Edited by
Lamberto Duò, Marco Finazzi, and
Franco Ciccacci

Magnetic Properties of Antiferromagnetic Oxide Materials

Related Titles

Kronmüller, H., Parkin, S. (eds.)

Handbook of Magnetism and Advanced Magnetic Materials

2007

ISBN: 978-0-470-02217-7

Rodriguez, J. A., Fernández-García, M. (eds.)

Synthesis, Properties, and Applications of Oxide Nanomaterials

2007

ISBN: 978-0-471-72405-6

Miller, J. S., Drillon, M. (eds.)

Magnetism: Molecules to Materials V

2005

ISBN: 978-3-527-30665-7

Waser, R., Böttger, U., Tiedke, S. (eds.)

Polar Oxides

Properties, Characterization, and Imaging

2005

ISBN: 978-3-527-40532-9

Morrish, A. H.

The Physical Principles of Magnetism

2001

ISBN: 978-0-7803-6029-7

Bozorth, R. M.

Ferromagnetism

2001

ISBN: 978-0-7803-1032-2

Edited by Lamberto Duò, Marco Finazzi, and Franco Ciccacci

Magnetic Properties of Antiferromagnetic Oxide Materials

Surfaces, Interfaces, and Thin Films

WILEY-VCH Verlag GmbH & Co. KGaA

The Editors

Lamberto Duò
Marco Finazzi
Franco Ciccacci
Dipartimento di Fisica
Politecnico di Milano
20133 Milano
Piazza Leonardo DA Vinci 32
Italy

Cover
Spieszdesign, Neu-Ulm, Germany

Library of Congress Card No.: applied for

British Library Cataloguing-in-Publication Data
A catalogue record for this book is available from the British Library.

Bibliographic information published by the Deutsche Nationalbibliothek
The Deutsche Nationalbibliothek lists this publication in the Deutsche Nationalbibliografie; detailed bibliographic data are available on the Internet at http://dnb.d-nb.de.

Cover Design Adam Design, Weinheim
Typesetting Laserwords Private Limited, Chennai, India
Printing and Binding Strauss GmbH, Mörlenbach

Printed in the Federal Republic of Germany
Printed on acid-free paper

ISBN: 978-3-527-40881-8

Contents

Magnetic Properties of Antiferromagnetic Oxide Materials.
Edited by Lamberto Duò, Marco Finazzi, and Franco Ciccacci

ISBN: 978-3-527-40881-8

Preface

This book is devoted to antiferromagnetic oxides, both in the form of surfaces and thin films, and in the form of interfaces and multilayers with other magnetic or nonmagnetic materials. Films and multilayers with a thickness of a few nanometers are important examples of low-dimensionality systems that can exhibit very different physical properties with respect to their bulk counterparts. This field is testifying a great experimental effort in the production of artificial structures with original magnetic properties. An example of relevant scientific and technological interest is the rapid development in the last years of spintronics, a discipline that aims to exploit the intrinsic spin of electrons and its associated magnetic moment, in addition to its fundamental electronic charge, in solid-state devices. The actual trend toward miniaturized magnetic devices requires new investigations of low-dimensional magnetic systems from a fundamental point of view. While confined systems based on ferromagnetic materials have been the subject of intensive study, the investigation of the magnetic properties of antiferromagnetic surfaces, thin films, and interfaces has received much less attention, mostly because of the experimental difficulties in addressing the local magnetic properties of antiferromagnetic materials. Nevertheless, the last two decades have witnessed a rapid development of experimental techniques characterized by a good sensitivity to such properties. As a consequence, systems in which a ferromagnet interacts with an antiferromagnetic partner have aroused an increasing interest in the scientific community, which is motivated by the discovery of their unique magnetic properties (e.g., exchange coupling, modification of the magnetic domain structure, and spin orientation) of potential interest for applications. In particular, the phenomenon of exchange bias [1], consisting in the breaking of the time reversal symmetry and in the onset of unidirectional magnetic anisotropy in a field-cooled system where a ferromagnetic material is coupled to an antiferromagnetic partner, is widely employed in the technology of reading heads for modern hard disks.

It is by now widely accepted that the key to understanding such phenomena is the spin structure at the antiferromagnet–ferromagnet interface and its dynamical changes. In this respect, transition-metal oxides represent an important class of antiferromagnetic materials because of their high Néel temperature and chemical stability. They are also large band-gap insulators so that their magnetic properties can be accurately described in terms of magnetic moments localized on the

Magnetic Properties of Antiferromagnetic Oxide Materials.
Edited by Lamberto Duò, Marco Finazzi, and Franco Ciccacci

ISBN: 978-3-527-40881-8

cations, interacting via well-understood short-range interactions. For these reasons, transition-metal oxides occupy a privileged space in the scientific literature and are often considered as model antiferromagnetic systems. Moreover, the interest in oxide antiferromagnets goes beyond the phenomena that can be observed when they interact with a ferromagnetic system. In fact, complex behaviors such as spin reorientation might be induced in antiferromagnetic oxides even in the absence of bias exchange with a ferromagnetic partner. Antiferromagnetic transition-metal oxides are also important in connection with at least three classes of oxide materials that have a great technological relevance (but that will not be discussed in this volume), namely, high-T_c superconductor cuprates, colossal magnetoresistance manganites, and multiferroic materials, where a widely accepted general idea is that in these materials the interplay of charge and spin fluctuations is a central issue. In particular, the paradigm model describing this interplay in cuprates is the Anderson's superexchange theory, which also governs the magnetic properties of antiferromagnetic insulators, such as NiO or CoO, often considered as archetype materials for the antiferromagnetic parent compounds of high-T_c cuprates.

This book gives the first comprehensive account of these topics, bringing together experimental and theoretical methods. It is focused on the study of the magnetic behavior of spatially confined antiferromagnetic transition-metal-oxide systems when their dimensions are scaled down to nanometric level, with particular emphasis on the growth and the magnetic characterization through different experimental methods and theoretical modeling approaches.

This volume is designed along the following guidelines. The tutorial Chapter 1 outlines the magnetic properties of low-dimensional antiferromagnetic transition -metal-oxide systems and contains an overview of the significant physical phenomena that intervene in determining their origin. Chapter 2 extensively discusses the techniques employed to fabricate multilayer and thin film structures. This chapter highlights the close relationship existing between the structure and magnetism in low-dimensional antiferromagnetic oxides. Chapter 3 discusses one of the most widespread experimental techniques giving access to the properties of antiferromagnetic materials, namely, X-ray absorption and dichroism. This chapter represents a general introduction to X-ray absorption and how it is measured. It discusses the selection rules that give rise to the polarization dependence of X-ray absorption and how information about the magnetic properties of matter can be obtained from them. Chapter 4 deals with the low-dimensional antiferromagnetic transition-metal oxides as such, describing samples consisting of thin epitaxial layers on nonmagnetic substrates, which are therefore not influenced by magnetic underlayers. The effects of symmetry-lowering and crystal field in noncubic environments are addressed both experimentally, discussing X-ray absorption and dichroism data, and theoretically, within an atomic multiplet approach. Chapter 5 is devoted to the phenomenon of exchange bias. The central topic of this chapter is the so-called "domain state model" in which exchange bias emerges as a consequence of a net magnetization in the volume and also at the interface of the antiferromagnet stabilized by nonmagnetic defects. Chapter 6 provides a theoretical discussion of the phenomena (exchange coupling, exchange bias, spin reorientation, magnetic domain structure, etc.) occurring at the interface between a ferromagnetic and an antiferromagnetic material. Phase diagrams describing the behavior of bilayers and trilayers are obtained as a function of the relevant parameters. Chapter 7 focuses on the experimental investigation of Fe_3O_4-based ferrimagnet–antiferromagnet multilayers comprising NiO or CoO as model systems for addressing interface morphology and chemical structure, anisotropy, interlayer exchange coupling, and how these mechanisms affect the overall magnetic properties of the multilayers. Finally, Chapter 8 presents an extensive X-ray microscopy study of the micromagnetic structure in exchange-coupled interfaces constituted by a ferromagnetic metal and an antiferromagnetic oxide.

To conclude, we would like to dedicate the last part of this preface to thank the authors of the various chapters of this book for their cooperation. We appreciate their patience in accepting our reminders for keeping to time and their willingness in following the editors' suggestions. We would also like to thank Edmund Immergut, Wiley-VCH in New York, for his friendly help in first proposing to us to start on this book, in enabling us to overcome our initial hesitations, and in guiding us in the first steps of the project up to the acceptance of the same by Wiley-VCH, and Anja Tschörtner, Wiley-VCH Ed. Physics Department in Berlin, for her continuous and extraordinary assistance during the preparation of this book.

Finally, we would like to express our gratitude towards our friends and colleagues, who are too numerous to list here, to whom we are much indebted for inspiring discussions and help.

Milano, October 2009

Lamberto Duò
Marco Finazzi
Franco Ciccacci

Reference

1. Meiklejohn, W.H. and Bean, C.P. (1957) New magnetic anisotropy. *Phys. Rev.*, **105**, 904.

List of Contributors

Salvatore Altieri
Università di Modena e
Reggio Emilia
Dipartimento di Fisica
Via G. Campi 213/a
I-41100 Modena
Italy

Julie A. Borchers
National Institute of
Standards and Technology
Gaithersburg
MD 20899
USA

Franco Ciccacci
Politecnico di Milano
LNESS – Dipartimento di Fisica
Piazza Leonardo da Vinci 32
20133 Milano
Italy

Lamberto Duò
Politecnico di Milano
LNESS – Dipartimento di Fisica
Piazza Leonardo da Vinci 32
20133 Milano
Italy

Marian Fecioru-Morariu
RWTH Aachen University
Physikalisches Institut (IIA)
Templergraben 55
52056 Aachen
Germany

and

Oerlikon Solar AG
Hauptstrasse la
9477 Trübbach
Switzerland

Marco Finazzi
Politecnico di Milano
LNESS – Dipartimento di Fisica
Piazza Leonardo da Vinci 32
20133 Milano
Italy

Gernot Güntherodt
RWTH Aachen University
Physikalisches Institut (IIA)
Templergraben 55
52056 Aachen
Germany

Magnetic Properties of Antiferromagnetic Oxide Materials.
Edited by Lamberto Duò, Marco Finazzi, and Franco Ciccacci

ISBN: 978-3-527-40881-8

Tjipke Hibma
University of Groningen
Zernike Institute for
Advanced Materials
Nijenborgh 4
NL-9747 AG Groningen
The Netherlands

Maurits W. Haverkort
Max Planck Institute for Solid
State Research
Heisenbergstrasse 1
D-70569
Stuttgart
Germany

Paola Luches
National Research Centre on
Nanostructures and Biosystems at
Surfaces
CNR-INFM
S3, Via G. Campi 213/a
I-41100 Modena
Italy

Alexander I. Morosov
Electronics and Automation
(Technical University)
Moscow State Institute of
Radioengineering
pr. Vernadskogo 78
Moscow 19454
Russia

Ulrich Nowak
Universität Konstanz
Fachbereich Physik
Universitätsstrasse 10
78457 Konstanz
Germany

Hendrik Ohldag
Stanford Synchrotron
Radiation Lightsource
SLAC National Accelerator Center
2575 Sand Hill Road
Mailstop 69
Menlo Park 94025
CA
USA

Maurizio Sacchi
Synchrotron Soleil
Boîte Postale 48
F-91142 Gif-sur-Yvette
Paris
France

Alexander S. Sigov
Electronics and Automation
(Technical University)
Moscow State Institute of
Radioengineering
pr. Vernadskogo 78
Moscow 19454
Russia

Sergio Valeri
Università di Modena e
Reggio Emilia
Dipartimento di Fisica
Via G. Campi 213/a
I-41100 Modena
Italy

and

National Research Centre on
Nanostructures and Biosystems at
Surfaces
CNR-INFM
S3, Via G. Campi 213/a
I-41100 Modena
Italy

Pieter Jan van der Zaag
Philips Research Laboratories
High Tech Campus 12a
5656 AE Eindhoven
The Netherlands

Jan Vogel
CNRS and UJF
Institut Néel
B.P. 166
38042 Grenoble
France

1
Low-Dimensional Antiferromagnetic Oxides : An Overview

Marco Finazzi, Lamberto Duò, and Franco Ciccacci

1.1
Introduction

In the last two decades, the availability of experimental techniques endowed with high sensitivity with respect to the magnetic properties of antiferromagnetic (AFM) materials has motivated a large amount of studies dedicated to the investigation of low-dimensional AFM systems consisting of small particles or films deposited onto either nonmagnetic or ferromagnetic (FM) substrates. Similar to the well-known FM materials, such confined AFM systems are in fact characterized by magnetic properties that, because of interface or size effects, can be considerably different from the ones observed in the bulk [1]. Examples range from the stabilization of exotic AFM ordering to the onset of uniaxial anisotropy in low-dimensional AFM samples. Moreover, systems comprising AFM–FM interfaces represent a world of their own, thanks to their rich phenomenology related to interface exchange coupling.

Finite-size effects in both FM and AFM materials reflect deviations from bulk properties associated with the reduction of the sample dimensions. So-called "intrinsic" effects occur in material systems for which one or more sample dimensions, for example, the thickness of a layer or diameter of a particle, is comparable with the intrinsic correlation length scale of the property being considered. Strongly correlated systems such as AFM oxides are characterized by very short correlation lengths, so intrinsic finite-size effects can be observed only in ultrathin films or nanoparticles. In addition, "surface-related" finite-size effects might be caused by the competition between the properties of atoms in the core of a particle or layer and those at the surface, possibly originating from the reduced coordination number. As an example, surface spins often possess higher magnetocrystalline anisotropy than the ones in the sample volume because of the reduced symmetry. "Chemical" or "structural" effects may also arise due to phenomena such as surface segregation, relaxation, or reconstruction. Of course, the environment (the material surrounding the particle or the film substrate) can also dramatically alter the properties of interface atoms through hybridization, strain, or chemical interdiffusion, not to mention the crucial role of exchange in determining the magnetism of systems where an AFM material interacts with an FM partner.

Magnetic Properties of Antiferromagnetic Oxide Materials.
Edited by Lamberto Duò, Marco Finazzi, and Franco Ciccacci

ISBN: 978-3-527-40881-8

The high degree of correlation between the magnetic, chemical, structural, and morphologic features obviously makes the preparation and the characterization of high-quality samples a crucial point in any study involving low-dimensional systems (see 2). In this respect, AFM transition-metal (TM) monoxides are often regarded as a privileged reference. The reasons for this choice are manifold. First, these oxides can be grown as high-quality thin films on appropriate substrates and are characterized by a high chemical and mechanical stability. Second, their AFM ordering temperature (Néel temperature T_N) is relatively high: to cite two relevant cases, $T_N = 523$ K for bulk NiO and $T_N = 291$ K for bulk CoO. In the latter case, the proximity of T_N to room temperature represents an additional advantage in realizing exchange-biased systems, where field cooling from above to below T_N is required. Another important feature of TM oxides is their insulating nature resulting from strong inter- and intra-atomic electronic correlations. Their magnetic properties arise as a consequence of the short-range superexchange interaction mediated by the oxygen bonds [2]. Because of the absence of itinerant magnetism associated with conduction electrons, the only long-range magnetic interaction is represented by the dipole–dipole interaction, which can be neglected in many cases. Therefore, from the magnetic point of view, TM oxides can be described in the frame of the Heisenberg or Ising formalism as ensembles of well-localized spins with near-neighbor interactions. Finally, we would like to mention that AFM TM oxides are also considered as model systems for the AFM parent compounds of high T_c cuprates since, in the latter, the interplay between charge and magnetic ordering is described by the Anderson's superexchange theory, which also governs the magnetic properties of AFM insulators such as NiO or CoO.

In this introductory chapter, we present a survey of the peculiar magnetic phenomena observed in low-dimensional systems based on AFM TM oxides, such as surfaces, thin films, interfaces with magnetic or nonmagnetic materials, and multilayers. We also give an overview of the significant physical phenomena that intervene in determining their origin. The chapter is organized as follows: in Section 1.2 we address finite-size effects on the value of the Néel temperature for AFM oxide particles and thin films. Section 1.3 is dedicated to AFM thin-film magnetic anisotropy and how it is influenced by interaction with a nonmagnetic substrate, while interlayer magnetic coupling and micromagnetic structure at AFM–FM interfaces and multilayers are examined in Sections 1.4 and 1.5, respectively. Finally, Section 1.6 concludes the chapter by discussing applications.

1.2
Finite-Size Effects on the Magnetic Ordering Temperature

The reduction of the critical magnetic ordering temperature T_{order} is a typical finite-size effect in both FM ($T_{\text{order}} = T_C =$ Curie temperature) and AFM ($T_{\text{order}} = T_N$) low-dimensional systems. It can be seen as a consequence of the sample asymptotically approaching the conditions at which the Mermin–Wagner theorem applies as its size is progressively reduced. This theorem states that, because of fluctuations, continuous symmetry cannot be spontaneously broken (i.e., the sample cannot develop long-range FM or AFM order) at finite temperature in systems with short-range interactions in dimensions $d \leq 2$ [3]. Experiments have evidenced drastic reductions in the magnetic transition temperature T_N for a variety of AFM oxides in low-dimensional geometries such as CuO nanoparticles [4–7], NiO thin films [8, 9], CoO nanoparticles [10] and films [11, 12], and Co_3O_4 nanoparticles [13–15], nanotubes [16], and nanowires [17].

In a simple mean-field approach, which assumes T_{order} to be proportional to the exchange energy density of the particle or thin film, the reduction of the magnetic transition temperature is due to the decrease in the total exchange energy associated with the reduced number of neighboring atoms. In this picture, the system environment is not passive and might contribute to defining the total exchange energy, for instance, by inducing surface magnetic anisotropy via a strain field or by modifying the magnitude of the interface moments and the strength of their mutual coupling through hybridization or other interactions (see below). According to the mean-field model, the variation of T_{order} with respect to the bulk value $T_{order}(\infty)$ is expected to be proportional to the inverse of the particle size or of the film thickness D. However, this prediction is not consistent with the experimental results, and the reduction in the ordering temperature T_{order} with size is better described in terms of scaling theories [18], according to which the correlation length of the fluctuations of the AFM order parameter diverges logarithmically with the reduced temperature $(T - T_{order})/T_{order}$ as the temperature T approaches the magnetic ordering transition T_{order}. For a system with size D much larger than a characteristic length ξ describing the spatial extent of the spin–spin coupling, this yields a fractional decrease of T_{order} that follows a power-law curve [19]:

$$\frac{T_{order}(\infty) - T_{order}(D)}{T_{order}(\infty)} = \left(\frac{\xi + a}{2D}\right)^{\lambda} \tag{1.1}$$

where λ is the (constant) shift exponent and a is the lattice spacing. For $D \ll \xi$, T_{order} is expected to vary *linearly* with respect to D [19]:

$$T_{order}(D) = T_{order}(\infty)\frac{D - a}{2\xi} \tag{1.2}$$

The details of the measured dependence of T_{order} upon D show large differences over the experimentally investigated systems, suggesting that environment effects play a significant role. Moreover, the value of λ is model-dependent: as anticipated above, one obtains $\lambda = 1$ in mean-field theory [18], while $\lambda = 1.4$ or 1.6 for a system described by either the three-dimensional Heisenberg [20] or Ising [21] Hamiltonian, respectively. In practice, ξ and λ are considered as adjustable parameters that have to be fitted to the experimental data: λ has been found to be close to 1.1 for Co_3O_4 nanoparticles [15], while $\lambda \approx 1.5$ for CuO thin films [11], ξ being of the order of a few nanometers in both cases. A more recent model explicitly considers the disordering effect of the lattice thermal vibrations on the spin–spin coherence length at the transition temperature. It predicts a T_{order} dependence on D of the form [22]

$$\frac{T_{order}(D)}{T_{order}(\infty)} = e^{-\left(\frac{1-\alpha}{D/D_0 - 1}\right)} \tag{1.3}$$

In the previous expression, $D_0 = 2(3 - d)a$ (with $d = 0$ for particles, $d = 1$ for nanorods, and $d = 2$ for thin films), while $\alpha = \sigma_s^2(D)/\sigma_v^2(D)$, σ_s and σ_v corresponding to the root-mean square average amplitude of the oscillations of atoms at the surface (or interface) and in the volume at the transition temperature, respectively. In this model, T_N is again expected to depend on the total system–environment interface exchange energy.

The models discussed above can be generalized to low-dimensional systems in interaction with *magnetic* environments. In this case, the total exchange energy density of the low-dimensional sample might be even higher than in the bulk because of interface exchange coupling. In this case T_N might increase as the sample size is reduced, as observed in NiO/CoO and Fe_3O_4/CoO multilayers [8, 23]. Similarly, it has been demonstrated that the magnetic coupling of Co nanoparticles embedded in a CoO matrix leads to a marked improvement in the thermal stability of the moments of the FM nanoparticles, with an increase of almost 2 orders of magnitude in the temperature at which superparamagnetism sets in compared to similar particles in a nonmagnetic medium [24]. An intriguing and not yet understood size-related effect is the dependence of the blocking temperature T_B upon the AFM layer thickness in exchange-biased AFM–FM multilayers. T_B is defined as the temperature above which the system does not display any bias, and in Fe_3O_4/CoO multilayers it is observed to *decrease* by reducing the CoO layer thickness while T_N, as mentioned above, *increases* [23].

To further highlight the active role even a nonmagnetic substrate might have in determining the value of T_N in a thin film, we conclude this section by discussing the case of NiO ultrathin films epitaxially deposited on MgO(001) and Ag(001) single-crystal substrates [25]. Stoichiometric and high-quality NiO films were grown by atomic-oxygen-assisted reactive deposition and capped *in situ* with a protecting MgO film consisting in 25 monolayers (MLs) that avoided NiO film contamination by the residual gas inside the ultrahigh-vacuum chamber and prevented possible oxygen loss during thermal cycling. NiO and MgO have the same rock-salt crystal structure with a lattice constant of 4.2 and 4.1 Å, respectively, corresponding to a tiny lattice misfit of 0.2%. Ag has a face-centered cubic (fcc) structure with a lattice constant equal 4.09 Å and compared to NiO has a lattice misfit of about 3%. Nevertheless, misfit dislocations are avoided by keeping the film thickness below the critical thickness for strain relaxation (about 30 ML for NiO/Ag) [26]. The magnetic properties of the NiO layer have been investigated with X-ray magnetic linear dichroism (see Chapter 3) performed at the Ni L_2 edge. By plotting the dichroic signal (L_2 ratio) as a function of temperature, as done in Figure 1.1, one obtains a direct measure of the long-range order parameter and of the Néel temperature of the material. Figure 1.1 reports results measured for 3-ML- and 30-ML-thick NiO films on Ag(001) and for a 3-ML-thick NiO film on MgO(001). For the 30-ML NiO/Ag film, an ordering temperature $T_N = 535$ K in thus measured, which is close to the bulk value of 523 K [27]. Apparently, the 30-ML NiO/Ag film is already thick enough to act as the bulk oxide and is not affected any longer by the underlying Ag substrate [28, 29]. Figure 1.1 also suggests that there is no magnetic order in the 3-ML NiO/MgO sample in the measured temperature range, denoting

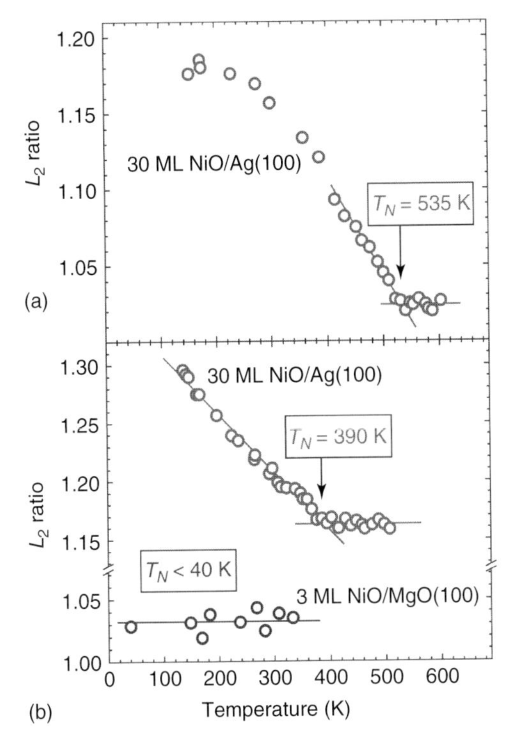

Figure 1.1 Temperature-dependent Ni linear dichroism signal (L_2 ratio) of (a) 30-ML NiO/Ag(100) and (b) 3-ML NiO/Ag(100) and 3-ML NiO/MgO(100). Reprinted figure with permission from [25]. © 2009, by the American Physical Society.

a strong finite-size-induced reduction of T_N that can be interpreted as discussed earlier in this section. Conversely, the Néel temperature of the 3-ML NiO/Ag film is $T_N = 390$ K, at least a factor of 10 higher than in the 3-ML NiO/MgO sample. It is important to notice that the NiO/Ag and NiO/MgO films are essentially identical as far as the NiO part is concerned (same thickness, crystal quality, interface roughness), except for the different value of the lattice mismatch with respect to the substrate. The origin of such a strongly different T_N value in NiO/MgO and NiO/Ag must thus be external to the NiO film itself and should be looked for in the different types of interactions at the NiO/MgO and NiO/Ag interfaces. To this purpose, we have to consider how the superexchange interactions in the NiO films can be modified by the presence of the substrate. The value of the superexchange coupling constant J in a system subjected to charge fluctuations described by the Hubbard energy U and the cation–anion charge-transfer energy Δ [30] is given by the following expression [31–33]:

$$J = \frac{2t^2}{\Delta^2}\left(\frac{1}{U} + \frac{1}{\Delta}\right) \tag{1.4}$$

with t being the anion 2p-cation 3d transfer integral. The value of J can be altered in three ways by the interaction with the substrate, resulting in a consequent modification of T_N.

The first mechanism refers to the prediction that a medium with a high dielectric polarizability should provide an effective screening for various charge excitations in a nearby located material [34]. Such a screening effect on the band gap, the Hubbard energy U, and the charge-transfer energy Δ has indeed been experimentally confirmed in C_{60}/Ag [35] and MgO/Ag [28, 29, 36]. According to this expectation, the observed higher T_N value for NiO/Ag compared to NiO/MgO would be associated to the conducting nature of the Ag substrate with respect to the insulating character of MgO, leading, in agreement with Eq. (1.4), to a higher superexchange coupling constant J as a result of the reduction of U and Δ by image-charge screening.

A second explanation considers the different strain state induced in the NiO overlayer by the Ag or MgO substrate. By modifying the in-plane interatomic distance, strain influences the overlap between adjacent orbitals and the value of the transfer integral t, affecting the value of J and hence T_N. We recall that the application of hydrostatic pressure is known to enhance T_N in TM oxides [37–40]. However, the theoretical [38] and experimental [39, 40] dependence of J and T_N on the lattice parameter variation is too weak to justify the 10-fold increase of T_N in NiO/Ag with respect to NiO/MgO as due to the different strain state. Moreover, the strain in an epitaxial layer is nonisotropic and one could expect that the lattice-spacing effect on T_N will be smaller than the one in an isotropically compressed system, since, in the first case, the change in the interatomic spacing along the surface normal is generally opposite to that in the film plane. Indeed, it has been experimentally shown that for NiO and CoO films on MgO uniaxial strain up to 2% has a negligible effect on T_N [8].

Finally, interface hybridization might represent a third mechanism according to which the substrate might be capable of influencing T_N. In this respect, we note that density-functional band-structure calculations on both free-standing and Ag(001)-supported ultrathin NiO films find that the superexchange constant J of a 3-ML-thick film should hardly be affected by a nearby substrate [41, 42]. These arguments leave screening effects as the only possible phenomenon that is able to account for the large value of the Néel temperature in 3-ML NiO/Ag thin films, suggesting at the same time an effective method to counterbalance finite-size-related reduction of critical temperatures in oxide systems.

1.3
AFM Anisotropy

As mentioned in the previous section, even a nonmagnetic substrate might play an active role in determining the value of T_N in a thin film. In this section, we discuss how the substrate can also influence the magnetic anisotropy of an AFM overlayer. We limit the discussion to nonmagnetic substrates, leaving AFM–FM coupled layers to the next section.

Well known and long-studied in FM systems, magnetic anisotropy is a phenomenon that has been unveiled only quite recently in AFM thin films. For instance, while the surface magnetic structure of cleaved NiO(001) single crystals is said to be bulk terminated [43], NiO(001) thin layers (less than 20 ML) epitaxially grown on MgO(001) exhibit an out-of plane uniaxial anisotropy [9]. On the other hand, the AFM anisotropy of epitaxial NiO/Ag(001) is found to depend on the NiO thickness: for 30-ML-thick NiO/Ag films, AFM domains with easy axis closer to the surface normal are favored, while 3-ML NiO/Ag are characterized by in-plane AFM anisotropy [44]. CoO films display local moments with magnitude and orientation strongly dependent on the strain induced by the substrate: the magnetic moments in CoO/MnO(001) are oriented out of plane while those in CoO/Ag(001) are in-plane [45]. Hereafter, we discuss the possible sources of AFM magnetic anisotropy.

1.3.1 Magnetocrystal Anisotropy

While a 3d isolated ion is described by a Hamiltonian that has spherical symmetry, the same ion embedded in a crystal is subjected to the crystal field resulting from the interactions of the electrons belonging to each ion with the surrounding atoms. If this crystal field is strong enough, the orbital degeneracy is completely removed and the ground state is an orbital singlet (the orbital momentum is "quenched") [46]. In these conditions, spin represents the only contribution to the total magnetic moment of the ion, and is, to first approximation, completely decoupled from the lattice. In other words, the system does not develop any magnetic anisotropy. However, if the crystal field is not too large, the spin–orbit interaction, which is proportional to $\hat{\mathbf{L}} \cdot \hat{\mathbf{S}}$ ($\hat{\mathbf{L}}$ = total orbital momentum operator; $\hat{\mathbf{S}}$ = total spin operator), prevents the quenching of the orbital momentum and couples the spin to the lattice, establishing magnetic anisotropy [46]. In a cubic ionic 3d compound, the crystal field can be considered as a small perturbation (of the order of 10^4 K) with respect to the Coulomb interactions between the electrons occupying the d-shell, but is considerably larger than the spin–orbit interaction (about 100 K). Sizable anisotropy can thus emerge only when the symmetry of the lattice and the degeneracy of the ground state are further reduced by a small perturbation. The above-mentioned anisotropy observed in CoO strained films can be explained in this frame (see also Chapter 4): the tetragonal distortion imposed by the substrate further splits the partially occupied t_{2g} orbitals that constitute the ground state resulting from the application of a cubic crystal-field to the 3d shell. The sign of the splitting depends on the type of tetragonal strain (compressive or extensive) and determines whether the CoO spins will preferentially align in-plane or out of plane [45].

1.3.2 Dipolar Anisotropy

Long-range magnetic dipole–dipole interactions have been suggested to play an important role in determining the bulk AFM structure of TM monoxides such as MnO and NiO, where the magnetocrystal anisotropy is expected to vanish [47].

In MnO, this happens because the d-shell electronic configuration of the Mn^{2+} ion is $3d^5$, corresponding to an orbital singlet ground state even in the isolated ion. The case of NiO is different: the Ni^{2+} 3d fundamental state is characterized by completely filled t_{2g} orbitals and partially occupied e_g orbitals. Now, $\hat{\mathbf{L}} \cdot \hat{\mathbf{S}}$ does not have matrix elements coupling different e_g states, implying that the spin–orbit interaction is also negligible for NiO.

The fcc lattice formed by the cations in TM monoxides can be viewed as a combination of four simple cubic lattices. In each sublattice, the moments are forced by the superexchange interaction to align ferromagnetically within {111} planes, while adjacent {111} planes are coupled antiferromagnetically. The superexchange interaction does not couple the magnetic moments of cations in different simple cubic sublattices since these are connected by 90° oxygen bonds. The relative orientations of the four lattices with respect to each other and to the crystallographic axes are instead imposed by anisotropic interactions. In MnO and NiO, the larger source of anisotropy is provided by the magnetic dipole interaction, which favors a collinear alignment between the sublattices, resulting in the so-called type-II AFM order. These are characterized by AFM domains (*T*-domains) in which the moments form FM foils parallel to one of the four equivalent {111} planes of the fcc lattice [47], with the spin aligned in the plane of the foil [47]. This type of AFM ordering is further stabilized by a rhombohedral distortion in the direction perpendicular to the foils caused by magnetostriction. Inside each (111) foil, the spin is driven to align along one of the three equivalent $[11\bar{2}]$ directions (giving three equivalent so-called *S*-domains) by sources of smaller anisotropy (see Figure 1.2).

The removal of translation symmetry and the possible presence of substrate -induced strain reduce the symmetry of the bulk and cause AFM anisotropy. Since in an AFM material there is no net magnetization, one can apply first-order

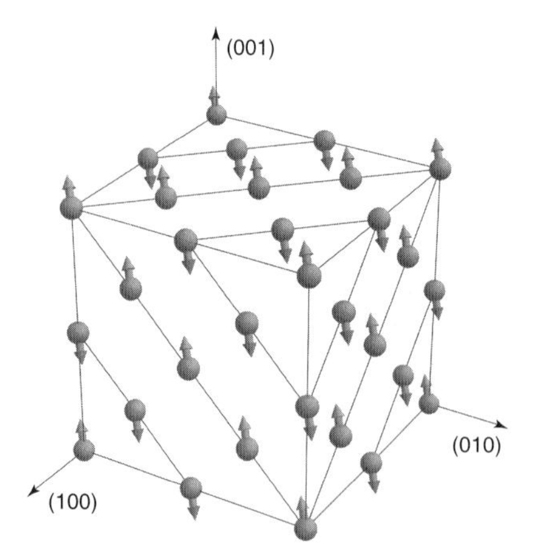

Figure 1.2 Collinear arrangement of magnetic cations in MnO and NiO. A domain with (111) ferromagnetic foils and moments parallel or antiparallel to the $[11\bar{2}]$ direction is shown.

perturbation theory [48] to evaluate the effects of the AFM film finite thickness τ and of the tetragonal strain $\varepsilon = (c/a) - 1$ (c and a being the out-of-plane and in-plane lattice parameters, respectively). Inclusion of tetragonal strain into a bulk AFM MnO or NiO lattice energetically favors S-domains with main spin component *perpendicular* ($S_{\perp}$) to the strain axis for $\varepsilon < 0$ (compression) and S-domains with main spin component *parallel* ($S_{||}$) to the strain axis for $\varepsilon > 0$ (expansion). In uniformly strained films, this effect coexists with a thickness-dependent perturbation to the total dipolar anisotropy, which stabilizes $S_{||}$ domains and in-plane anisotropy [48]. At variance with the well-known shape anisotropy of FM thin films, which is proportional to the film volume, this AFM shape anisotropy is proportional to τ^{-1} [48]. In the original work [25], the thickness-dependent AFM anisotropy experimentally observed in NiO/Ag(001) films was indeed interpreted as being a consequence of AFM dipolar anisotropy.

1.4 Interlayer Coupling in AFM–FM Bilayers and Multilayers

AFM surfaces are said to be compensated when the surface magnetization of the AFM material is null. In TM monoxides, which are characterized by an AFM order similar to that shown in Figure 1.2, {001} and {011} planes are nominally compensated, while the {111} surfaces are totally uncompensated. The latter, however, exposes the same chemical species and a net electric charge (TM oxides are highly ionic compounds). For this reason, the {111} surfaces are unstable and tend to reconstruct. At compensated AFM–FM interfaces, the exchange interaction is strongly frustrated. Frustration can be partially released by the presence of defects (chemical interdiffusion, atomic steps, missing atoms, dislocations), which make the interface partially uncompensated, at least on a local scale. Defects are instead a source of frustration at nominal totally uncompensated interfaces. There is a general consensus in the scientific community about the importance of considering this interplay between frustration and defects to understand magnetic properties such as exchange bias, interlayer coupling, and micromagnetic structure of AFM–FM bilayers and multilayers. We omit to discuss exchange bias in this introductory chapter, since this argument is extensively treated in this book (see Chapters 5–7). We concentrate instead, in Section 1.4.1, on interlayer coupling, while the micromagnetic structure at AFM–FM interfaces is the subject of Section 1.4.2.

1.4.1 AFM–FM Interface Coupling

Micromagnetic calculations based on energy minimization show that the ground-state configuration of an ideal magnetically compensated AFM–FM interface corresponds to a perpendicular orientation of the bulk FM moments relative to the AFM magnetic easy axis direction, an arrangement known as *spin-flop state* [49]. This configuration is stable since the magnetic moments both

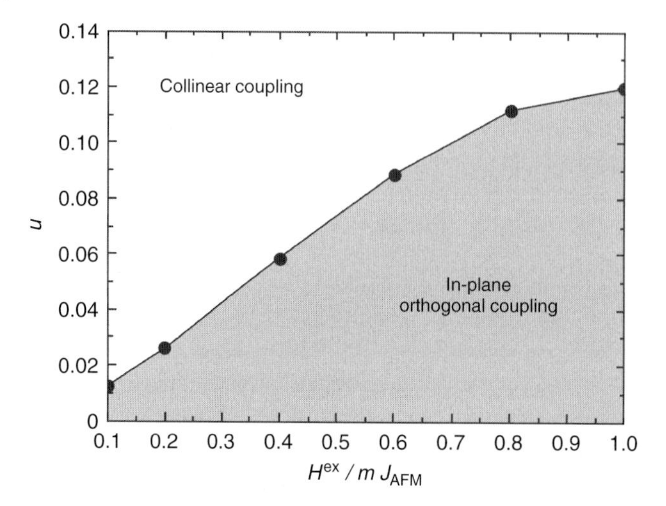

Figure 1.3 Diagram describing the regions in the parameter space for which the AFM anisotropy axis is either collinear or orthogonal to the FM magnetization, obtained for an AFM–FM bilayer with the characteristics of NiO/Fe(001). H^{ex} represents the magnitude of the effective exchange field of magnetically active defects at the AFM–FM interface, while u is the fraction of interface atomic sites occupied by such defects: $u = 0$ for an ideal compensated interface; $u = 1$ for a totally uncompensated interface. J_{AFM} and m are the values of the superexchange coupling constant and of the local magnetic moments in the AFM material, respectively. Reprinted figure with permission from [51]. © 2004, by the American Physical Society.

in the FM and in the AFM layer exhibit a small canting from the 90° coupling that vanishes away from the interface but induces a small magnetization in the AFM layer. The spin-flop ground state was also suggested to lead to exchange bias [49], a result that was later confuted by a more accurate model describing the spin dynamics in terms of moment precession rather than energy minimization [50]. The presence of defects can, however, destroy the spin-flop ground state, inducing a collinear coupling between the AFM and FM moments across the interface. The reason why this happens can be captured by a mean-field calculation where the perturbation associated with the defects is simulated by an effective exchange field of magnitude H^{ex} acting on the atoms belonging to a compensated AFM–FM interface reproducing the characteristics of the NiO/Fe(001) bilayer [51, 52]. The result is summarized in Figure 1.3, which shows the type of average AFM–FM coupling (collinear rather than perpendicular) between the two layers as a function of H^{ex} and of the fraction u of magnetically active defects at the interface. Monte Carlo simulations find AFM spins aligning collinearly with the FM moments above T_N, with a transition from collinear to perpendicular alignment of the FM and AFM spins at a lower temperature in the case of rough interfaces [53]. Besides interface defects, also volume defects such as dislocations inside the AFM material can influence the coupling between the FM magnetization and the AFM anisotropy axis by disrupting the collinear alignment of the spins inside the AFM layer [54].

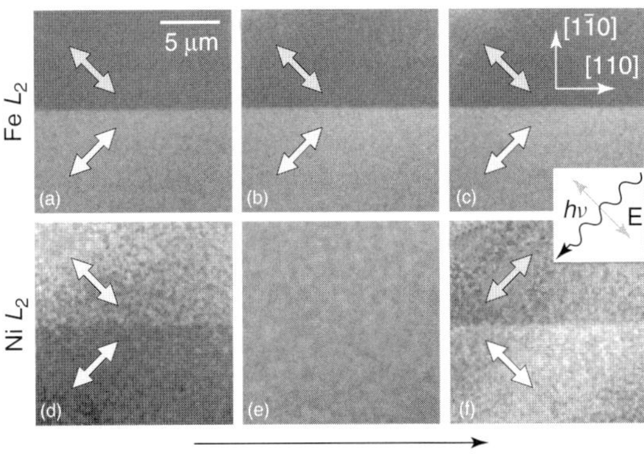

Figure 1.4 Photoemission electron microscopy images excited by linear X-rays and showing the magnetic contrast obtained at the Fe $L_{2,3}$ (top) and Ni L_2 (bottom) absorption edges on an NiO/Fe(001) wedged thin film. The field of view shown in each panel straddles a domain wall separating two Fe domains with in-plane magnetization perpendicular to each other. The arrows indicate the direction of the magnetization in the Fe substrate and of the NiO film easy axis. The reported crystallographic directions refer to the Fe substrate. The NiO thickness increases from the left to the right, with an average value in the left, central, and right panels equal to about 12, 18, and 24 Å, respectively. The NiO thickness variation in each image is about 3 Å. The direction of the NiO easy axis has been rotated by 90° with respect to the one previously reported in [58] (see [59]). Reprinted figure with permission from [58]. © 2006, by the American Physical Society.

Because of the presence of frustration between competitive exchange interactions, the AFM–FM interface coupling in real systems can strongly depend on the interface preparation conditions. Moreover, strain-induced magnetoelastic effects should also be considered (see Chapter 8). For instance, thin FM metal films (Fe, Co) on NiO(001) exhibit perpendicular FM–AFM coupling [55–57], while the coupling is collinear for thin NiO films on Fe(001) [51, 58]. Actually, the original articles [51, 56–58] based on X-ray dichroism indicated the type of coupling to be the opposite of the one reported above. This confusion originated from a misinterpretation of the linear dichroism spectra, which was corrected in later work [59]. The coupling in NiO/Fe(001) is also NiO thickness-dependent: the anisotropy axis is parallel to the Fe substrate magnetization when the NiO thickness is less than about 15 Å, but rapidly becomes perpendicular parallel to the Fe magnetization for a NiO coverage higher than 25 Å, as displayed in Figure 1.4 [58].

A thorough evaluation of exchange and magnetoelastic effects has been conducted for epitaxial NiO thin films on single magnetite (Fe_3O_4) crystals [60]. Magnetite is a *ferrimagnetic* material that has a lattice parameter in the (001) plane that is almost exactly twice (mismatch only 0.5%) that of NiO [61]. As a consequence, in an epitaxially grown system, the surface spins of the Fe_3O_4 interact only, to first approximation, with one of the two uncompensated sublattices of the (001) NiO interface plane. This situation would therefore correspond to an uncompensated NiO/Fe_3O_4(001) interface where collinear coupling would be expected to be energetically favored. Instead, a spin-flop coupling is observed, stabilized by magnetostrictive deformations induced by the magnetite substrate [60]. Collinear coupling was indeed found for the (111) and (110) interfaces, attributed to a strain-induced AFM stacking asymmetry in the NiO. All three interfaces display an uncompensated AFM magnetization, with the largest value observed for the (110) interface, while the (111) and (001) interfaces exhibit only 10% of that value [60].

The interface coupling in $CoO/Fe_3O_4(001)$ superlattices (see chapter 7) is also found to be perpendicular [62]. This result, however, has not been explained as being a consequence of magnetoelastic effects, but by rather considering [63] the role of anisotropic exchange, a phenomenon first studied by Dzyaloshinsky and Moriya [64, 65]. By coupling the direction of the spin to the crystal axes, the spin–orbit interaction is responsible for single-ion magnetic anisotropy, as outlined in Section 1.3.1. The same effect also introduces an anisotropic contribution to the isotropic exchange Hamiltonian. This antisymmetric Dzyaloshinsky–Moriya term can be written as $\mathbf{D}(\mathbf{S}_1 \times \mathbf{S}_2)$, with $\mathbf{S}_1$ and $\mathbf{S}_2$ being the spin operators of two neighboring magnetic ions, and $\mathbf{D}$ a vector that vanishes when the crystal field around each ion has inversion symmetry with respect to the center of their mid-point [46]. Therefore, the Dzyaloshinsky–Moriya anisotropic exchange interaction, which favors a perpendicular coupling, cannot *a priori* be neglected for magnetic ions lying on opposite sides of an interface, where inversion symmetry is broken.

A strong AFM–FM interface coupling might also affect the type of order that is stabilized in the AFM material, as demonstrated in a study conducted on Co/CoO core/shell nanoparticles obtained by oxidation of Co nanospheres. In such a system, Co–CoO interfaces are highly crystalline and oxidation leads to the decompensation of the (100) CoO surface, resulting in a strong core-shell coupling [66]. Polarized-neutron diffraction finds both $(\frac{1}{2}, \frac{1}{2}, \frac{1}{2})$ and (100) AFM modulation, as in bulk CoO [67], corresponding to a stacking of alternate FM foils with the normal parallel to either the [111] direction (type-II AFM order, as in Figure 1.2) or to the [100] direction (type-I AFM order). While in bulk CoO the $(\frac{1}{2}, \frac{1}{2}, \frac{1}{2})$ neutron diffraction peak prevails [67], oxidation of Co nanoparticles hugely enhances the (100) peak [66].

1.4.2
Coupling between FM Layers Separated by an AFM Oxide Spacer

The insulating nature of TM AFM oxides excludes any intervention of the Ruderman–Kittel–Kasuya–Yosida (RKKY) [68–70] interaction in determining the relative orientation of the magnetization in FM layers in FM–AFM multilayers. The RKKY interaction, in fact, couples the magnetic moments in FM layers through the conduction electrons of a nonmagnetic metal spacer. Instead, when the spacer is an AFM insulator, FM interlayer coupling can be described in terms of Slonczewski's proximity magnetism model [71], which has been developed for uncompensated interfaces and requires strong AFM–FM interfacial coupling, as compared with the domain-wall energy in the AFM material. For perfectly flat interfaces, the expected coupling between the FM layers is either parallel or antiparallel

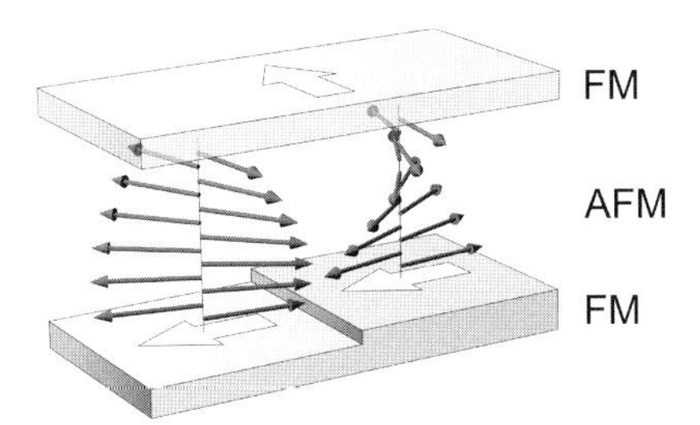

Figure 1.5 An artist's impression of the winding spin structures expected to form in the AFM spacer separating two FM slabs according to Sloncewski's proximity magnetism model. The magnetization in the FM layers is indicated by white arrows. Thin gray arrows indicate the spin direction in the AFM planes.

depending on the number of AFM planes. AFM–FM interface roughness results in a lateral modulation of the number of AFM layers, and therefore in a competition between parallel and antiparallel coupling. According to Slonczewski, the perpendicular coupling through an AFM spacer should then be described by an energy term of the form $C[\theta^2 + (\theta - \pi)^2]$, where θ is the angle between the magnetization of the two FM layers and C the AFM-thickness-dependent coupling strength. Therefore, the Slonczewski coupling energy displays a minimum at $\theta = \pi/2$, accounting for perpendicular interlayer alignment. The particular expression of the coupling energy term depends on the AFM spins forming winding structures, which are schematically shown in Figure 1.5.

The proximity magnetism model has recently been generalized to the case where the interfacial coupling and AFM domain-wall energy densities are comparable [72]. In this "extended" proximity model, the FM–FM coupling energy can be approximated, to lowest order, by a combination of a biquadratic contribution, proportional to $\cos^2\theta$, and a term proportional to $\sin^2(2\theta)$. The main difference between Sloncewski's model and its extended version is the behavior of the FM layers at high applied fields, as the magnetizations of the two FM layers become parallel only asymptotically for the Slonczewski coupling term, while in the extended proximity model saturation of the magnetizations of both FM layers in the direction of the applied field is obtained at finite fields. This behavior is related to the existence of local minima at $\theta = 0$ and $\theta = \pi$ in the coupling energy term of the extended model. The existence of these additional minima stems from the fact that, for a weak AFM–FM coupling, the AFM spins can rearrange by unwinding the twisted AFM magnetic structure predicted by the proximity model, confining the frustration at one of the AFM–FM interfaces [72]. The presence of the $\sin^2(2\theta)$ term in the expression of the coupling energy is explained by the inequivalence between $\pm\pi$ and 0 energy minima, which is a result of the different arrangements of the AFM local moments in different coupling configurations [72].

Perpendicular FM interlayer coupling has indeed been observed in Fe_3O_4/NiO/Fe_3O_4(001) trilayers [73], which appear to be well described by Slonczewski's model since, as discussed above, the NiO/Fe_3O_4(001) interface is magnetically uncompensated while the magnetocrystal anisotropy in NiO is weak. These two characteristics thus fulfill the conditions of strong interfacial coupling compared to the AFM domain-wall energy density, which is approximately proportional to $\sqrt{JK}$ [74], J and K being the exchange coupling constant and the anisotropy constant of the material, respectively. On the other hand, Fe/NiO/Fe(001) trilayers are better described by the extended proximity magnetism model [75]. Despite that Fe_3O_4 is also formed at the interface obtained by depositing NiO on Fe(001) [76], a different oxide, namely FeO, is expected to form when Fe is deposited on NiO(001) to complete the trilayer [77]. The AFM–FM coupling in the Fe/NiO/Fe trilayer can thus be expected to be much lower than in Fe_3O_4/NiO/Fe_3O_4, at least at one interface.

Both proximity models predict a perpendicular FM–FM interlayer coupling only when the AFM spacer thickness is below a critical value t_c, which is expected to be of the same order of magnitude as the width of domain walls (δ_W) in the bulk AFM material. Indeed, a transition between perpendicular to parallel FM–FM coupling in zero applied magnetic field is observed at similar values of t_c for Fe_3O_4/NiO/Fe_3O_4 ($t_c \approx 5$ nm) and Fe/NiO/Fe ($t_c \approx 4$ nm) trilayers. However, the values measured for t_c are considerably smaller than the experimental value of δ_W in bulk NiO single crystals, which has been reported to vary between 134 and 184 nm [78]. Such a large discrepancy is not surprising since proximity models assume a coherent rotation of spins in AFM planes parallel to the interfaces of the multilayer, and do not take into account the three-dimensional structure at the AFM domain walls originating at defects in the AFM volume, such as vacancies and dislocations, or at the interface, such as atomic steps. As discussed in 6, these defects are expected to reduce the coupling energy between the FM layers, with the result that the value of t_c might be considerably smaller than that of δ_W in the bulk AFM material.

1.5
Micromagnetic Structure at AFM–FM Interfaces

It has been shown that in FM films grown on top of an AFM material, the magnetic domains and domain walls tend to be small compared to the case of otherwise similar FM films grown on nonmagnetic substrates. In some cases, domain sizes of the order of a few micrometers have been observed [56, 57, 79–83]. Investigating the origin of the phenomena that contribute to the formation of small magnetic domains is obviously of primary importance in information technology [84]. Chapter 8 will thoroughly address the micromagnetic structure of AFM–AF systems by discussing magnetic-contrast soft X-ray microscopy. In this section, we focus instead on Fe/NiO/Fe trilayers, which display magnetic domains of the Fe overlayer that can be even smaller, with a minimum lateral size of about 20 nm, as shown in Figure 1.6 [85]. The magnetic contrast observed in Figure 1.6b,c is due to domains exhibiting opposite magnetization as a result of the Fe–Fe coupling across the NiO spacer, whose thickness has been chosen so as to obtain a collinear

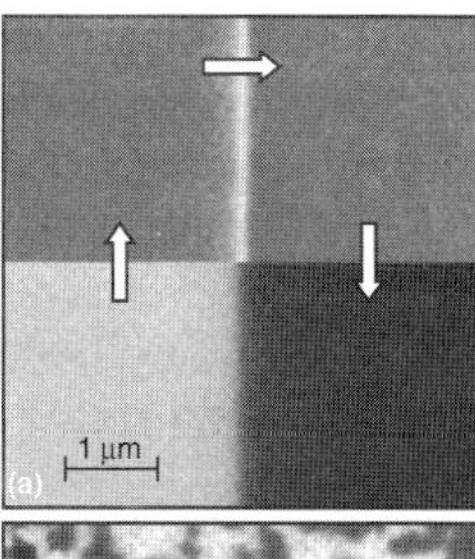

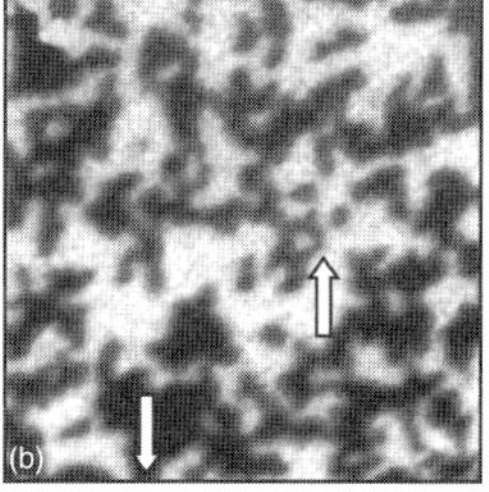

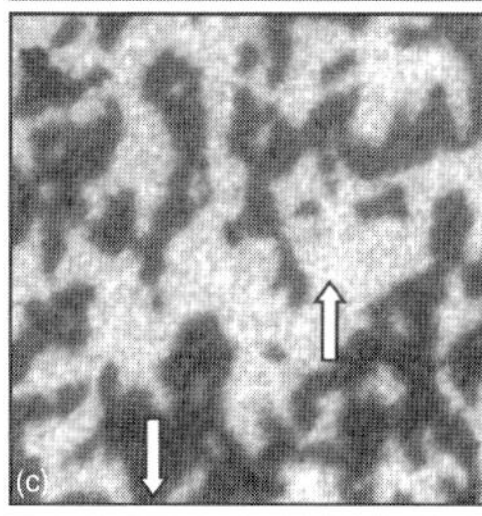

Figure 1.6 Spin-polarized low-energy electron microscopy images of magnetic domains in: (a) Fe(001) substrate. In the upper part of the image the spin of the primary electron beam is perpendicular to the domain wall visible in the field of view. In the bottom part the spin is parallel. (b) Fe/NiO/Fe with $t_{NiO} = 4.5$ nm and $t_{over} = 1.3$ nm. (c) Same sample area (b) with t_{over} increased to 6 nm. In (b) and (c), the spin of the primary beam is parallel to the vertical axis of the figure. The field of view in the three images is 4×4 μm^2. A domain coarsening going from panel (b) to (c) is visible. The direction of the magnetization in domains appearing dark or bright is indicated by the arrows. In panel (a), the 180° domain walls are sufficiently wide ($\delta \approx 240$ nm) to reveal the spin structure of the wall in the upper half of panel (a), obtained by using a 90° spin polarization of the primary electron beam of the microscope. Reprinted figure with permission from [85]. © Copyright (2007) by the American Physical Society.

FM–FM coupling. Similar domains are also obtained for thinner NiO layers, corresponding to a perpendicular coupling in zero applied field (see Section 1.4.2).

Before discussing the conditions that allow the stabilization of such small magnetic domains, it is worth mentioning the mechanisms that determine the width δ_W of a domain in a bulk material. In fact, a domain cannot be significantly smaller than δ_W, otherwise the system would gain energy by simply suppressing it, that is, by allowing the moments inside the domain to reorient in the same direction as the magnetization of the surrounding material. Approximately, δ_W is given by the relation [74, 84]

$$\delta_W \approx ab\sqrt{\frac{J}{K}} \tag{1.5}$$

where J and K are the exchange coupling and the magnetocrystal anisotropy constants, respectively, while a is the atomic lattice spacing and b is a dimensionless parameter, which depends on the details of the spin structure within the wall, for example, $b \approx 10$ in bulk Fe [86]. A large exchange interaction with respect to the magnetocrystal anisotropy favors thick domain walls, since the local moments would exert a strong torque on those of the neighboring atoms.

Conversely, a large magnetic anisotropy leads to narrow domain walls since the moments of each atom tend to align parallel with the crystallographic directions. Generally, domain walls are quite larger than the lattice spacing since the anisotropy is much smaller than the exchange energy. For example, in bulk Fe $K \approx 4$ μeV and $J \approx 100$ meV per atom [86–89], corresponding to a δ_W value of the order of several hundred nanometers, as observed also in Figure 1.6a. This bulk value for δ_W has to be compared with the value $\delta_W < 20$ nm (20 nm corresponding to the lateral resolution of the instrument) measured in Figure 1.6b,c on a Fe/NiO/Fe trilayer. Such a large decrease in δ_W cannot be traced back to a possible increment in the value of the single-ion anisotropy constant K even though the low coordination number and the presence of strain could, in principle, induce a large magnetocrystal anisotropy at surfaces or interfaces (see section 1.3.1). In fact seems unreasonable that this mechanism could produce the 100-fold raise of K that would be required to justify the observed value of δ_W. The stabilization of small magnetic domains in the FM overlayer can be instead associated with the AFM–FM interface exchange energy.

One of the first models describing the statistical properties of the magnetic domains at AFM–FM interfaces is due to Malozemoff [90]. In his model, the presence of interface roughness gives rise to a random field acting on the interface spins. The magnetically softer (either AFM or FM) material breaks up into domains whose size is determined by the competition between the exchange interaction and an additional uniaxial in-plane anisotropy. This random field model was the first attempt at including the role of defects to explain why the exchange bias energy density in AFM–FM systems is much smaller than what should be expected from the value of the exchange interaction between AFM and FM atoms across the surface. This model, however, considers a domain structure that resembles a chessboard, where all the domains have a lateral dimension equal to L. This leads to an average interface energy density scaling as $1/L$, that is, to an interfacial exchange coupling energy on the domain footprint scaling as L. On inspection of Figure 1.6, it is seen that the domains in the Fe overlayer show instead fractal morphology, with a fractal dimension of about 1.6, very close to the value expected for an Ising system with random fluctuations [85]. The assimilation of the overlayer to an Ising system seems justified by the strong coupling with the second Fe layer, which results, as explained in Section 1.4.2, in a strong uniaxial anisotropy of the overlayer. Models reproducing the fractal structure of the domains [91] show that the exchange coupling energy on a *minimum* stable domain footprint should instead scale as L^2. Moreover, the structure of domain walls at AFM–FM interfaces is determined by frustration of the exchange interaction rather than by the balance between exchange and magnetocrystal anisotropy, as explained in 6. The presence of frustration is the driving mechanism that justifies the stabilization of much smaller domains at AFM–FM interfaces than in the bulk of the single constituent materials [92]. By combining all these elements, one realizes that the minimum domain size depends on the balance between the exchange energy e_{DW} contained in a domain wall encircling a circular FM domain of diameter equal to L and the AFM–FM interface exchange energy $e_{f,af}$. The domain is stable if $e_{DW} \propto L$ equals $e_{f,af} \propto L^2$ [85]. If $e_{f,af} > e_{DW}$ the domain expands and if $e_{f,af} < e_{DW}$ the domain shrinks. The scaling laws for e_{DW} and $e_{f,af}$ allow determining of the minimum domain size [85]. By explicitly including in the model the dependence of the exchange energy per unit domain-wall length, one can also reproduce the observed coarsening of the domains, also visible from the comparison of Figure 1.6b,c, which have been obtained on the same sample area after successive Fe depositions [85]. The coarsening is a consequence of e_{DW} increasing with the Fe thickness (a higher domain wall obviously requires more energy). The higher negative pressure exerted on the domain by its wall has to be compensated by an increase of the domain footprint to provide the necessary AFM–FM exchange energy to prevent the domain from collapsing.

The importance of frustration over anisotropy is further confirmed by magnetic-contrast X-ray photoemission microscopy studies on Fe/CoO/Fe(001) trilayers. FM/AFM/FM trilayers in which the AFM material is either NiO or CoO are characterized by interfaces with similar morphology and chemical quality, but NiO and CoO exhibit large differences in the respective values of the magnetocrystal anisotropy: CoO is characterized by $K = 2 \times 10^5$ erg cm^{-3}, while $K = 3.3 \times 10^2$ erg cm^{-3} for NiO [48, 93]. Despite the large difference in the magnetocrystal anisotropy (3 orders of magnitude), the morphology and the minimum domain size in the topmost layer in Fe/NiO/Fe and Fe/CoO/Fe trilayers are very similar [94].

1.6 Applications

The most common technological application of AFM materials is in spin valves based on the giant magnetoresistance (GMR) [95, 96] and tunnel magnetoresistance (TMR) [97, 98] effects, which are currently used as reading heads for magnetic storage media or as memory elements in magnetic random-access memories (MRAMs). Such devices consist of two conducting FM materials that exhibit high or low electrical resistance depending on the relative alignment of the magnetic layers. An AFM layer is used to pin the harder FM reference layer by exchange bias, so that the magnetization of only the softer layer in the spin valve is reversed by the application of an external magnetic field. Similarly, the most advanced disk media are antiferromagnetically coupled, making use of interfacial exchange to effectively increase the stability of small magnetic particles whose behavior would otherwise be superparamagnetic.

For most spin-valve sensors, metallic antiferromagnets like NiMn, PtMn, or IrMn were and still are used. Although there were spin-valves using NiO in the early days of the discovery of GMR effect, their use was discontinued for thermal stability reasons. Though the idea of using Co-ferrite as a bias material to replace PtMn in current-in-plane sensors was mooted, this was not implemented. Thus, the number of AFM oxide applications for magnetic recording is limited. Nonetheless, although the AFM pinning layer in a spin-valve does not directly contribute to the magnetoresistance and rather constitutes a parasitic resistance, a considerably enhanced in-plane magnetoresistance as compared to conventional all-metal spin-valves has been observed in Co/Cu/Co and $Ni_{80}Fe_{20}/Cu/Ni_{80}Fe_{20}$ spin-valves confined within insulating AFM NiO layers [99, 100]. Such an improvement is because spin-polarized electrons are reflected back in the FM layer at the NiO/metal interface.

However, as recording densities increase and critical dimensions shrink accordingly, the industry has moved from a current-in-plane to a current-perpendicular-to-plane sensor geometry. Lower sensor cross sections require materials with low resistance–area products like metals to obtain adequate signal-to-noise ratios. For current-in-plane sensors, oxides were useful as they do not shunt the current; however, these sensors have already been abandoned. At present, current-perpendicular-to-plane sensors with conducting antiferromagnets are employed. The same is true also for MRAMs (Stefan Maat, San Jose Research Center - Hitachi Global Storage Technologies, private communication).

1.7 Conclusions

AFM oxide materials in low-dimensional geometries, either in nonmagnetic or magnetic environments, display a rich variety of magnetic behaviors. They are very interesting materials to investigate the fundamental physics of finite-size effects expressed by magnetic systems. Despite the limited applications in actual technology, AFM oxides represent very important reference and model systems for studying the interface coupling phenomena that are ultimately exploited in devices such as spin-valves.

This book provides an extensive discussion of the complex and intriguing phenomena observed in such systems, with particular emphasis on the growth and the magnetic characterization through different experimental methods and theoretical modeling approaches.

References

1. Finazzi, M., Duò, L., and Ciccacci, F. (2009) Magnetic properties of interfaces and multilayers based on thin antiferromagnetic oxide films. *Surf. Sci. Rep*, **64**, 139.
2. Anderson, P.W. (1950) Antiferromagnetism. Theory of superexchange interaction. *Phys. Rev*, **79**, 350.
3. Mermin, N.D. and Wagner, H. (1966) Absence of ferromagnetism or antiferromagnetism in one- or two-dimensional isotropic Heisenberg models. *Phys. Rev. Lett*, **17**, 1133.
4. Punnoose, A., Magnone, H., Seehra, M.S., and Bonevich, J. (2001) Bulk to nanoscale magnetism and exchange bias in CuO nanoparticles. *Phys. Rev. B*, **64**, 174420.
5. Punnoose, A. and Seehra, M.S. (2002) Hysteresis anomalies and exchange bias in 6.6 nm CuO nanoparticles. *J. Appl. Phys*, **91**, 7766.
6. Stewart, S.J., Multigner, M., Marco, J.F., Berry, F.J., Hemando, A., and Gonzalez, J.M. (2004) Thermal dependence of the magnetization of antiferromagnetic copper(II) oxide nanoparticles. *Solid State Commun*, **130**, 247.
7. Zheng, X.G., Xu, C.N., Nishikubo, K., Nishiyama, K., Higemoto, W., Moon, W.J., Tanaka, E., and Otabe, E.S. (2005) Finite-size effect on Néel temperature

in antiferromagnetic nanoparticles. *Phys. Rev. B*, **72**, 014464.

8. Abarra, E.N., Takano, K., Hellman, F., and Berkowitz, A.E. (1996) Thermodynamic measurements of magnetic ordering in antiferromagnetic superlattices. *Phys. Rev. Lett*, **77**, 3451.
9. Alders, D., Tjeng, L.H., Voogt, F.C., Hibma, T., Sawatzky, G.A., Chen, C.T., Vogel, J., Sacchi, M., and Iacobucci, S. (1998) Temperature and thickness dependence of magnetic moments in NiO epitaxial films. *Phys. Rev. B*, **57**, 11623.
10. Sako, S., Ohshima, K., Sakai, M., and Bandow, S. (1996) Magnetic property of CoO ultrafine particle. *Surf. Rev. Lett*, **3**, 109.
11. Ambrose, T. and Chien, C.L. (1996) Finite-size effects and uncompensated magnetization in thin antiferromagnetic CoO layers. *Phys. Rev. Lett*, **76**, 1743.
12. Tang, Y.J., Smith, D.J., Zink, B.L., Hellman, F., and Berkowitz, A.E. (2003) Finite size effects on the moment and ordering temperature in antiferromagnetic CoO layers. *Phys. Rev. B*, **67**, 054408.
13. Gangopadhyay, S., Hadjipanayis, G.C., Sorensen, C.M., and Klabunde, K.J. (1993) Exchange anisotropy in oxide passivated Co fine particles. *J. Appl. Phys*, **73**, 6964.
14. Resnick, D.A., Gilmore, K., Idzerda, Y.U., Klem, M.T., Allen, M., Douglas, T., Arenholz, E., and Young, M. (2006) Magnetic properties of Co_3O_4 nanoparticles mineralized in Listeria innocua Dps. *J. Appl. Phys*, **99**, 08Q501.
15. He, L., Chen, C., Wang, N., Zhou, W., and Guo, L. (2007) Finite size effect on Néel temperature with Co3O4 nanoparticles. *J. Appl. Phys*, **102**, 103911.
16. Wang, R.M., Liu, C.M., Zhang, H.Z., Chen, C.P., Guo, L., Xu, H.B., and Yang, S.H. (2004) Porous nanotubes of Co3O4: synthesis, characterization, and magnetic properties. *Appl. Phys. Lett*, **85**, 2080.
17. Salabas, E.L., Rumplecker, A., Kleitz, F., Radu, F., and Schuth, F. (2006) Exchange anisotropy in nanocasted Co3O4 nanowires. *Nano. Lett*, **6**, 2977.
18. Fisher, M.E. and Barber, M.N. (1972) Scaling theory for finite-size effects in the critical region. *Phys. Rev. Lett*, **28**, 1516.
19. Zhang, R. and Willis, R.F. (2001) Thickness-dependent curie temperatures of ultrathin magnetic films: effect of the range of spin-spin interactions. *Phys. Rev. Lett*, **86**, 2665.
20. Chen, K., Ferrenberg, A.M., and Landau, D.P. (1993) Static critical behavior of three-dimensional classical Heisenberg models: a high-resolution Monte Carlo study. *Phys. Rev. B*, **48**, 3249.
21. Ferrenberg, A.M. and Landau, D.P. (1991) Critical behavior of the three-dimensional Ising model: A high-resolution Monte Carlo study. *Phys. Rev. B*, **44**, 5081.
22. Lang, X.Y., Zheng, W.T., and Jiang, Q. (2006) Size and interface effects on ferromagnetic and antiferromagnetic transition temperatures. *Phys. Rev. B*, **73**, 224444.
23. van der Zaag, P.J., Ijiri, Y., Borchers, J.A., Feiner, L.F., Wolf, R.M., Gaines, J.M., Erwin, R.W., and Verheijen, M.A. (2000) Difference between blocking and Néel temperatures in the exchange biased Fe_3O_4/CoO system. *Phys. Rev. Lett*, **84**, 6102.
24. Skumryev, V., Stoyanov, S., Zhang, Y., Hadjipanayis, G., Givord, D., and Nogués, J. (2003) Beating the superparamagnetic limit with exchange bias. *Nature*, **423**, 850.
25. Altieri, S., Finazzi, M., Hsieh, H.H., Haverkort, M.W., Lin, H.-J., Chen, C.T., Frabboni, S., Gazzadi, G.C., Rota, A., Valeri, S., and Tjeng, L.H. (2009) Image charge screening: A new approach to enhance magnetic ordering temperatures in ultrathin correlated oxide films. *Phys. Rev. B*, **79**, 174431.
26. Giovanardi, C., di Bona, A., Altieri, S., Luches, P., Liberati, M., Rossi, F., and Valeri, S. (2003) Structure and morphology of ultrathin NiO layers on Ag(001). *Thin Solid Films*, **428**, 195.
27. Slack, G.A. (1960) Crystallography and domain walls in antiferromagnetic NiO crystals. *J. Appl. Phys*, **31**, 1571.

28. Altieri, S., Tjeng, L.H., Voogt, F.C., Hibma, T., and Sawatzky, G.A. (1999) Reduction of Coulomb and charge-transfer energies in oxide films on metals. *Phys. Rev. B*, **59**, R2517.
29. Altieri, S., Tjeng, L.H., Voogt, F.C., Hibma, T., Rogojanu, O., and Sawatzky, G.A. (2002) Charge fluctuations and image potential at oxide-metal interfaces. *Phys. Rev. B*, **66**, 155432.
30. Zaanen, J., Sawatzky, G.A., and Allen, J.W. (1985) Band gaps and electronic structure of transition-metal compounds. *Phys. Rev. Lett*, **55**, 418.
31. Anderson, P.W. (1959) New approach to the theory of superexchange interactions. *Phys. Rev*, **115**, 2.
32. Zaanen, J. and Sawatzky, G.A. (1987) The electronic structure and superexchange interactions in transition-metal compounds. *Can. J. Phys*, **65**, 1262.
33. Jefferson, J.H. (1988) Theory of superexchange in antiferromagnetic insulators. *J. Phys. C: Solid State Phys*, **21**, L193.
34. Duffy, D.M. and Stoneham, A.M. (1983) Conductivity and 'negative U' for ionic grain boundaries. *J. Phys. C*, **16**, 4087.
35. Hesper, R., Tjeng, L.H., and Sawatzky, G.A. (1997) Strongly reduced band gap in a correlated insulator in close proximity to a metal. *Europhys. Lett*, **40**, 177.
36. Altieri, S., Tjeng, L.H., and Sawatzky, G.A. (2001) Ultrathin oxide films on metals: new physics and new chemistry? *Thin Solid Films*, **400**, 9.
37. Bloch, D. (1966) The 10/3 law for the volume dependence of superexchange. *J. Phys. Chem. Solids*, **27**, 881.
38. Zhang, W.-B., Hu, Y.-L., Han, K.-L., and Tang, B.-Y. (2006) Pressure dependence of exchange interactions in NiO. *Phys. Rev. B*, **74**, 054421.
39. Sidorov, V.A. (1998) Differential thermal analysis of magnetic transitions at high pressure: Néel temperature of NiO up to 8 GPa. *Appl. Phys. Lett*, **72**, 2174.
40. Massey, M.J., Chen, N.H., Allen, J.W., and Merlin, R. (1990) Pressure dependence of two-magnon Raman scattering in NiO. *Phys. Rev. B*, **42**, 8776.
41. Casassa, S., Ferrari, A.M., Busso, M., and Pisani, C. (2002) Structural, magnetic, and electronic properties of the NiO monolayer epitaxially grown on the (001) Ag surface: an ab initio density functional study. *J. Phys. Chem. B*, **106**, 12978.
42. Cinquini, F., Giordano, L., Pacchioni, G., Ferrari, A.M., Pisani, C., and Roetti, C. (2006) Electronic structure of NiO/Ag(100) thin films from DFT+U and hybrid functional DFT approaches. *Phys. Rev. B*, **74**, 165403.
43. Hillebrecht, F.U., Ohldag, H., Weber, N.B., Bethke, C., Mick, U., Weiss, M., and Bahrdt, J. (2001) Magnetic moments at the surface of antiferromagnetic NiO(100). *Phys. Rev. Lett*, **86**, 3419.
44. Altieri, S., Finazzi, M., Hsieh, H.H., Lin, H.-J., Chen, C.T., Hibma, T., Valeri, S., and Sawatzky, G.A. (2003) Magnetic dichroism and spin structure of antiferromagnetic NiO(001) films. *Phys. Rev. Lett*, **91**, 137201.
45. Csiszar, S.I., Haverkort, M.W., Hu, Z., Tanaka, A., Hsieh, H.H., Lin, H.-J., Chen, C.T., Hibma, T., and Tjeng, L.H. (2005) Controlling orbital moment and spin orientation in CoO layers by strain. *Phys. Rev. Lett*, **95**, 187205.
46. Yosida, K. (1998) *Theory of Magnetism*, Series in Solid-State Sciences, Springer.
47. Keffer, F. and O'Sullivan, W. (1957) Problem of spin arrangements in MnO and similar antiferromagnets. *Phys. Rev*, **108**, 637.
48. Finazzi, M. and Altieri, S. (2003) Magnetic dipolar anisotropy in strained antiferromagnetic films. *Phys. Rev. B*, **68**, 054420.
49. Koon, N.C. (1997) Calculations of exchange bias in thin films with ferromagnetic/antiferromagnetic interfaces. *Phys. Rev. Lett*, **78**, 4865.
50. Schulthess, T.C. and Butler, W.H. (1998) Consequences of spin-flop coupling in exchange biased films. *Phys. Rev. Lett*, **81**, 4516.

51. Finazzi, M., Portalupi, M., Brambilla, A., Duò, L., Ghiringhelli, G., Parmigiani, F., Zacchigna, M., Zangrando, M., and Ciccacci, F. (2004) Magnetic anisotropy of NiO epitaxial thin films on Fe(001). *Phys. Rev. B*, **69**, 014410.
52. Finazzi, M. (2004) Interface coupling in a ferromagnet/antiferromagnet bilayer. *Phys. Rev. B*, **69**, 064405.
53. Tsai, S.-H., Landau, D.P., and Shulthess, T.C. (2003) Effect of interfacial coupling on the magnetic ordering in ferro-antiferromagnetic bilayers. *J. Appl. Phys*, **93**, 8612.
54. Finazzi, M., Biagioni, P., Brambilla, A., Duò, L., and Ciccacci, F. (2005) Disclinations in thin antiferromagnetic films on a ferromagnetic substrate. *Phys. Rev. B*, **72**, 024410.
55. Matsuyama, H., Haginoya, C., and Koike, K. (2000) Microscopic imaging of Fe magnetic domains exchange coupled with those in a NiO(001) surface. *Phys. Rev. Lett*, **85**, 646.
56. Ohldag, H., Scholl, A., Nolting, F., Anders, S., Regan, T.J., Hillebrecht, F.U., and Stöhr, J. (2001) Spin reorientation at the antiferromagnetic NiO(001) surface in response to an adjacent ferromagnet. *Phys. Rev. Lett*, **86**, 2878.
57. Ohldag, H., Regan, T.J., Stöhr, J., Scholl, A., Nolting, F., Lüning, J., Stamm, C., Anders, S., and White, R.L. (2001) Spectroscopic identification and direct imaging of interfacial magnetic spins. *Phys. Rev. Lett*, **87**, 247201.
58. Finazzi, M., Brambilla, A., Biagioni, P., Graf, J., Gweon, G.-H., Scholl, A., Lanzara, A., and Duò, L. (2006) Interface coupling transition in a thin epitaxial antiferromagnetic film interacting with a ferromagnetic substrate. *Phys. Rev. Lett*, **97**, 097202.
59. Arenholz, E., van der Laan, G., Chopdekar, R.V., and Suzuki, Y. (2007) Angle-dependent Ni^{2+} x-ray magnetic linear dichroism: interfacial coupling revisited. *Phys. Rev. Lett*, **98**, 197201.
60. Krug, I.P., Hillebrecht, F.U., Haverkort, M.W., Tanaka, A., Tjeng, L.H., Gomonay, H., Fraile-Rodríguez, A., Nolting, F., Cramm, S., and Schneider, C.M. (2008) Impact of interface orientation on magnetic coupling in highly ordered systems: a case study of the low-indexed Fe_3O_4/NiO interfaces. *Phys. Rev. B*, **78**, 064427.
61. Borchers, J.A., Erwin, R.W., Berry, S.D., Lind, D.M., Ankner, J.F., Lochner, E., Shaw, K.A., and Hilton, D. (1995) Long-range magnetic order in Fe_3O_4/NiO superlattices. *Phys. Rev. B*, **51**, 8276.
62. Ijiri, Y., Borchers, J.A., Erwin, R.W., Lee, S.-H., van der Zaag, P.J., and Wolf, R.M. (1998) Perpendicular coupling in exchange-biased Fe_3O_4/CoO superlattices. *Phys. Rev. Lett*, **80**, 608.
63. Ijiri, Y., Schulthess, T.C., Borchers, J.A., van der Zaag, P.J., and Erwin, R.W. (2007) Link between perpendicular coupling and exchange biasing in Fe_3O_4/CoO multilayers. *Phys. Rev. Lett*, **99**, 147201.
64. Dzyaloshinsky, I. (1957) Thermodynamic theory of weak ferromagnetism in antiferromagnetic substances. *Sov. Phys. JETP*, **5**, 1259.
65. Moriya, T. (1960) Anisotropic superexchange interaction and weak ferromagnetism. *Phys. Rev*, **120**, 91.
66. Inderhees, S.E., Borchers, J.A., Green, K.S., Kim, M.S., Sun, K., Strycker, G.L., and Aronson, M.C. (2008) Manipulating the magnetic structure of Co core/CoO shell nanoparticles: implications for controlling the exchange bias. *Phys. Rev. Lett*, **101**, 117202.
67. Tomiyasu, K., Inami, T., and Ikeda, N. (2004) Magnetic structure of CoO studied by neutron and synchrotron x-ray diffraction. *Phys. Rev. B*, **70**, 184411.
68. Ruderman, M.A. and Kittel, C. (1954) Indirect exchange coupling of nuclear magnetic moments by conduction electrons. *Phys. Rev*, **96**, 99.
69. Kasuya, T. (1956) A theory of metallic ferro- and antiferromagnetism on Zener's model. *Prog. Theor. Phys*, **16**, 45.
70. Yosida, K. (1957) Magnetic properties of Cu-Mn alloys. *Phys. Rev*, **106**, 893.
71. Slonczewski, J.C. (1995) Overview of interlayer exchange theory. *J. Magn. Magn. Mater*, **150**, 13.

72. Xi, H. and White, R.M. (2000) Coupling between two ferromagnetic layers separated by an antiferromagnetic layer. *Phys. Rev. B*, **62**, 3933.
73. van der Heijden, P.A.A., Swüste, C.H.W., de Jonge, W.J.M., Gaines, J.M., van Eemeren, J.T.W.M., and Schep, K.M. (1999) Evidence for roughness driven 90° coupling in Fe_3O_4yNiOyFe_3O_4 trilayers. *Phys. Rev. Lett*, **82**, 1020.
74. Hubert, A. and Schäfer, R. (1998) *Magnetic Domains: The Analysis Of Magnetic Microstructures*, Springer, Berlin.
75. Brambilla, A., Biagioni, P., Portalupi, M., Zani, M., Finazzi, M., Duò, L., Vavassori, P., Bertacco, R., and Ciccacci, F. (2005) Magnetization reversal properties of Fe/NiO/Fe(001) trilayers. *Phys. Rev. B*, **72**, 174402.
76. Finazzi, M., Brambilla, A., Duò, L., Ghiringhelli, G., Portalupi, M., Ciccacci, F., Zacchigna, M., and Zangrando, M. (2004) Chemical effects at the buried NiO/Fe(100) interface. *Phys. Rev. B*, **70**, 235420.
77. Regan, T.J., Ohldag, H., Stamm, C., Nolting, F., Lüning, J., Stöhr, J., and White, R.L. (2001) Chemical effects at metal/oxide interfaces studied by x-ray-absorption spectroscopy. *Phys. Rev. B*, **64**, 214422.
78. Weber, N.B., Ohldag, H., Gomonaj, H., and Hillebrecht, F.U. (2003) Magnetostrictive domain walls in antiferromagnetic NiO. *Phys. Rev. Lett*, **91**, 237205.
79. Nolting, F., Scholl, A., Stöhr, J., Seo, J.W., Fompeyrine, J., Siegwart, H., Locquet, J.-P., Anders, S., Lüning, J., Fullerton, E.E., Toney, M.F., Scheinfein, M.R., and Padmore, H.A. (2000) Direct observation of the alignment of ferromagnetic spins by antiferromagnetic spins. *Nature*, **405**, 767.
80. Scholl, A., Nolting, F., Seo, J.W., Ohldag, H., Stöhr, J., Raoux, S., Locquet, J.-P., and Fompeyrine, J. (2004) Domain-size-dependent exchange bias in Co/$LaFeO_3$. *Appl. Phys. Lett*, **85**, 4085.
81. Welp, U., Velthuis, S.G.E., Felcher, G.P., Gredig, T., and Dahlberg, E.D. (2003) Domain formation in exchange biased Co/CoO bilayers. *J. Appl. Phys*, **93**, 7726.
82. Blomqvist, P., Krishnan, K.M., and Ohldag, H. (2005) Direct imaging of asymmetric magnetization reversal in exchange-biased Fe/MnPd bilayers by x-ray photoemission electron microscopy. *Phys. Rev. Lett*, **94**, 107203.
83. Chopra, H.D., Yang, D.X., Chen, P.J., Brown, H.J., Swartzendruber, L.J., and Egelhoff, W.F. Jr. (2000) Nature of magnetization reversal in exchange-coupled polycrystalline NiO-Co bilayers. *Phys. Rev. B*, **61**, 15312.
84. Dennis, C.L., Borges, R.P., Buda, L.D., Ebels, U., Gregg, J.F., Hehn, M., Jouguelet, E., Ounadjela, K., Petej, I., Preibeanu, I.L., and Thornton, M.J. (2002) The defining length scales of mesomagnetism: a review. *J. Phys.: Condens. Matter*, **14**, R1175.
85. Rougemaille, N., Portalupi, M., Brambilla, A., Biagioni, P., Lanzara, A., Finazzi, M., Schmid, A.K., and Duò, L. (2007) Exchange-induced frustration in Fe/NiO multilayers. *Phys. Rev. B*, **76**, 214425.
86. Lilley, B.A. (1950) Energies and widths of domain boundaries in ferromagnetics. *Phil. Mag*, **41**, 792.
87. Stoner, E.C. (1950) Ferromagnetism: magnetization curves. *Rep. Prog. Phys*, **13**, 83.
88. Kittel, C. (1949) Physical theory of ferromagnetic domains. *Rev. Mod. Phys*, **21**, 541.
89. Stearns, M.B. (1986) in *3d, 4d, and 5d Elements, Alloys and Compounds*, Landolt-Börnstein, New Series, Group III, Vol. 19, Pt. a(ed. H.P.J.Wijn), Springer, Berlin, p. 34.
90. Malozemoff, A.P. (1987) Random-field model of exchange anisotropy at rough ferromagnetic-antiferromagnetic interfaces. *Phys. Rev. B*, **35**, 3679.
91. Esser, J., Nowak, U., and Usadel, K.D. (1997) Exact ground-state properties of disordered Ising systems. *Phys. Rev. B*, **55**, 5866.
92. Morosov, A.I. and Sigov, A.S. (2004) New type of domain walls: domain

walls caused by frustrations in multilayer magnetic nanostructures. *Phys. Solid State*, **46**, 395.

93. Carey, M.J., Berkowitz, A.E., Borchers, J.A., and Erwin, R.W. (1993) Strong interlayer coupling in CoO/NiO antiferromagnetic superlattices. *Phys. Rev. B*, **47**, 9952.

94. Brambilla, A., Sessi, P., Cantoni, M., Finazzi, M., Rougemaille, N., Belkhou, R., Vavassori, P., Duò, L., and Ciccacci, F. (2009) Frustration-driven micromagnetic structure in Fe/CoO/Fe thin film layered systems. *Phys. Rev. B*, **79**, 172401.

95. Grünberg, P., Schreiber, R., Pang, Y., Brodsky, M.B., and Sowers, H. (1986) Layered magnetic structures: evidence for antiferromagnetic coupling of Fe layers across Cr interlayers. *Phys. Rev. Lett*, **57**, 2442.

96. Baibich, M.N., Broto, J.M., Fert, A., Nguyen Van Dau, F., Petroff, F., Eitenne, P., Creuzet, G., Friederich, A., and Chazelas, J. (1988) Giant magnetoresistance of (001)Fe/(001)Cr magnetic superlattices. *Phys. Rev. Lett*, **61**, 2472.

97. Miyazaki, T. and Tezuka, N. (1995) Giant magnetic tunneling effect in Fe/Al_2O_3/Fe junction. *J. Magn. Magn. Mater*, **139**, L231.

98. Moodera, J.S., Kinder, L.R., Wong, T.M., and Meservey, R. (1995) Large magnetoresistance at room temperature in ferromagnetic thin film tunnel junctions. *Phys. Rev. Lett*, **74**, 3273.

99. Swagten, H.J.M., Strijkers, G.J., Bloemen, P.J.H., Willekens, M.M.H., and de Jonge, W.J.M. (1996) Enhanced giant magnetoresistance in spin-valves sandwiched between insulating NiO. *Phys. Rev. B*, **53**, 9108.

100. Veloso, A., Freitas, P.P., Wei, P., Barradas, N.P., Soares, J.C., Almeida, B., and Sousa, J.B. (2000) Magnetoresistance enhancement in specular, bottom-pinned, Mn83Ir17 spin valves with nano-oxide layers. *Appl. Phys. Lett*, **77**, 1020.

2
Growth of Antiferromagnetic Oxide Thin Films

Sergio Valeri, Salvatore Altieri, and Paola Luches

2.1
Introduction

Because of the challenging fundamental electronic properties, of the potential applications in different areas ranging from catalysis to spin electronics and of their significant role in the chemical and physical processes taking place in the ecosystem, there is a steadily growing interest in the research on metal-oxides [1–4]. From the magnetic point of view, much of the past and current efforts have focused on binary metal-oxide (Me_xO_y) thin films containing 3d transition metals like Mn, Fe, Co, or Ni, because of their importance in magnetic devices technology.

Antiferromagnets, unlike ferromagnets, do not exhibit a net magnetization and cannot be readily controlled by, or coupled to, an external magnetic field. Thus, most of the studies of magnetism for technological applications were focused on ferromagnets. Antiferromagnets, however, are intrinsically very interesting magnetic systems due to the complex interplay between magnetic and structural order, which presents new challenges and can lead to novel applications. Antiferromagnetic (AF) films are also studied for their application in magnetic devices, where they can be incorporated into data storage systems.

The unique properties of AF materials are especially evident in systems of reduced dimensionality, as pointed out by a number of studies [3–8]. Therefore, special efforts have been addressed to the preparation and study of AF transition-metal oxide films. Phenomena influenced by the film thickness (i.e., by the vertical confinement) were mainly investigated, namely, the AF-superparamagnetic transition in ultrathin AF films, the thickness-dependent Néel temperature (T_N) in AF layers, and the coupling of ferro-, antiferro-, and nonmagnetic films in multilayers. Critical properties such as magnetic order, ordering temperature, and interfacial moments were found to depend on the stoichiometry, defectivity, and morphology of the films, on the extent of the crystalline order, and on the sharpness of the interfaces between the film and the substrate or between different films in multilayers, which are to a great extent determined by the preparation method [3, 4, 6, 7].

Magnetic Properties of Antiferromagnetic Oxide Materials.
Edited by Lamberto Duò, Marco Finazzi, and Franco Ciccacci

ISBN: 978-3-527-40881-8

The two main driving forces to develop AF oxide materials in the form of thin films and multilayers were on one side the progressive improvement of the fabrication procedures and on the other, the availability of more and more sophisticated and reliable characterization methods and apparatus. The first approach to the fabrication of ordered oxide layers was the controlled oxidation of bulk metal single crystals. Exposure of metal surfaces to oxygen leads to the moderately rapid formation of a thin (3–5 monolayers (MLs) thick) oxide overlayer. At room temperature (RT), the oxygen uptake stops once this limit is reached, because the formed oxide passivates the surface, while at higher temperatures the slow growth of a thicker oxide is typically observed [9–11]. The simultaneous impingement of argon ions on the substrate during oxygen exposures was found to enhance the oxidation process [12]. Direct (and ion assisted) oxidation of metal surfaces generally results in the presence of a rather imperfect oxide film, that might result from the large lattice mismatch between the oxide film and the substrate, in spite of the use of highly oriented metal surfaces as substrates.

Oxidation of thin metal films predeposited on suitable substrates has successively been explored to reduce the mismatch and improve the quality of the oxide films. Metal deposition, postoxidation, and annealing cycles can be repeated several times to overcome the diffusion limit and obtain several monolayer thick oxide films [13–15]. Thicker oxide layers can be produced by sputtering of bulk oxide targets or by reactive sputtering of pure metal targets in Ar^+–O_2 mixed atmosphere. Tens of nanometer thick NiO layers have also been epitaxially grown on MgO or Si substrates by radio frequency (RF) or direct current (DC) sputtering [16, 17]. Pulsed laser deposition was also exploited for the fabrication of very thick NiO and CoO films [18], and of FeO and Fe_2O_3 films [19].

However, in recent years the interesting and novel physical properties of thin and ultrathin films and multilayers claimed for a much better control of film thickness, morphology, and crystal quality. Since the crystal quality and morphology of ultrathin films can strongly influence their electronic and magnetic properties [3], new growth protocols have been proposed to ensure high-quality epitaxial growth, mainly based on the reactive deposition of metal layers on low mismatch substrates under a controlled oxygen atmosphere. Reactive deposition represents the most popular and successful method for metal-oxide thin film fabrication [1] and a number of selected examples will be discussed in the following sections of this work. This very flexible approach opened up the way to overlayer structural engineering via the mismatch-induced interfacial strain and to the modification of the oxide properties via the electronic interactions between the overlayer and the substrate atoms.

It is noteworthy to mention that a further step in the reactive deposition procedure is represented by the use of more reactive oxidizing agents like atomic oxygen or NO_2 [1, 20, 21]. It has been reported that this procedure significantly improves the film quality mainly in terms of interface sharpness, surface flatness, and substrate wettability even at the lowest coverages. Relevant recent examples refer to the preparation of NiO [22] and FeO [23] films.

As mentioned before, the availability of more and more sophisticated and reliable characterization methods and apparatus was a key factor in pushing both fundamental and technological activities on magnetic films in general and oxide AF films in particular. Pioneering characterization activity was mainly based on X-ray photoemission spectroscopy (XPS) and Auger electron spectroscopy (AES) as far as the composition and electronic properties of the metal-oxide AF films are concerned, while structure and growth morphology were mainly investigated by low-energy electron diffraction (LEED) and reflection high-energy electron diffraction (RHEED) [21], respectively. XPS technique was successfully complemented by inverse photoemission (IP) for the full description and complete understanding of the electronic states in correlated oxides [24, 25]. Species-specific structural characterization of oxide films was possible by exploiting forward focusing of either outgoing Auger electrons (Auger electron diffraction (AED)) [26] and photoelectrons (X-ray photoelectron diffraction (XPD)) [27] or ingoing primary electrons (primary beam diffraction modulated electron emission (PDMEE)) [28], while the growth morphology took advantage of the profile analysis (spot profile analysis (SPA)) of LEED spots [29]. Strain analysis, relaxation behavior, and interface studies in a number of metal-oxide films on metals or oxide substrates have been performed by using this set of techniques.

The advent of scanning probes opened up a wide range of exciting opportunities for the morphologic characterization of oxide films on metal substrates by imaging surfaces in real space. The growth and the local electronic structure of NiO and CoO oxide films have been studied from submonolayer coverage up to several layer thickness by scanning tunneling microscopy (STM) and scanning tunneling spectroscopy (STS) [30, 31].

A relevant step in the analytical capabilities was represented by the use of synchrotron facilities. Polarization-dependent X-ray absorption spectroscopy (XAS) has allowed us to obtain an accurate description of the local atomic environment of metal atoms in metal-oxides, including the exact determination of the in-plane and out-of-plane strains in oxide layers [32, 33]. Specular X-ray reflectivity (XRR) has allowed a very accurate film thickness and interface roughness evaluation [34, 35]. Grazing incidence X-ray scattering (GIXS) has provided very precise details on the atomic structure, roughness, relaxation, reconstruction, growth mode, and interface morphology of a number of oxide thin films [36].

The study of magnetism in AF thin films has posed a real experimental challenge due to the magnetically compensated nature of the systems. Conventional optical, X-ray, and neutron techniques [37] were implemented by the use of X-ray magnetic *linear* dichroism (XMLD) spectroscopy [3, 38]. XMLD spectroscopy can be applied to all *uniaxial* magnetic systems, for example, antiferromagnets. It has been shown that XMLD spectroscopy in conjunction with photoelectron emission microscopy (PEEM) is capable of imaging the detailed AF domain structure of a surface or interface [39, 40]. In comparison to other imaging techniques of AF domains, XMLD/PEEM spectromicroscopy offers elemental specificity, surface sensitivity, and improved spatial resolution. The first unambiguous surface images with AF contrast were obtained by XMLD microscopy for NiO(100) grown on MgO(100) [41].

Modeling of metal-oxide films fully supported the experimental investigations and in a number of cases it provided a major help in designing more focused experiments. However, theoretical investigations also present some difficulties. First, it is not trivial to select a quantum-mechanical *ab initio* approach capable of describing at the same level of accuracy both a metal and a magnetic insulator; furthermore, a periodic model is necessary to take into account both short- and long-range Coulomb effects, which play a fundamental role in the characteristics of the electronic structure of the interface.

A quantum-mechanical *ab initio* approach based on the use of a periodic model and of a hybrid-exchange Hamiltonian has been adopted to simulate the properties of ultrathin layers of nickel oxide epitaxially grown on Ag(001) [42]. Density-functional theory (DFT) methods fail in describing the electronic structure of transition-metal oxides, when these are dominated by the on-site Coulomb repulsion between d electrons. To overcome this shortcoming, an approach that combines DFT with a Hubbard Hamiltonian for the Coulomb repulsion and exchange interaction has been used to investigate the relative stability of various FeO/Pt(111) interface structures, taking also into account different magnetic configurations [43] and helping in the interpretation of STM images [43, 44]. The structure and thermodynamic stability of other AF oxides, such as the (0001) surfaces of hematite, as a function of temperature and oxygen pressure, were investigated by *ab initio* DFT with the generalized gradient approximation [45]. Molecular dynamics calculations were carried out to understand the formation of misfit dislocations and the interfacial structural features at the buried interface of epitaxially grown α-Fe_2O_3(0001)/α-Al_2O_3(0001) system. The calculations show that misfit dislocations form in the Al_2O_3 substrate and terminate at the interface, in agreement with the experimental observations [46]. The surface termination of Fe_2O_3(0001) prepared on α-Al_2O_3(0001) has been studied by a comparison of XPD experiments with quantum-mechanical scattering theory [47]. The stability of epitaxial MnO(100) and MnO(111) layers on Pd(100) surface as a function of lattice strain has been examined by *ab initio* DFT. All calculations were performed using the "Vienna" *ab initio* simulation package within the spin-polarized generalized gradient approximation and are based on the projector augmented-wave method. It was suggested that the growth of MnO(111) layers is energetically preferred over MnO(100) due to the epitaxial stabilization at the metal-oxide interface [48].

This chapter is organized as follows. In Section 2.2, we discuss the most recent studies on NiO films. Section 2.3 is dedicated to the most relevant results obtained on CoO films. Section 2.4 deals with the relatively less studied AF binary oxides, such as MnO, FeO, and α-Fe_2O_3. In Sections 2.5 and 2.6 some special topics, such as the oxide–substrate interface and films exposing polar surfaces, are discussed. Finally, Section 2.7 reports the conclusions and shows some trends for future studies in this field.

2.2 Nickel Oxide

2.2.1 Ultrathin NiO Layers

The recent discovery of unprecedented properties in metal-supported subnanometric oxide films [6, 49–52] has fostered growing interest for ML oxide layers on metal substrates. This is a very subtle issue which requires the control at the atomic scale of chemical and structural interface abruptness in complex heterostructures, through atomic-layer control of the growth processes. Among the 3d oxides, monolayer NiO and CoO on Ag(001), Pd(001), Pt(111), Ir(001), and Rh vicinal surfaces are the most studied systems [1, 15, 21, 22, 29, 31, 53–68]. For a film thickness larger than 2 ML, epitaxial and stoichiometric oxide layers can be easily prepared on various metallic and insulating substrates by standard O_2-based reactive deposition [1]. Nevertheless, even for these simple rock-salt-oxide monolayers, controlling and understanding the growth properties at the 3d oxide–metal interface remains a challenging problem. Indeed, there is a general consensus in the literature that monolayer films of 3d oxide deposited on metal substrates by O_2-based reactive deposition result in a very puzzling growth behavior characterized by complex atomic reconstruction, nonuniform oxide layer thickness, and multiple oxidation states with properties that drastically depend on the particular growth parameters such as O_2 pressure and/or deposition/annealing temperature [1, 15, 29, 31, 53–63, 66, 67] and by the particular nature of the oxidizing gas used [22].

The case of 1-ML NiO/Ag(001) is somehow a special one since it has been extensively studied by several groups and it has also recently been used to address the general question of why a single monolayer of a 3d oxide can be easily grown as a chemically and structurally well defined single epitaxial phase on insulating substrates but not so on metallic substrates [22]. For a coverage up to 1 ML, the O_2-reactive deposition of Ni on Ag(001) at substrate temperatures in the range of 300–500 K and O_2 pressure in the 10^{-6} mbar range results in the formation of a characteristic 90°-rotated two-domain (2×1) structure (see STM data in Figure 2.1) [1, 29, 53–56] which evolves into a (1×1) structure as the film thickness is increased to 2 ML or higher [54, 55]. The (2×1) superstructure of 1-ML NiO/Ag(001) is a peculiar oxide–metal interface phase, which is one atom high and wets the Ag(001) surface [1, 29, 53–56]. Postgrowth annealing treatments at 500 K either under ultra-high-vacuum (UHV) or under O_2 atmosphere result in the disappearance of the (2×1) superstructure accompanied by a dramatic restructuring of the deposited NiO film, which condenses into double-layer NiO islands with a height of two atomic planes (Figure 2.1) [53, 55, 56], for which the opening of a bulk-like conductivity gap has been observed by STS [53]. The low sticking coefficient of O_2 on Ag(001), the solubility of nickel atoms in the silver subsurface, and the high mobility of silver atoms even at relatively low temperatures are believed to play an important role in the formation of the (2×1) phase [55]. To explain its origin, a model of the atomic structure was proposed, which is based

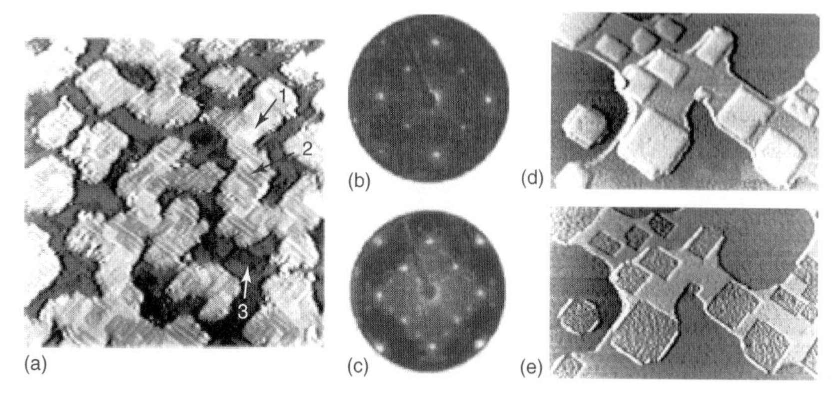

Figure 2.1 (a) 30 × 30 nm^2 STM image of two-thirds ML NiO deposited at RT with O_2 on Ag(001) showing Ag islands of monoatomic height on the substrate (1), islands with (2 × 1) periodicity mainly protruding from the substrate (2), but also incorporated within the substrate (3); (b) LEED pattern of clean Ag(100) surface; (c) LEED pattern of the (2 × 1)−1 ML NiO deposited on Ag(100) with O_2 at 463 K; (d) 50 × 100 nm^2 STM image showing the formation of double-layer NiO/Ag islands obtained after annealing at 450 K the two-thirds ML NiO deposited at RT, measured with $V = 5$ V and $I = 2$ nA; (e) same as (d) measured with $V = 1$ V. (Reprinted from [53, 54] with permission.)

on the formation of rows of Ni and O atoms with different possible registries [53]. Tensor LEED calculations indicated that the atomic structure of the (2 × 1) phase is similar to that of a NiO(111) polar surface consisting of coplanar and alternating Ni and O layers, where the (2 × 1) symmetry is a result of the in-plane displacement of the nickel and oxygen atoms compared to those in ideal NiO(111) surface [55].

While there is a universal consensus about the formation of the (2 × 1) phase for O_2-reactive deposition of Ni on Ag(001), the chemical composition of this monolayer phase is more controversial. Some reports claim that the (2 × 1)−1 ML NiO/Ag(001) is chemically stoichiometric [53, 55, 56]. However, low-energy ion scattering (LEIS) and XPS studies clearly showed the presence of nonoxidized metallic nickel atoms in the 1-ML NiO/Ag(001) film deposited by O_2-reactive deposition [54, 55], and a connection between the existence of the (2 × 1) phase and the presence of metallic nickel was established [54]. In particular, it was shown that by substantially increasing the O_2/Ni ratio during the reactive deposition of the 1-ML NiO(001)/Ag(001), an almost complete disappearance of the (2 × 1) superstructure and a simultaneous strong reduction of the fraction of nonoxidized nickel can be achieved [55]. This behavior could be traced back to the substantial solubility of nickel atoms in Ag which might be limited by increasing the oxidation probability for nickel atoms impinging on the Ag substrate surface [55].

The O_2-reactive deposition of monolayer amount of Ni on the Pt(111) metal surface has been studied by STM and LEED as a function of substrate temperature and O_2 pressure in the range 400–550 K and 10^{-10}–10^{-6} mbar, respectively [57].

It was found that one-dimensional network-like Ni–O structures and islands grow at 400 K under an O_2 atmosphere of 10^{-6} mbar. Depending on the combination of substrate temperature and O_2 pressure, Ni–O structures with (7×1) and (4×2) reconstructions were observed and found to transform reversibly into a (2×2) reconstruction within a narrow O_2 pressure range of $1.5–2.0 \times 10^{-6}$ mbar. In analogy with the known (2×2) reconstruction observed for oxygen adsorbed on a Ni(111) surface, the Ni–O (2×2) phase was assigned to the adsorption of 0.25 ML of oxygen on the Pt(111)-supported Ni(111) monolayer, while the Ni–O (7×1) and (4×2) were tentatively interpreted as due to the partial oxidation of the Ni(111) monolayer to NiO(001). All the observed Ni–O structures were stable only under high O_2 pressure. Indeed, a reduction of the oxygen pressure down to the 10^{-10} mbar range led to oxygen desorption, leaving metallic Ni(111) monolayer islands on the Pt(111) surface [57]. The atomically reconstructed Ni–O phases observed on Pt(111) bear similarities with those reported for the oxidation of a monolayer amount of Ni deposited on Rh(15 15 13) and (5 5 3) vicinal surfaces, although in the latter case the prepared nickel oxide structures turned out to be more stable under UHV conditions [58, 59]. When deposited on the rhodium vicinal surfaces, the nickel atoms decorate the step edges, and upon dosing 15 Langmuir (1 L $= 1.33 \times 10^{-6}$ mbar s) of O_2 at 573 K uniaxially ordered quasi-one-dimensional monolayer Ni–O structures are formed exhibiting (2×1) and (6×1) symmetries [58, 59] with an oxygen-deficient chemical composition.

The growth of 1-ML NiO on a metal substrate which exhibits a high reactivity toward oxygen and high intermixing with nickel atoms presents additional experimental complications, which call for more elaborated growth solutions. This is the case of 1-ML NiO/Pd(001) for which it was shown that oxygen can act either as an inhibitor or as a promoter of NiO epitaxial growth, depending on the dose of metallic Ni deposited on the Pd(001) surface [60–63]. For this system, two distinct growth procedures were developed, namely, RT O_2-reactive deposition and postoxidation methods, consisting of repeated cycles of UHV deposition of a predetermined amount of metallic nickel atoms followed by annealing at 440 K under O_2 atmosphere in the 10^{-6} mbar range [60–63]. The postoxidation method turned out to be effective for the growth of epitaxial NiO film only if the initial dose of Ni evaporated on the clean Pd(001) substrate exceeded a critical value corresponding approximately to two NiO equivalent monolayers. In this case, however, the obtained oxide overlayer was strongly oxygen deficient and poorly long-range ordered. For a Ni dose smaller than the critical one, a polycrystalline NiO film is obtained due to a substantial oxidation of the Pd substrate promoted by the presence of Ni. The O_2-reactive deposition method starting from the bare Pd(001) surface yielded a NiO layer with improved quality. To avoid both Pd(001) oxidation and Ni–Pd intermixing, the deposition of the NiO monolayer either by reactive deposition or by postannealing methods was performed on an oxygen preadsorbed $(\sqrt{5} \times \sqrt{5})$-R27^0 O/Pd(001) surface obtained by dosing the Pd surface with 600 L of O_2 at 570 K. The use of a starting oxygen-preadsorbed Pd(001) surface prevents the surface oxidation during film growth and warrants the best long-range ordered oxide monolayer. The monolayer NiO film obtained in this way has a very peculiar

$c(4 \times 2)$-defective structure due to the formation of a regular Ni vacancy array with a strongly two-dimensional wetting character and a Ni_3O_4 chemical composition, which was also directly proved by atomically resolved STM experiments (Figure 2.2) [58, 59]. The $c(4 \times 2)$-Ni_3O_4/Pd(001) monolayer is a very peculiar phase which is stabilized by the oxide–metal interface and does not have a counterpart at the surface of bulk NiO.

An important step forward in the preparation of 1-ML NiO/Ag(001) with highly controlled chemical composition, crystal structure, and morphology has been reported in a work where the reactive deposition of monolayer nickel oxide was performed using *atomic* rather than molecular oxygen at a substrate temperature of 463 K [22]. The activated oxygen has often been used to improve the oxidation conditions of growing oxide films in plasma-based deposition methods [1]. In those applications, however, the fraction of atomic oxygen is only limited and in general not even quantitatively known. On the contrary, in [22] the experimental conditions were chosen in such a way that the growth of the 1-ML NiO/Ag(001) was entirely controlled by a collimated beam of atomic oxygen produced by a UHV-compatible

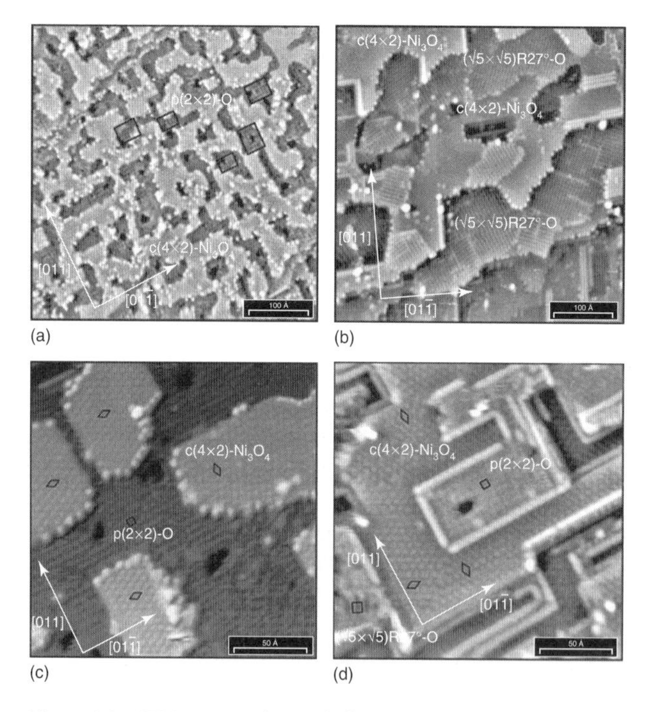

Figure 2.2 STM images obtained after O_2-reactive deposition of Ni on Pd(100) under different conditions: (a) 0.5-ML Ni deposited at RT with $P_{O2} = 2 \times 10^{-6}$ mbar and then annealed at 570 K under an oxygen pressure of $P_{O2} = 5 \times 10^{-7}$ mbar; (b) 0.5-ML Ni deposited at 523 K under $P_{O2} = 2 \times 10^{-6}$ mbar; (c) 0.25-ML Ni deposited as in (a); (d) 0.25-ML Ni deposited as in (b). The observed reconstructions in the different areas are indicated in the figures. (Reprinted from [61] with permission.)

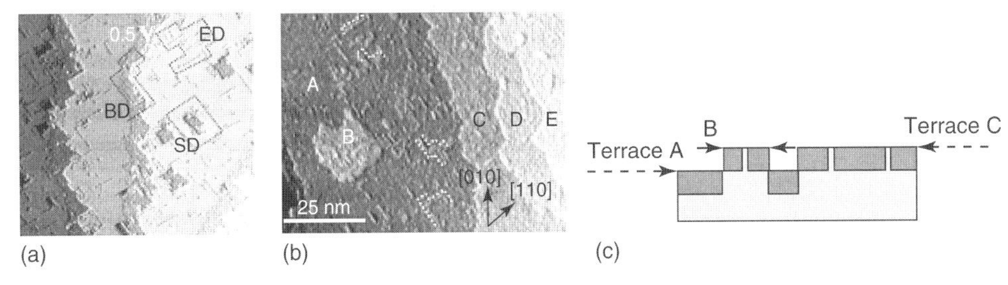

Figure 2.3 Monolayer NiO(001) deposited at 463 K by *atomic* oxygen on Ag(001). (a) 150 × 150 nm^2 STM image of 0.3-ML NiO(001)/Ag(001). The ED and SD dashed contours highlight regions covered by one-dimensional (ED) and two-dimensional (BD) islands appearing as depressions with edges parallel to <110> crystal directions; the BD dashed contour highlights the border of the Ag step edge, which after NiO deposition appears strongly reoriented along <110>, and the nearby down-hill terrace region, which appears as a depression area after the NiO deposition. (b) STM image of 0.85-ML NiO(001) covering the Ag(001) surface with uniform monoatomic thickness; uncovered Ag areas appear as one-dimensional interconnected protrusions with fractional coverage of about 0.15 ML. (c) Side-view sketch of (b) with monoatomic NiO(100) islands (dark gray) and monoatomic Ag terraces, steps, and ridges (light gray). (Reprinted from [22] with permission.)

O_2-thermal cracker. *In situ* XPS/AES, LEED, and voltage-dependent STM analysis showed that the NiO films grown by atomic oxygen on Ag(001) have the same stoichiometry as cleaved NiO, a highly uniform monoatomic thickness, and a (1 × 1) symmetry at any coverage in the 0.2–1.0-ML range (Figure 2.3) [22]. The combined XPS and STM study showed that the (1 × 1)–1 ML NiO/Ag(001) phase has a very peculiar Ni 2p XPS lineshape, well distinguishable from that of 2–3-ML and thicker NiO/Ag(001) films (Figure 2.4), as expected according to theoretical DFT + U calculations [64], thus providing a spectroscopic tool to characterize and identify the (1 × 1) monoatomic NiO layer on Ag(001). The atomic-oxygen-reactive deposition of NiO/Ag(001) also revealed new features at the submonolayer coverage, showing that NiO monoatomic islands nucleate both on flat silver terraces and at Ag step edges forming two-dimensional and one-dimensional nickel oxide monoatomic nanostructures with polar edges oriented along the <110> crystal directions. Moreover, the monoatomic NiO islands nucleated at the Ag step edges turned out to induce a dramatic reorientation of the silver step edge itself, which, after the NiO film growth, exhibits a preferential <110> orientation [22]. The stabilization of a well-stoichiometric (1 × 1)–1 ML NiO/Ag(001) phase with an excellent crystal structure and highly uniform monoatomic thickness was ascribed to the strong oxidizing power of the atomic oxygen used in the reactive deposition [22]. It was proposed that in general the complex growth properties of monolayer 3d oxides on metal substrates may be a result of the competition between two main opposite processes involving 3d adatoms on a metal surface exposed to an oxidizing gas, namely, an oxidation process due to the gas molecules and a reduction process due to electron transfer and/or hybridization from the underneath metal surface. On the basis of this idea, it was suggested that under extreme oxidizing conditions such as those

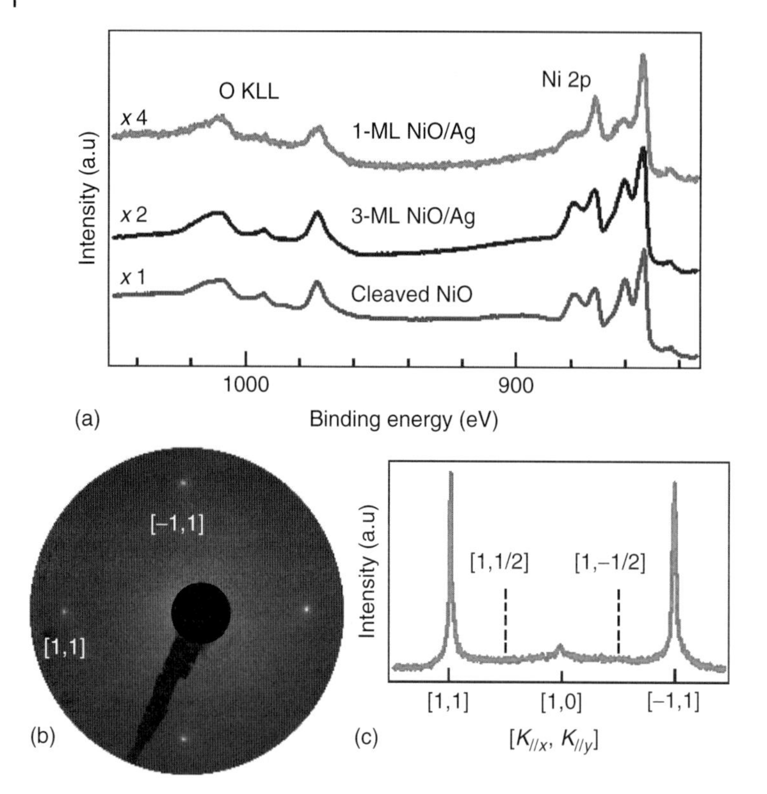

Figure 2.4 Monolayer NiO(001) deposited at 463 K by *atomic* oxygen on Ag(001): (a) Ni 2p XPS and O KLL Auger spectra of 1- and 3-ML NiO(001)/Ag(001), and cleaved NiO; (b) LEED pattern of 1-ML NiO(001)/Ag(001) at electron beam energy $E = 94$ eV; (c) intensity profile across the [1, 1] and [−1, 1] spots in (b) showing the absence of fractional order spots. (Reprinted from [22] with permission.)

provided by atomic oxygen it could be possible to make the oxidation much more efficient than the reduction process, with the consequent formation of a well-defined ionic crystal phase stabilized by Madelung potential and image potential interactions at the oxide–metal interface, thus preventing the formation of a reconstructed and off-stoichiometric oxide monolayer [22]. According to this model, among the 3d transition metals Ni is the species with the highest reluctance toward oxidation when supported on an electron donor reservoir such as a metal substrate, as shown by the ratio $R = E_A/E_I$ between the electron affinity (E_A) and the 3d ionization potential (E_I) that provides a quantitative measure of the difficulty to oxidize 3d atoms on metal surfaces (for Sc, Ti, V, Cr, Fe, Co, Ni, and Cu, R values are 0.2, 0.1, 0.7, 0.4, 0.2, 0.6, 2.3, and 0.5, respectively). Therefore, on the basis of the results found for the atomic-oxygen-reactive deposition of 1-ML NiO/Ag(001), it was suggested [22] that similarly nonreconstructed and atomically flat monolayers of other 3d oxides on a metal substrate might be obtained by reactive deposition methods based on atomic oxygen or other strongly oxidizing species such as NO_2 or ozone. These types of growth methods were extensively used for oxide film growth on oxide substrates [21, 65], but have not yet been systematically explored for oxide–metal interfaces.

2.2.2
Thick NiO Films

The most direct way to obtain NiO thin films on a metal substrate is to oxidize a Ni single crystal. A large number of studies in the past used this method (see, for example, [9] and references therein). However, because of the large lattice mismatch (19%) between Ni and its oxide, the crystal quality of the films obtained in this way was not satisfactory [9]. The growth of good-quality NiO films has recently involved the use of substrates with lower lattice mismatch, and it is typically performed in molecular-beam epitaxy (MBE) systems by reactive deposition, that is, by the evaporation of Ni under an O_2 (or other oxidizing agents) partial pressure in the 10^{-8}–10^{-6} mbar range. The typical Ni evaporation rates used are of the order of 1 ML min^{-1}. The substrate is typically kept at a moderately high temperature (450–570 K) during the growth, in order to increase the mobility of the atoms when they reach the surface and obtain epitaxial films with uniform thickness. The use of different substrates leads to films with different magnetic properties, both due to epitaxy, that is, to the different strain status of the films [3, 4, 7], and to the different electronic properties of the film–substrate interface [6].

The most widely used metal substrate for the growth of good-quality epitaxial NiO films is Ag(001), because of its low mismatch (2%) with respect to the NiO lattice. LEED, angle resolved ultraviolet photoelectron spectroscopy (ARUPS), and AES showed the better quality of NiO films obtained by reactive deposition on Ag(001) at RT, compared to the ones prepared by oxidation of Ni overlayers predeposited on the same substrate [69]. NiO growth on Ag(001) has been shown to proceed in a layer-by-layer fashion by electron reflectivity measurements [29]. The films have been found to be stoichiometric in the 3–50-ML range [34, 70], and to have the rock-salt structure – with oxygen atoms on the top of the substrate Ag atoms – as shown by the multiple-scattering simulation of XAS data [32, 71–73]. A combined IP and valence-band (VB) photoemission study proved that NiO films on Ag(001) present a VB electronic structure closer to the ideal one than a NiO single crystal [24]. This result allows us to infer that the acceptor-like defects present in cleaved NiO(100) samples, which are responsible for the pinning of the Fermi level close to the valence-band maximum, are not present in the grown thin films [24]. The morphology of NiO films on Ag(001) has been characterized by STM [6, 70], scanning electron microscopy (SEM), and high-resolution transmission electron microscopy (HRTEM) [6]. As clearly seen in Figure 2.5a,b, the NiO layer uniformly covers the Ag substrate, supporting the idea of a very regular layer-by-layer growth mode. The NiO film surface shows very small islands of irregular shape with an average diameter of 4–7 nm. The root mean square roughness is in the 0.1 nm range. The interlayer distances within the NiO film can be directly measured by HRTEM [6] (Figure 2.5c). The characterization of the strain in AF oxide films is a very relevant

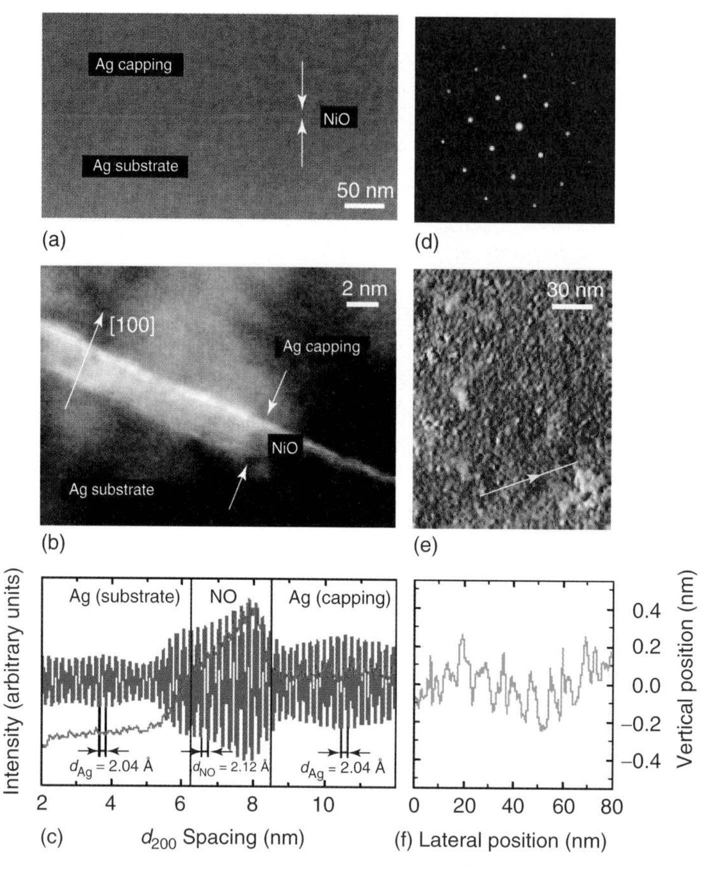

Figure 2.5 Ag/NiO/Ag cross-sectional lamella: (a) SEM image; (b) HRTEM image; (c) HRTEM intensity line scan through Ag/NiO/Ag interfaces; (d) selected area diffraction (SAD) pattern. (e) STM image of uncapped 3-ML NiO(100)/Ag(100) at $V = 3$ V and $I = 0.1$ nA. (f) STM intensity line scan along the yellow line in panel (e). (Reprinted from [6] with permission.)

issue, since it has been shown to deeply influence the magnetic properties, in particular the preferential orientation of the AF domains, shown to be in- or out-of-plane depending on the sign of the epitaxial deformation of the film structure [3, 4, 6, 7]. In NiO/Ag(001) films domains with in-plane orientation have been shown to be preferentially occupied as a consequence of the compressive in-plane strain [4, 7]. PDMEE and LEED were used to determine that at the first stages of the growth the films have a strained rock-salt structure, due to the epitaxial mismatch with the Ag substrate, with an in-plane contraction of 2% and an out-of-plane expansion consistent with the assumption of the bulk Poisson ratio [70]. The tetragonal distortion persists up to approximately 10 ML and it is gradually released for larger thickness until the structure relaxes to one of the bulk NiO at approximately 20 ML. On this system, XAS has allowed us to have a more accurate determination of the film strain as a function of thickness and to exclude significant intermixing with the substrate [73]. The results of the quantitative analysis of the data measured on the Ni and O K-edges in the extended energy range using multiple-scattering simulations indicate that the 3-ML film has both in-plane and out-of-plane interatomic distances compatible with the ones expected for perfect pseudomorphism on Ag(001), within the experimental accuracy. At 10 ML, the strain is partially released and the 50-ML film has a completely relaxed structure with interatomic distances very similar to the ones of bulk NiO. The mechanism for strain relaxation has been investigated in detail by Wollschläger *et al.* in a SPA–LEED study [29]. A high-resolution image of the specular electron beam on a 10-ML NiO film on Ag(001) is shown in Figure 2.6. Besides the specular peak additional streaky satellites along the <100> directions are clearly visible, together with weaker satellites with a fourfold symmetry. The behavior of the satellites positions, which move toward the central spot both for increasing film thickness and decreasing scattering phase, allows us to ascribe them to scattering from mosaics with a slightly tilted angle with respect to the (001) surface. These mosaics are formed during growth for a film thickness larger than 5 ML, that is, when the epitaxial strain starts to relax. The average angle of the tilted surfaces decreases with increasing film thickness. The average defect distance has been estimated from the position of one of the satellites, which does not depend on the scattering phase, to be approximately 15 nm [29]. The atomic scale structure of the metal–oxide interface is also a crucial issue in systems of this kind and will be dealt with in detail in Section 2.5. Here, we would like only to mention that for the NiO/Ag(001) system the metal–oxide interface distance has been determined by means of two experimental techniques, PDMEE and XAS, finding a value of 2.37 ± 0.05 Å,

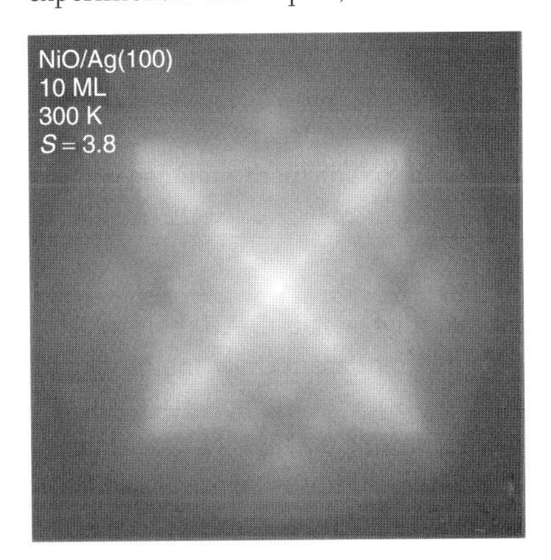

Figure 2.6 High-resolution recording of the specular (00) beam after depositing 10-ML NiO on Ag(100) for a scattering phase $S = 3.8$, with $S = dk_{\perp}/2\pi$, where d is the distance between the layers and $k_{\perp}$ is the component of the scattering vector **k** perpendicular to the surface. Additional satellites are clearly visible in direction to the {11} spots (equivalent to crystallographic <001> directions). The broadening of the satellites from the NiO film in the <001> direction and their sharpness in the perpendicular direction point to anisotropic features. (Reprinted from [29] with permission.)

significantly expanded with respect to both NiO and Ag interlayer distance [74]. The measured value is in good agreement with the one obtained by *ab initio* simulations [42, 74].

Among the other metal substrates used for the growth of NiO films, Au has a lattice mismatch with NiO comparable to the one of Ag. However, to the best of our knowledge, only the Au(111) surface has been used for the epitaxial growth of NiO films, giving (111)-oriented films. This topic is discussed in detail in Section 2.6.

The NiO/Pd(001) system is characterized by a larger lattice mismatch (7.3%) compared to Ag and Au substrates. A systematic study of this system by STM and LEED has been performed by Schoiswohl *et al.* [75, 76]. Stoichiometric (001)-oriented three-dimensional islands are obtained by reactive deposition at RT on top of the interfacial $c(4 \times 2)$ wetting layer (see Section 2.2.1). The epitaxial strain is completely released at 10–12 ML. The mechanism for strain release has been shown to involve, also for this system, the formation of mosaics, whose angle with respect to the surface decreases both with film thickness and with thermal treatments. Postdeposition annealing at 570–670 K in oxygen gives a significantly improved film morphology with larger islands, flat island surfaces, and improved structural order. The NiO/Pd(001) system, however, shows a low thermal stability due to the alloying tendency of Ni and Pd. After postdeposition annealing treatments at 670 K a long-range periodic superstructure appears (Figure 2.7). This periodic modulation can be due to two different effects. The first is the formation of a Moiré pattern, caused by the interference of tunneling processes between the closely related lattices of the overlayer and the substrate. The second effect is a periodic height modulation related to the presence of ordered defects, such as a dislocation network at the interface or other complex distortions caused by the

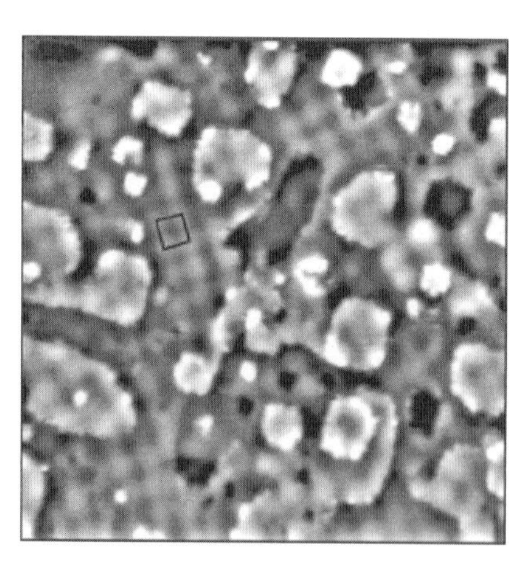

Figure 2.7 STM image illustrating the superstructure formed after annealing at 670 K a 6-ML-thick NiO film grown on Pd(001). 750×750 Å^2, $V = 3$ V, $I = 0.1$ nA; the superstructure unit cell is indicated in the image. (Reprinted from [76] with permission.)

lattice mismatch. It is interesting to note that patterned oxides of this kind can be used to induce the growth of ordered arrays of nanostructures [77, 78].

Since the lattice mismatch between NiO and MgO is less than −1%, it is possible to grow epitaxial NiO films also on this substrate. RHEED, XRR, and X-ray diffraction (XRD) studies have been reported by Hibma and coworkers [21, 35]. Considerable surface roughness has been observed for reactive growth in O_2 at substrate temperatures between 300 and 770 K, while films of superior quality have been obtained using NO_2 as an oxidizing agent. In the latter case, the NiO films are shown to grow layer-by-layer if the MgO substrate is kept between 470 and 670 K, while for lower growth temperatures an increase of surface disorder is observed [21]. The NiO layers are found to be fully strained up to a thickness of about 300 ML and still partially strained up to 800 ML [35]. The onset of the relaxation process appears at a thickness much larger compared to the theoretically predicted one, probably due to the presence of kinetic barriers during the growth process. It is interesting to note that for NiO films on MgO(001) the magnetic properties are different with respect to the ones of NiO films on Ag(001) due to the film structure and to the different interface. At variance with NiO films on Ag(001), in which the orientation of the AF domains is found to be preferentially in plane, for NiO films on MgO the orientation is mainly out-of-plane, due to the strain in the opposite direction [3]. Furthermore, the absence of an image charge compensation in NiO/MgO films leads also to different T_N in the two cases [6]. Epitaxial growth of NiO on the other two low-Miller-index surfaces of MgO, that is, MgO(110) and MgO(111), has also been studied [16]. The experiments give evidence for a flat NiO surface only on MgO(001), while on the other substrates the surface is rough and the films surface consists of 100 facets forming either wires, on the (110) surface, or tetrahedrons, on the (111) surface, which lower the surface energy.

Another substrate used for the epitaxial growth of NiO films is Fe(001) [25, 79]. This system deserves particular attention since it represents an epitaxial system made of an AF film deposited on a ferromagnetic (FM) substrate, which has significant applications related to the exchange-bias process. NiO grows with its [80] direction parallel to Fe[100] direction, that is, 45° rotated in the surface plane (the resulting lattice mismatch is only 3%). Compared to the other substrates mentioned above, Fe is much more reactive toward oxidation. To avoid an uncontrolled oxidation of the substrate, a passivated surface, namely, Fe(001)-$p(1\times1)$O, very stable to further oxidation, was used. The stoichiometry of the NiO films on Fe is confirmed by XPS (Figure 2.8) and the epitaxial orientation, by LEED. Spin-resolved IP measurements give a further proof of the good quality of the obtained NiO films. The Fe L_{23} XAS before and after the deposition of 20-ML NiO clearly indicate that the Fe substrate is partially oxidized by the reactive deposition of NiO, forming a thick Fe_3O_4 phase (see Section 2.5). Annealing to temperatures above 600 K gives rise to a strong interfacial intermixing, with the formation of a FeO layer and the diffusion of metallic Ni in the Fe substrate (Figure 2.8).

Also $Fe_3O_4(100)$ $(\sqrt{2} \times \sqrt{2})R45^\circ$ has been used as a substrate for NiO film growth [81]. In this case, in order not to oxidize the substrate to Fe_2O_3, the first 5 ML of NiO were grown one by one by depositing enough metallic Ni to be oxidized into 1-ML

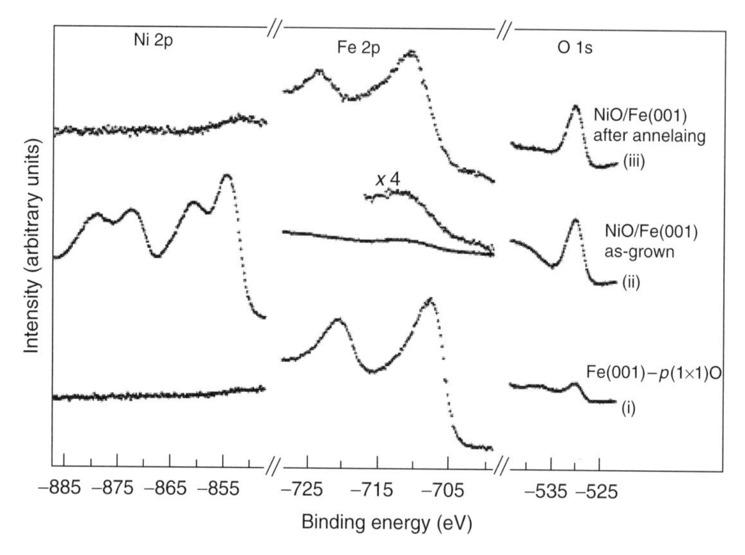

Figure 2.8 NiO films on Fe(001): XPS data related to Ni 2p, Fe 2p, and O 1s core levels, respectively. From bottom up, the spectra refer to: (i) the Fe(001)-$p(1 \times 1)$O substrate, (ii) an as-grown 30-Å-thick NiO/Fe(001) film, and (iii) the same film after 3 minutes annealing at 900 K. Intensities have not been rescaled. Note that the small bump, present in case (c), in the Ni 2p profile at ~852-eV binding energy is due to Fe 2s emission. In the Fe 2p spectra a shift of about 2.8 eV is evidenced comparing the substrate and the interface lineshapes. (Reprinted from [25] with permission.)

NiO and by subsequently exposing it to molecular oxygen. Films thicker than 5 ML were grown by reactive deposition. Even by using this procedure, some oxidation of the substrate at the interface could not be ruled out. Stoichiometric (011)-oriented films are obtained. Some three-dimensional clustering on the thin-film surface is observed.

2.3
Cobalt Oxide

2.3.1
Ultrathin CoO Layers

Beside NiO, CoO is the most studied 3d oxide system in the ML range. The chemical composition, atomic structure, and morphology of monolayer Co–O phases deposited by molecular oxygen in the pressure range of 10^{-6} mbar on a metal substrate dramatically depend on the particular preparation conditions in a much more complex way than in the case of 1-ML NiO on metals [31, 66, 67]. Monolayer amount of Co reactively deposited on Ag(001) at 300 K appears in STM images as a very rough film with irregularly shaped islands on four different height levels, involving also the presence of squared Ag islands and vacancies formed by partial removal of silver atoms from the substrate. Under these deposition conditions, Co–O structures with metallic character and hexagonal lattice periodicity of 0.3 nm were detected, which, however, were not found after annealing or high-temperature deposition [31, 66, 67]. In Co monolayer films deposited at 390 K or annealed at 400 K after RT deposition, Ag islands and vacancies are not visible in STM images, which instead reveal the appearance of two distinct types of Co–O islands, namely, round-shaped islands and square-shaped islands with edges oriented parallel to the <110> crystal direction. Similar islands were also found for RT-reactive deposition of Co monolayer amount on Ag(001) followed by annealing at 470 K [31, 66, 67]. In the latter case, three different types of Co–O related features were observed on the Ag(001) surface: double- (multi-) layer CoO(001), double-layer CoO(111) with threefold symmetry and a periodicity of 1.7 nm, and single-layer CoO(111) islands (Figure 2.9). The

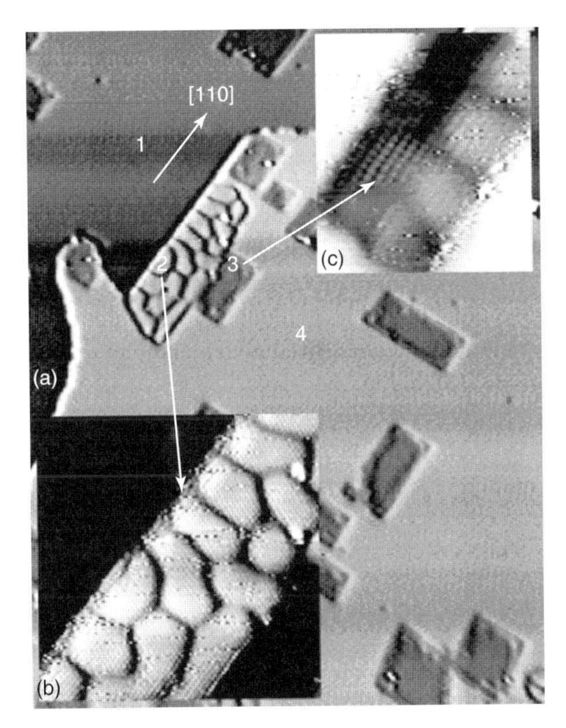

Figure 2.9 STM images of 0.5 ML of Co reactively deposited on Ag(100) at 400 K and 7 $\times$ 10^{-6} mbar O_2, and subsequently postannealed at 460 K (area = 109 $\times$ 86 nm^2, V = −3 V, I = 0.1 nA): region 1 and 4 correspond to the clean Ag(001) substrate; region 2 shows monolayer CoO(111) developed at Ag(100) step edges; region 3 shows CoO(001) multilayer islands appearing as squared depressions. The inset (b) (area = 18 $\times$ 18 nm^2, V = −0.1 V, I = 0.1 nA), obtained after zooming in, shows the atomic corrugation of CoO(111). In the inset (c) (area = 9 $\times$ 8 nm^2, V = −0.1 V, I = 0.25 nA), the tip conditions have changed for a few scanning lines, allowing the recognition of the atomic corrugation of CoO(001). (Reprinted from [67] with permission.)

double- (multi-) layer CoO(001) islands appeared in STM images either as embedded or floating on top of the Ag surface. The former have squared shape with edges aligned parallel to the <110> crystal directions, while the latter have more rounded shape, which was interpreted as due to the coexistence of both <110> and <100> edge orientations. The CoO(111) islands have edges aligned parallel to <110> directions and turned out to transform into double- (multi-) layer CoO(001) islands upon further prolonged annealing at temperatures higher than 470 K. In an attempt to identify the chemical nature of the CoO(001) and the CoO(111) islands, detailed voltage-dependent STM images and STS studies were performed [31, 66, 67]. This study showed that the double- (multi-) layer CoO(001) islands have an apparent topographic height that strongly changes with voltage and a band gap which matches the known band gap of bulk CoO, thus showing that these types of islands correspond to condensation of cobalt-oxide multilayers. On the contrary, the CoO(111) islands did not show voltage-dependent topographic height changes and exhibited metallic I–V curves which suggest that these islands may be more likely related to a metastable Co–O adsorbed phase rather than to stable real cobalt-oxide phase. In this study, it was not possible to conclude whether the atomic structure of the CoO(111) islands corresponds to that of (111)-oriented layers in CoO or to that of Co_3O_4 spinel-like structure. These works show that even after high-temperature annealing, O_2-reactive deposition of Co monolayer amount on Ag(001) does not result in the formation of a single monolayer of CoO oxide phase, but CoO oxide film with a minimum thickness of two atomic planes or more is always achieved [31, 66, 67].

In another STM and LEED work [15], monolayer CoO films were prepared either by O_2-reactive deposition or by Co deposition and postoxidation cycles on both the clean (1×1)-Ir(100) surface and the O-(2×1)-Ir(100) iridium surface precovered with half a monolayer of oxygen. It was shown that (sub)-monolayer deposition of Co on the Ir surface results in the formation of a CoO phase in a layer-by-layer fashion with three levels of topographic height and with a strong buckling of 0.5 Å. Moreover, the LEED pattern exhibited a peculiar $c(10 \times 2)$ quasi-hexagonal spot arrangement (Figure 2.10) which could be related to the hexagonal arrangement of ions in (111)-oriented layers of the two stable phases of bulk cobalt oxide, namely, the fcc-type rock-salt CoO and the cubic spinel Co_3O_4 [31]. In this work no chemical analysis was performed to determine the stoichiometric composition of the deposited CoO/Ir film. Nevertheless, on the basis of the comparison between detailed STM and LEED quantitative analysis and structural modeling of the rock-salt CoO and the cubic spinel Co_3O_4 structures, it was concluded that the atomic structure of the cobalt-oxide films grown on the iridium surface matches that of the hexagonal layer of polar CoO(111) accommodated on the square lattice of Ir. Small patches of Co_3O_4 were also detected in monolayer cobalt-oxide films deposited under more oxygen-rich conditions or on the oxygen preadsorbed Ir surface. The stability of the prepared monolayer CoO(111)/Ir(100) was explained on the basis of both the observed strong buckling and the presence of possible oxygen vacancies imaged in atomically resolved STM data, although other stabilization

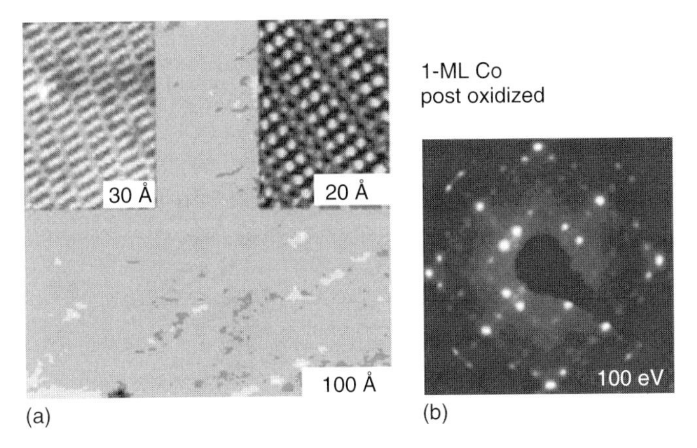

Figure 2.10 STM images recorded at three different blow ups (a), and $c(10 \times 2)$ LEED pattern (b) after full oxidation of 1-ML Co deposited on Ir(100)−(1 × 1). (Reprinted from [15] with permission.)

mechanisms of the polar cobalt-oxide surface such as impurity doping could not be excluded [31]. The stabilization of an impurity-free and unreconstructed monolayer CoO(111) polar film on a metal surface was reported for Co monolayer deposition on the Pt(111) surface and O_2 postoxidation [68].

2.3.2
Thick CoO Films

Cobalt-oxide films have been obtained by direct oxidation of hcp cobalt single crystals (see for example [82] and references therein). Depending on the experimental conditions, either CoO, Co_3O_4, or a mixture of the two phases can be obtained. By oxidation at 300–500 K under mild oxidizing conditions CoO is the major component obtained. For (111)-oriented CoO films, obtained by oxidation of the Co(0001) phase, due to the large lattice mismatch with the hcp substrate (20%), the structural quality is not very good. The Co(11$\bar{2}$0) surface has a lower lattice mismatch (5% along the [0001] direction and −2% along [11$\bar{2}$0] direction) and allows the growth of CoO films with better quality.

Some studies have also dealt with the oxidation of fcc Co films grown on Cu(001). The oxidation starts at exposures slightly larger than 10 L, after the chemisorption stage, and gives stoichiometric CoO films [83]. Angle- and spin-resolved photoemission measurements indicate that at high enough (>7 L) O_2 exposures at RT AF CoO is formed, while at low temperatures (150 K) the film is at least partially oxidized to Co_3O_4 [84].

The oxidation of 10-ML-thick body centered tertragonal (bct) Co films grown on Fe(001) has also been studied by PDMEE, XPS, and AES [14, 85, 86]. After oxidation under an oxygen partial pressure of 10^{-7} mbar at 300 K the authors observed the formation of CoO films with a tetragonally distorted rock-salt structure, rotated in plane by 45° with respect to the unit cell of the Co film. The distortion reduces if the oxide layer thickness increases. The oxide formation induces a further strain in the unreacted underlying bct Co film. The interface between the CoO film and the underlying unreacted Co film is not chemically sharp and it includes the presence of a significant amount of Co sites with lower oxidation state (Figure 2.11). Oxidation at 500 K gives oxide films of larger thickness and reduces the relative

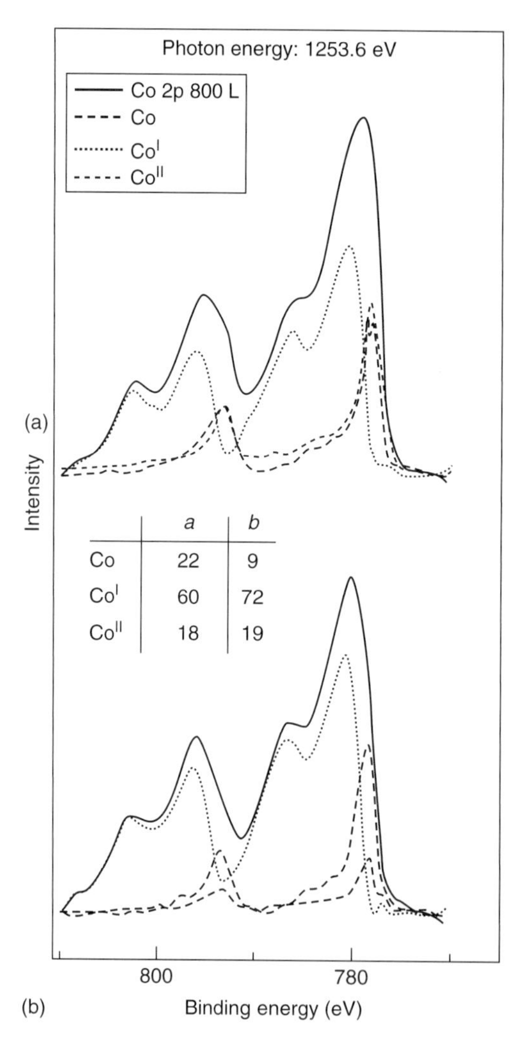

Figure 2.11 Oxidation of a 10-ML-thick bct Co film grown on Fe(001): Co 2p XPS spectrum of the 800 L O_2 exposed cobalt surface, in both normal (a) and grazing (b) take-off geometries. Contributions of metallic Co (Co), of Co in the CoO phase (Co^{I}) and Co in the interfacial understoichiometric phase (Co^{II}) are shown and their relative intensities are reported in the inset. (Reprinted from [85] with permission.)

thickness of the interfacial understoichiometric region and the stoichiometric CoO phase. The thermal stability of such films in the 300–620 K range has also been studied [87]. The CoO layers are stable in terms of composition and thickness for moderate annealing (up to 1 hour and 470 K). At higher temperatures, the oxide film thickness is reduced and the amount of Co atoms in low coordination sites is increased.

Good-quality CoO films have also been obtained on Fe(001) substrates by Co deposition under oxygen atmosphere. Even though the CoO/Fe interface has been shown to be significantly oxidized, the CoO films show the right stoichiometry, as proved by XPS, and above 5 ML also the expected epitaxial orientation, 45° rotated in plane, as shown by LEED [88]. Epitaxial CoO films have also been grown on $Fe_3O_4(001)$. The two materials have fcc oxygen sublattices with unit cells differing only by 1.4%. In order not to oxidize the Fe_3O_4 substrate, in the ultrathin limit the growth was performed by deposition of enough Co to be oxidized into 1-ML CoO and subsequent exposure to 10 L of O_2; for the growth of films with thickness above 5 ML reactive deposition was used. The growth has been shown to proceed layer-by-layer and the films have been shown to be stoichiometric in the 1–20-ML range [89].

A study of reactive growth of CoO films by evaporation of metallic Co under NO_2 partial pressure on a cleaved MgO(001) substrate has been performed by Peacor and Hibma [21]. The analysis of RHEED specular beam intensity oscillations has shown that the growth at 700 K proceeds in a layer-by-layer mode. Growth at 500 K instead induces the formation of the Co_3O_4 phase after the first four CoO layers (Figure 2.12). The obtained results indicate that the relative influence of the

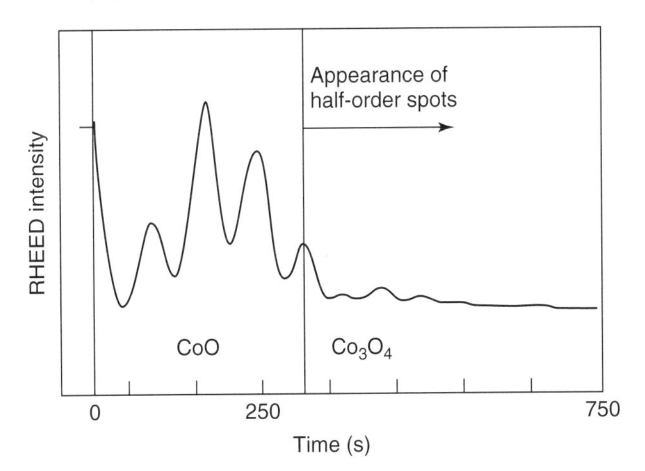

Figure 2.12 RHEED intensity oscillations of CoO deposited on *ex situ* cleaved MgO at 473 K using NO as an oxidizing agent. The sudden decrease in period length of one-fourth, which is accompanied by the appearance of half-order spots in the RHEED pattern, indicates a change in growth from CoO to Co_3O_4. (Reprinted from [21] with permission.)

NO_2 oxidizing power and the substrate/overlayer interface determines the relative abundance of the two cobalt-oxide phases.

Another substrate used for the growth of CoO films by reactive growth is Ag(001) [53, 90–92]. The group of Neddermeyer showed by STM that the growth of CoO films proceeds in a three-dimensional mode. In this study, the Ag(001) substrate was kept at 460 K during growth and the obtained films were annealed to 670 K in oxygen after the growth in order to improve the film quality [90, 91]. More recently, CoO films on Ag(001) were studied by grazing incidence X-ray diffraction (GIXRD), which proved that the films have the rock-salt structure with O atoms sitting on top of the substrate Ag atoms [92]. Such measurements also allowed us to evaluate the interface distance, which is found to be 2.19 ± 0.03 Å for a 1-ML-thick film and 2.32 ± 0.07 Å for a 4-ML-thick film. The in-plane lattice parameter of the CoO films has been found to gradually evolve from a value corresponding to pseudomorphism on Ag to a value close to the bulk one at 23 ML. The radial scans around the (2, −2) peak of Ag along the (h, $-h$, 0.08) direction as a function of CoO thickness also show the presence of two satellites, one at each side of the Ag peak (Figure 2.13), corresponding to diffraction from a periodic structure with a period larger than the crystal one. Following a work by Renaud *et al.* [93], the satellites were assigned to diffraction from the dislocation network, which develops within the CoO film in order to release the strain. The expected period for the dislocation network (9.37 nm), calculated using the Ag and CoO lattice parameters, coincides

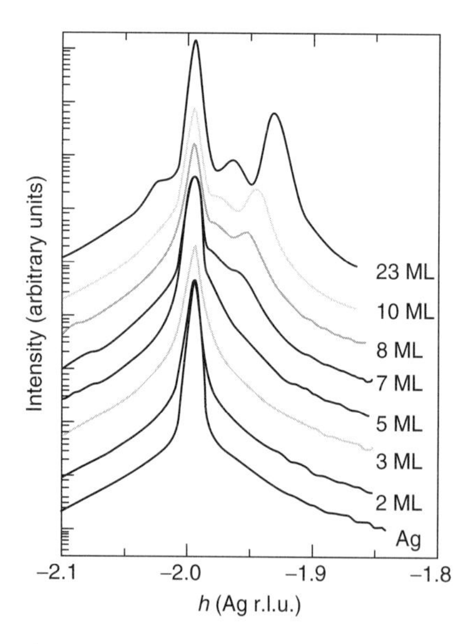

Figure 2.13 GIXRD radial scan along the (h, $-h$, 0.08) direction around the (2, −2) Bragg peak of Ag for films of CoO of different thicknesses grown on Ag(001). (Reprinted from [92] with permission.)

within the experimental accuracy with the period measured on the radial scans (9.2 nm).

MnO(001) represents a substrate with lattice mismatch of opposite sign compared to Ag(001) for the growth of CoO films [5]. As in the case of NiO, also for CoO it is interesting to note that the film strain and its sign largely determine the magnetic anisotropy, the magnitude, and the orientation of magnetic moments close to the interface [5].

2.4 Other Oxides

2.4.1 MnO(001)

Among the 3d transition-metal monoxides, MnO plays a particular role due to the high spin ground state of the $3d^5$ configuration of Mn^{2+} ions. Therefore, MnO could be an excellent model system to investigate many kinds of spin-dependent or magnetic interactions between electrons in the surface of an AF solid. Mn monoxide is also long known as a materialwhose specific electronic and magnetic structure constitutes a challenge for *ab initio* calculations due to the half-filled d-states of Mn [94]. In practice, most experimental investigations on crystalline MnO(001) surfaces suffer from the bad structural quality of the surface of cleaved bulk crystals. In contrast to, for example, NiO, in fact, the cleavage of a MnO single crystal along the (001) plane usually leads to a surface structure, which consists of arrays of pyramids with (111) surfaces. As a consequence, oxide surfaces of a high structural quality can be only prepared by the growth of thin films on suitable substrates, and this is one of the main interests in preparing epitaxially grown MnO films in the monolayer and multilayer range. Again in contrast to NiO, there is a large variety of Mn oxides, among which those with Mn in higher ionization states are the most stable ones, for example, Mn_2O_3. The reactive deposition of the less stable AF MnO phase is therefore challenging. A constant evaporation of Mn is very difficult to handle due to its high vapor pressure. Moreover, the oxygen pressure is a critical parameter owing to the large variety of Mn oxides. If both parameters are properly tuned, it is possible to prepare MnO films with a structural order comparable to that of the corresponding crystal substrate.

Epitaxial growth of MnO surfaces of high structural order has been reported on Ag(001) substrate. As electronic fingerprints, the characteristics of the Mn 2p line profile and especially the strength of the Mn 3s splitting, both measured by XPS, were used, whereas the bulk structure of the films was examined by XPD [95]. The structural order of the surface was studied by LEED [96]. Epitaxial growth of MnO(100) films, 20–30-ML thick, has also been reported on a Pd(100) single-crystal surface [48]. The main goal of this work was to examine the influence of lattice mismatch on the epitaxial orientation of the growing oxide films. The reactive deposition, followed by a brief postannealing step in UHV up to 870 K, results in the growth of a well-ordered MnO(100) layer, as shown by SPA–LEED. The surface lattice constant determined from the separation of the LEED spots has been found to be identical with the (100) in-plane lattice parameter of bulk MnO crystals. Atomic force microscopy (AFM) operated in the frequency-modulation mode shows that the MnO(100) surface is atomically flat and consists of terraces with lateral dimensions of up to 500 Å, which are separated by monatomic steps running predominantly along the main azimuthal substrate crystallographic directions. A number of interface-stabilized intermediate oxide phases have been detected for Mn oxide films thinner than 5 ML. MnO layers between 3 and 10 ML have a strained lattice as a result of the lattice mismatch at the Pd(100)/MnO(100) interface.

2.4.2 FeO

FeO (wüstite) is an AF insulator with a rock-salt phase which is not stable in the bulk at temperatures below 850 K. Mainly, the FeO(111) surface has been the object of investigation (see for example [1] and references therein). Heteroepitaxial FeO films were first produced by postoxidation of iron films deposited on Pt(111) and Pt(100) [97], and it was observed that FeO(111) films grew layer-by-layer on both substrates. After that, several studies have been performed to investigate the influence of the thickness of the film on the resulting phase. The most successful work has involved growth on Pt (111) and sapphire substrates. The lattice mismatch of the oxygen sublattice of FeO(111) on Pt(111) is 9.75%. Quantitative LEED studies on FeO(111) films prepared by deposition and postoxidation of Fe showed that a (8×8) superstructure at 1 ML evolves as a (2×2) reconstructed surface at 8 ML [98]. A different LEED and STM study [99] reported that up to 2 ML the oxide grows as FeO(111). At higher coverage the film changes to $Fe_3O_4(111)$ and subsequent annealing in O_2 at temperatures above 1070 K converts the film to α-$Fe_2O_3(0001)$. The low-coverage FeO(111) layers consist of hexagonal close-packed iron–oxygen bilayers laterally expanded with respect to the bulk FeO structure and are oxygen terminated, resulting in a Moiré pattern that is clearly seen in the STM images. The corrugation of this pattern in thin FeO films on Pt(111) was reported to depend on variations of the surface potential within the Moiré unit cell, variations that can be induced by very subtle differences in geometry [100] (Figure 2.14). Well ordered and oriented superstructure formation was observed for FeO(111) films on Ru(0001) up to 4 ML [101]. The formation of Moiré patterns was also reported in a STM, XPS, and ion-scattering spectroscopy study of FeO formation on Au(111) surfaces by postoxidation of predeposited Fe [102]. On Pt(100) substrate the formation of the FeO(111) phase has been reported for very thin films, consisting of a buckled bilayer terminated by oxygen atoms over a Fe layer [103].

A study of epitaxial Fe films on Ag(111) was performed by two different procedures [104]. A 10-ML Fe deposition followed by oxidation resulted in poorly ordered FeO(111) films. By the second procedure, consisting in the sequential production

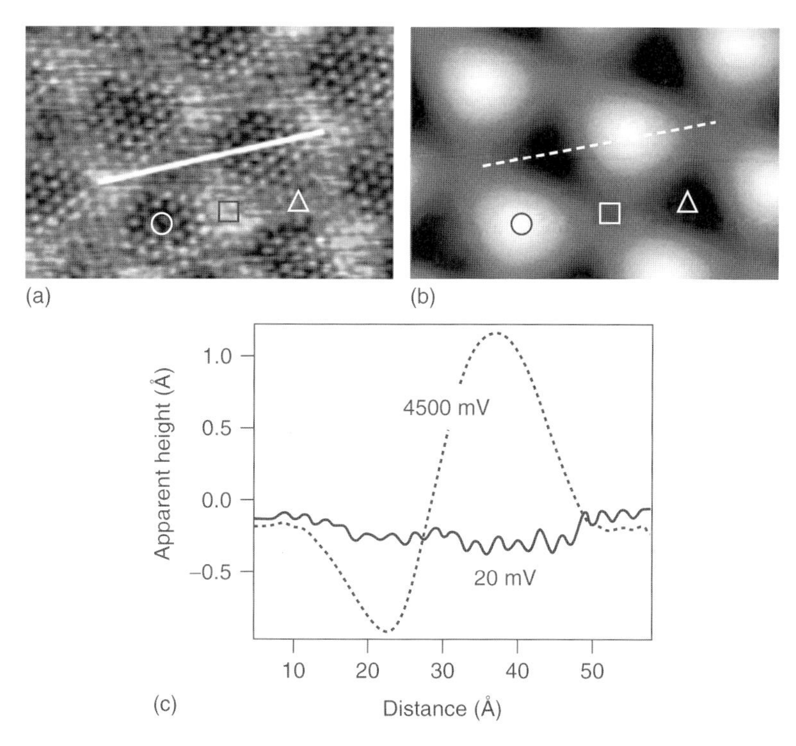

Figure 2.14 75 × 50 Å^2 STM images of FeO thin film on Pt(111), at sample bias voltages of (a) 20 mV and (b) 4500 mV. (c) Line profiles from the STM images across the Moiré unit cell. I = 0.3 nA. (Reprinted from [100] with permission.)

of submonolayer Fe films followed by oxidation, a much better crystallographic order has been obtained up to about 10-Å thickness.

The growth, structure, and morphology of ultrathin oxide layers formed on a Fe(110) single-crystal surface by oxidation with atomic or molecular oxygen as well as by reactive Fe deposition have been investigated by AES, LEED, and grazing ion-scattering [11, 23] (Figure 2.15). A well-ordered FeO(111) film with low defect density was only obtained with atomic oxygen. Independent of the preparation method, long-range structural order is poor if the oxide film thickness exceeds three to five layers. This is attributed to the relatively large mismatch between FeO(111) and Fe (110) [11]. The FeO(111) surface was reported to be ferromagnetically ordered even above a bulk T_N of 198 K, and the magnetic surface layer was found to be antiferromagnetically coupled with the underlying FM Fe(110) substrate through paramagnetic FeO. The FM order at the FeO(111) surface was ascribed to reconstruction [23]. Also the FeO(001) phase has been investigated, although to a minor extent with respect to the (111) one. LEED investigation of the surface of an FeO(001) thin film prepared by reactive deposition on Ag(001)

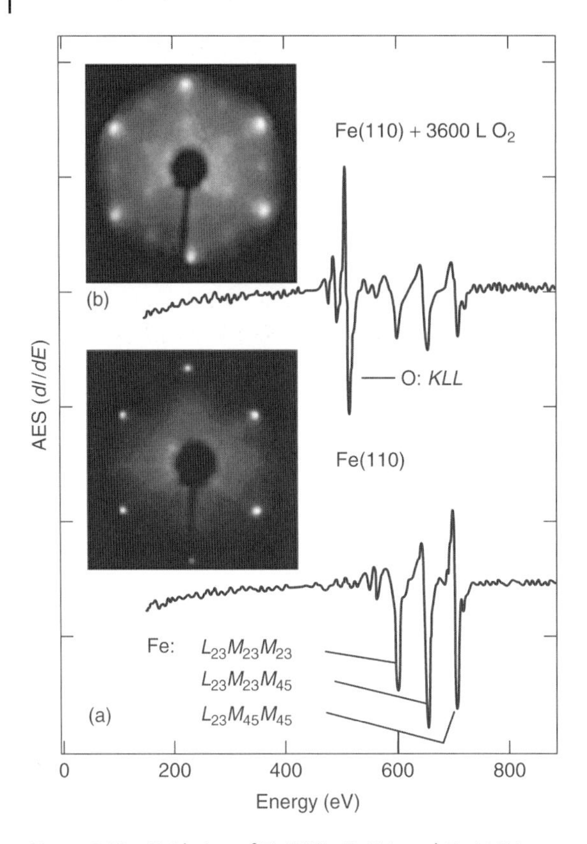

Figure 2.15 Oxidation of Fe(110): O KLL and Fe LMM Auger spectra at 3-keV primary electron energy and LEED patterns (a) for clean Fe(110) and (b) for Fe(110) exposed to 3600 L of O_2 at 570 K. The LEED pattern is obtained at a beam energy of 93 eV in (a) and 60 eV in (b). (Reprinted from [23] with permission.)

showed that the surface has termination structure close to the bulk one with a very small rumple on the first layer, in agreement with the structure of the other rock-salt oxides [105].

Polycrystalline FeO films were deposited on kapton, glass strips, and carbon pieces, by reactive DC magnetron sputtering in a mixture of Ar and O_2 gases. An AF-type behavior has been found at low temperatures. This observation is in contrast to the bulk AF properties of FeO and has been explained by the formation of defect clusters in FeO film [19].

2.4.3 α-Fe_2O_3

Hematite (α-Fe_2O_3) in its bulk form is an AF material with a T_N of 95 K. The compound can be classified as a charge-transfer insulator. However, the electronic and magnetic properties of thin films can be substantially different from those of the bulk because of the large contribution of surface and interface effects. As for the FeO case, a number of groups have worked on the growth of thin, crystalline hematite films, using Pt(111) or sapphire substrates and evaporating single or multiple monolayers of Fe followed by annealing in O_2 at temperatures of 870–1070 K, or by reactive deposition (see [1] and references therein). The oxygen sublattice of the (0001) plane of α-Fe_2O_3 aligns with Pt(111) and with α-Al_2O_3(0001), which have a reasonably low lattice mismatch with hematite (4.9 and 5.8%, respectively). The disadvantage is that just one epitaxial orientation is allowed for the film, that is, (0001)Fe_2O_3||(111)Pt and (0001)Fe_2O_3||Al_2O_3(0001). The primary advantage of using a Pt(111) substrate is that Fe_2O_3/Pt systems typically show a large enough conductivity to be studied by electron spectroscopies and STM. Thin films of α-Fe_2O_3 were obtained by further oxidizing magnetite layers, previously prepared on Pt(111) by repeated cycles of iron deposition and subsequent oxidation [106, 107]. It turned out that the oxidation has to be performed at temperatures above 1070 K and oxygen pressures higher than 10^{-3} mbar [108].

Clean, stoichiometric, epitaxial thin films of α-Fe_2O_3, up to 100 ML in thickness with flat and defect-free surfaces, have been synthesized by oxygen-plasma-assisted MBE using polished α-Al_2O_3(0001) and MgO(001) single crystals as substrates [109], and characterized by RHEED, LEED, and XPS/XPD. The overall quality of the oxide layers was found to be critically dependent on the growth conditions and on the choice of the substrate. In particular, the oxygen partial pressure and the substrate temperature were reported to be relevant parameters [108, 109]. A significant improvement of crystal quality was obtained by using a Pt(111) buffer layer predeposited on the basal-plane sapphire [80]. α-Fe_2O_3 films were also prepared by evaporating Fe from a Knudsen cell onto an α-Al_2O_3(0001) substrate, and simultaneously oxidizing the metal particles with a NO_2 flux coming from a small buffer volume. The RHEED patterns suggest that layer-by-layer growth of α-Fe_2O_3(0001) occurs for the first few monolayers. Subsequently, the growth mode changes to three-dimensional growth [110].

The early stages of the film growth were investigated by noncontact atomic force microscopy (nc-AFM) and a variety of different techniques. The interfacial layer was found to be badly distorted but commensurate with the substrate [111]. The interfacial disordering has been investigated by ion scattering and HRTEM supported by molecular dynamics calculations and ascribed to misfit dislocations associated with lattice mismatch between the substrate and the film [46] (Figure 2.16). Comparison of XPD experiments with quantum-mechanical scattering theory enabled the spacing of the first four surface layers to be precisely determined and revealed that the oxide surface is Fe terminated [112]. However, surface structure and termination of hematite were reported to strictly depend on ambient oxygen gas pressure applied during the preparation procedure [108] (Figure 2.17). Using STM and infrared reflection absorption spectroscopy it has been observed that the surface of α-Fe_2O_3 films grown on Pt(111) exhibits ferryl (Fe = O) groups, which may coexist with domains of

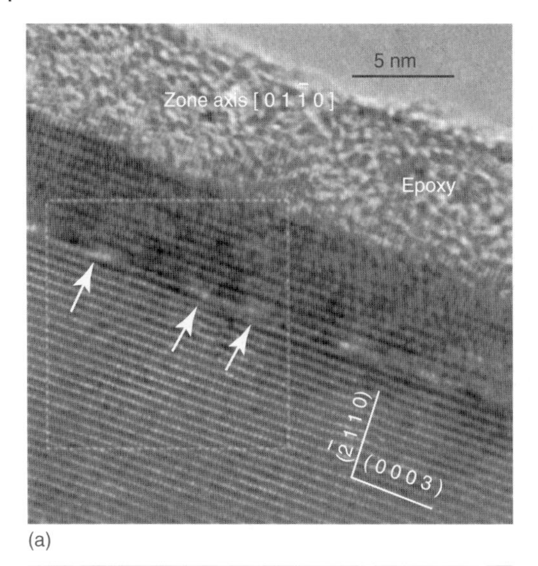

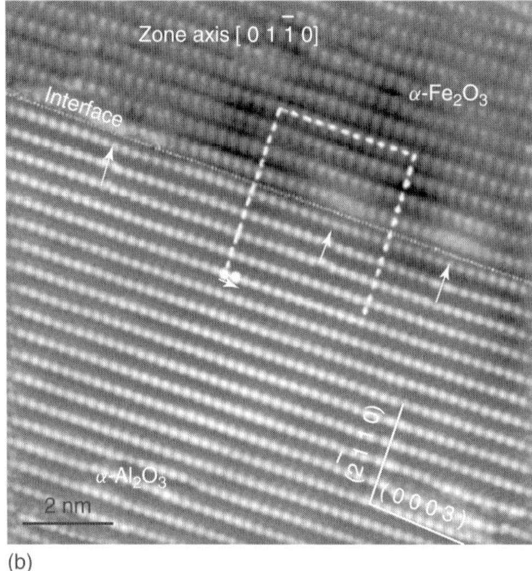

Figure 2.16 (a) HRTEM micrograph of the misfit dislocations at the Fe_2O_3/Al_2O_3 interface imaged from the $[01\bar{1}0]$ zone axis, in which three misfit dislocations are present in the box marked by dashed lines. (b) Magnified and Fourier filtered image of the region in (a) showing an extra $(\bar{2}110)$ lattice planes on the α-Al_2O_3 side as indicated by the white arrows. (Reprinted from [46] with permission.)

the Fe-terminated surface [113]. The close similarity to the results on the (0001) surfaces of Cr_2O_3 and V_2O_3 strongly suggests that the metal oxygen double bond termination under certain oxygen pressure conditions is the most stable for the close-packed surfaces of transition-metal oxides with the corundum structure.

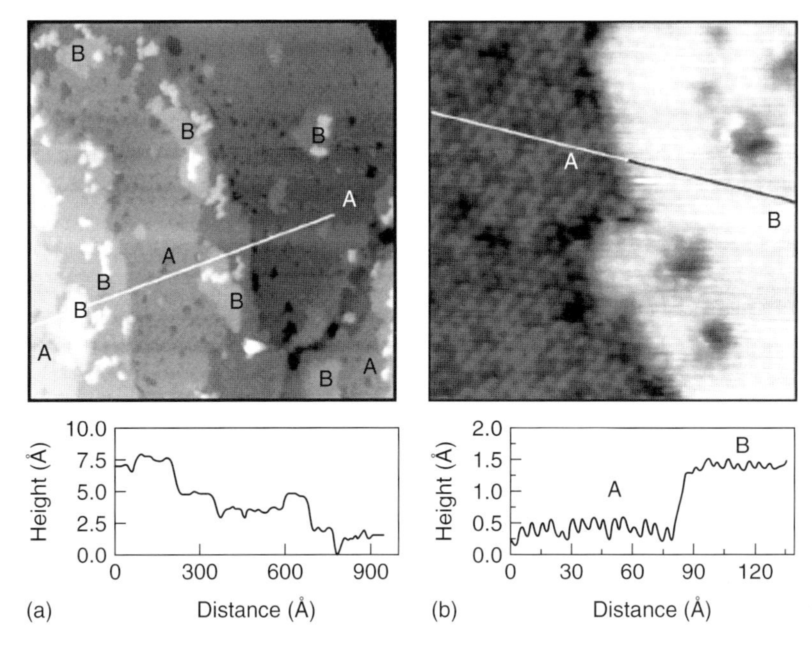

Figure 2.17 Large scale 1000 × 1000 Å (a) and atomic resolution 120 × 120 Å (b) STM images of the α-Fe_2O_3(0001) surface prepared on Pt(111) at 1003 K in 10^{-1} mbar O_2 with profile lines shown below. Region A and B are separated by an apparent height difference of ~1.2 Å, equivalent regions are separated by ~2.4 Å in height, as expected for equivalent surface terminations of α-Fe_2O_3(0001). Structure A exhibits a larger atomic corrugation than structure B. (a) $V = 1.5$ V, $I = 0.9$ nA; (b) $V = 1.3$ V, $I = 1.25$ nA. (Reprinted from [108] with permission.)

Hematite thin films have also been prepared on Si substrates using *n*-butylferrocene and oxygen in a low-pressure metallorganic chemical vapor deposition reactor [114]. The growth rates were studied in the 673–873 K temperature range, while the structure and morphology were characterized by XRD and SEM, and chemical states by XPS. Crystalline films were only obtained for deposition at higher temperatures. RF magnetron sputtering was used to prepare hematite thin films on stainless steel and Si(001) single-crystal substrates [115]. By XRD and AFM analysis the films were found to be polycrystalline α-Fe_2O_3 with a morphology consisting of columnar grains of random orientation.

2.5 Oxide–Substrate Interface

The atomic scale details of the interface between an oxide film and the substrate on which it is grown represent an important issue to be investigated, since they largely determine the epitaxial orientation, the structural and morphological quality, and, as a consequence, the electronic and magnetic properties of the film. Even in the case of morphologically and chemically sharp interfaces, the electronic properties at the oxide/substrate interface are significantly different from the ones of the oxide bulk phases [116]. For this reason, a large number of studies have recently been devoted to the characterization of the first few atomic layers on both sides of the interface.

A theoretical and experimental investigation of the atomic structure of the noninteracting NiO/Ag(001) interface has been performed by *ab initio* calculations, PDMEE, and XAS studies [42, 74]. The intensity angular distributions (IADs) of the Auger Ag MNN emission as a function of the incidence angle of the primary exciting beam, reported in Figure 2.18a, show that the [110] forward focusing

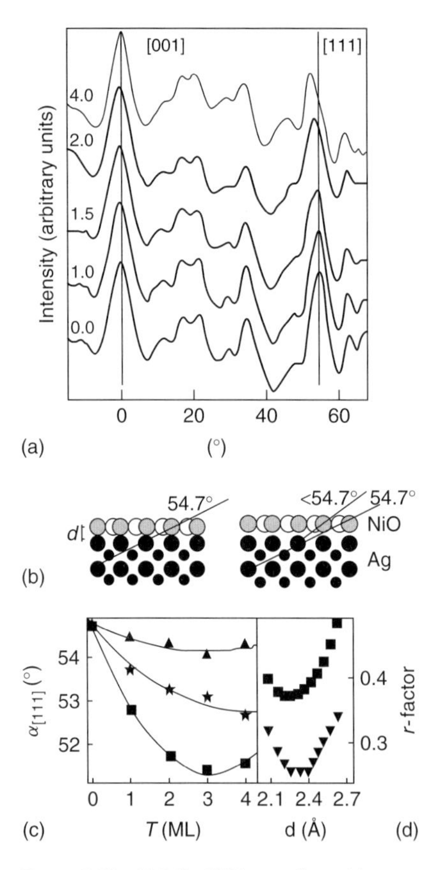

Figure 2.18 NiO/Ag(001) interface: (a) Intensity angular distribution of Ag MNN Auger electrons along the [80] azimuth for increasing NiO thickness (T) (from bottom to top $T = 0.0$, 1.0, 1.5, 2.0, and 4.0 ML). (b) Schematic representation of the NiO/Ag(001) system (O-on-top model) along the [80] azimuth for unexpanded (interfacial distance $d = 2.045$ Å, left) and expanded ($d > 2.045$ Å, right) interfacial distances. (c) Simulated [110] peak position versus T, for $d = 2086$, 2.30, and 2.50 Å, ▲, *, and ■ symbols, respectively. (d) r factors versus d for the 4-ML-thick film assuming either the O-on-Ag (▼) or the Ni-on-Ag (■) configurations. (Reprinted from [74] with permission.)

feature shifts to lower angular positions with increasing NiO thickness. This shift is the consequence of a value of the interface distance (d) significantly larger than that expected by extending the Ag crystal periodicity along the [001] direction ($c/2 = 2045$ Å), as evidenced by the two schemes in Figure 2.18b. The PDMEE patterns have been simulated on the basis of single scattering cluster calculations. Figure 2.18c reports the simulated α_{111} value as a function of NiO thickness, for three different values of d in the O-on-top model. The 4-ML IAD simulation for both O- and Ni-on-top models (Figure 2.18d) allowed us to establish that the O-on-Ag configuration is the preferred one and to quantify the interface distance as $d = 2.3 \pm 0.1$ Å. Polarization-dependent Ni K-edge XAS has then been used to improve the accuracy of d determination. The multiple-scattering fit of the XAS spectrum collected in grazing angle geometry on a 3-ML NiO film has been obtained with a Ni–Ag contribution at a distance of 3.13 ± 0.04 Å, implying a d value of 2.37 ± 0.05 Å. To validate the picture emerging from PDMEE and XAS experimental data, *ab initio* calculations on the same system have been performed [42, 74] using a hybrid-exchange Hamiltonian, because of its satisfactory performance in describing the bulk properties of both Ag and NiO. The z coordinates of all atoms of the oxide have been optimized, while the in-plane lattice parameter of the overlayer has been forced to coincide with the optimized one for the silver substrate. The presence of an oxide layer above the first one has the effect of reducing the corrugation of the first oxide layer, because of the attraction exerted by O anions in the second layer on Ni cations, and of reducing the interaction energy between the metal and the oxide. The calculated d value for a 2-ML-thick film (2.40 Å) is in excellent agreement with the Ni K-edge XAS value.

A very different situation is found when more reactive substrates, such as Fe(001), are used. In some cases, in order to avoid complex interface situations, the growth of oxide films using e-beam evaporation of bulk oxides has been used (see for example [117] for the MgO/Fe system). For the growth of NiO films on Fe, a passivated surface, Fe(001)-$p(1 \times 1)$O, very stable to further oxidation, was used [25, 79]. Nevertheless, the Fe L_{23} XAS and X-ray magnetic circular dichroism (XMCD) spectra before and after the deposition of 20-ML NiO clearly indicate that the Fe substrate is partially oxidized by the reactive deposition of NiO, forming a thick Fe_3O_4 phase (Figure 2.19). No relevant Fe–Ni mixing is detected. As expected, annealing at temperatures above 603 K gives rise to a strong interfacial intermixing with the formation of a FeO layer and the diffusion of metallic Ni in the Fe substrate [79].

It has to be noted that when Fe is deposited on NiO a very different situation is encountered. In particular, Fe is found to be oxidized mainly in the surface plane forming a buckled Fe–O-like layer [118], NiO is reduced, and the metallic Ni diffuses into the Fe film forming a bct alloy [119].

Also the CoO/Fe(001) interface, even if a passivated Fe(001)-$p(1 \times 1)$O surface is used, is dominated by the presence of iron oxides with a different valence character, mostly FeO and Fe_3O_4, involving several layers at the interface [88]. On the opposite interface, Fe/CoO(001), the reduction of CoO to metallic Co has been observed [88].

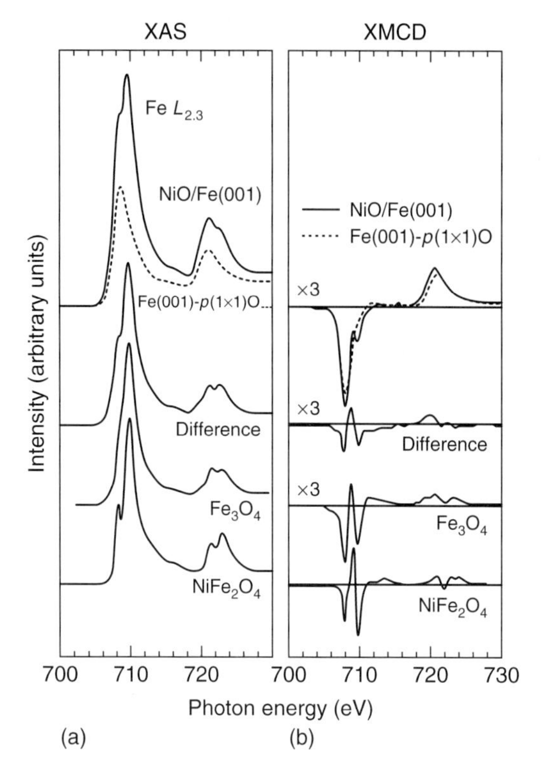

Figure 2.19 Comparison between the Fe $L_{2,3}$ XAS (a) and XMCD (b) spectra from 20-ML NiO/Fe(001) and Fe(001)-*p*(1 × 1)O. After normalization, the spectra obtained on Fe(001)-*p*(1 × 1)O have been multiplied by a factor equal to 0.5. The difference spectra are compared to the Fe_3O_4 and to the $NiFe_2O_4$ XAS and XMCD. (Reprinted from [79] with permission.)

These results indicate that the atomic scale details of the interface between two materials, in particular between an AF oxide and a metal, and all the consequent properties do not depend only on the thermodynamic properties of the two materials coming into contact, but also on nonequilibrium kinetic parameters, which strongly depend on the experimental conditions used and of course on the growth sequence. This issue is very relevant because it evidences that in the view the applications, that is, in the field of magnetoelectronic devices, it is important to choose not only the right materials but also the best growth procedures to obtain the desired functionality.

2.6
Polar-Oxide Surfaces

A very promising class of systems in view of the applications is represented by polar-oxide surfaces (for a recent review see, e.g., [120]). The specificity of these systems derives from the combined effect of surface orientation and termination, which results in a macroscopic polarization along the direction normal to the surface and in an electrostatic instability [120, 121]. The mechanisms for polarity compensation may involve either a change of the surface stoichiometry (vacancy formation, adsorption of foreign species, faceting) or a change of the surface electronic structure (partial or total filling of surface states, metallization). Polar-oxide ultrathin films can also be stabilized by the substrate on which they are synthesized, without any depolarization effect from the surface. These systems, with unprecedented structural and electronic properties, are very challenging for nanomaterials engineering.

Concerning the AF films dealt with in this chapter, the (111) surfaces of rock-salt oxides such as NiO, CoO, FeO are polar.

Different substrates have been used for the growth of NiO(111) thin films. Earlier works report the direct oxidation of the Ni(111) surface (see, e.g., [122] and references therein). In this case the formation of thick triangular-shaped (111)-oriented islands has been shown to occur for high oxygen exposures [123] and the surface has been found to be stabilized by a surface monolayer of nickel hydroxide [122]. NiO growth on Au(111) at RT has been shown by LEED and STM to induce the formation of a three-domain NiO(001) crystallite structure with an average diameter of approximately 5 nm [124]. By increasing the growth temperature to 570 K, instead, the formation of $p(2 \times 2)$-reconstructed NiO(111) film up to 6 ML and of faceted structures for larger coverages are observed [124]. Barbier *et al.* have compared the NiO(111) single-crystal $p(2 \times 2)$ surface structure and that of a 5-ML NiO(111) film on Au(111) [125]. Both were shown to locally exhibit the theoretically predicted octopolar reconstruction. However, while the single crystal exhibits a single Ni termination with double steps, the thin film exhibits both possible terminations (O and Ni) with single steps. These surfaces were found to be nonreactive with respect to hydroxylation. The situation is more complex when a Cu(111) substrate is used, as shown by the LEED patterns in [126], which have been interpreted as being originated either by a α-Ni_2O_3 hexagonal phase, or by a structural distortion of the NiO(111)$(\sqrt{3} \times \sqrt{3})R30^\circ$ structure. The use of a Pt(111) substrate for the growth of NiO films of thickness above 1 ML, instead, shows the formation of three-dimensional islands with an average diameter of 40 nm [57]. Annealing such islands at 850 K is reported to induce the formation of (001) facets [57].

Thin CoO(111) films were obtained by the oxidation of Co(0001). The observed (1×1) LEED pattern has been tentatively assigned to the presence of hydroxyl groups on the surface, which could not be removed without damaging the film [127]. Using Au(111) as a substrate the stabilization of a well-ordered bulk-like CoO(111) layer was shown to be due to the presence of OH groups on the surface [128]. (111)-oriented CoO islands have also been found to be formed by reactive evaporation of Co under O_2 atmosphere on the Ag(001) surface [31]. In this case, the islands reveal a Moiré-like modulation with hexagonal structure and no surface reconstruction. A rather flat, unreconstructed polar CoO(111) surface was grown on a Pt(111) substrate. In this case, electron energy loss spectroscopy (EELS) and ultraviolet photoemission spectroscopy (UPS) suggest the existence of low-energy excitations originating from finite density of states at the Fermi level, possibly ascribed to the shifted O-derived band in the CoO(111) surface layer [68]. CoO(111) films have also been obtained by oxidation of thin (up to 4 ML) Co films epitaxially grown on the (1×1) phase of Ir(001) [15]. LEED and STM measurements for this system showed the presence of a slight lattice distortion giving a $c(10 \times 2)$ coincidence structure with the substrate. The mechanism for polarity compensation for this system is probably the formation of oxygen vacancies in the top layer, although surface hydroxylation or other mechanisms could not be excluded. Very recently, the same authors showed that reactive deposition of Co under O_2 atmosphere on Ir(001) gives CoO films in the 5–50-ML range whose bulk rock-salt structure is terminated by a thin slab of CoO in the wurtzite-like phase. The film surface is found to be metallic, in order to compensate polarity [129].

An interesting case in which polarity may be responsible for drastic changes in the magnetic properties is observed in FeO(111) films obtained by oxidation of a Fe(011) surface. In this system, in fact, a FM surface ordering is observed above the bulk T_N (198 K) due to a complex surface reconstruction to compensate polarity [23, 130]. The results obtained for FeO films grown on Pt(111) are also very interesting. Only low-thickness (below 2.5 ML) FeO(111) films could be stabilized before three-dimensional Fe_3O_4 islands started to form [97, 131]. The bilayer film is oxygen terminated, and it shows a slightly expanded in-plane lattice spacing [132]. The interplanar distance is significantly (50%) contracted with respect to the bulk value, in order to reduce the out-of-plane electric dipole [132]. The lattice mismatch between FeO and Pt gives rise to a Moiré superstructure with 25-Å periodicity. As shown in Figure 2.20, the corrugation of STM images has been related to the variation of the surface potential within the Moiré unit cell [78, 100, 133].

2.7 Conclusions and Perspectives

We have summarized the advances made in the growth and characterization of AF oxide thin films in the last decades. The most studied material is NiO, appealing for the applications due to its high T_N. Relatively less studied are CoO films, which can be prepared on various substrates as well. Fewer studies are oriented toward the growth of FeO, Fe_2O_3, and MnO thin films, probably due to the experimental difficulties in obtaining the desired stoichiometry, given the large number of oxidation states of Fe and Mn. In general, we have shown that a rich variety of phases can be obtained in the monolayer and submonolayer range, while the growth procedures to prepare stoichiometric films of larger thickness are well established. A number of issues have been raised, such as the importance of the interface and properties connected with polarity.

The large number of available studies testifies the increasing importance of this field of research. The use of AF oxides in the form of metal-supported ultrathin films has in fact allowed us to avoid the charging problems encountered when bulk samples are studied using charged probes and to apply electron spectroscopies and

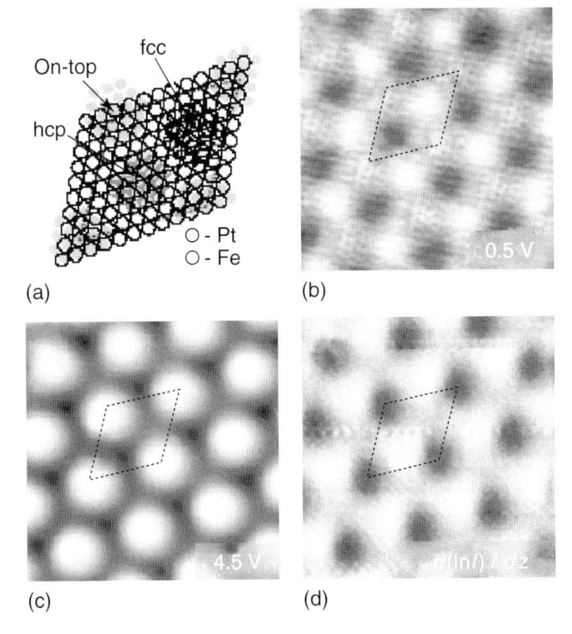

Figure 2.20 (a) Structural model of the Moiré cell formed by FeO film and Pt(111). (b) STM topographic images of FeO/Pt(111) taken at 0.5 V and (c) 4.5 V sample bias. The current was set to 0.1 nA. (d) Apparent-barrier-height image of FeO/Pt(111) taken at 3.5 V sample bias. The current was set to 0.1 nA, the z modulation to 1.0 Å. All images are 100×100 Å^2 in size. (Reprinted from [78] with permission.)

STM to characterize bulk-like materials at the atomic scale. On the other side, the use of oxides in the form of ultrathin films grown on various substrates has allowed us to suitably tune the electronic and magnetic properties of the material through strain- and interface-related effects.

Relevant future trends can be envisaged along different directions.

1) The possibility of engineering to some extent the surface structure of thin and ultrathin oxide layers to create oxide templates suitable for self-assembling of ordered arrays of metal nanoclusters has been demonstrated, for example, for FeO ultrathin films on Pt(111). A coincidence structure of 2.6-nm periodicity has been obtained. Clusters nucleated on the surface after depositing 0.2-Å Fe. The clusters are well confined and repeat the surface superstructure. FeO grown on Rh(111) instead forms a coincidence structure similar to FeO/Pt(111) but with smaller periodicity of 1.4 nm due to the large film/substrate mismatch. The possibility of tuning the size and spatial order of metal clusters by an appropriate choice of the oxide template is therefore foreseen [134]. On a lower (atomic) scale, the distinct spatial distribution of single Au atoms on FeO/Pt(111) has been traced back to a varying surface potential on this polar, ultrathin oxide film [78].

2) Surface engineering of thick oxide film can be obtained by the interfacial misfit dislocation network formed at the interface between the oxide film and the underlying substrate. The misfit between an epitaxial, 5-nm-thick CoO film and an Ag(001) substrate is accommodated by the underlying interfacial dislocations network that is buried at the interface. This induces a periodic displacement field on the oxide surface, which provides a network of sites for preferential nucleation and growth of Ni nanoparticles. The achievement of the control of the self-assembling process of metal cluster on the surface of a fairly thick oxide film represents a step forward in the production of real devices by self-organized growth [77].
3) A critical issue in the application of magnetic oxides in nanotechnologies is the strong reduction of the relevant critical ordering temperatures due to well-known finite-size effects. Interfacing with a suitable substrate has been proposed as a new method to compensate these effects and enhance magnetic ordering temperatures. The T_N of 3-ML NiO(100) film grown by atomic-oxygen-assisted reactive deposition on MgO(100), measured by using the magnetic linear dichroic effect in Ni L_2 XAS spectra, was found to be lower than 40 K, i.e, much smaller than the bulk value (523 K), due to finite-size effects. On the contrary, a T_N of 390 K was measured for a 3-ML NiO(100) film on Ag(100), showing that the dramatic finite-size effects can be almost counterbalanced by the proximity of the metal [6, 51]. Interfacing oxide films with polarizable media can therefore be a practical way toward designing oxide nanostructures with improved critical temperatures for various forms of magnetic ordering. Control of epitaxial strain is an alternative route toward the improvement of critical temperatures in magnetic oxides, as already suggested for superconductor oxides [135]. Epitaxial strain associated with the mismatch between the oxide film and the substrate can be tuned, for example, by a suitable choice of the substrate itself and/or by the addition of buffer layer(s) between the overlayer and the substrate.
4) Finally, within the field of future data storage technology, nanoparticles consisting of FM metal core and an AF metal-oxide shell (core–shell nanoparticles) are also of increasing interest [136]. These particles can be described as an ultrathin AF oxide film on a FM metal substrate. Co–CoO core–shell nanoparticles (similarly to Co–CoO bilayers) are also an archetypal system for basic studies of exchange bias, given the good growth properties of Co, the convenient, that is, near RT, T_N of CoO, and the large exchange-bias field exhibited by these materials. From the technological side, exchange-bias effect in Co–CoO nanoparticles has been investigated in order to beat the superparamagnetic limit and to improve the thermal stability in nanometric magnetic recording elements [137].

Acknowledgments

We wish to thank G. Pacchioni for having motivated our interest toward oxide films starting from the collaboration in the context of INFM PRA-ISADORA project in

2000. A number of stimulating cooperations within the Physics Department of the University and the CNR–INFM S[3] Center in Modena have supported our activities in the field. We also acknowledge the fruitful discussions with the colleagues involved in the FP6 STREP project GSOMEN. Finally, we would like to thank the editors F. Ciccacci, L. Duò, and M. Finazzi for giving us the opportunity to write this chapter.

References

1. Chambers, S.A. (2000) Epitaxial growth and properties of thin film oxides. *Surf. Sci. Rep.*, **39**, 105–180.
2. Freund, H.-J., Kuhlenbeck, H., and Staemmler, V. (1996) Oxide surfaces. *Rep. Prog. Phys.*, **59**, 283–347.
3. Alders, D., Tjeng, L.H., Voogt, F.C., Hibma, T., Sawatzky, G.A., Chen, C.T., Vogel, J., Sacchi, M., and Iacobucci, S. (1998) Temperature and thickness dependence of magnetic moments in NiO epitaxial films. *Phys. Rev. B*, **57** (18), 11623–11631.
4. Altieri, S., Finazzi, M., Hsieh, H.H., Lin, H.-J., Chen, C.T., Hibma, T., Valeri, S., and Sawatzky, G.A. (2003) Magnetic dichroism and spin structure of AF NiO(001) films. *Phys. Rev. Lett.*, **91**, 137201.
5. Csiszar, S.I., Haverkort, M.W., Hu, Z., Tanaka, A., Hsieh, H.H., Lin, H.-J., Chen, C.T., Hibma, T., and Tjeng, L.H. (2005) Controlling orbital moment and spin orientation in CoO layers by strain. *Phys. Rev. Lett.*, **95**, 187205.
6. Altieri, S., Finazzi, M., Hsieh, H.-H., Haverkort, M.W., Lin, H.-J., Chen, C.T., Frabboni, S., Gazzadi, G.C., Rota, A., Valeri, S., and Tjeng, L.H. (2009) Image charge screening: a new approach to enhance magnetic ordering temperatures in ultrathin correlated oxide films. *Phys. Rev. B*, **79**, 174431.
7. Krishnakumar, S.R., Liberati, M., Grazioli, C., Veronese, M., Turchini, S., Luches, P., Valeri, S., and Carbone, C. (2007) Magnetic linear dichroism studies of in situ grown NiO thin films. *J. Magn. Magn. Mater.*, **310**, 8–12.
8. Finazzi, M, Duò, L., Ciccacci, F. (2007) Magnetic properties of interfaces and multilayers based on thin antiferromagnetic oxide films, *Surf. Sci. Rep.*, **62**, 337–371.
9. Bäumer, M., Cappus, D., Kuhlenbeck, H., Freund, H.-J., Wilhelmi, G., Brodde, A., and Neddermeyer, H. (1991) The structure of NiO(100) films grown on Ni(100) as determined by low-energy electron diffraction and scanning tunnelling microscopy. *Surf. Sci.*, **253**, 116–128.
10. Muller, F., Steiner, P., Straub, Th., Reinicke, D., Palm, S., de Masi, R., and Hufner, S. (1999) Full hemispherical intensity maps of crystal field transitions in NiO(001) by angular resolved electron energy loss spectroscopy. *Surf. Sci.*, **442**, 485–497.
11. Busch, M., Gruyters, M., and Winter, H. (2006) FeO(111) formation by exposure of Fe(110) to atomic and molecular oxygen. *Surf. Sci.*, **600**, 2778–2784.
12. de Jesus, J.C., Pereira, P., Carrazza, J., and Zaera, F. (1996) Influence of argon ion bombardment on the oxidation of nickel surfaces. *Surf. Sci.*, **369**, 217–230.
13. Sambi, M., Sensolo, R., Rizzi, G.A., Petukhov, M., and Granozzi, G. (2003) Growth of NiO ultrathin films on Pd(100) by post-oxidation of Ni films: the effect of pre-adsorbed oxygen. *Surf. Sci.*, **537**, 36–54.
14. Gazzadi, G.C., Borghi, A., di Bona, A., and Valeri, S. (1998) Epitaxial growth of CoO on the (001) surface of bct cobalt. *Surf. Sci.*, **402-404**, 632–635.
15. Giovanardi, C., Hammer, L., and Heinz, K. (2006) Ultrathin cobalt oxide films on Ir(100)-(1x1). *Phys. Rev. B*, **74**, 125429.

16. Warot, B., Snoeck, E., Baulès, P., Ousset, J.C., Casanove, M.J., Dubourg, S., and Bobo, J.F. (2001) Epitaxial growth of NiO layers on MgO(001) and MgO(110). *Appl. Surf. Sci.*, **177**, 287–291.
17. Hotovy, I., Bùc, D., Hascik, S., and Nennewitz, O. (1998) Characterization of NiO thin films deposited by reactive sputtering. *Vacuum*, **50** (1-2), 41–44.
18. Tanaka, M., Mukai, M., Fujimori, Y., Kondoh, M., Tasaka, Y., Baba, H., and Usami, S. (1996) Transition metal oxide films prepared by pulsed laser deposition for atomic beam detection. *Thin Solid Films*, **281-282** (1-2), 453–456.
19. Dimitrov, D.V., Hadjipanayis, G.C., Papaefthymiou, V., and Simopoulos, A. (1997) Unusual magnetic behavior in sputtered FeO and α-Fe2O3 thin films. *J. Vac. Sci. Technol. A*, **15**, 1473–1477.
20. Sacchi, M., Hague, C.F., Gota, S., Guiot, E., Gautierc Soyer, M., Pasquali, L., Mrowka, S., Gullikson, E.M., and Underwood, J.H. (1999) Resonant scattering of polarized soft X-rays for the study of magnetic oxide layers. *J. Electron Spectrosc. Relat. Phenom.*, **101–103**, 407–412.
21. Peacor, S.D. and Hibma, T. (1994) Reflection high-energy electron diffraction study of the growth of NiO and CoO thin films by molecular beam epitaxy. *Surf. Sci.*, **301**, 11–18.
22. Rota, A., Altieri, S., and Valeri, S. (2009) Growth of oxide-metal interfaces by atomic oxygen: monolayer of NiO(001) on Ag(001). *Phys. Rev. B*, **79**, 161401(R).
23. Mori, K., Yamazaki, M., Hiraki, T., Matsuyama, H., and Koike, K. (2005) Magnetism of a FeO(111)/Fe(110) surface. *Phys. Rev. B*, **72**, 014418.
24. Portalupi, M., Duò, L., Isella, G., Bertacco, R., Marcon, M., and Ciccacci, F. (2001) Electronic structure of epitaxial thin NiO(100) films grown on Ag(100): towards a firm experimental basis. *Phys. Rev. B*, **64**, 165402.
25. Duò, L., Portalupi, M., Marcon, M., Bertacco, R., and Ciccacci, F. (2002) Epitaxial thin NiO films grown on Fe(001) and the effect of temperature. *Surf. Sci.*, **518**, 234–242.
26. Marre, K., Neddermeyer, H., Chassé, A., and Rennert, P. (1996) Auger electron diffraction from NiO(100) layers on Ag(100). *Surf. Sci.*, **357-358**, 233–237.
27. Muller, F., de Masi, R., Steiner, P., Reinicke, D., Stadtfeld, M., and Hufner, S. (2000) EELS investigation of thin epitaxial NiO/Ag(001) films: surface states in the multilayer, monolayer and submonolayer range. *Surf. Sci.*, **459**, 161–172.
28. Valeri, S. and di Bona, A. (1997) Modulated electron emission by scattering-interference of primary electrons. *Surf. Rev. Lett.*, **4**, 141–160.
29. Wollschläger, J., Erdös, D., Goldbach, H., Hopken, R., and Schröder, K.M. (2001) Growth of NiO and MgO films on Ag(100). *Thin Solid Films*, **400**, 1–8.
30. Bertrams, Th. and Neddermeyer, H. (1996) Growth of NiO(100) layers on Ag(100): characterization by scanning tunneling microscopy. *J. Vac. Sci. Technol. B*, **14**, 1141–1144.
31. Hagendorf, Ch., Shantyr, R., Meinel, K., Schindler, K.-M., and Neddermeyer, H. (2003) Scanning tunneling microscopy and spectroscopy investigation of the atomic and electronic structure of CoO islands on Ag(001). *Surf. Sci.*, **532-535**, 346–350.
32. Luches, P., Groppo, E., Prestipino, C., Lamberti, C., Giovanardi, C., and Boscherini, F. (2003) Ni atomic environment in epitaxial NiO layers on Ag(001). *Nucl. Instrum. Methods Phys. Res., Sect. B*, **200**, 371–375.
33. Sanchez-Agudo, M., Yubero, F., Fuentes, G.G., Gutierrez, A., Sacchi, M., Soriano, L., and Sanz, J.M. (2000) Study of the growth of ultrathin films of NiO on Cu(111). *Surf. Interface Anal.*, **30**, 396–400.
34. Luches, P., Altieri, S., Giovanardi, C., Moia, T.S., Valeri, S., Bruno, F., Floreano, L., Morgante, A., Santaniello, A., Verdini, A., Gotter, R., and Hibma, T. (2001) Growth structure and epitaxy of ultrathin NiO films on Ag(001). *Thin Solid Films*, **400** (1-2), 139–143.

35. James, M.A. and Hibma, T. (1999) Thickness-dependent relaxation of NiO(001) overlayers on MgO(001) studied by X-ray diffraction. *Surf. Sci.*, **433-435**, 718–722.
36. Renaud, G. (1998) Oxide surfaces and metal/oxide interfaces studied by grazing incidence X-ray scattering. *Surf. Sci. Rep.*, **32**, 1–90.
37. Roth, W.L. (1960) Neutron and optical studies of domains in NiO. *J. Appl. Phys.*, **31**, 2000.
38. Thole, B.T., van der Laan, G., and Sawatzky, G.A. (1985) Strong magnetic dichroism predicted in the M4,5 x-ray absorption spectra of magnetic rare-earth materials. *Phys. Rev. Lett.*, **55**, 2086–2088.
39. Stöhr, J., Padmore, H.A., Anders, S., Stammler, T., and Scheinfein, M.R. (1998) Principles of x-ray magnetic dichroism spectromicroscopy. *Surf. Rev. Lett.*, **5** (6), 1297–1308.
40. Spanke, D., Solinus, V., Knabben, D., Hillebrecht, F.U., Ciccacci, F., Gregoratti, L., and Marsi, M. (1998) Evidence of in-plane AF domains in ultrathin NiO films. *Phys. Rev. B*, **58**, 5201–5204.
41. Stöhr, J., Scholl, A., Regan, T.J., Anders, S., Lüning, J., Scheinfein, M.R., Padmore, H.A., and White, R.L. (1999) Images of AF structure of a NiO(100) surface by means of x-ray magnetic linear dichroism spectromicroscopy. *Phys. Rev. Lett.*, **83**, 1862–1865.
42. Casassa, S., Ferrari, A.M., Busso, M., and Pisani, C. (2002) Structural, magnetic, and electronic properties of the NiO monolayer epitaxially grown on the (001) Ag surface: an ab Initio density functional study. *J. Phys. Chem. B*, **106** (50), 12978–12985.
43. Giordano, L. and Pacchioni, G. (2007) Interplay between structural, magnetic, and electronic properties in a FeO/Pt(111) ultrathin film. *Phys. Rev. B*, **76**, 075416.
44. Galloway, H.C., Sautet, P., and Salmeron, M. (1996) Interplay between structural, magnetic, and electronic properties in a FeO/Pt(111) ultrathin film. *Phys. Rev. B*, **54** (16), R11145–R11148.
45. Bergermayer, W. and Schweiger, H. (2004) Ab initio thermodynamics of oxide surfaces: O2 on Fe2O3(0001). *Phys. Rev. B*, **69** (19), 195409.
46. Maheswaran, S., Thevuthasan, S., Gao, F., Shutthanandan, V., Wang, C.M., and Smith, R.J. (2005) Misfit dislocations at the single-crystal Fe2O3/Al2O3 interface. *Phys. Rev. B*, **72**, 075403.
47. Thevuthasan, S., Kim, Y.J., Yi, S.I., Chambers, S.A., Morais, J., Denecke, R., Fadley, C.S., Liu, P., Kendelewicz, T., and Brown, G.E. Jr. (1999) Surface structure of MBE-grown α-Fe2O3(0001) by intermediate-energy X-ray photoelectron diffraction. *Surf. Sci.*, **425**, 276–286.
48. Allegretti, F., Franchini, C., Bayer, V., Leitner, M., Parteder, G., Xu, B., Fleming, A., Ramsey, M.G., Podloucky, R., Surnev, S., and Netzer, F.P. (2007) Epitaxial stabilization of MnO(111) overlayers on a Pd(100) surface. *Phys. Rev. B*, **75**, 224120.
49. Sterrer, M., Risse, T., Pozzoni, U.M., Giordano, L., Heyde, M., Rust, H.-P., Pacchioni, G., and Freund, H.-J. (2007) Control of the charge state of metal atoms on thin MgO films. *Phys. Rev. Lett.*, **98**, 96107.
50. Pacchioni, G., Giordano, L., and Baistrocchi, M. (2005) Charging of metal atoms on ultrathin MgO/Mo(100) films. *Phys. Rev. Lett.*, **94**, 226104.
51. Altieri, S., Tjeng, L.H., Voogt, F.C., Hibma, T., and Sawatzky, G.A. (1999) Reduction of Coulomb and charge-transfer energies in oxide films on metals. *Phys. Rev. B*, **59**, R2517–R2520.
52. Altieri, S., Tjeng, L.H., Voogt, F.C., Hibma, T., Rogojanu, O., and Sawatzky, G.A. (2002) Charge fluctuations and image potential at oxide-metal interfaces. *Phys. Rev. B*, **66**, 155432.
53. Sebastian, I., Bertrams, T., Meinel, K., and Neddermeyer, H. (1999) Scanning tunnelling microscopy on the growth and structure of NiO(100) and

CoO(100) thin films. *Faraday Discuss.*, **114**, 129–140.

54. Giovanardi, C., di Bona, A., and Valeri, S. (2004) Oxygen-dosage effect on the structure and composition of ultrathin NiO layers reactively grown on Ag(001). *Phys. Rev. B*, **69**, 075418.
55. Caffio, M., Cortigiani, B., Rovida, G., Atrei, A., and Giovanardi, C. (2004) Early stages of NiO growth on Ag(001): a study by LEIS, XPS, and LEED. *J. Phys. Chem. B*, **108**, 9919–9926.
56. Caffio, M., Atrei, A., Cortigiani, B., and Rovida, G. (2006) STM study of the nanostructures prepared by deposition of NiO on Ag(001). *J. Phys.: Condens. Matter*, **18**, 2379–2384.
57. Hagendorf, C., Shantyr, R., Neddermeyer, H., and Widdra, W. (2006) Pressure-dependent Ni-O phase transitions and Ni oxide formation on Pt(111): an in situ STM study at elevated temperatures. *Phys. Chem. Chem. Phys.*, **8**, 1575–1583.
58. Schoiswohl, J., Mittendorfer, F., Surnev, S., Ramsey, M.G., Andersen, J.N., and Netzer, F.P. (2006) Chemical reactivity of Ni-Rh nanowires. *Phys. Rev. Lett.*, **97**, 126102.
59. Parteder, G., Allegretti, F., Wagner, M., Ramsey, M.G., Surnev, S., and Netzer, F.P. (2008) Growth and oxidation of Ni nanostructures on stepped Rh surfaces. *J. Phys. Chem. C*, **112**, 19272–19278.
60. Agnoli, S., Sambi, M., Granozzi, G., Atrei, A., Caffio, M., and Rovida, G. (2005) A LEED I-V structural determination of the c(4x2) Ni3O4/Pd(100) monolayer phase: an ordered array of Ni vacancies. *Surf. Sci.*, **576**, 1–8.
61. Agnoli, S., Sambi, M., Granozzi, G., Schoiswohl, J., Surnev, S., Netzer, F.P., Ferrero, M., Ferrari, A.M., and Pisani, C. (2005) Experimental and theoretical study of a surface stabilized monolayer phase of Nickel oxide on Pd(100). *J. Phys. Chem. B*, **109**, 17197–17204.
62. Agnoli, S., Orzali, T., Sambi, M., Granozzi, G., Schoiswohl, J., Surnev, S., and Netzer, F.P. (2005) Reactive growth of NiO ultrathin films on Pd(100): a multitechnique approach. *J. Electron Spectrosc. Relat. Phenom.*, **144-147**, 465–469.
63. Agnoli, S., Barolo, A., Finetti, P., Pedona, F., Sambi, M., and Granozzi, G. (2007) Core and valence band photoemission study of strained ultrathin NiO films on Pd(100). *J. Phys. Chem. B*, **111**, 3736–3743.
64. Cinquini, F., Giordano, L., Pacchioni, G., Ferrari, A.M., Pisani., C., and Roetti, C. (2006) Electronic structure of NiO/Ag(100) thin films from DFT + U and hybrid functional DFT approaches. *Phys. Rev. B*, **74**, 165403.
65. Voogt, F.C., Fujii, T., Smulders, P.J., Niesen, L., James, M.A., and Hibma, T. (1999) NO2-assisted molecular-beam epitaxy of Fe3O4, Fe3-δ O4, and γ-Fe2O3 thin films on MgO(100) *Phys. Rev. B*, **60**, 11193.
66. Sebastian, I. and Neddermeyer, H. (2000) Scanning tunneling microscopy on the atomic and electronic structure of CoO thin films on Ag(100). *Surf. Sci.*, **454-456**, 771–777.
67. Shantyr, R., Hagendorf, C., and Neddermeyer, H. (2004) Scanning tunneling microscopy and spectroscopy studies on structural and electronic properties of thin films of Co oxides and oxide precursor states on Ag(100). *Thin Solid Films*, **464-465**, 65–75.
68. Entani, S., Kiguchi, M., and Koichiro, S. (2004) Fabrication of polar CoO(111) thin films on Pt(111). *Surf. Sci.*, **566-568**, 165–169.
69. Marre, K. and Neddermeyer, H. (1993) Growth of ordered thin films of NiO on Ag(100) and Au(111). *Surf. Sci.*, **287/288** (2), 995–999.
70. Giovanardi, C., di Bona, A., Altieri, S., Luches, P., Liberati, M., Rossi, F., and Valeri, S. (2003) Structure and morphology of ultrathin NiO layers on Ag(001). *Thin Solid Films*, **428** (1-2), 195–200.
71. Luches, P., Groppo, E., D'Addato, S., Lamberti, C., Prestipino, C., Valeri, S., and Boscherini, F. (2004) NiO and MgO ultrathin films by polarization dependent XAS. *Surf. Sci.*, **566-568** (1), 84–88.
72. Groppo, E., Prestipino, C., Lamberti, C., Luches, P., Giovanardi, C., and Boscherini, F. (2003) Growth of NiO on Ag(001): atomic environment,

strain, and interface relaxations studied by polarization dependent extended x-ray absorption fine structure. *J. Phys. Chem. B*, **107**, 4597–4606.

73. Groppo, E., Prestipino, C., Lamberti, C., Carbone, R., Boscherini, F., Luches, P., Valeri, S., and D'Addato, S. (2004) O K-edge x-ray absorption study of ultrathin NiO epilayers deposited in situ on Ag(001). *Phys. Rev. B*, **70** (16), 165408.
74. Lamberti, C., Groppo, E., Prestipino, C., Casassa, S., Ferrari, A.M., Pisani, C., Giovanardi, C.C., Luches, P., Valeri, S., and Boscherini, F. (2003) Oxide/metal interface distance and epitaxial strain in the NiO/Ag(001) system. *Phys. Rev. Lett.*, **91** (4), 046101.
75. Schoiswohl, J., Agnoli, S., Xu, B., Surnev, S., Sambi, M., Ramsey, M.G., Granozzi, G., and Netzer, F.P. (2005) Growth and thermal behaviour of NiO nanolayers on Pd(100). *Surf. Sci.*, **599** (1-3), 1–13.
76. Schoiswohl, J., Zheng, W., Surnev, S., Ramsey, M.G., Granozzi, G., Agnoli, S., and Netzer, F.P. (2006) Strain relaxation and surface morphology of nickel oxide nanolayers. *Surf. Sci.*, **600** (5), 1099–1106.
77. Torelli, P., Soares, E.A., Renaud, G., Gragnaniello, L., Valeri, S., Guo, X.X., and Luches, P. (2008) Self-organized growth of Ni nanoparticles on a cobalt-oxide thin film induced by a buried misfit dislocation network. *Phys. Rev. B*, **77**, 081409(R).
78. Nilius, N., Rienks, E.D.L., Rust, H.-P., and Freund, H.-J. (2005) Self-organization of gold atoms on a polar FeO(111) surface. *Phys. Rev. Lett.*, **95**, 066101.
79. Finazzi, M., Brambilla, A., Duò, L., Ghiringhelli, G., Portalupi, M., Ciccacci, F., Zacchigna, M., and Zangrando, M. (2004) Chemical effects at the buried NiO/Fe(001) interface. *Phys. Rev. B*, **70** (23), 235420.
80. Gao, Y. and Chambers, S.A. (1997) Heteroepitaxial growth of α-Fe2O3, γ-Fe2O3 and Fe3O4 thin films by oxygen-plasma-assisted molecular beam epitaxy. *J. Cryst. Growth*, **174**, 446–454.
81. Wang, H.-Q., Gao, W., Altman, E.I., and Henrich, V.E. (2004) Studies of the electronic structure at the Fe3O4– NiO interface. *J. Vac. Sci. Technol. A*, **22** (4), 1675–1681.
82. Klingenberg, B., Grellner, F., Borgmann, D., and Wedler, G. (1997) Low-energy electron diffraction and X-ray photoelectron spectroscopy on the oxidation of cobalt (11-20). *Surf. Sci.*, **383**, 13–24.
83. Gonzalez, L., Miranda, R., Salmerón, M., Vergés, J.A., and Ynduráin, F. (1981) Experimental and theoretical study of Co adsorbed on the surface of Cu: reconstructions, charge density waves, surface magnetism, and oxygen absorption. *Phys. Rev. B*, **24** (6), 3245–3254.
84. Clemens, W., Vescovo, V., Kachel, T., Carbone, C., and Heberardt, W. (1992) Spin-resolved photoemission study of the reaction of O2 with fcc Co(100). *Phys. Rev. B*, **46** (7), 4198–4204.
85. Valeri, S., Borghi, A., Gazzadi, G.C., and di Bona, A. (1999) Growth and structure of cobalt oxide on (001) bct cobalt film. *Surf. Sci.*, **423** (2-3), 346–356.
86. Luches, P., Giovanardi, C., Moia, T., Valeri, S., Bruno, F., Floreano, L., Gotter, R., Verdini, A., Morgante, A., and Santaniello, A. (2002) Epitaxy of ultrathin CoO on films studied by XPD and GIXRD. *Surf. Rev. Lett.*, **9** (2), 937–941.
87. Borghi, A., di Bona, A., Bisero, D., and Valeri, S. (1999) Structural and compositional stability structure of cobalt oxide grown on (001) bct Co. *Appl. Surf. Sci.*, **150** (1-4), 13–18.
88. Brambilla, A., Sessi, P., Cantoni, M., Duò, L., Finazzi, M., and Ciccacci, F. (2008) Epitaxial growth and characterization of CoO/Fe(001) thin film layered structures. *Thin Solid Films*, **516** (21), 7519–7524.
89. Wang, H.-Q., Altman, E.I., and Henrich, V.E. (2008) Interfacial properties between CoO(100) and Fe3O4(100). *Phys. Rev. B*, **77** (8), 085313.
90. Sebastian, I. and Neddermeyer, H. (2000) Scanning tunnelling microscopy on the atomic and electronic structure

of CoO thin films on Ag(100). *Surf. Sci.*, **454-456**, 771–777.

91. Sindhu, S., Heiler, M., Schindler, K.-M., Widdra, W., and Neddermeyer, H. (2004) Growth mechanism and angle-resolved photoemission spectra of cobalt oxide (CoO) thin films on Ag(100). *Surf. Sci.*, **566-568**, 471–475.
92. Torelli, P., Soares, E.A., Renaud, G., Valeri, S., Guo, X.X., and Luches, P. (2007) Nanostructuration of CoO film by misfit dislocations. *Surf. Sci.*, **601**, 2651–2655.
93. Renaud, G., Guénard, P., and Barbier, A. (1998) Misfit dislocation network at the Ag/MgO(001) interface: a grazing incidence X-ray-scattering study. *Phys. Rev. B*, **58** (11), 7310–7318.
94. Bayer, V., Franchini, C., and Podloucky, R. (2007) Ab initio study of the structural, electronic, and magnetic properties of Mn(100) and Mn(111). *Phys. Rev. B*, **75**, 035404.
95. Muller, F., de Masi, R., Reinicke, D., Steiner, P., Hufner, S., and Stowe, K. (2002) Epitaxial growth of MnO/Ag(001) films. *Surf. Sci.*, **520**, 158–172.
96. Soares, E.A., Paniago, R., de Carvalho, V.E., Lopes, E.L., Abreu, G.J.P., and Pfannes, H.-D. (2005) Quantitative low-energy electron diffraction analysis of MnO (100) films grown on Ag(100). *Phys. Rev. B*, **73**, 035419.
97. Vurens, G.H., Maurice, V., Salmeron, M., and Somorjai, G.A. (1992) Growth, structure and chemical properties of FeO overlayers on Pt(100) and Pt(111). *Surf. Sci.*, **268**, 170–178.
98. Weiss, W. and Somorjai, G.A. (1993) Preparation and structure of 1-8 monolayer thick epitaxial iron oxide films grown on Pt(111). *J. Vac. Sci. Technol. A*, **11** (4), 2138–2144.
99. Ritter, M., Ranke, W., and Weiss, W. (1998) Growth and structure of ultrathin FeO films on Pt(111) studied by STM and LEED. *Phys. Rev. B*, **57**, 7249.
100. Rienks, E.D.L., Nilius, N., Rust, H.-P., and Freund, H.-J. (2005) Surface potential of a polar oxide film: FeO on Pt(111). *Phys. Rev. B*, **71**, 241404(R).
101. Ketteler, G. and Ranke, W. (2002) Self-assembled periodic Fe3O4 nanostructures in ultrathin FeO(111) films on Ru(0001). *Phys. Rev. B*, **66**, 033405.
102. Khan, N.A. and Matranga, C. (2008) Nucleation and growth of Fe and FeO nanoparticles and films on Au(111). *Surf. Sci.*, **602**, 932–942.
103. Ritter, M., Over, H., and Weiss, W. (1997) Structure of epitaxial iron oxide films grown on Pt(100) determined by low energy electron diffraction. *Surf. Sci.*, **371**, 245–254.
104. Waddill, G.D. and Ozturk, O. (2005) Epitaxial growth of iron oxide films on Ag(111). *Surf. Sci.*, **575**, 35–50.
105. Lopes, E.L., Abreu, G.J.P., Paniago, R., Soares, E.A., de Carvalho, V.E., and Pfan, H.-D. (2007) Atomic geometry determination of FeO(001) grown on Ag(001) by low energy electron diffraction. *Surf. Sci.*, **601**, 1239–1245.
106. Schedel-Niedrig, Th., Weiss, W., and Schlogl, R. (1995) Electronic structure of ultrathin ordered iron oxide films grown onto Pt(111). *Phys. Rev. B*, **52**, 17449–17460.
107. Weiss, W. (1997) Structure and composition of thin epitaxial iron oxide films grown onto Pt(111). *Surf. Sci.*, **377-379**, 943–947.
108. Shaikhutdinov, Sh.K. and Weiss, W. (1999) Oxygen pressure dependence of the α-Fe2O3(0001) surface structure. *Surf. Sci.*, **432** (3), L627–L634.
109. Kim, Y.J., Gao, Y., and Chambers, S.A. (1997) Selective growth and characterization of pure, epitaxial α-Fe2O3(0001) and Fe3O4(011) films by plasma-assisted molecular beam epitaxy. *Surf. Sci.*, **371**, 358–370.
110. Fujii, T., Alders, D., Voogt, F.C., Hibma, T., Thole, B.T., and Sawatzky, G.A. (1996) In situ RHEED and XPS studies of epitaxial thin α-Fe2O3(0001) films on sapphire. *Surf. Sci.*, **366**, 579–586.
111. Yi, S.I., Liang, Y., Thevuthasan, S., and Chambers, S.A. (1999) Morphological and structural investigation of the early stages of epitaxial growth of α-Fe2O3(0001) on α-Al2O3(0001) by oxygen-plasma-assisted MBE. *Surf. Sci.*, **443**, 212–220.

112. Thevuthasan, S., Kim, Y.J., Yi, S.I., Chambers, S.A., Morais, J., Denecke, R., Fadley, C.S., Liu, P., Kendelewicz, T., and Brown, G.E. Jr. (1999) Surface structure of MBE-grown α-Fe2O3(0001) by intermediate-energy X-ray photoelectron diffraction. *Surf. Sci.*, **425**, 276–286.
113. Lemire, C., Bertarione, S., Zecchina, A., Scarano, D., Chaka, A., Shaikhutdinov, S., and Freund, H.-J. (2005) Ferryl (Fe = O) termination of the hematite α-Fe2O3(0001) surface. *Phys. Rev. Lett.*, **94**, 166101.
114. Singh, M.K., Yang, Y., and Takoudis, C.G. (2008) Low-pressure metallorganic chemical vapour deposition of Fe2O3 thin films on Si(100) using n-butylferrocene and oxygen. *J. Electrochem. Soc.*, **155**, D616–D623.
115. Uribe, J.D., Osorio, J., Barrero, C.A., Girata, D., Morales, A.L., Devia, A., Gomez, M.E., Ramirez, J.G., and Gancedo, J.R. (2006) Hematite thin films: growth and characterization. *Hyperfine Interact.*, **169**, 1355–1362.
116. Altieri, S., Tjeng, L.H., and Sawatzky, G.A. (2000) Electronic structure and chemical reactivity of oxide-metal interfaces: MgO(100)/Ag(100). *Phys. Rev. B*, **61** (24), 16948–16955.
117. Meyerheim, H.L., Popescu, R., Jedrecy, N., Vedpathak, M., Sauvage-Simkin, M., Pinchaux, R., Heinrich, B., and Kirschner, J. (2002) Surface X-ray diffraction analysis of the MgO/Fe(001) interface: evidence for an FeO layer. *Phys. Rev. B*, **65** (14), 144433.
118. Luches, P., Bellini, V., Colonna, S., Di Giustino, L., Manghi, F., Valeri, S., and Boscherini, F. (2006) Iron oxidation, interfacial expansion, and buckling at the Fe/NiO(001) interface. *Phys. Rev. Lett.*, **96**, 106106.
119. Benedetti, S., Luches, P., Liberati, M., and Valeri, S. (2004) Chemical reactions and interdiffusion at the Fe/NiO(0 0 1) interface. *Surf. Sci.*, **572** (2-3), L348–L354.
120. Goniakowski, J., Finocchi, F., and Noguera, C. (2008) Polarity of oxide surfaces and nanostructures. *Rep. Prog. Phys.*, **71** (1), 016501.
121. Noguera, C. (2000) Polar oxide surfaces. *J. Phys.: Condens. Matter*, **12** (31), R367–R410.
122. Kitakatsu, N., Maurice, V., and Marcus, P. (1998) Local decomposition of NiO ultra-thin films formed on Ni(111). *Surf. Sci.*, **411** (1-2), 215–230.
123. Hildebrandt, S., Hagendorf, Ch., Doege, T., Jeckstiess, Ch., Kulla, R., Neddermeyer, H., and Uttich, Th. (2000) Real time scanning tunneling microscopy study of the initial stages of oxidation of Ni(111) between 400 and 470 K. *J. Vac. Sci. Technol. A*, **18** (3), 1010–1015.
124. Ventrice, C.A., Bertrams, Th., Hannemann, H., Brodde, A., and Neddermeyer, H. (1994) Stable reconstruction of the polar (111) surface of NiO on Au(111). *Phys. Rev. B*, **49** (8), 5773–5776.
125. Barbier, A., Mocuta, C., Kuhlenberg, H., Peters, K.F., Richter, B., and Renaud, G. (2000) Atomic structure of the polar NiO(111)-p(2×2) surface. *Phys. Rev. Lett.*, **84** (13), 2897–2900.
126. Stanescu, S., Boeglin, C., Barbier, A., and Deville, J.-P. (2004) Epitaxial growth of ultra-thin NiO films on Cu(111). *Surf. Sci.*, **549** (2), 172–182.
127. Hassel, M. and Freund, H.-J. (1995) NO on CoO(111)/Co(0001): hydroxyl assisted adsorption. *Surf. Sci.*, **325** (1-2), 163–168.
128. Sindhu, S., Heiler, M., Schindler, K.-M., and Neddermeyer, H. (2003) A photoemission study of CoO-films on Au(111). *Surf. Sci.*, **541** (1-3), 197–206.
129. Meyer, W., Hock, D., Biedermann, K., Gubo, M., Müller, S., Hammer, L., and Heinz, K. (2008) Coexistence of rocksalt and wurtzite structure in nanosized CoO films. *Phys. Rev. Lett.*, **101**, 016103.
130. Koike, K. and Furukawa, T. (1996) Evidence for FM Order at the FeO(111) Surface. *Phys. Rev. Lett.*, **77** (18), 3921–3924.
131. Ranke, W., Ritter, M., and Weiss, W. (1999) Crystal structures and growth mechanism for ultrathin films of ionic compound materials: FeO(111) on Pt(111). *Phys. Rev. B*, **60** (3), 1527–1530.

132. Kim, Y.J., Westphal, C., Ynzunza, R.X., Galloway, H.C., Salmeron, M., Van Hove, M.A., and Fadley, C.S. (1997) Interlayer interactions in epitaxial oxide growth: FeO on Pt(111). *Phys. Rev. B*, **55** (20), R13448–R13551.

133. Giordano, L., Pacchioni, G., Goniakowski, J., Nilius, N., Rienks, E.D.L., and Freund, H.-J. (2007) Interplay between structural, magnetic, and electronic properties in a FeO/Pt(111) ultrathin film. *Phys. Rev. B*, **76**, 075416.

134. Berdunov, N., Mariotto, G., Balakrishnan, K., Murphy, S., and Shvets, I.V. (2006) Oxide templates for self-assembling arrays of metal nanoclusters. *Surf. Sci.*, **600**, L287–L290.

135. Locquet, J.-P., Perret, J., Fompeyrine, J., Machler, E., Seo, J.W., and van Tendeloo, G. (1998) Doubling the critical temperature of La1.9Sr0.1CuO4 using epitaxial strain. *Nature*, **394**, 453–456.

136. Iglesias, O., Labarta, A., and Batlle, X. (2008) Exchange bias phenomenology and models of core/shell nanoparticles. *J. Nanosci. Nanotechnol.*, **8**, 2761–2780.

137. Skumryev, V., Stoyanov, S., Zhang, Y., Hadjipanayis, G., Givord, D., and Nogués, J. (2003) Beating the superparamagnetic limit with exchange bias. *Nature*, **423**, 850–853.

3
Dichroism in X-ray Absorption for the Study of Antiferromagnetic Materials

Jan Vogel and Maurizio Sacchi

Since the first predictions and experiments on polarization-dependent X-ray absorption in magnetic materials around 1980, X-ray dichroism has developed as an important tool for the study of element-specific magnetic properties in bulk materials, thin films, and nanostructures. Many experimental techniques exist to measure the magnetic properties of materials, but only a few can provide element-selective information on magnetic multilayers or on antiferromagnetic (AF) materials. One of these techniques is X-ray magnetic dichroism in X-ray absorption, the dependence of the absorption spectra on the polarization of the incoming X-rays due to the local symmetry of the magnetic and electronic properties. The element selectivity is obtained by tuning the X-ray energy to an absorption edge of the element of interest. The energy of these edges is strongly related to the binding energy of the excited core electron, characteristic for each element. The most frequently used absorption edges are the $L_{2,3}$-edges of the 3d transition metals (TM) (dominated by $2p \rightarrow 3d$ excitations), the $M_{4,5}$-edges of the rare earths (RE) ($3d \rightarrow 4f$ excitations) and the K-edges of the TM ($1s \rightarrow 4p$ excitations). The $L_{2,3}$ and $M_{4,5}$-edges, involving direct transitions to the magnetic 3d and 4f levels, are located in the soft X-ray range (500–2000 eV) and exhibit very high X-ray absorption cross sections. This allows very thin layers (a fraction of a monolayer) or very diluted materials (less than 1% atomic fraction) to be measured. Two main types of X-ray dichroism can be discerned. The most commonly used is X-ray magnetic circular dichroism or XMCD, the difference in absorption of left circularly polarized (LCP) and right circularly polarized (RCP) X-rays. XMCD sum rules can be used to obtain accurate information on the spin and orbital contributions to the magnetic moment in an element-selective way [1, 2]. XMCD measures the projection of the spin and orbital moments on the incoming X-ray direction, and is therefore a very powerful tool for the study of ferro- or ferrimagnetic thin films. However, it does not provide information on AF order. This information can be obtained with X-ray linear dichroism or XLD, the difference in absorption of linearly polarized X-rays with the polarization vector parallel and perpendicular to the quantization axis. Other than to AF order, XLD is sensitive also to ferro(ferri)magnetic order and to crystalline electric fields with lower than cubic symmetry, making it sometimes hard to discern between the different origins. Historically, the first prediction of X-ray dichroism

Magnetic Properties of Antiferromagnetic Oxide Materials.
Edited by Lamberto Duò, Marco Finazzi, and Franco Ciccacci

ISBN: 978-3-527-40881-8

was made by Erskine and Stern [3] but the first experimental proof was given by Van der Laan and coworkers [4] 10 years later. They used the huge linear magnetic dichroism in RE $M_{4,5}$-edges predicted by Thole *et al.* [5] to study the magnetic order in Tb–Fe garnets. The first experimental proof of XMCD was provided by Schütz *et al.* [6] for the K-edge of Fe. An exponential increase in the use of this technique started when sum rules were derived that relate the XMCD signal to the orbital and spin part of the magnetic moment for each element separately [1, 2]. This made it possible to quantitatively relate the orbital magnetic moment and its anisotropy to magnetocrystalline anisotropy [7]. It was also used to demonstrate an increase in the orbital magnetic moment, due to a decrease in quenching by the crystalline electric field, going from bulk materials to thin layers to small clusters [8]. The element selectivity was exploited to show the small proximity-induced magnetic moments in otherwise nonmagnetic materials like Cu [9] or Pd [10].

Linear X-ray dichroism was first used to reveal magnetic effects [4] and crystalline electric field effects [11, 12] on the ground state of RE ions. A first example of linear X-ray dichroism induced by AF order was shown for hematite (Fe_2O_3) using the Morin transition where the magnetic moments rotate by 90° [13]. It was later used to investigate spin–spin correlations [14] and the thickness dependence of the Néel temperature in thin NiO layers [15]. This last paper gave rise to a large number of studies using XLD for the study of magnetism in bulk and thin-layer NiO, using either spectroscopy or magnetic imaging using photoemission electron microscopy (PEEM). Many of these examples are treated in this book.

In this chapter, we first give a theoretical basis for X-ray absorption and dichroism. After a general introduction to X-ray absorption and how it is measured, we discuss the selection rules that give rise to the polarization dependence of X-ray absorption. Approaches to dichroism using multiplet theory and band structure effects are both treated. In the next part, we discuss sum rules for both circular and linear X-ray dichroism. We then turn to the application of linear dichroism to magnetic and crystal-field-induced effects in ferromagnetic, ferrimagnetic, and AF materials. We conclude with a discussion of the most recent developments in this field.

3.1 X-ray Absorption and X-ray Dichroism

X-ray absorption spectroscopy (XAS) using synchrotron radiation is a well-established technique providing information on the electronic, structural, and magnetic properties of matter. In X-ray absorption, a photon is absorbed by the atom, giving rise to the transition of an electron from a core state to an empty state above the Fermi level. The absorption cross section depends on the photon energy and on the measured element. To excite an electron from a given core level, the photon energy has to be equal or higher than the binding energy of this core level. A new absorption channel becomes thus accessible when the photon energy is scanned from below to above this core-level energy. The energies of the absorption edges therefore correspond to the core-level energies, which are characteristic for each element, making X-ray absorption an element-selective technique. In the dipole approximation, the cross section for X-ray absorption can be written as

$$W_{\mathrm{fi}} \propto \sum_{q} |\langle \Phi_{\mathrm{f}} | \hat{e}_q \cdot \vec{r} | \Phi_{\mathrm{i}} \rangle|^2 \delta_{E_{\mathrm{f}} - E_{\mathrm{i}} - \hbar\omega} \tag{3.1}$$

where Φ_i and Φ_f are the initial and final states of the absorbing atom, the δ-function accounts for the conservation of energy, $\vec{r}$ is the position operator of the excited electron, and $\hat{e}_q$ is the polarization vector of the incoming X-rays. The origin of X-ray dichroism lies in the dependence of the absorption cross section on this polarization vector. The X-rays can be RCP ($q = -1$), LCP ($q = +1$), or linearly polarized with the polarization vector along the quantization axis ($q = 0$). Linearly polarized X-rays with the electric vector perpendicular to the quantization axis can be represented as a coherent superposition of circularly polarized waves with opposite helicity ($q = \pm 1$). In the dipole approximation, by absorbing a photon, electrons can only make transitions that do not change their spin ($\Delta s = 0$) and change their orbital number by 1 ($\Delta l = \pm 1$). An electron that is initially in the 1s-level can undergo a transition to empty p-levels, 2p electrons can go to empty s or d-levels, and so on.

In this chapter, we mainly consider $L_{2,3}$ edges in TM (dominated by 2p to 3d transitions) and $M_{4,5}$ edges in RE (dominated by 3d to 4f transitions). The $L_{2,3}$ edges are particularly important for TM-based AF oxides. Two extreme approaches for describing the X-ray absorption process can be considered. The first one, called *one-electron approximation*, considers only the electron that is excited from a core level to an empty state above the Fermi level. The interaction with the other electrons and the core-hole is neglected in the first approximation. In that case, the absorption spectrum should look like the partial empty density of states. Band structure calculations and multiple scattering calculations can be used to simulate the X-ray absorption spectra. These calculations give quite accurate results for the X-ray absorption spectra from deep core levels to delocalized empty levels, like, for instance, the K-edge absorption spectra of TM. They have also been used for the $L_{2,3}$ edges of metallic TMs, but in that case correlation effects have to be taken into account to obtain accurate results for the shape of the dichroic spectra. When intra-atomic electron–electron and electron–hole interactions are more important than the interaction with the environment, the basis for simulations of the X-ray absorption and dichroism spectra is given by atomic multiplet calculations. In that case, the electron–electron and electron–hole correlations are taken into account explicitly. The environment can then be simulated by taking into account crystal-field and charge-transfer effects on the atomic multiplets [16]. The origin of X-ray dichroism in these two approaches is explained in Sections 3.1.1 and 3.1.2.

3.1.1
X-ray Magnetic Circular Dichroism in the One-Electron Approximation

In general, band-structure effects play a fundamental role in the shape of spectra of metallic TM, and a one-electron approach is often used to treat the excitations.

Table 3.1 Basis states of the $2p_{1/2}$ and $2p_{3/2}$ levels and the corresponding spherical harmonics.

	Spherical harmonic
$\lvert \frac{1}{2}, \frac{1}{2}\rangle$	$-\frac{1}{\sqrt{3}} Y_{10}^{\uparrow} + \frac{\sqrt{2}}{\sqrt{3}} Y_{11}^{\downarrow}$
$\lvert \frac{1}{2}, -\frac{1}{2}\rangle$	$-\frac{\sqrt{2}}{\sqrt{3}} Y_{1-1}^{\uparrow} + \frac{1}{\sqrt{3}} Y_{10}^{\downarrow}$
$\lvert \frac{3}{2}, \frac{3}{2}\rangle$	$Y_{11}^{\uparrow}$
$\lvert \frac{3}{2}, \frac{1}{2}\rangle$	$\frac{\sqrt{2}}{\sqrt{3}} Y_{10}^{\uparrow} + \frac{1}{\sqrt{3}} Y_{11}^{\downarrow}$
$\lvert \frac{3}{2}, -\frac{1}{2}\rangle$	$\frac{1}{\sqrt{3}} Y_{1-1}^{\uparrow} + \frac{\sqrt{2}}{\sqrt{3}} Y_{10}^{\downarrow}$
$\lvert \frac{3}{2}, -\frac{3}{2}\rangle$	$Y_{1-1}^{\uparrow}$

A semiclassical approach to X-ray dichroism at the $L_{2,3}$-edges, taking into account relativistic effects (spin–orbit (SO) coupling) in the initial p-state of the electron but not in the d-state, was given in 1975 by Erskine and Stern [3]. After the first experimental proof of XMCD in the $L_{2,3}$-edges by Chen *et al.* [17], this model was refined by Smith *et al.* [18] and Stöhr [19] to include SO coupling in the d-band.

The relativistic basis states $\lvert j, m_j\rangle$ for the p-levels are given in Table 3.1, as well as their representation using spherical harmonics Y_{l,m_l}. It is quite straightforward to calculate the transition probabilities from p to d levels using these spherical harmonics. For example, if the 3d-level is not SO split, the transition from the $\lvert \frac{1}{2}, \frac{1}{2}\rangle$ state in the $2p_{1/2}$ manifold to the 3d-states, for LCP light, is given by $\Phi = \lvert\langle 3d \rvert \frac{(x+iy)}{r} \lvert 2p_{1/2}(\frac{1}{2}, \frac{1}{2})\rangle\rvert^2$. If we assume that the final states can be treated incoherently, that is, the final state is an incoherent sum of the different $Y_{l,m_l}(l = 2)$ states, then this transition can be written as

$$\Phi = \left|\left\langle Y_{2,1}^{\uparrow} \left| \frac{(x+iy)}{r} \right| \frac{1}{\sqrt{3}} Y_{1,0}^{\uparrow} \right\rangle\right|^2 R^2 + \left|\left\langle Y_{2,2}^{\downarrow} \left| \frac{(x+iy)}{r} \right| \frac{\sqrt{2}}{\sqrt{3}} Y_{1,1}^{\downarrow} \right\rangle\right|^2 R^2 \tag{3.2}$$

Here we have used the dipole selection rules $\Delta m_l = +1$ for LCP light, and $\Delta s = 0$ (constant spin). R is the integral $R = \int r^2 R_p(r) R_d(r)$, where $R_p(r)$ and $R_d(r)$ are the 2p and 3d radial wave functions. The dipole matrix elements for transitions from a state with angular momentum l to a state $l + 1$ are given in the Bethe and Salpeter equation (BS), Chapter 4 [20]:

$$\left\langle l+1, m_l+1 \left| \frac{(x+iy)}{r} \right| l, m_l \right\rangle = \sqrt{\frac{(l+m_l+2)(l+m_l+1)}{2(2l+3)(2l+1)}} \tag{3.3}$$

$$\left\langle l+1, m_l-1 \left| \frac{(x-iy)}{r} \right| l, m_l \right\rangle = \sqrt{\frac{(l-m_l+2)(l-m_l+1)}{2(2l+3)(2l+1)}} \tag{3.4}$$

For the above-mentioned example with $\frac{(x+iy)}{r}$, this gives $\Phi = \frac{1}{3}\lvert \mathrm{BS}(l = 1, m_l = 0)\rvert^2 R^2 + \frac{2}{3}\lvert \mathrm{BS}(l = 1, m_l = 1)\rvert^2 R^2 = \frac{1}{3}\frac{1}{5} R^2 + \frac{2}{3}\frac{2}{5} R^2 = \frac{1}{3} R^2$. It can be easily calculated that in the case of a nonmagnetic material where the number of empty spin-up and spin-down states is equal, the integrated absorption is the same, at both edges, for LCP and RCP light. If the spin-up states are completely filled, as is the case of strong ferromagnets like Ni and Co, only spin-down electrons can undergo a transition and only the $Y^{\downarrow}$-parts of the wavefunctions have to be taken into account. In that case, the intensity of the L_2-edge using LCP light is given by

$$\left|\left\langle Y^{\downarrow}_{2,2}\left|\frac{(x+iy)}{r}\right|\frac{\sqrt{2}}{\sqrt{3}}Y^{\downarrow}_{1,1}\right\rangle\right|^2 R^2 + \left|\left\langle Y^{\downarrow}_{2,1}\left|\frac{(x+iy)}{r}\right|\frac{1}{\sqrt{3}}Y^{\downarrow}_{1,0}\right\rangle\right|^2$$

$$R^2 = \frac{2}{3}\frac{2}{5}R^2 + \frac{1}{3}\frac{1}{5}R^2 = \frac{1}{3}R^2 \tag{3.5}$$

while RCP light gives $\frac{2}{3}\frac{1}{15}R^2 + \frac{1}{3}\frac{1}{5}R^2 = \frac{1}{9}R^2$. At the L_3-edge, LCP light gives $(\frac{1}{3}\frac{2}{5} + \frac{2}{3}\frac{1}{5} + \frac{1}{15})R^2 = \frac{1}{3}R^2$, while RCP light gives $(\frac{1}{3}\frac{1}{15} + \frac{2}{3}\frac{1}{5} + \frac{2}{5})R^2 = \frac{5}{9}R^2$. The total absorption at the L_2-edge is $\frac{4}{9}R^2$ and at the L_3-edge $\frac{8}{9}R^2$, giving the expected branching ratio $L_3 : L_2 = 2 : 1$ for unpolarized light in the absence of SO coupling in the 3d-band [21]. The difference LCP − RCP is $\frac{2}{9}R^2$ at the L_2 and $-\frac{2}{9}R^2$ at the L_3-edge, giving a branching ratio for the dichroism of $L_3 : L_2 = -1 : 1$.

A more intuitive idea of the origin of XMCD can be obtained if we look separately at the spin-up and spin-down parts of the wave functions, using a two-step model [19]. In the case of a final state without spin splitting LCP light excites $\frac{1}{3}\frac{1}{5}R^2 + \frac{2}{3}\frac{1}{15}R^2 = \frac{1}{9}R^2$ spin-up and $\frac{2}{3}\frac{2}{5}R^2 + \frac{1}{3}\frac{1}{5}R^2 = \frac{1}{3}R^2$ spin-down electrons at the L_2-edge. So at the L_2-edge, LCP (RCP) light excites 75% (25%) spin-up and 25% (75%) spin-down electrons. At the L_3-edge, 37.5% (62.5%) spin-up and 62.5% (37.5%) spin-down electrons are excited by LCP (RCP) light. In a nonmagnetic material, the absorption of LCP and RCP light is the same, but as soon as there is an unbalance in the number of available empty spin-up and spin-down states, the absorption of the two polarizations will be different, with a difference that is opposite at the L_2 and L_3-edges.

3.1.1.1 Spin-Orbit Coupling in the *d*-Band

For a complete relativistic description of X-ray dichroism in the one-electron model, SO coupling has to be included also in the d-band. SO coupling will split the d-band in $d_{5/2}$ and $d_{3/2}$ where the latter has the lowest energy. A partly filled band will therefore contain more empty $d_{5/2}$-states, which will favor L_3-edge absorption with respect to the L_2-edge ($2p_{1/2} \rightarrow d_{5/2}$ transitions are dipole-forbidden). The $L_3 : L_2$ branching ratio will therefore increase, both in the total X-ray absorption and in the XMCD. In particular, this means that the integrated XMCD signal will not be zero anymore in the presence of SO coupling in the 3d-band. It has been shown by Thole *et al.* [1] that the integrated XMCD intensity is actually proportional to the 3d orbital moment, as discussed in Section 3.2.

3.1.1.2 Core-Hole and Other Many-Body Effects

Deviations from the statistical 2 : 1 value of the $L_3 : L_2$ branching ratio can occur when the energy of the 2p SO coupling is reduced, as it happens going toward lighter 3d-elements, and becomes comparable to the energy of the Coulomb interactions between two 3d electrons and between a 3d electron and the 2p core-hole. In light 3d TM, the branching ratio often seems closer to 1, regardless of the SO coupling in the 3d-shell. Actually, this observation expresses a more fundamental problem, that is, that one can no longer define an L_3 and an L_2-edge, since none of the transitions correspond to a pure $2p_{3/2}$ or $2p_{1/2}$ core-hole in the final state. This has important consequences for a sum rule, which permits to calculate the spin magnetic moment from dichroism spectra, as discussed in Section 3.2.2.

Even for the TM at the end of the series, more intensity is present at the onset of the absorption edge than would be expected from band structure. The presence of a core-hole in the final state actually attracts electronic states to the Fermi level, shifting absorption intensity to the edges.

Many-body effects play a role in the ground state as well. In Figure 3.1, we show the absorption and dichroism at the $L_{2,3}$-edges of Ni-metal. While the satellite at 6 eV in the L_3-edge absorption spectrum could be reproduced by one-electron calculations [22], this is not the case for the satellite at 4 eV in the XMCD. Several groups [23, 24] used a many-body configuration interaction approach, in which the electronic state of Ni is described as a superposition of different $3d^n$ configurations. Using an Anderson impurity model, the ground state of an Ni atom can be described by a superposition of $3d^8$, $3d^9\underline{L}$ and $3d^{10}\underline{L}^2$ configurations with different weights (where $\underline{L}$ denotes a hole in orbitals on neighboring sites). With this model, the experimentally observed L_3 to L_2 ratios could be reproduced, as well as the satellites in the XMCD curve, which were shown to be due mainly to the contribution of the ^3F term in the $3d^8$ ground-state multiplet [23]. Using the same type of calculations, van der Laan showed [25] that the peak asymmetry in the 2p circular dichroism

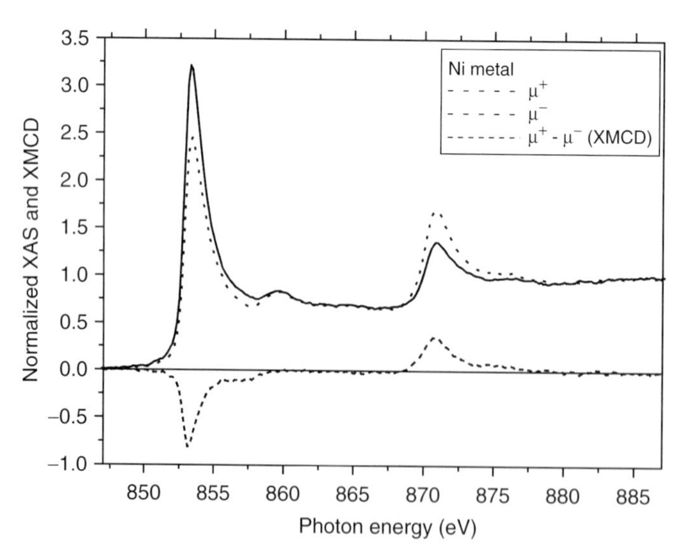

Figure 3.1 Spectra of Ni metal taken with left (μ^+) and right (μ^-) circularly polarized X-rays, together with the difference or XMCD curve.

signal of magnetic 3d metals strongly depends on the d count and is inversely proportional to the spin polarization.

3.1.2 XMCD in the Strongly Correlated Limit: Multiplet Effects

The first experimental proof of X-ray dichroism was given for the $M_{4,5}$ absorption edges of RE materials (or lanthanides) [4]. These results were obtained after the theoretical prediction of large magnetic effects on the polarization dependence of these absorption edges, dominated by the 3d $\rightarrow$ 4f transitions. The 4f states in the lanthanides are strongly localized and partly screened from the environment by the delocalized 5d and 6s electrons. The electronic levels of the lanthanide ions are primarily determined by Coulomb interactions, followed by SO interactions. In first approximation, the 3d $\rightarrow$ 4f transitions can thus be described in a purely atomic multiplet model. Magnetic and crystal-field effects can then be treated as small perturbations. For the $L_{2,3}$-edges of ionic TM like TM-oxides crystal-field effects and the (chemical) interaction with the environment are much more important and cannot be treated as small perturbations anymore.

In the case of (quasi)atomic transitions, the wavefunctions in Eq. (3.1) can be written using spherical harmonics characterized by the quantum numbers J, M and J', M' for the initial and final states, respectively. In this case, the absorption cross section is given by

$$W_{\text{fi}} = \sum_q \left| \left\langle J' M_{J'} \middle| \hat{e}_q \cdot \vec{r} \middle| J M_J \right\rangle \right|^2 = \sum_q \begin{pmatrix} J & 1 & J' \\ -M_J & q & M_{J'} \end{pmatrix}^2 |\langle J' || \hat{e}_q \cdot \vec{r} || J \rangle|^2 \tag{3.6}$$

The matrix element has been separated in an angular part (the squared 3J-symbol) and a radial part, using the Wigner-Eckarts theorem. The radial part $|\langle J' || \hat{e}_q \cdot \vec{r} || J\rangle|^2$ gives the spectral intensity of the transition. The dependence of the absorption on the light polarization comes from the dependence of the 3J-symbol on M and q. The 3J-symbol is nonzero only if $|J - 1| \leq J' \leq J + 1$ and $q = \Delta M = M_J - M_{J'} = 0, \pm 1$. Linearly polarized light with its polarization vector parallel to the quantization axis (defined, for instance, by an external or internal magnetic field) gives rise to transitions with $q = 0$, while $q = +1$ ($q = -1$) transitions are induced by left (right) circularly polarized light propagating parallel to the quantization axis.

In spherical symmetry, the ground state given by Hund's rules is $(2J + 1)$-degenerate and all M_J ($-J \leq M_J \leq J$) levels are equally occupied. It can be shown that in this case the light polarization does not have an influence on the absorption spectra [26]. A magnetic field (internal or external) can lift this degeneracy and split the ground state in Zeeman levels with different values of M_J and at low enough temperature these different levels will be unequally occupied. The simplest case is given by the $M_{4,5}$-edges of a Yb^{3+} ion ($3d^{10}4f^{13}$ in the initial state) in a magnetic field [27]. This field will induce a Zeeman splitting of the $^2F_{7/2}$ ground state in 15 levels with $M_J = -\frac{7}{2}, -\frac{5}{2}, -\frac{3}{2}, \ldots, \frac{7}{2}$ and energy $\mu_B\, gHM_J$, where μ_B

Table 3.2 Values of the 3J-symbols as a function of ΔJ and q.

$\Delta J/q$	-1	0	$+1$
-1	$\frac{J(J-1)-(2J-1)M+M^2}{2J(2J+1)(2J-1)}$	$\frac{J^2-M^2}{J(2J+1)(2J-1)}$	$\frac{J(J-1)+(2J-1)M+M^2}{2J(2J+1)(2J-1)}$
-0	$\frac{J(J+1)-M-M^2}{2J(2J+1)(J+1)}$	$\frac{M^2}{2J(2J+1)(J+1)}$	$\frac{J(J+1)+M-M^2}{2J(2J+1)(J+1)}$
$+1$	$\frac{(J+1)(J+2)+(2J+3)M+M^2}{2(2J+3)(2J+1)(J+1)}$	$\frac{(J+1)^2-M^2}{2(2J+3)(2J+1)(J+1)}$	$\frac{(J+1)(J+2)-(2J+3)M+M^2}{2(2J+3)(2J+1)(J+1)}$

is the Bohr magneton, g is the Landé factor for the considered RE and H is the magnetic field strength. At zero Kelvin, only the level lying lowest in energy, $M_J = -\frac{7}{2}$, will be occupied. The final state after absorption has the configuration $3d^9 4f^{14}$, with only one possible energy level characterized by the term symbol $^2D_{5/2}$, with $-\frac{5}{2} \leq M_J \leq \frac{5}{2}$. Therefore, at zero Kelvin, only transitions with $\Delta M = +1$ are possible and only LCP light, or linearly polarized light with the polarization vector perpendicular to the magnetic field will be absorbed. At higher temperatures, the other M_J-levels will be occupied according to a Boltzmann distribution, and the other transitions (first $\Delta M = 0$ for $M_J = -\frac{5}{2}$, then also $\Delta M = -1$ for $M_J = -\frac{3}{2}$) become possible. However, a dependence of the absorption on the polarization will persist up to temperatures for which $kT \gg \mu_0 gH$.

Using Table 3.2, one can see that the difference between the transitions for $q = -1$ and $q = +1$ (circular dichroism) is proportional to $\langle M_J \rangle$ and thus to the magnetic moment ($|M| = \langle M_J \rangle \mu_0 g_{\alpha J}$). The difference between $q = 0$ and $q = \pm 1$ (linear dichroism) is proportional to $\langle M_J^2 \rangle - \frac{1}{3}J(J+1)$ or to the deviation from the statistical value of $\langle M_J^2 \rangle$. Linear dichroism is thus sensitive to magnetic order, but it cannot give the sign, or the direction, of magnetic moments. However, it can be very useful for the study of antiferromagnetism, as shown in Section 3.5. Moreover, in contrast to XMCD, linear dichroism is also sensitive to crystalline electric fields with a less than cubic symmetry. This often limits the interest in linear dichroism for the study of the magnetism of RE, since for many magnetic materials based on RE, the RE ion is in a low-symmetry crystalline environment.

3.1.2.1 **Ligand Field Atomic Multiplet Calculations**

When the effect of the environment (neighboring atoms, magnetic fields) cannot be considered a small perturbation with respect to intra-atomic effects, the purely atomic model mentioned above cannot be used anymore. With the exception of the case of RE, crystal or ligand fields have to be treated on the same footing as SO coupling and intra-atomic Coulomb interactions. This is especially the case for compounds of 3d TMs, which represent a class of materials of large interest for magnetic dichroism studies. Owing to the importance of the 3d electrons in the chemical bonding in these compounds, the ground and excited states ($2p^6 3d^n$ and $2p^5 3d^{n+1}$) at the $L_{2,3}$ absorption edges must be calculated taking the crystal field into account explicitly [16].

In a first approximation, the crystal-field model considers the TM as an isolated atom surrounded by a distribution of point charges that represent the neighboring atoms in a solid or molecule. This simple model has been used with success to explain optical spectra [28] as well as X-ray absorption spectra [29]. Many of the features of these spectra are mainly determined by symmetry considerations, which has made the use of group theory in the crystal-field model very efficient. Group theory is used in crystal-field theory to translate, or branch, the results obtained in atomic symmetry to cubic symmetry and then to any point groups with lower symmetry (see, e.g., [30]). The use of group theory has allowed to considerably simplify the calculations in the crystal-field model and thus to decrease the calculation times. With current computer speeds, it becomes possible to perform *ab initio* calculations in any geometry, without using group theory considerations (see, e.g. [31]).

The crystal field model was initially developed by Yamaguchi *et al.* [32] and later generalized by Thole, van der Laan, and coworkers [5, 33, 34]. The crystal-field Hamiltonian contains intra-atomic interactions as well as perturbation terms. Typical parameters that can be adjusted to fit experimental results are reduction of the intra-atomic Slater integrals (to mimic hybridization), the local exchange field, and the symmetry and strength of the crystal field.

3.1.2.2 **Charge-Transfer Effects**

Charge-transfer (CT) effects are the effects of charge fluctuations in the initial and final states. In the atomic multiplet and crystal field multiplet models, a single electronic configuration is used to describe ground state as well as final states. In order to take charge fluctuations into account, one can perform calculations for different electronic configurations and combine them coherently, that is, summing amplitudes (with given weights) then taking the square modulus, rather than summing the intensities pertaining to each contribution. The ground state of Ni^{2+} in NiO, for instance, can be presented as a weighted sum of $3d^8$ and $3d^9\,\underline{L}$ configurations, where $\underline{L}$ represents a hole on the ligand site. The energy difference between these two configurations is determined by the charge-transfer energy Δ. For NiO, this means that an electron has been moved from the oxygen 2p-valence band to the Ni 3d-band. In many cases, two configurations will be enough to explain the spectral shapes, but, in particular, for high valence states it can be important to include more configurations. For TM spectra for which a $3d^{N+2}\,\underline{L}^2$ configuration is included, its energy is $2\,\underline{\Delta} + U_{dd}$, where U_{dd} is the correlation energy between two 3d-electrons [35], induced by Coulomb and direct exchange interactions. U_{dd} is the energy difference one obtains when an electron is transferred from one metal site to another, that is, a transition $3d^N + 3d^N \rightarrow 3d^{N+1} + 3d^{N-1}$. The number of interactions between two $3d^N$ configurations (N^2) is one more than the number of interactions between $3d^{N+1}$ and $3d^{N-1}$ ($(N+1)(N-1) = N^2 - 1$), implying that this energy difference is equal to the correlation energy between two 3d-electrons. The main effects of CT on the X-ray absorption spectral shape are the formation of small satellites and the contraction of the multiplet structures.

The formation of small satellites or even the absence of visible satellite structures is a special feature of X-ray absorption spectroscopy. Its origin is the fact that X-ray absorption is a neutral spectroscopy and the local charge of the final state is equal to the charge of the initial state. This implies that the screening in initial and final states is similar, leading to weak charge-transfer satellites.

The contraction of the multiplet structure due to CT can be understood assuming two multiplet states split by an energy δ. They both mix with a CT state that is positioned at an energy Δ above the lowest energy multiplet state I, and thus at $\Delta - \delta$ above the second multiplet state II. Assuming that the hopping terms are the same for states I and II, CT will lead to the largest energy gain for the state having an energy closest to CT. State II will thus gain more energy by CT than state I, leading to a reduction of δ and thus to an apparent compression of the multiplet.

For further reading on multiplet calculations for X-ray absorption, the reader is referred to books by Cowan [36] and by Kotani and De Groot [16]. In Chapter 4, extensive use of multiplet calculations is made.

3.2
Sum Rules for X-ray Dichroism

The integrated intensities of the absorption and XMCD spectra can be used to obtain quantitative magnetic information on the absorbing atom. In the beginning of the 1990s, sum rules were derived that directly relate these experimental intensities to the ground-state expectation values of the orbital [1] ($\langle L_z \rangle$) and spin [2] ($2\langle S_z \rangle$) magnetic moments. The subscript z indicates that the components of the total momenta projected on the propagation direction of the X-rays are measured. Given the element and symmetry specificity of X-ray absorption, these sum rules provide element- and symmetry-specific information, that is, at the $L_{2,3}$ edges of TMs the local 3d momenta are measured, while the $M_{4,5}$ edges of RE provide the 4f momenta.

3.2.1
Orbital Moment

In Section 3.1.1.1, it was mentioned that in the simple one-electron model the orbital moment is proportional to the total integrated intensity of the dichroism curve. The mathematical derivation hereof was given by Thole *et al.* using graphical methods for angular momentum algebra [1] while alternative derivations were given by other authors [37]. The general form of the orbital sum rule is given below:

$$\langle L_z \rangle = \frac{2l\,(l+1)(4\,l+2-n)}{l\,(l+1)+2-c\,(c+1)} \times \frac{\int\limits_{j^+ + j^-} d\omega(\mu^+ - \mu^-)}{\int\limits_{j^+ + j^-} d\omega(\mu^+ + \mu^- + \mu^0)} \tag{3.7}$$

In this formula l is the orbital quantum number of the valence state, c that of the core state, n is the number of electrons in the valence shell, j^+ (j^-) corresponds to the $c + \frac{1}{2}$ ($c - \frac{1}{2}$) absorption edge and μ^+, μ^-, and μ^0 are the absorption coefficients for LCP light, RCP light, and light that is linearly polarized with the polarization vector parallel to the quantization axis. For $L_{2,3}$-edges, $l = 2$ ($3d$) and $c = 1$ ($2p$), giving

$$\langle L_z \rangle = 2(\Delta L_3 + \Delta L_2) \times \frac{(10 - n)}{\int_{L_3+L_2} d\omega(\mu^+ + \mu^- + \mu^0)} \tag{3.8}$$

In the absence of XLD (which is usually very small in metallic TM systems [38]), μ^0 will be the average of μ^+ and μ^- and $\int_{L_3+L_2} d\omega(\mu^+ + \mu^- + \mu^0)$ can be obtained from the experimental spectra subtracting a proper background accounting for transitions other than 2p → 3d (like p → s). The number of electrons in the valence shell (n) is usually obtained from band-structure calculations. A procedure to use the sum rules on experimental spectra has been given, for example, by Chen *et al.* [39].

The orbital sum rule has been used to show an increase in the orbital moment at interfaces [40], surfaces [41], for impurities [42], and small clusters [8]. It has also provided a first experimental proof [7] for the relation between the magnetocrystalline anisotropy and the anisotropy in the orbital moment predicted by theory [43].

3.2.2
Spin Moment

The sum rule for the spin moment was given in 1993 by Carra *et al.* [2]. Its general form is

$$\langle S_z \rangle = \frac{3c\,(4l + 2 - n)}{l(l+1) - 2 - c(c+1)} \times \left[\frac{\int_{j^+} d\omega(\mu^+ - \mu^-) - [(c+1)/c] \int_{j^-} d\omega(\mu^+ - \mu^-)}{\int_{j^+ + j^-} d\omega(\mu^+ + \mu^- + \mu^0)} - \frac{l(l+1)[l(l+1) + 2c(c+1) + 4] - 3c(c-1)^2(c+2)^2}{6lc(l+1)(4l+2-n)} \langle T_Z \rangle \right] \tag{3.9}$$

At the $L_{2,3}$ edges this gives

$$\langle S_z \rangle = \frac{3}{2}(\Delta L_3 - 2\,\Delta L_2) \frac{(10 - n)}{\int_{L_3+L_2} d\omega(\mu^+ + \mu^- + \mu^0)} - 3.5 \langle T_Z \rangle \tag{3.10}$$

$\langle T_Z \rangle$ is the expectation value of the magnetic dipole operator $\mathbf{T} = \sum_i (\mathbf{s}_i - 3\mathbf{r}_i(\mathbf{r}_i \cdot \mathbf{s}_i)/r_i^2)$. It may be noted that in the sum rule for the spin moment the difference between the integrated intensities over L_3 and L_2-edges is needed, unlike for the orbital sum rule where the sum over the two edges (and thus the total integrated intensity) is used. In order to determine these two integrated intensities unambiguously, the 2p SO coupling, which induces the separation between the L_3 and L_2-edges, should be much larger than the Coulomb interactions in the final state. This is not the general case for 3d-TM, but at the end of the series, where the SO coupling is largest, the error becomes relatively small ($\leq$5% for Ni [2]).

The term $\langle T_Z \rangle$ is due to an anisotropy in the spin moment, and can be induced by an anisotropic charge distribution around the atom (corresponding to a charge quadrupole moment [44, 45]) or by the SO interaction. The first contribution is zero in bulk systems with cubic symmetry, but can be enhanced at surfaces and interfaces [46], where the cubic symmetry is broken. The contribution due to the SO interaction is small in 3d metals (small SO coupling) but larger in 4d and 5d metals. It is especially large in 4f metals (RE) where it is the dominant contribution, but in this case an analytical expression for T can be found [2, 47].

Though T does not contribute to the magnetic moment $L + 2S$, it contributes to the magnetocrystalline anisotropy energy (MAE) [45]: $\delta E \approx -\frac{1}{4}\xi\widehat{\mathbf{S}} \cdot [\langle \mathbf{L}^{\downarrow} \rangle - \langle \mathbf{L}^{\uparrow} \rangle] + \frac{\xi^2}{\Delta E_{ex}} \left[\frac{21}{2}\widehat{\mathbf{S}} \cdot \langle T \rangle + 2\langle (L_\zeta S_\zeta)^2 \rangle \right]$. The orbital magnetic moment as measured with XMCD is given by $\langle \mathbf{L}^{\downarrow} \rangle + \langle \mathbf{L}^{\uparrow} \rangle$, which is equivalent to $\langle \mathbf{L}^{\downarrow} \rangle - \langle \mathbf{L}^{\uparrow} \rangle$ only if $\langle \mathbf{L}^{\uparrow} \rangle = 0$, that is, when the spin-up band is filled (like in Ni and Co metal).

3.2.3
Sum Rule for Linear Dichroism

In the general case, the MAE is not proportional to the orbital moment anisotropy but to the anisotropy in the SO interaction, as has been shown by Van der Laan [48]. He developed a sum rule that relates the SO anisotropy to the XLD in X-ray absorption. The two major methods to measure XLD, the difference between absorption spectra with $\mathbf{P} \perp \mathbf{M}$ and $\mathbf{P} \parallel \mathbf{M}$, are by (i) keeping the magnetization direction fixed and turning the linear polarization vector or (ii) keeping the linear polarization vector fixed and turning $\mathbf{M}$. The latter may be difficult when the magnetic anisotropy (induced by magnetostatic and magnetocrystalline effects) is large, but the dependence of the dichroism on the SO anisotropy will be enhanced. In AF materials, it is even harder to rotate the magnetization vector away from the easy axis, but this can happen naturally by the exchange coupling at a ferromagnetic/AF interface, for example, [49]. In this case, the dependence of the absorption spectra on the angle between magnetization and crystallographic axis has to be taken into account explicitly [50].

Van der Laan showed [48] that the anisotropy in the SO interaction $\langle \lambda_a \rangle$ can be obtained by measuring the difference in branching ratio B ($L_3/(L_2 + L_3)$ for the $L_{2,3}$ edges) for $\mathbf{P} \perp \mathbf{M}$ and $\mathbf{P} \parallel \mathbf{M}$:

$$\langle \lambda_a \rangle \approx \frac{\Delta B \cdot \langle n_h \rangle}{1 - B_0} \qquad (3.11)$$

where B_0 is the statistical value for the branching ratio, which is 2/3 for the $L_{2,3}$-edges. Using this formula, Dhesi *et al.* [51] have given the following formula for the MAE

$$\mathrm{MAE} = \frac{\zeta \lambda_a}{2} = \frac{\zeta n_h}{2A} \frac{\Delta I_{L_3} - 2\Delta I_{L_2}}{I_{L_3} + I_{L_2}} \tag{3.12}$$

where n_h is the number of holes in the 3d-band, ζ is the radial part of the SO interaction and $I_{L_{2,3}}$ and $\Delta I_{L_{2,3}}$ are the integrated intensities over the $L_{2,3}$-edges of the XAS and XLD spectra, respectively. Dhesi *et al.* used the element specificity of this technique to determine separately the MAE of Co and Fe in Co/Fe bilayers [51]. In metallic 3d TM, the SO anisotropy, and thus the XLD, is generally small, smaller than 10% even for materials with a strong MAE like thin Co layers with perpendicular magnetic anisotropy [7, 48]. Also, in AF oxides the integrated signal over the L_3 and L_2 edges is generally small, but large effects can be observed *within* one SO split edge like is the case, for instance, at the Ni L_2 edge of NiO.

3.3 Experimental Determination of X-ray Absorption

Measuring the absorption coefficient over the so-called soft X-ray range of photon energies (roughly, 50–2000 eV) presents peculiar technical aspects that have to be considered carefully. Normally, a measurement of the intensity impinging on the sample, $I_0(E)$, and of the intensity transmitted through it, $I(E)$, allows one to determine the energy-dependent linear absorption coefficient $\mu(E) : \mu(E) = ln[I_0(E)/I(E)]/t$, t being the sample thickness traversed by the X-rays. In the soft X-ray range, though, this approach is seldom used because the very high absorption coefficients require samples and substrates that must be extremely thin and, at the same time, homogeneous. One prefers to rely upon indirect methods that measure secondary products of the absorption process, typically electrons and photons, the former being by far the more common of the two.

Collecting the total electron yield (TEY) or, equivalently, the sample drain current is a simple way of measuring a soft X-ray absorption spectrum. In spite of its widespread use, a careful analysis of the limitations of interpreting TEY measurements in terms of absorption spectra is not always carried out and the use of TEY data often relies on the implicit but unsupported assumption that they simply measure the absorption. Another common notion about TEY is that it has a large probing depth d (values in excess of 20 nm are often cited). This assumption comes from the fact that TEY deals mainly with very low energy electrons (a few electronvolts) which have, according to the so-called universal curve [52, 53], a very large inelastic mean free path (IMFP) in solids. We performed several experiments [54, 55] in order to directly test the reliability of this assumption. The outcome is that if 10 nm can represent a correct estimate of d for a limited number of materials (e.g., Si_3N_4), such value has no universal applicability. Our results, supported by the work of other authors [56, 57], show, for instance, that iron and nickel in their metallic form have $d = 1.5$ nm and $d = 1.8$ nm, respectively, while values as low as 1 nm have been measured in RE [55]. Finally, a specific study dealing with the determination of TEY probing depth in hematite and magnetite iron oxides [58] led to values of $d = 3.5$ nm and $d = 4.5$ nm, respectively. A first conclusion that could be drawn from these studies is that TEY is not as bulk sensitive as it was claimed in the past: a probing depth of a few nanometers is not always sufficient for neglecting surface effects safely. Moreover, the TEY probing depth is strongly material dependent and, to make things worse, no reliable model has been found yet for an accurate prediction of its value in a given material.

The knowledge of the exact value of d is very important in the evaluation of angular-dependent saturation effects in TEY-detected absorption spectra. It is particularly important to account for these effects in XLD experiments, where often it is required to compare spectra obtained in different beam-to-sample geometries. Saturation occurs when the measured TEY signal is no longer proportional to the absorption cross section and the intensity of prominent peaks is reduced or "saturated." The TEY intensity for an infinitely thick sample can be written as

$$I(\alpha, E) = \frac{Ad}{d + \lambda(E) \sin\alpha} \tag{3.13}$$

where A is the number of electrons produced per absorbed photon, α is the angle of incidence of the X-rays defined relative to the sample surface, d is the probing depth, and $\lambda(E) = 1/\mu(E)$ is the absorption length. When the condition $[d << \lambda(E)$ $\sin\alpha]$ is satisfied for a given α, the measured yield is inversely proportional to the absorption length and thus proportional to the absorption coefficient:

$$I(\alpha, E) \approx \frac{Ad}{\lambda(E) \sin\alpha} \propto \frac{1}{\lambda(E)} = \mu(E) \tag{3.14}$$

The proportionality constant is α-dependent, since the TEY intensity $I(\alpha, E)$ varies as $\frac{1}{\sin\alpha}$ due to the variation of the length of the photon path within the active thickness d. Rewriting Eq. (3.7) as

$$I(\alpha, E) \sin\alpha = \frac{A}{\csc\alpha + \lambda(E)/d} \tag{3.15}$$

$I(\alpha, E)\sin\alpha$ represents a normalized intensity per absorbing atom that is α-independent as long as $d/\lambda(E)$ is much smaller than $\sin\alpha$. This condition will be satisfied better for photon energies corresponding to weak absorption and/or for high values of α (normal incidence of the photons). When changing the angle toward grazing incidence, the approximation $d < \lambda(E) \sin\alpha$ progressively loses its validity. Figure 3.2 illustrates how angular-dependent saturation sets in at the absorption maxima of the L_3 and L_2 edges of Fe in Fe_3O_4. Similar data are shown in Figure 3.6(a) for the Ni-L_3-edge in an NiO film. As expected, the departure from a constant value of $I(\alpha, E)\sin\alpha$ appears sooner (i.e. closer to normal incidence) at L_3 than at L_2, since the former has a much stronger absorption coefficient. Equation (3.8) indicates that, for a given angle, saturation effects depend on λ/d. As both quantities change when moving from pure metals to oxides, saturation effects are expected to be, for a given element and absorption resonance, different for metallic or oxide compounds [59]. However, a clear trend for different materials (oxides vs metals) has not been identified yet. Finally, it is worth mentioning that the shorter

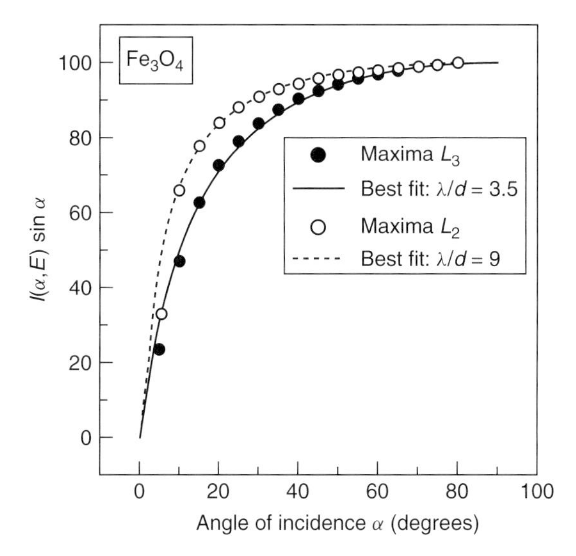

Figure 3.2 Experimental values of $I(\alpha, E)\sin\alpha$ as a function of α for a 240-Å-thick Fe_3O_4 film. Solid and open circles represent photon energies corresponding to the maxima of the Fe L_3 and L_2 edges, respectively. Both intensity curves are normalized to 100 for $\alpha = 90^\circ$. Curve fittings with the expression of Eq. (3.8) yield $\lambda/d = 3.5$ for the L_3 edge (solid line) and $\lambda/d = 9$ for the L_2 edge (dashed line).

than predicted probing depth of TEY is essential for a meaningful association between $I(\alpha, E)$ and $\mu(E)$. A d of some 20 nm would be, for many materials, of the same order of the minimum absorption length; under these conditions, saturation effects would be unavoidable even at normal incidence [57], let alone as a function of α. This is one of the major problems affecting the total fluorescence yield detection mode, a photon-in photon-out process where the penetration of the incoming photons and the probing depth of the outgoing photons are usually of the same order of magnitude, leading to $\lambda d \sim 1$.

Another problem of using TEY on AF oxides is related to the insulating nature of most oxides. At the absorption edges, the absorption cross section, and thus the number of extracted secondary electrons, is particularly high. This can lead to charging effects that are strongest at the maximum of the absorption, leading to an apparent decrease in the measured absorption intensity. This effect is particularly strong for oxide layers deposited on insulating substrates, like MgO. A reduction of the incoming X-ray intensity and/or partly covering the surface with a conducting layer may then be necessary to obtain reliable results.

3.4 Linear X-ray Dichroism in Rare-Earth Compounds

Since applications of X-ray magnetic dichroism to RE-based AF oxides are scarce, we use other compounds to illustrate the potentials and limitations of the technique, keeping in mind that AF order would produce exactly the same XLD, but no XMCD. As shown in Section 3.1.2, within the atomic approach, linear and circular dichroism measure $(3\langle M_J^2\rangle - J(J+1))$ and $\langle M_J\rangle$, respectively. While $\langle M_J\rangle$ different from zero can be related immediately to the presence of local magnetic order, the interpretation of a nonzero $(3\langle M_J^2\rangle - J(J+1))$ can be more subtle, since it can be of nonmagnetic origin too. AF order, where $\langle M_J\rangle = 0$, can be studied by XLD only; it is therefore important to be aware of contributions other than magnetic that will influence the outcome of linear dichroism experiments.

3.4.1 Fe_xTb_{1-x} Amorphous Thin Films

One of the first applications of magnetic dichroism to study thin films concerned Fe_xTb_{1-x} amorphous layers with perpendicular magnetization [60, 61]. RE/TM amorphous films present peculiar magnetic properties, which have attracted great interest in both fundamental and applied research [62]. Binary and ternary alloys with different RE/TM ratios, when prepared in the form of thin films, can show a large magnetic anisotropy with the easy axis perpendicular to the film plane. Their application as high-density recording media was proposed many years ago, triggering a large effort to understand the physical mechanism underlying and controlling the macroscopic magnetic properties. Apparently, the simplified picture of a "disordered" amorphous material does not apply: the large magnetic anisotropy perpendicular to the film plane is driven by a structural anisotropy, that is, the RE–RE, TM–TM and RE–TM distances are different in and perpendicular to the film plane [63, 64].

Macroscopic magnetic measurements show that the net magnetic moments of the RE and TM sublattices are antiparallel, the RE moments exhibiting a canted structure [62] around the net magnetization direction. The size of the perpendicular magnetic anisotropy depends on the film composition [65] and on the details of the film preparation, like the temperature at which it was deposited [66].

In Figure 3.3, the spectra of the M_5 edge of Tb at two different temperatures are reported, taken with the linear polarization vector of the light perpendicular (normal photon incidence, filled circles) and almost parallel (open circles) to the direction of the total magnetization of the sample, that is, the surface normal. The specific sample considered in Figure 3.3 has $x = 0.18$. The difference curves between the two orientations are also reported, multiplied by an (arbitrary) factor 3, and compared to the theoretical dichroism obtained from atomic multiplet calculations.

In the case of a pure Zeeman splitting, at very low temperature, only the $M_J = -6$ level should be occupied, leading to a saturation value for $\sqrt{\langle M_J^2\rangle}$ of 6. However, the value found for the sample in Figure 3.3 is only 4.85 at 35 K. Actually, the pure Zeeman splitting is not a very accurate description of the real situation. The presence of other perturbing fields, like crystalline electric fields (CEFs), leads to the occurrence of complex magnetic structures and to a mixing of the $|M_J\rangle$ atomic wave functions in building up the perturbed level scheme. In a semiclassical picture, for the RE–TM

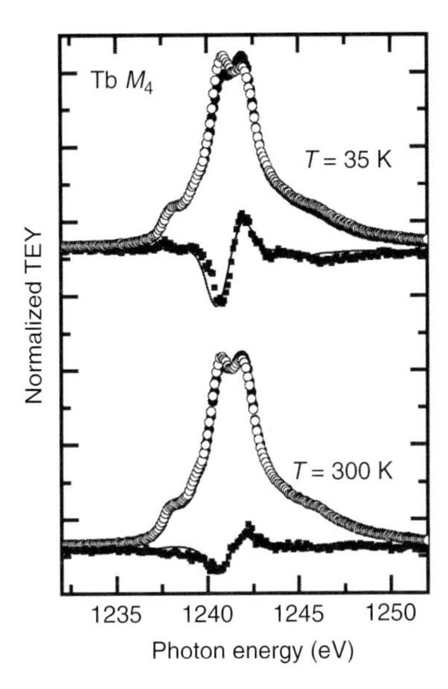

Figure 3.3 Linear dichroism at the Tb M_5 absorption edge for $Fe_{82}Tb_{18}$ at 35 K and RT (300 K). Spectra are for normal (•) and 10° incidence (°). The difference is multiplied by a factor 3. Lines are the results of atomic calculations, using $\langle M_J \rangle$ as a fitting parameter.

alloys this can be represented by a situation where the Tb moments are lying on a cone making an angle θ with the net magnetization direction. The $\langle M_J^2 \rangle$ along this direction is then given by $\sqrt{\langle M_J^2 \rangle} = \sqrt{\langle M_J^2 \rangle_{\max}} * \cos\theta \Rightarrow \theta = \arccos\sqrt{\langle M_J^2 \rangle / 36}$. This gives an average angular aperture θ around the axis perpendicular to the plane of 35° for $x = 0.18$. In principle, XLD could be related solely to CEF effects. Complementary measurements using circularly polarized light are then necessary to be sure that there is a net magnetic moment in the Tb sublattice, as was done for the given Fe_xTb_{1-x} sample [61].

The first example of XLD at the RE $M_{4,5}$ edges of ascertained nonmagnetic origin refers to ultrathin RE layers grown on Si(111) and Cu(001) [11, 67]. An atom adsorbed on a substrate experiences a strongly asymmetric environment with a finite charge density on one side and vacuum on the other. The corresponding CEF shows a dominant axis of symmetry perpendicular to the surface, which basically determines the splitting of the free atom degenerate levels into a new perturbed scheme. Various techniques can give information about CEF splitting, and notably inelastic neutron scattering (INS), but, when a surface layer is to be studied, most of them fail because of lack of sensitivity. However, thanks to the large cross sections involved, measurements at the $M_{4,5}$ edges of RE can be easily performed on layers with submonoatomic thickness (less than 10^{12} at.cm^{-2}).

We have used XLD to investigate the crystal-field-induced splitting on Er ($^4I_{15/2}$ ground state) [68] and Dy ($^6H_{15/2}$) ions deposited on the Si(111) 7 × 7 surface [11].

The two sets of data could be explained coherently within a simplified model supposing a CEF of $C_{\infty v}$ symmetry, a result that gave experimental evidence to the relation between crystal fields and XLD. Recently, the use of XLD in the analysis of CEF has found renewed interest, with applications to single crystals [69].

These examples of soft X-ray dichroism in RE compounds illustrate how this technique can produce results that are relatively easy to interpret in terms of local and element-selective magnetic and crystal-field properties. XLD, in particular, provides information about ferromagnetic and AF order, as well as about crystal-field effects. On one hand, this makes it more versatile than circular dichroism, which is sensitive to FM order only. On the other hand, disentangling these contributions of different origin is not always trivial and can make the interpretation less straightforward.

3.5 Magnetic Dichroism in TM Oxides

Applications of X-ray magnetic dichroism are more frequent for TM-oxides than for RE-oxides. The interpretation scheme presented in Section 3.1.2 allows one to treat SO coupling as well as magnetic and crystal-field perturbations on the same footing. Calculations are more complex and rely on a higher number of parameters, but they also offer a more direct way to account for the interplay between electronic and magnetic properties. The first example of soft X-ray magnetic linear dichroism applied to TM oxides was reported by Kuiper *et al.* in 1993 [13], investigating the spin-reorientation (Morin) transition in hematite. Absorption data were collected as a function of temperature around the transition temperature of 263 K, without changing the relative orientation of the photon wave vector and the polarization vector with respect to the crystal axes. Since no crystallographic transition occurs in Fe_2O_3 around 263 K, these measurements provided the first clear evidence of magnetic XLD in an AF TM-oxide.

The experiment of Kuiper *et al.* aimed at a proof of principle, that is, at demonstrating the interest and the potential of the technique, since the magnetic properties of Fe_2O_3 single crystals were well known in advance, but its role was fundamental in inspiring further applications.

3.5.1 Magnetic Linear Dichroism in Thin NiO Films on MgO

One of the first applications in which XLD was used to obtain quantitative information on AF materials was on thin NiO layers on MgO. The research program started within a collaboration between the Universities of Groningen and Nijmegen in the Netherlands. The peculiar characteristics of these AF samples, whose thickness could be of a few atomic layers only, made them quite difficult to analyze using conventional techniques and X-ray dichroism proved very powerful for studying them. Measurements on NiO started in 1991 at LURE, analyzing the temperature and thickness dependence of XLD at the Ni-2p resonances [70]. The analysis of experimental data found a strong support in the multiplet theory developed by Thole, De Groot, and coworkers at about the same time, which predicted that the L_2 edge of $3d^9$ Ni^{2+} in NiO would be very sensitive to magnetic order and, moreover, would provide dichroism measurements with an unquestionable, internally referenced, signature for magnetic analysis.

The magnetic structure of NiO has been widely studied. The Bravais lattice of NiO is simple cubic, with the Ni and O occupying face-centered cubic (fcc) sublattices. Various techniques have been used to reveal the magnetic structure of bulk NiO to be of type-II fcc, that is, ferromagnetic sheets of spins in {1,1,1} planes, which are antiferromagnetically stacked along the $\langle 1, 1, 1\rangle$ directions. In total, there are 12 distinct domains possible: four principal so-called T domains, corresponding to the four possible $\langle 1, 1, 1\rangle$ directions, which are themselves divided into three S-domains corresponding to three possible $\langle 1, 1, \overline{2}\rangle$ spin directions. If all these domains were present in the measured thin layers, no magnetic XLD would be observed, since $\langle M_J^2\rangle$ would be equal in every direction. XLD could, however, still be caused by CEF effects. In the bulk of NiO, the cubic symmetry inhibits CEF-induced XLD, but an eventual effect can originate from the surface and from the interface with MgO, where the cubic symmetry is broken. A rather larger XLD induced by crystal fields was observed, for instance, for a monolayer of NiO sandwiched between Ag(100) and MgO(100) [71].

3.5.1.1 **Calculations**

Initial theoretical studies of the $L_{2,3}$ edges of NiO with cluster and multiplet calculations including the crystal field showed a good overall agreement with experimental spectra [4]. The Ni ($2p^5$ $3d^9$) final states are dominated by the 2p SO coupling, which splits the spectrum into two parts, roughly corresponding to states with a $2p_{3/2}$ and a $2p_{1/2}$ core-hole (the L_3 and the L_2 edge, respectively). Both edges consist of a multiplet, whose detailed shape is determined by the crystal and magnetic fields and the final-state pd-Slater integrals F^2_{pd}, G^1_{pd}, and G^3_{pd}. F^k_{pd} and G^k_{pd} represent the direct and exchange parts of the electrostatic interaction between the p and d holes. Since divalent Ni in the final state contains only one d and one p hole, dd and pp correlations do not have to be considered [72]. The calculations presented in Figures 3.4 and 3.5 were performed with a version of Cowan's program [36], modified by Thole *et al.* [5] to include crystal-field effects in any desired point group. The spectra are calculated for O_h symmetry, using a value of $10\ Dq = 1.65\ eV$. The Hartree–Fock Slater integrals were scaled down to 80% of their atomic values to account for hybridization effects in the solid state [72]. Magnetic effects were introduced using an exchange field acting on the 3d-orbitals. This leads to an exchange term in the Hamiltonian equal to $H_{ex} = n\langle S_z\rangle J$, where n is the number of nearest neighbors (6 for Ni in NiO), $\langle S_z\rangle$ is the spin–spin correlation function, and J is the superexchange. With neutron scattering a value for J of 19 meV has been obtained for bulk NiO [73, 74], a value which has also been used for the calculations shown here. In fact, small changes in the line shapes

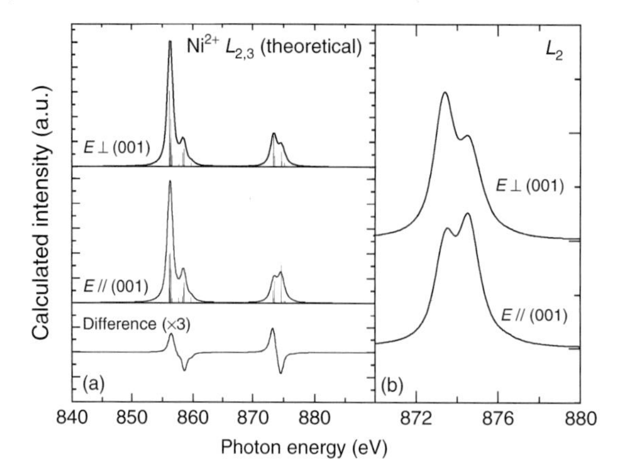

Figure 3.4 (a) Theoretical $L_{2,3}$ absorption spectra for Ni^{2+} at 0 K, with superexchange parameter $J = 19$ meV and the magnetic moments along the $[1,1,\bar{2}]$ equivalent directions. The spectra are for perpendicular (top) and parallel (middle) orientations of the [0,0,1] direction and the electric vector of the linearly polarized light. (b) The L_2 edge in more detail.

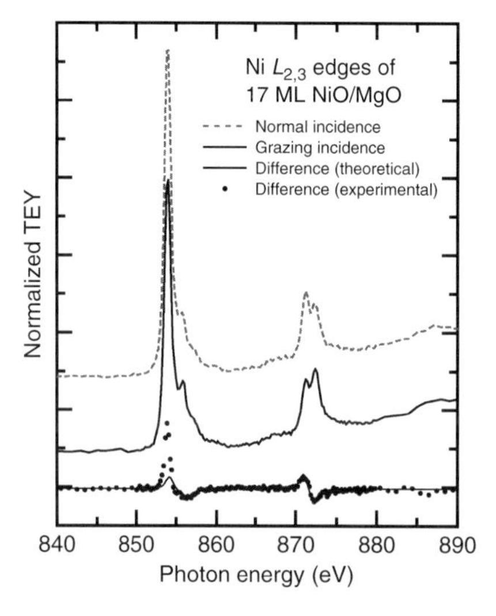

Figure 3.5 Experimental Ni $L_{2,3}$ edges of 17 ML NiO/MgO, at RT, with the linearly polarized light at normal and 15° incidence. The experimental difference is compared to the theoretical curves of Figure 3.4.

of the $L_{2,3}$ edges were observed for different thicknesses of the NiO layers, due to modifications in the superexchange parameter J (see [14]). These changes, however, do not influence the interpretation of the data presented here.

The calculated spectra for parallel and perpendicular orientations of light polarization vector and net magnetization axis are shown in Figure 3.4. To account for lifetime effects, the theoretical line spectra were broadened with a Lorentzian of $\delta = 0.4\,\text{eV}$ for the L_3 part, and of $\delta = 0.50\,\text{eV}$ and $\delta = 0.54\,\text{eV}$ for the first and second peak of the L_2 part.

The increased broadening at higher energy originates from the decreased lifetime due to the opening of Coster–Kronig Auger decay channels. The spectra were further convoluted with a Gaussian of $\delta = 0.1\,\text{eV}$ to simulate the limited experimental resolution. See [14] for more details.

3.5.1.2 Sample Preparation

The NiO films were grown on MgO(001) substrates, and details about the preparation can be found in [75]. Single-crystal MgO samples were cleaved ex situ, yielding platelets of $\sim 5 \times 10\,\text{mm}^2$. They were annealed for 4–16 hours, at 750°C, in an ultrahigh vacuum (UHV) system, to remove hydrocarbon contamination. The NiO layers were grown in an UHV molecular beam epitaxy system with a base pressure of less than 10^{-10} torr. Ni metal was evaporated from alumina crucibles at 1300 to 1375°C, in a NO—NO_2 pressure of 1×10^{-7} torr.

During preparation, the quality of the films was monitored by reflection high-energy electron diffraction, observing layer-by-layer growth. Low-energy electron diffraction (LEED) was used to check the surface structure of the films, while X-ray photoelectron spectroscopy (XPS) measurements were performed to check if the films were well oxidized. The resulting samples were very smooth, well oxidized, and did not contain any nitrogen impurities. See Chapter 4 for more details.

3.5.1.3 Experiment

Five NiO samples of different thicknesses (3, 6, 12, 17 and 50 ML, 1 ML $\simeq$ 4 Å) were used for this experiment. The samples were clamped on a copper plate, attached to the cold finger of a liquid nitrogen cooled cryostat. Heating wires inside the cryostat allowed to stabilize at any temperature in the range $\sim$80 K $\div$ 650 K, the temperature being controlled by two thermocouples, one of them placed at the sample position and the other close to the heat source. The sample could be rotated by 360°, with an accuracy of $\pm 1°$, around the vertical axis, perpendicular to both the propagation and the electric vector of the incoming light. With this mounting, for every sample, at least three spectra, corresponding to different orientations of the polarization vector and the surface normal (the [001] direction of the substrate), were taken, at different temperatures.

In a second experiment, the 17-ML sample was mounted on a different sample holder, where an additional azimuthal rotation of 360° could be performed, with an accuracy of 2°, around the normal to the sample surface.

3.5.1.4 Results

In Figure 3.5 two spectra are reported, taken on the 17-ML sample at room temperature (RT), with the polarization vector perpendicular (normal incidence) and almost parallel (grazing incidence) to the [001] direction of the substrate. The difference between the two is compared to the calculated one from Figure 3.4.

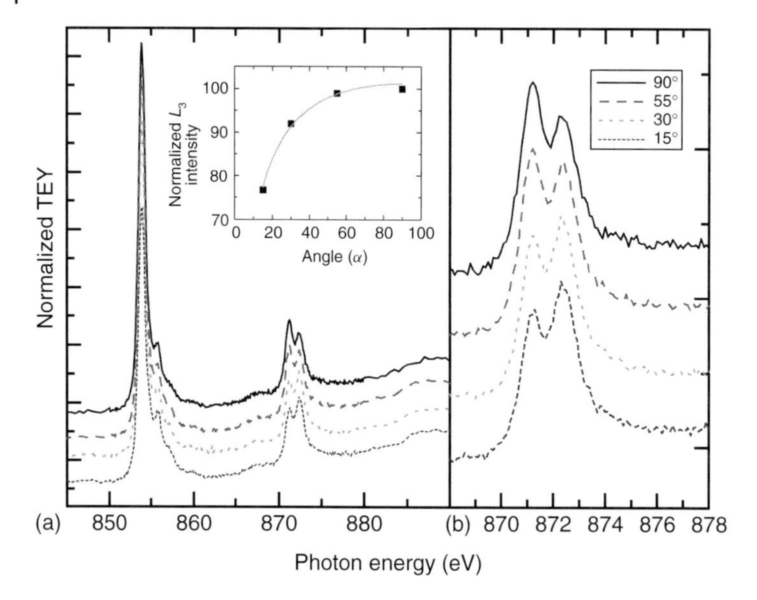

Figure 3.6 (a) Ni $L_{2,3}$ absorption edges of 17-ML NiO/MgO(100) taken at RT for different angles of incidence. Inset: L_3 normalized intensity, fitted by using Eq. (3.6). (b) The L_2 part of the same spectra.

It is evident that experiment and calculation match well, apart from the energy region where the maximum L_3 absorption occurs. A more detailed angular dependence for the L_3 and L_2 edges is given in Figure 3.6; while the L_2 edge exhibits large changes in line shape and relative intensities of the two main structures, the variations in the L_3 edge are dominated by saturation effects (see Section 3.3). This is shown in the inset of Figure 3.6(a), where the maximum intensity versus angle is reported. These experimental points were fitted using Eq. (3.6), giving a value of λ/d of 8 ± 1, which is much lower than the values found for Ni metal ($\lambda/d = 18 \pm 2$) [76] and Dy ($\lambda/d = 11 \pm 1$) [55]. This can partly be explained by the larger value of the escape depth d expected for insulators [54], but for very grazing angles also the increased reflectivity at the L_3-edge has to be taken into account [77].

Because of the possible occurrence of saturation and other angular-dependent effects, we decided to concentrate on the L_2 edge only. The cross section at L_2 is much smaller than at the L_3 and the two peaks which build it up have more or less equal intensity. The above-mentioned effects, which make the interpretation of the changes in the L_3 absorption rather hazardous, have therefore much less influence on the shape of the L_2 edge. In Figure 3.6(b), which shows the L_2 part of the spectra, the change in the relative intensity of the two peaks upon changing the angle is clearly observable.

Figure 3.7(a) compares the angular-dependent spectra, at RT, for layers of different thicknesses, showing that below 12 ML there is essentially no dichroism. This result confirms that the lower symmetry at the surface (C_{4v}) is not the main

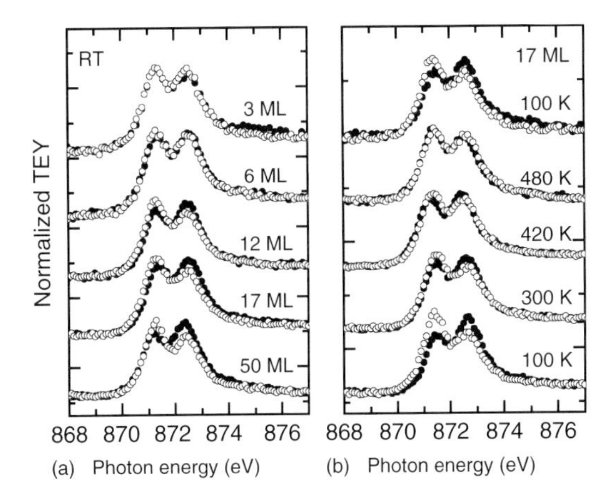

Figure 3.7 (a) Polarization dependence of the L_2 edge of Ni, at RT, for NiO layers of different thicknesses. (b) Polarization dependence of the L_2 edge of Ni for 17 ML of NiO on MgO(001), at different temperatures.

source of XLD in the NiO thin layers, since in that case the effect should decrease with increasing layer thickness.

In Figure 3.7(b), the polarization-dependent spectra are reported for 17 ML, taken at different sample temperatures. The dichroism disappears at a temperature between 420 and 480 K. The temperature changes are completely reversible, as confirmed by the top curve in Figure 3.7(b), taken after a complete thermal cycle. For 50 ML, the dichroism vanishes at 500 K, that is, very close to the Néel temperature of bulk NiO (525 K). Therefore, the presence of dichroism can be associated safely with the presence of antiferromagnetic ordering in the NiO layers.

Assuming the magnetic origin of the dichroism, the direction of the magnetic moments can be found by comparing the experimental spectra with calculations for different directions of the exchange field J. As mentioned above, in bulk NiO, there are 12 distinct magnetic domains possible. For S-domains with spins along the $[\bar{2},1,1]$, $[1,\bar{2},1]$, and $[1,1,\bar{2}]$ equivalent directions, the average $\langle M_J^2 \rangle$ is along the [1,0,0], [0,1,0], and [0,0,1] axes, respectively. In bulk NiO these axes are equivalent, making a preferential ordering of the magnetic moments along one of them unlikely. This is however not the case for the thin layers, where the breaking of the cubic symmetry at the surface and the interface with the MgO makes the [0,0,1] direction inequivalent to the [0,1,0] and [1,0,0]. Though not directly influencing the absorption spectra, the reduced symmetry in the thin layers can still have an effect by causing a preferential aligning of the magnetic moments along an axis of high symmetry.

In the present case, where the scattering plane is the (100) plane of the NiO(001) film and the light is p-polarized, an $\langle M_J^2 \rangle$ along the [1,0,0] axis does not result in an angular dependence of the spectra. The calculations show that for the [0,1,0] axis the angular dependence should be reversed with respect to the one observed in the experiment. Apparently, only the $[1,1,\bar{2}]$ domains are present in the thin

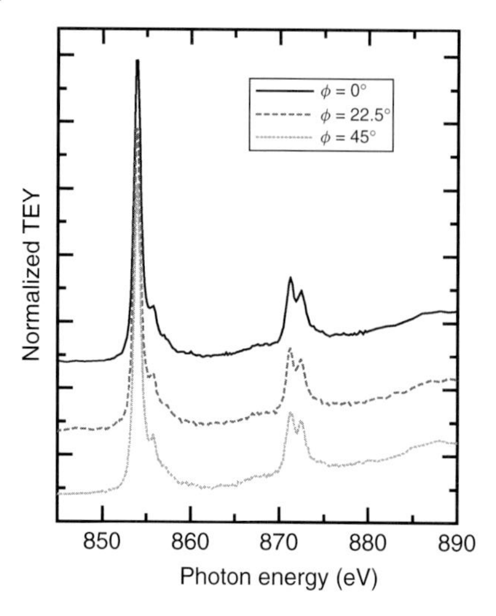

Figure 3.8 Spectra taken at normal incidence, for the 17-ML sample at RT, for different azimuthal angles ϕ between the polarization vector of the light and the [010] axis.

films, resulting in a maximum value of $\langle M_J^2 \rangle$ along the [0,0,1] axis, perpendicular to the surface. This is confirmed by the three spectra in Figure 3.8, which are taken on the 17-ML sample, at normal incidence, but with different angles between the polarization vector of the light and the [0,1,0] direction. The spectra are very similar, showing the absence of azimuthal dependence around the [0,0,1] direction. Figure 3.4(a) shows the theoretical $L_{2,3}$ spectra for normal and 15° incidence if only the $[1,1,\bar{2}]$, $[1,\bar{1},\bar{2}]$, $[\bar{1},1,\bar{2}]$, and $[\bar{1},\bar{1},\bar{2}]$ (and opposite) spin directions are assumed to be present, in equal proportions, to have the resultant $\langle M_J^2 \rangle$ along the [0,0,1] axis. In Figure 3.4(b), the L_2 edges of the same spectra are drawn. A comparison with the spectra of Figure 3.6 shows that the agreement is good.

Apparently, the stabilization of a net spin axis along the surface normal, as observed in the thin NiO layers on MgO(001) presented here, depends on the epitaxial strain in the NiO layer. Altieri *et al.* [78] showed that the AF S-domains in thin NiO layers can have their main spin component either in- or out-of-plane, depending on the lattice mismatch with the substrate (see Chapter 4). See also [79] for a general review on magnetic properties of low-dimensional AF oxides.

A summary of the results of the angular- and temperature-dependent measurements is reported in Figure 3.9. The x-axis represents the NiO film thickness in monolayers and the y-axis represents the sample temperature during the measurement. A filled square in the graph means that the sample at that temperature showed dichroism, and an open square means that it did not. Half-filled squares mean the observed difference was just above the noise level, and hence doubtful. Even taking into account the difficulty to determine exactly where the dichroism vanishes, a semiquantitative trend for the Néel temperature T_N versus the number of NiO layers can be drawn.

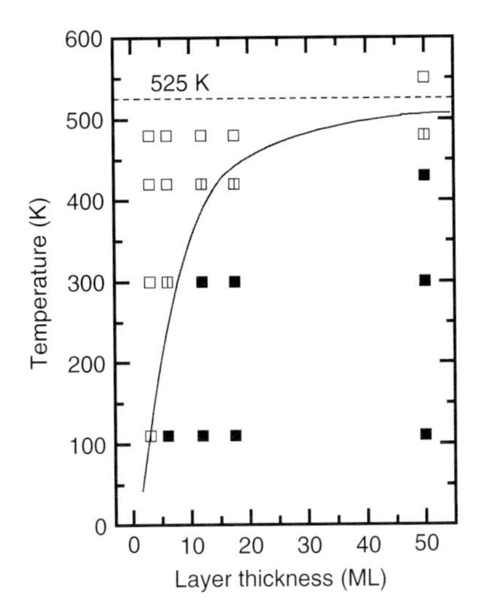

Figure 3.9 Observed dichroism as a function of temperature for different layer thicknesses. The symbols are explained in the text.

For the calculations of the angular-dependent absorption mentioned above, the magnetic moments were assumed to be along the $[\bar{2}11]$ equivalent directions, like in bulk NiO. This can be justified for the rather thick NiO layers we investigated. In that case, the absorption spectra depend only on the angle between these spin directions and the polarization vector of the incoming X-rays. Obviously, the angular dependence will be different when the spin directions are not along the $[\bar{2}11]$ but along other crystallographic axes. Arenholz *et al.* [50] have shown that, in general, the angle of the linear polarization vector with both the crystallographic axes and the spin direction have to be taken into account.

3.6 Conclusions

In this chapter, we have discussed the basics of circular and linear dichroism in X-ray absorption, and the way it can be used to obtain information on ferromagnetic, ferrimagnetic, and AF order with element selectivity. Especially, it was shown that XLD is useful for the study of AF materials, but that care has to be taken when using it. This was illustrated using XLD in RE materials, showing that linear dichroism can be induced not only by ferromagnetic order but also by nonmagnetic crystal-field effects. The X-ray linear dichroism in thin NiO layers on MgO(001) was then shown to be of AF origin using the thickness and temperature dependence of the effect. It was used to show that the preferential spin directions in these layers are oriented perpendicular to the plane, and to determine the thickness-dependent Néel temperature. Many other examples of the use of X-ray linear dichroism for the study of AF oxides are given in this book, using either spectroscopy (Chapter 4) or magnetic imaging (Chapter 8).

References

1. Thole, B.T., Carra, P., Sette, F. and van der Laan, G. (1992) X-ray circular dichroism as a probe of orbital magnetization. *Phys. Rev. Lett.*, **68** (12), 1943.
2. Carra, P., Thole, B.T., Altarelli, M., and Wang, X. (1993) X-ray circular dichroism and local magnetic fields. *Phys. Rev. Lett.*, **70** (5), 694.
3. Erskine, J.L. and Stern, E.A. (1975) Calculation of the $M_{2,3}$ magneto-optical absorption spectrum of ferromagnetic nickel. *Phys. Rev. B*, **12** (11), 5016.
4. van der Laan, G., Thole, B.T., Sawatzky, G.A., Goedkoop, J.B., Fuggle, J.C., Esteva, J.-M., Karnatak, R., Remeika, J.P., and Dabkowska, H.A. (1986) Experimental proof of magnetic x-ray dichroism. *Phys. Rev. B*, **34** (9), 6529–6531.
5. Thole, B.T., van der Laan, G., and Sawatzky, G.A. (1985) Strong magnetic dichroism predicted in the $M_{4,5}$ x-ray absorption spectra of magnetic rare-earth materials. *Phys. Rev. Lett.*, **55** (19), 2086–2088.
6. Schütz, G., Wagner, W., Wilhelm, W., Kienle, P., Zeller, R., Frahm, R., and Materlik, G. (1987) Absorption of circularly polarized x rays in iron. *Phys. Rev. Lett.*, **58** (7), 737–740.
7. Weller, D., Stöhr, J., Nakajima, R., Carl, A., Samant, M.G., Chappert, C., Mégy, R., Beauvillain, P., Veillet, P., and Held, G.A. (1995) Microscopic origin of magnetic anisotropy in Au/Co/Au probed with x-ray magnetic circular dichroism. *Phys. Rev. Lett.*, **75** (20), 3752–3755.
8. Gambardella, P., Rusponi, S., Veronese, M., Dhesi, S.S., Grazioli, C., Dallmeyer, A., Cabria, I., Zeller, R., Dederichs, P.H., Kern, K., Carbone, C., and Brune, H. (2003) Giant magnetic anisotropy of single cobalt atoms and nanoparticles. *Science*, **300**, 1130.
9. Samant, M.G., Stöhr, J., Parkin, S.S.P., Held, G.A., Hermsmeier, B.D., Herman, F., Van Schilfgaarde, M., Duda, L.-C., Mancini, D.C., Wassdahl, N., and Nakajima, R. (1994) Induced spin polarization in Cu spacer layers in Co/Cu multilayers. *Phys. Rev. Lett.*, **72** (7), 1112.
10. Vogel, J., Fontaine, A., Cros, V., Petroff, F., Kappler, J.-P., Krill, G., Rogalev, A., and Goulon, J. (1997) Structure and magnetism of Pd in Pd/Fe multilayers studied by x-ray magnetic circular dichroism at the Pd $L_{2,3}$-edges. *Phys. Rev. B*, **55** (6), 3663.
11. Sacchi, M., Sakho, O., and Rossi, G. (1991) Strong dichroism in the dy 3*d* − > 4f x-ray absorption at Dy/Si(111) interfaces. *Phys. Rev. B*, **43**, 1276.
12. Vogel, J. and Sacchi, M. (1996) An x-ray dichroism study of magnetic and crystal field effects in thin Rare Earth overlayers. *Surf. Sci.*, **365**, 831.
13. Kuiper, P., Searle, B.G., Rudolf, P., Tjeng, L.H., and Chen, C.T. (1993) X-ray magnetic dichroism of antiferromagnet Fe_2O_3: The orientation of magnetic moments observed by Fe 2*p* x-ray absorption spectroscopy. *Phys. Rev. Lett.*, **70** (10), 1549.
14. Alders, D., Vogel, J., Levelut, C., Peacor, S.D., Hibma, T., Sacchi, M., Tjeng, L.H., Chen, C.T., Vand der Laan, G., Thole, B.T., and Sawatzky, G.A. (1995) Magnetic x-ray dichroism study of the nearest-neighbor spin-spin correlation-function and long range magnetic order parameter in antiferromagnetic NiO. *Europhys. Lett.*, **32**, 259.
15. Alders, D., Tjeng, L.H., Voogt, F.C., Hibma, T., Sawatzky, G.A., Chen, C.T., Vogel, J., Sacchi, M. and Iacobucci, S. (1998) Temperature and thickness dependence of magnetic moments in NiO epitaxial films. *Phys. Rev. B*, **57** (18), 11623.
16. De Groot, F.M.F. and Kotani, A. (2008) *Core Level Spectroscopy of Solids*, Taylor & Frances.

17. Chen, C.T., Sette, F., Ma, Y., and Modesti, S. (1990) Soft-x-ray magnetic circular dichroism at the $L_{2,3}$ edges of nickel. *Phys. Rev. B*, **42** (11), 7262.
18. Smith, N.V., Chen, C.T., Sette, F., and Mattheiss, L.F. (1992) Relativistic tight-binding calculations of x-ray absorption and magnetic circular dichroism at the L_2 and L_3 edges of nickel and iron. *Phys. Rev. B*, **46** (2), 1023.
19. Stöhr, J. and Wu, Y. (1994) *New Directions in Research with Third-Generation Soft X-Ray Synchrotron Radiation Sources*, Vol. 254, chapter Magnetic X-ray Dichroism, NATO ASI Series, Series E: Applied Sciences, Kluwer academic, Dordrecht. pp. 221–250.
20. Bethe, H.A. and Salpeter, E.E. (1977) *Quantum Mechanics of One-and Two-Electron Atoms*, Plenum, New York.
21. Thole, B.T. and van der Laan, G. (1987) Systematics of the relation between spin-orbit-splitting in the valence band and the branching ratio in x-ray absorption spectra. *Europhys. Lett.*, **4**, 1083.
22. Chen, C.T., Smith, N.V., and Sette, F. (1991) Exchange, spin-orbit, and correlation effects in the soft-x-ray magnetic-circular-dichroism spectrum of nickel. *Phys. Rev. B*, **43** (8), 6785.
23. Jo, T. and Sawatzky, G.A. (1991) Ground state of ferromagnetic nickel and magnetic circular dichroism in Ni 2p core x-ray-absorption spectroscopy. *Phys. Rev. B*, **43** (10), 8771.
24. van der Laan, G. and Thole, B.T. (1992) Electronic correlations in Ni 2p and 3p magnetic x-ray dichroism and x-ray photoemission of ferromagnetic nickel. *J. Phys.: Condens. Matt.*, **4**, 4181.
25. van der Laan, G. (1997) Line shape of 2p magnetic-x-ray-dichroism spectra in 3d metallic systems. *Phys. Rev. B*, **55** (13), 8086.
26. Sacchi, M. and Vogel, J. (2001) *Magnetism and Synchrotron Radiation*, Chapter X-ray Dichroism in Absorption, Springer, Berlin, p. 287.
27. Goedkoop, J.B., Thole, B.T., van der Laan, G., Sawatzky, G.A., de Groot, F.M.F., and Fuggle, J.C. (1988) Calculations of magnetic x-ray dichroism in the 3d absorption spectra of rare-earth compounds. *Phys. Rev. B*, **37** (4), 2086.
28. Sugano, S., Tanabe, Y., and Kamimura, H. (1970) *Multiplets of Transition-Metal Ions in Crystals*, Academic Press, New York, London.
29. de Groot, F.M.F. (2001) High resolution x-ray emission and x-ray absorption spectroscopy. *Chem. Rev.*, **101**, 1779.
30. Butler, P.H. (1981) *Point Group symmetry, Applications, Methods and Tables*, Plenum, New York.
31. Ikeno, H., de Groot, F.M.F., Stavitski, E., and Tanaka, I. (2009) Multiplet calculations of $L_{2,3}$ x-ray absorption near-edge structures for 3d transition-metal compounds. *J. Phys.: Condens. Matt.*, **21**, 104208.
32. Yamaguchi, T., Shibuya, S., and Sugano, S. (1982) *J. Phys. C*, **14**, 2625.
33. Thole, B.T., van der Laan, G. and Butler, P.H. (1988) Spin-mixed ground-state of Fe Phthalocyanine and the temperature-dependent branching ratio in x-ray absorption spectroscopy. *Chem. Phys. Lett.*, **149**, 295.
34. van der Laan, G. and Thole, B.T. (1991) Strong magnetic x-ray dichroism in 2p absorption spectra of 3d transition-metal ions. *Phys. Rev. B*, **43** (16), 13401.
35. Zaanen, J., Sawatzky, G.A., and Allen, J.W. (1985) Band gaps and electronic structure of transition-metal compounds. *Phys. Rev. Lett.*, **55** (4), 418.
36. Cowan, R.D. (1981) *The Theory of Atomic Structure and Spectra*, University of California Press, Berkeley.
37. Altarelli, M. (1993) Orbital-magnetization sum rule for x-ray circular dichroism: a simple proof. *Phys. Rev. B*, **47** (2), 597.
38. Schwickert, M.M., Guo, G.Y., Tomaz, M.A., O'Brien, W.L., and Harp, G.R. (1998) X-ray magnetic linear dichroism in absorption at the *L* edge of metallic Co, Fe, Cr, and V. *Phys. Rev. B*, **58** (8), R4289.
39. Chen, C.T., Idzerda, Y.U., Lin, H.-J., Smith, N.V., Meigs, G., Chaban, E., Ho, G.H., Pellegrin, E., and Sette, F. (1995) Experimental confirmation of the x-ray magnetic circular dichroism sum rules for iron and cobalt. *Phys. Rev. Lett.*, **75** (1), 152.

40. Wu, Y., Stöhr, J., Hermsmeier, B.D., Samant, M.G., and Weller, D. (1992) Enhanced orbital magnetic moment on Co atoms in Co/Pd multilayers: a magnetic circular x-ray dichroism study. *Phys. Rev. Lett.*, **69** (15), 2307.
41. Tischer, M., Hjortstam, O., Arvanitis, D., Hunter Dunn, J., May, F., Baberschke, K., Trygg, J., Wills, J.M., Johansson, B., and Eriksson, O. (1995) Enhancement of orbital magnetism at surfaces: Co on Cu(100). *Phys. Rev. Lett.*, **75** (8), 1602.
42. Vogel, J. and Sacchi, M. (1996) Magnetic moments in as-deposited and annealed Ni layers on Fe(001): an x-ray-dichroism study. *Phys. Rev. B*, **53** (6), 3409.
43. Bruno, P. (1989) Tight-binding approach to the orbital magnetic moment and magnetocrystalline anisotropy of transition-metal monolayers. *Phys. Rev. B*, **39** (1), 865.
44. Dürr, H.A. and van der Laan, G. (1996) Magnetic circular x-ray dichroism in transverse geometry: importance of noncollinear ground state moments. *Phys. Rev. B*, **54** (2), R760.
45. van der Laan, G. (1998) Microscopic origin of magnetocrystalline anisotropy in transition metal thin films. *J. Phys.: Condens. Matter*, **10** (14), 3239.
46. Wu, R. and Freeman, A.J. (1994) Limitation of the magnetic-circular-dichroism spin sum rule for transition metals and importance of the magnetic dipole term. *Phys. Rev. Lett.*, **73** (14), 1994.
47. Carra, P., Konig, H., Thole, B.T., and Altarelli, M. (1993) Magnetic x-ray dichroism - general features of dipolar and quadrupolar spectra. *Physica B*, **192**, 182.
48. van der Laan, G. (1999) Magnetic linear x-ray dichroism as a probe of the magnetocrystalline anisotropy. *Phys. Rev. Lett.*, **82** (3), 640.
49. Ohldag, H., Scholl, A., Nolting, F., Anders, S., Hillebrecht, F.U., and Stöhr, J. (2001) Spin reorientation at the antiferromagnetic NiO(001) surface in response to an adjacent ferromagnet. *Phys. Rev. Lett.*, **86** (13), 2878.
50. Arenholz, E., van der Laan, G., Chopdekar, R.V., and Suzuki, Y. (2007) Angle-dependent Ni^{2+} x-ray magnetic linear dichroism: interfacial coupling revisited. *Phys. Rev. Lett.*, **98** (19), 197201.
51. Dhesi, S.S., van der Laan, G., and Dudzik, E. (2002) Determining element-specific magnetocrystalline anisotropies using x-ray magnetic linear dichroism. *Appl. Phys. Lett.*, **80** (9), 1613–1615–.
52. Powell, C.J. (1974) Attenuation lengths of low-energy electrons in solids. *Surf. Sci.*, **44** (1), 29–46.
53. Seah, M.P. and Dench, W.A. (1979) Quantitative electron spectroscopy of surfaces: a standard data base for electron inelastic mean free paths in solids. *Surf. Interface Anal.*, **1**, 2–11.
54. Abbate, M., Goedkoop, J.B., de Groot, F.M.F., Grioni, M., Fuggle, J.C., Hoffman, S., Petersen, H., and Sacchi, M. (1992) Probing depth of soft x-ray absorption-spectroscopy measured in total-electron-yield mode. *Surf. Interf. Anal.*, **18**, 65.
55. Vogel, J. and Sacchi, M. (1994) Experimental estimate of absorption length and total electron yield (TEY) probing depth in Dysprosium. *J. Electron Spectr. Related Phen.*, **67** (1), 181.
56. O'Brien, W.L. and Tonner, B.P. (1994) Orbital and spin sum rules in x-ray magnetic circular dichroism. *Phys. Rev. B*, **50** (17), 12672–12681.
57. Nakajima, R., Stöhr, J., and Idzerda, Y.U. (1999) Electron-yield saturation effects in *L*-edge x-ray magnetic circular dichroism spectra of Fe, Co, and Ni. *Phys. Rev. B*, **59** (9), 6421–6429.
58. Gota, S., Gautier-Soyer, M., and Sacchi, M. (2000) Fe 2p absorption in magnetic oxides: quantifying angular-dependent saturation effects. *Phys. Rev. B*, **62** (7), 4187–4190.
59. Jones, R.G. and Woodruff, D.P. (1982) Sampling depths in total electron yield and reflectivity sexafs studies in the soft x-ray region. *Surf. Sci.*, **114** (1), 38–46.
60. Sacchi, M., Kappert, R.J.H., Fuggle, J.C., and Marinero, E.E. (1991) Magnetic ordering in Tb_xFe_{1-x} amorphous films: an application of x-ray dichroism with linearly polarized light. *Appl. Phys. Lett.*, **59** (7), 872.
61. Vogel, J., Sacchi, M., Kappert, R.J.H., Fuggle, J.C., Goedkoop, J.B., Brookes,

N.B., van der Laan, G., and Marinero, E.E. (1995) Magnetic properties of Fe and Tb in Tb_xFe_{1-x} amorphous films studied with soft x-ray circular and linear dichroism. *J. Magn. Magn. Mater.*, **150** (3), 293.

62. Rhyne, J.J. (1979) *Handbook on the Physics and Chemistry of Rare Earths*, vol. 2 chapter 16, North-Holland Publishing Company, Amsterdam, pp. 259–294.
63. Harris, V.G., Aylesworth, K.D., Das, B.N., Elam, W.T., and Koon, N.C. (1992) Structural origins of magnetic anisotropy in sputtered amorphous Tb-Fe films. *Phys. Rev. Lett.*, **69** (13), 1939–1942.
64. Yan, X., Hirscher, M., Egami, T., and Marinero, E.E. (1991) Direct observation of anelastic bond-orientational anisotropy in amorphous $Tb_{26}Fe_{62}Co_{12}$ thin films by x-ray diffraction. *Phys. Rev. B*, **43** (11), 9300–9303.
65. Miyazaki, T., Hayashi, K., Yamaguchi, S., Takahashi, M., Yoshihara, A., Shimamori, T., and Wakiyama, T. (1988) Magnetization, Curie temperature and perpendicular magnetic anisotropy of evaporated Fe-Rare Earth amorphous alloy films. *J. Magn. Magn. Mater.*, **75** (3), 243.
66. Hellman, F. and Gyorgy, E.M. (1992) Growth-induced magnetic anisotropy in amorphous Tb-Fe. *Phys. Rev. Lett.*, **68** (9), 1391–1394.
67. Kappert, R.J.H., Vogel, J., Sacchi, M., and Fuggle, J.C. (1993) Linear-dichroism studies of thin Dy overlayers on Ni(110) and Cu(110) substrates. *Phys. Rev. B*, **48** (4), 2711.
68. Castrucci, P., Yubero, F., Vicentin, F.C., Vogel, J., and Sacchi, M. (1995) Surface crystal field at the Er/Si(111) interface studied by soft-x-ray linear dichroism. *Phys. Rev. B*, **52** (19), 14035–14039.
69. Hansmann, P., Severing, A., Hu, Z., Haverkort, M.W., Chang, C.F., Klein, S., Tanaka, A., Hsieh, H.H., Lin, H.-J., Chen, C.T., Fåk, B., Lejay, P., and Tjeng, L.H. (2008) Determining the crystal-field ground state in rare earth heavy fermion materials using soft-x-ray absorption spectroscopy. *Phys. Rev. Lett.*, **100** (6), 066405.
70. Sacchi, M. (1993) Soft X-ray linear dichroism, in *Proceedings of the VUV-10 Conference* (eds Y. Petroff, F. Wuilleumier, and I. Nenner), World Scientific Co., Singapore, pp. 431.
71. Haverkort, M.W., Csiszar, S.I., Hu, Z., Altieri, S., Tanaka, A., Hsieh, H.H., Lin, H.-J., Chen, C.T., Hibma, T., and Tjeng, L.H. (2004) Magnetic versus crystal-field linear dichroism in NiO thin films. *Phys. Rev. B*, **69** (2), 020408.
72. De Groot, F.M.F. (1991) X-ray absorption of transition metal oxides, PhD thesis, University of Nijmegen.
73. Hutchings, M.T. and Samuelsen, E.J. (1972) Measurement of spin-wave dispersion in nio by inelastic neutron scattering and its relation to magnetic properties. *Phys. Rev. B*, **6** (9), 3447.
74. Dietz, R.E., Parisot, G.I., and Meixner, A.E. (1971) Infrared absorption and raman scattering by two-magnon processes in NiO. *Phys. Rev. B*, **4** (7), 2302–2310.
75. Peacor, S. and Hibma, T. (1994) Reflection high-energy electron diffraction study of the growth of NiO and CoO thin films by molecular beam epitaxy. *Surf. Sci.*, **103**, 11.
76. Vogel, J. and Sacchi, M. (1994) Polarization and angular dependence of the $L_{2,3}$ absorption edges in Ni(110). *Phys. Rev. B*, **49** (5), 3230.
77. Alders, D., Hibma, T., Sawatzky, G.A., Cheung, K.C., van Dorssen, G.E., Roper, M.D., Padmore, H.A., van der Laan, G., Vogel, J., and Sacchi, M. (1997) Grazing incidence reflectivity and total electron yield effects in soft x-ray absorption spectroscopy. *J. Appl. Phys.*, **82** (6), 3120.
78. Altieri, S., Finazzi, M., Hsieh, H.H., Lin, H.J., Chen, C.T., Hibma, T., Valeri, S., and Sawatzky, G.A. (2003) Magnetic dichroism and spin structure of antiferromagnetic NiO(001) films. *Phys. Rev. Lett.*, **91** (13), 137201.
79. Finazzi, M., Duò, L., and Ciccacci, F. (2009) Magnetic properties of interfaces and multilayers based on thin antiferromagnetic oxide films. *Surf. Sci. Rep.*, **64** (4), 139–167.

4
Antiferromagnetic Oxide Films on Nonmagnetic Substrates

Tjipke Hibma and Maurits W. Haverkort

4.1
Introduction

The most prominent use of antiferromagnetic (AF) materials is as substrates for the biasing of magnetic moments in ferromagnetic materials. For this reason, AF thin layers are mostly studied in combination with ferromagnetic substrates or overlayers [1], with emphasis on the magnetic structure of the latter. In this chapter, however, we focus on the magnetic structure of AF 3d transition-metal (TM) monoxide materials. Therefore, we restrict ourselves to samples consisting of thin epitaxial layers on nonmagnetic substrates, not influenced by magnetic underlayers. For a rigorous understanding of the magnetic structure, it is necessary to figure out the full valence state electronic structure, that is, the effect of correlation on the behavior of TM d-electrons, the bonding between the TM ions and the neighboring oxygen ions, and the role of spin–orbit coupling (SOC) and exchange interaction. X-ray absorption spectroscopy (XAS), which is of great help in disentangling these interactions, has been treated extensively in the previous chapter. The use of linear polarized radiation, that is, the X-ray linear dichroism technique, is of particular significance for AF materials. This technique is sensitive to a noncubic charge density and to noncubic ligand fields. Through SOC, it is also sensitive to the anisotropic spin distribution in magnetically ordered materials, that is, the technique is capable of measuring the mean square magnetic moment, which is finite in AF materials. Also, the absolute spin direction can be determined. In bulk crystals of the AF oxides discussed in this chapter, the linear dichroism is often zero because the individual contributions of single domains are compensated by others related by cubic symmetry operations. To induce measurable effects, growth strain, which breaks the cubic symmetry and stabilizes a subset of magnetic domains, can be used.

Magnetic Properties of Antiferromagnetic Oxide Materials.
Edited by Lamberto Duò, Marco Finazzi, and Franco Ciccacci

ISBN: 978-3-527-40881-8

4.2
Electronic Structure of TM Oxides

4.2.1
Mott-Hubbard and Charge Transfer Insulators

The special position of TM compounds in solid state research is undoubtedly related to the correlated nature of the d-electrons. Simple independent electron approaches completely fail in describing even basic properties. One of the most disturbing discrepancies has been the failure to predict whether a TM compound will be a metal or an insulator. The first important step toward a satisfactory description has been the introduction of the Mott–Hubbard model [2, 3]. In this model, the correlation gap is linked to the large effective on-site Coulomb interaction U. The model was quite successful for the early TM metals but not for the others. Later on, it was recognized that the gap could also be of the charge transfer type. The parameter of interest here is Δ, the energy needed to transfer an electron from the ligand ion to a distant TM ion. Both parameters, U and Δ, play an equivalent role in the well-known classification scheme of Zaanen, Sawatzky, and Allen [4]. Roughly the early TM monoxides, TiO and VO, fall into the Mott–Hubbard regime, MnO, FeO, CoO, and NiO are AF insulators in the intermediate regime, and CuO is an AF charge transfer semiconductor. CrO is missing in this list, because it does not exist as a bulk compound for some reason [5].

The most important difference between a band insulator and a Mott–Hubbard/ charge transfer insulator is the highly localized nature of the valence orbitals, leaving many of the local degrees of freedom intact and bringing about a variety of subtle forms of charge, spin, and orbital ordering. The single band Mott–Hubbard model is not able to explain these fine details, even if charge transfer effects are included. This becomes particularly evident, if spectroscopic data have to be explained, in which, in addition to the ground state, excited states are involved. For a description of the detailed electronic structure of TM compounds, the full multiorbital character has to be taken into account and two types of interactions which are of comparable strength have to be treated first. One is the full multiorbital Coulomb interaction between the electrons of the TM ion, and the other is the bonding to the neighboring ions. Fortunately, the size of the calculation can be restricted considerably due to the localized character of the orbitals. For instance, for a proper description of the magnetic properties it is often sufficient to consider only the d-orbitals of a central TM ion and the p-orbitals of the neighboring oxygen-ligand ions. The same approximation holds for the interpretation of XAS at the $L_{2,3}$ edge of TM oxides. This is a valid approach, because the final state is strongly excitonic and can be well described within a TMO_6 cluster. Transport properties, on the other hand, obviously, cannot be understood on the basis of these cluster calculations, because the k-dependence is totally neglected.

4.2.2 Ligand Field Theory

The bonding between the d-metal ions and the environment can be included by introducing a so-called ligand field (LF). Originally, the LF was approximated by a collection of point charges and the theory was called the crystal-field theory (CFT) [6, 7]. The concept is very simple and with the aid of group theory a qualitatively correct energy level scheme for a particular arrangement of ligand ions or molecules can be fabricated. However, quantitatively the predictions made by CFT are quite wrong. A much more realistic approach is to calculate the LF parameters using molecular orbital theory to describe the TM ion ligand bonds. This approach is known as the ligand field theory (LFT) [8].

In practice, there are two extreme cases, the strong and the weak field limit. In the strong field limit, the d-electron correlations are less important and can be incorporated by introducing an average exchange correlation potential. An independent electron LFT treatment may be sufficient for a qualitative description of the local electronic structure of the TM ion. For a quantitative treatment, in particular in the weak field limit, however, it is vital to use a multiplet theory including all d-electron correlations in addition to the LF interaction. This is called the multiplet ligand field theory (MLFT). Because of their simplicity, independent electron LFT concepts are often used to explain results in a qualitative way, even if they were obtained using more elaborate theories like MLFT.

4.2.2.1 Independent Electron Ligand Field Theory

Group theory plays a very important role in LFT because it can be used to design one-electron orbital combinations allowed by symmetry. In a cubic TM oxide crystal, the divalent TM ion is in a site having octahedral O_h symmetry and the d-levels split into a threefold degenerate t_{2g} (d_{xy}, d_{xz} and d_{yz}) and twofold degenerate $e_g(d_{3z^2-r^2}$ and $d_{x^2-y^2})$ level.

Group theory can also be used to construct so-called symmetry adapted linear combinations (SALC's) of ligand orbitals. These combinations are formed out of either the σ or π type 2p orbitals of the oxygen ions. For the cubic case, the σ type SALC's can be of symmetry a_{1g}, e_g, and t_{1u}, and the π type SALC's can be of symmetry t_{1g}, t_{2g}, t_{1u}, and t_{2u}. Because only orbitals having the same symmetry can combine into new molecular orbitals, the possible combinations in the valence state region are 3d and σ orbitals of e_g symmetry on the one hand and 3d and π orbitals of t_{2g} symmetry on the other. For both, two new sets of orbitals can be formed, a bonding set having a high oxygen 2p- and an antibonding set having a high 3d-character. Because σ-overlap is larger than π-overlap, the upward shift of the e_g level is larger than the upward shift of the t_{2g} level. The energy difference between the two is called the LF splitting parameter Δ_O (or 10 D_q) (see Figure 4.1).

If the cubic crystal is distorted in one direction, the crystal structure becomes tetragonal and the local symmetry of the TM ion is D_{4h}. The e_g level splits into a nondegenerate $a_{1g}(d_{3z^2-r^2})$ and $b_{1g}(d_{x^2-y^2})$ level and the t_{2g} levels into a twofold degenerate $e_g(d_{xz}$ and $d_{yz})$ and a nondegenerate $b_{2g}(d_{xy})$ level. The corresponding

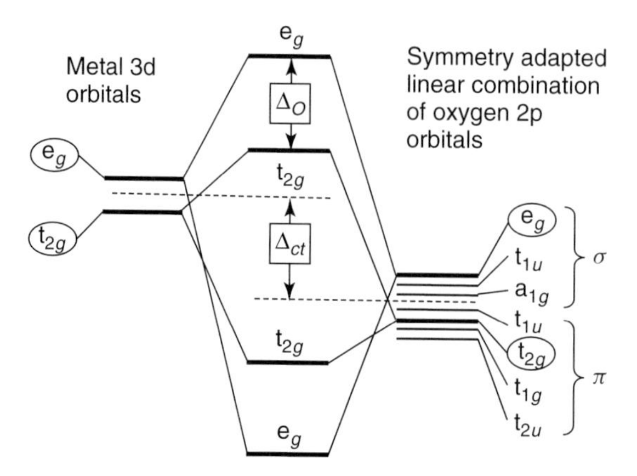

Figure 4.1 Molecular orbital energy scheme for an octahedral TMO_6 cluster. The metal 3d-orbitals, split into t_{2g} and e_g orbitals by an octahedral crystal potential, can only mix with symmetry adapted linear combinations of the 6 oxygen 2p orbitals having the same symmetry. The bonding orbitals have a predominant oxygen character and are completely filled. The antibonding orbitals have a high metal character and are only partly filled. The separation between the t_{2g} and e_g levels, which is called the LF splitting parameter Δ_O, increases as a result of the covalent bonding, because the overlap between the metal and oxygen orbitals is larger for the σ than for the π orbitals. The charge transfer energy Δ_{ct} is also indicated in this figure.

splitting parameters are Δ_{e_g} and $\Delta_{t_{2g}}$ respectively. If one of the cubic axes is shorter than the other two, the overlap between the TM and the oxygen is largest in the shorter z direction and the $d_{x^2-y^2}$ and d_{xy} levels are lower in energy. If the axis is longer, the order of the levels is reversed (Figure 4.2).

Once the LF energy level scheme has been established, the levels can be filled with the d-electrons of the TM ion. To minimize Coulomb repulsion, degenerate levels first take up a single electron keeping the spins parallel. For d^4-d^7 configurations, the filling order is different in the weak and strong LF case.

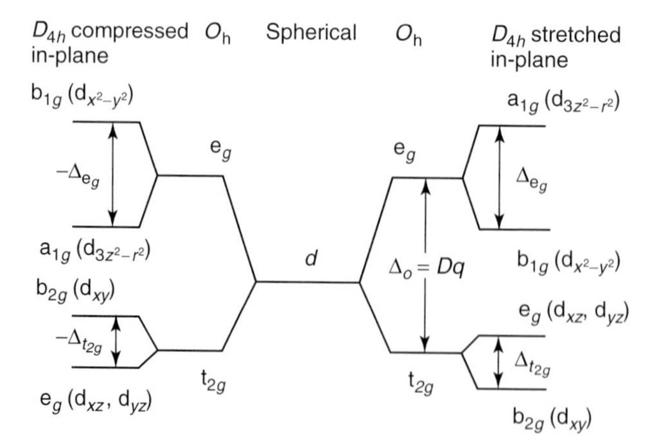

Figure 4.2 The LF splitting for a tetrahydral distortion with D_{4h} symmetry.

If the LF is stronger than the Coulomb repulsion, the t_{2g} levels are filled first with up to six electrons. If the LF is weaker, the fourth and the fifth electrons go into the e_g level with parallel spins, before the t_{2g} level is completely filled with paired electrons. Ions will have a low spin (LS) in a strong LF, and a high spin (HS) in a weak LF. For the the oxygen ligand, the LF for a divalent TM ion is quite weak and consequently it is a high spin case. The number of d-electrons for Mn^{2+}, Fe^{2+}, Co^{2+}, and Ni^{2+} is 5, 6, 7, and 8, respectively, and S is equal to $\frac{5}{2}$, 2, $\frac{3}{2}$, and 1.

4.2.2.2 **Multiplet Ligand Field Theory**

In most cases, Coulomb repulsion is of the same order of magnitude as the LF splitting and cannot be neglected. Fortunately, it can be incorporated in an accurate way for the XAS spectra of the highly correlated oxides we want to describe, because it is a good approximation to limit the calculation to the valence orbitals of the TM ion and its nearest-neighbor oxygen ions.

To keep the many-particle basis set small, one could use perturbation theory [9, 10]. In the first step, the electron microstates are sorted into groups having the same energy if Coulomb repulsion is included. Because SOC is of minor importance in 3d atoms, the Russell–Saunders coupling scheme can be applied. Groups of states having the same **S** and **L** are called *multiplets* and denoted by ^{2S+1}L. The electron repulsion energy of these multiplets can be expressed in terms of the so-called Slater integrals F^0, F^2, and F^4. F^0 represents the monopole part of the Coulomb repulsion and F^2 and F^4 the multipole part of the orbital dependent Coulomb interaction. (It may be noted that F^0, F^2 and F^4 can be related to the often-used Coulomb repulsion parameter U and Hunds-rule exchange J_H. However, where the Slater integrals are exactly defined, there is no common definition for U and J_H [11–13].) Next, the LF interaction is applied. It will split the states of a given **L** and allow for the mixing of different terms. For example, a spherical symmetric term with $L = 3$ will split in O_h symmetry into T_{1g}, T_{2g}, and A_g terms. The T_{1g} term then could mix with a T_{1g} term that branches from a spherically symmetric term with $L = 1$, and L is not a good quantum number anymore. Finally, the smaller interactions like SOC, exchange fields, and magnetic fields have to be included.

With modern computers such a perturbation theory approach is not needed anymore, as memory is sufficiently available, nowadays. It is more efficient to build the entire Hamiltonian as a large, but sparse, matrix and solve the eigensystem problem with the use of Lanczos methods. To understand the results of such calculations, it is often useful to work out, in retrospect, which atomic terms contribute most to the wave function.

4.2.3
Spin–Orbit Coupling in Cubic Symmetry

SOC is not important for most of the properties studied in TM compounds, because the SOC in 3d TM compounds is only of the order of 10 (Ti) to 100 (Cu) meV. At the same time, the LFs that split degenerate d-orbitals can be of the order of 1 eV and, therefore the orbital moment is quenched. SOC is only a small perturbation. However, if one wants to study magnetic anisotropies, the inclusion of SOC becomes very important. Magnetic anisotropy energies are often very small. Therefore, the small perturbation that SOC induces can still be the leading term for the energy of the magnetic anisotropy. The effect of SOC generally will be the creation of a small orbital momentum. The size of the induced orbital momentum may depend on the direction of the magnetization. For metallic systems, it has been shown that the directional dependent part of the orbital momentum scales with the magnetic anisotropy energy [14]. For the insulating compounds discussed in this chapter, this concept still holds, the difference being that the orbital moments in insulating compounds are often much larger than in metals.

The AF TM monoxides have a close to cubic symmetry and the splitting between the t_{2g} and e_g orbitals is much larger than the SOC. However, there is still orbital degeneracy within the t_{2g} and/or e_g subshell. The orbital degeneracy of the e_g orbitals does not couple to the spin directly. The e_g orbitals consist of a $d_{x^2-y^2}$ orbital, which is a linear combination of a d_2 and d_{-2} orbital, and a $d_{3z^2-r^2}$ orbital, which is the d_0 orbital. The SOC Hamiltonian can couple states with the same angular momentum and same spin, or states with $\Delta m = \pm 1$ and $\Delta s = \mp 1$. This is not possible within the e_g subshell. For t_{2g} electrons , however, SOC can lift the threefold orbital degeneracy, as shown in the next subsections. First the effect of SOC on a single electron in an open t_{2g} subshell is discussed followed by its effect on configurations with more than one local electron. For a single electron, one can neglect Coulomb repulsion and use the simpler language of LFT. Although the SOC Hamiltonian ($\sum_i \zeta l_i \cdot s_i$) is a one particle operator, LFT is not sufficient to explain SOC effects for systems with more than one electron locally. This does not work, because SOC mixes states with spin-up and spin-down, creating a multi-Slater-determinant ground state. In atomic physics, it has become customary to approximate the SOC operator by $\lambda \mathbf{S} \cdot \mathbf{L}$ with $\lambda = g\zeta$ and $g = \pm\frac{1}{2S}$ for a Hunds-rule ground state and more (−) or less (+) than half-filled shells [15]. For an open t_{2g} shell, a similar approximation is useful in order to obtain intuitive insight.

4.2.3.1 Single Electron in an Open t_{2g} Shell

A t_{2g} shell is threefold orbital degenerate and one can make linear combinations of the real t_{2g} orbitals to form complex orbitals that carry an orbital momentum of 1 μ_B. In Figure 4.3a, we show the linear combinations that lead to an orbital momentum of 1 μ_B in the z direction. By a cyclic permutation of the coordinates, one can also create an orbital momentum in the x-direction, by combining d_{xz} and d_{xy} orbitals, or in the y-direction, by combining d_{yz} and d_{xy} orbitals. In fact, one could create an orbital momentum of $1\mu_B$ in any direction by making the proper linear combination of these orbitals. A threefold degenerate set of orbitals with an orbital momentum of 1 μ_B is reminiscent of a p-shell. One could call the $d_{xy} = \imath\sqrt{1/2}(d_{-2} - d_2)$ orbital a t_0 orbital in analogy to the p_0 orbital. Acting with the operator l^+ on the t_0 orbital will yield the following: $\hbar 2\imath\sqrt{1/2}d_{-1} = \hbar\sqrt{2}t_1$, from which it follows that $t_1 = \imath d_{-1}$. Similarly, by acting with l^- on the t_0 orbital one gets $t_{-1} = \imath d_1$. This can be summarized by stating that the t_{2g} orbitals can

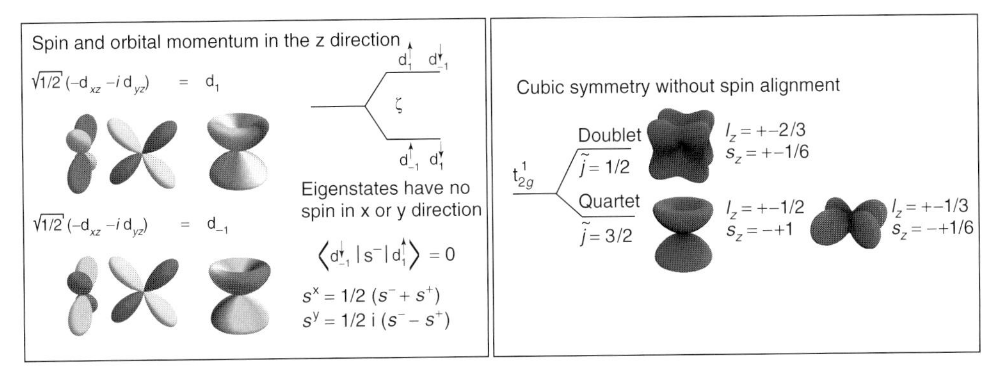

Figure 4.3 Eigenstates of the SOC Hamiltonian for a single t_{2g} electron. (a) The spin is assumed to be fixed in the z direction and the d_{xy} orbital has not been included. SOC recombines the d_{xz} and the d_{yz} orbital in order to create an orbital momentum of 1 μ_B (anti) parallel to the spin. (b) No spin alignment assumed. All three t_g orbitals recombine to a state with $\tilde{j} = \frac{3}{2}$ or $\tilde{j} = \frac{1}{2}$.

be mapped onto p orbitals with an orbital g factor of -1. The effective orbital momentum for the t_{2g} shell will be denoted by $\tilde{l}$. For a single t_{2g} electron, the $s = \frac{1}{2}$ couples with the effective orbital momentum $\tilde{l} = 1$ to $\tilde{j} = \frac{1}{2}$ or $\tilde{j} = \frac{3}{2}$. For a single p electron, the antiparallel alignment would be lower in energy than the parallel alignment. For a t_{2g} electron $g = -1$, and therefore, the $\tilde{j} = \frac{3}{2}$ state is lower in energy than the $\tilde{j} = \frac{1}{2}$ state. In Figure 4.3b, we show the eigenstates for the SOC operator within the t_{2g} subshell. One should note that although this state has a large unquenched orbital momentum, it does not have a magnetic anisotropy as the orbital momentum can point in any direction. This is different from the general situation where SOC is a perturbation and generates an orbital momentum, which depends on the spin direction and is related to the magnetic anisotropy.

4.2.3.2 d^6 and d^7 Configurations

For both a d^6 and a d^7 HS configuration, one has an open t_{2g} shell. The effective orbital momentum is $\tilde{L} = 1$ and the spin is $S = 2$ ($S = \frac{3}{2}$) for the d^6 (d^7) configuration. The spin couples with the effective orbital momentum to a $\tilde{J} = 1, 2, 3$ ($\frac{1}{2}, \frac{3}{2}, \frac{5}{2}$). The $\tilde{J} = 1$ ($\frac{1}{2}$) state is the ground state. These states cannot be represented by single Slater determinants, which complicates the explanation of physical properties. This can easily be seen if one realizes that the operator S^- acting on a fully spin polarized state with six electrons will create four different determinants with two spin-down electrons and four spin-up electrons. SOC mixes these determinants. However, it is often possible to go to extreme limits, that is, either extreme large crystal-field distortions or extreme large exchange fields, in order to retrieve wave functions that can be represented as a single Slater determinant. This helps in understanding the physics, and although it does not get the numerical values correct, it does show the correct trends.

4.3 Magnetic Structure

4.3.1 Magnetic Ordering of MnO, FeO, CoO and NiO

In the paramagnetic high-temperature state, MnO, FeO, CoO, and NiO all have a rocksalt or NaCl crystallographic structure. Below the Néel point, these 3d TM monoxides order antiferromagnetically into magnetic superstructures, which can be determined using neutron diffraction. From the classic papers by Shull, Strauser, and Wollan [16], the magnetic unit cell is found to be close to cubic with a doubled lattice parameter in order to accommodate the AF spin arrangement. In fact, the spins were found to be parallel on cubic {111} planes (so-called F sheets) but antiparallel on neighboring {111} planes (see Figure 4.4). The relative orientation of the spins with respect to each other is determined by exchange interactions. The main factor responsible for this particular (so-called type 2) AF arrangement is the 180° superexchange interaction. It couples next nearest-neighbor (nnn) spins. The nearest-neighbor (nn) interaction is much weaker and frustrated. This frustration is lifted by exchange striction. Assuming a collinear spin structure, a trigonal distortion restricts the number of possible spin structures to the four experimentally observed *T*-domains.

The absolute orientation of the spins within the {111} sheets, on the other hand, is determined by interactions coupling the spin to the lattice. Possible

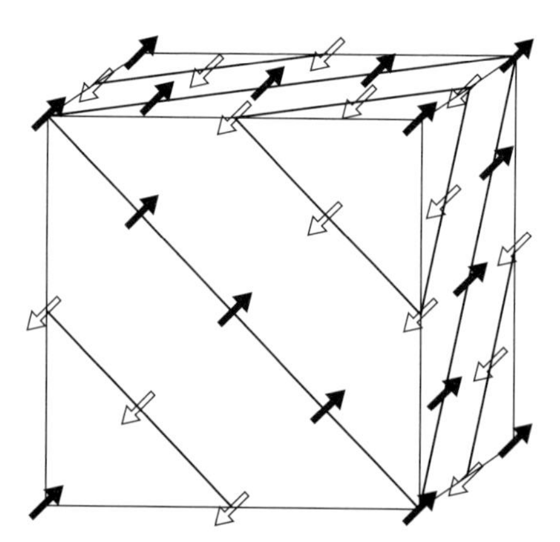

Figure 4.4 Basic spin ordering in the TM monoxides MnO, FeO, CoO and NiO. The spins are ferromagnetically ordered within {111} planes, and AF in neighboring {111} planes in all four compounds. The orientation of the spins with respect to the crystallographic axes, however, is different. In NiO and MnO, the spins are oriented in $\langle 11\bar{2} \rangle$ directions (as is actually shown in the above figure), in FeO in the [111] direction, and in CoO slightly tilted away from the cubic axes. All materials have structures that are slightly distorted to lower than cubic symmetries, which are consistent with these orientations.

contributions to this magnetic anisotropy are the dipole–dipole interaction and the single ion anisotropy. In the latter case, the spin is coupled to the lattice in two steps, the spin–orbit interaction couples the spin to the orbital momentum, and the crystal-field couples the orbital to the lattice. For closed shell configurations in cubic symmetry, the orbital momentum is quenched and spin–orbit interaction is small in contrast to open shell t_{2g} configurations. For the TM monoxides considered in this chapter, single ion anisotropy is small for Mn^{2+} (d^5) and Ni^{2+} (d^8), but it is large for Co^{2+} (d^7) and Fe^{2+} (d^6).

4.3.1.1 MnO and NiO

In both MnO and NiO, the trigonal distortion stabilized by exchange is the main distortion. It leads to an elongated rhombohedral unit cell with $\alpha < 60^\circ$. The orientation of the spins with respect to the lattice is determined by a detailed interplay between dipole–dipole interactions and single ion anisotropy. There is a relatively large force, both from dipole–dipole interactions and SOC, that restricts the spins to a direction in the {111} planes. The precise direction within these planes is a subtle and quite involved problem. Experimentally, a further reduction of the crystal symmetry to monoclinic is observed for both NiO [17] and MnO [18] and the spins are found to be oriented in one of the three possible $\{11\bar{2}\}$ directions perpendicular to the rhombohedral axis, thus generating 24 domains in total, namely 6 possible *S* domains (including spin reversed domains) for each of the 4 *T* domains.

4.3.1.2 FeO and CoO

In the paramagnetic phase, both CoO and FeO have a nearly cubic structure, but below the Néel temperature (T_N) the spin orientation induces a noncubic charge density, which distorts the crystal structure. For instance, for spins being oriented in the [001] direction one would expect a charge density as depicted in Figure 4.3a. For electrons (FeO, d^6) this would lead to a tetragonal elongation ($c > a$), and for holes (CoO, d^7) to a tetragonal contraction ($c < a$).

For CoO, there is still some debate about the precise spin direction. It is clear, however, that there is a relatively large tetragonal distortion, with ($c < a$) in accordance with the above example, accompanied by a smaller rhombohedral elongation lowering the total symmetry to monoclinic. The general consensus on the spin direction is that it is slightly tilted away from the tetragonal axis. [19–21].

For FeO, one finds the spin to be in the [111] direction. An accompanying orbital momentum can be associated with a $\sqrt{1/3}(\sqrt{1/2}(\imath - 1)d_{yz} + \sqrt{1/2}(-\imath - 1)d_{xz} + d_{xy})$ orbital, which has a larger charge density perpendicular to the [111] direction than parallel to it. Therefore, FeO shows a trigonal elongation. There are no other distortions found in FeO, which might be related to the fact that both the magnetostriction and the local single ion anisotropy are optimized by the same trigonal elongation.

4.4
X-ray Absorption Spectroscopy

One of the most powerful spectroscopic techniques to study the 3d valence state, including the magnetic structure, is 2p XAS. The subject has been discussed at length in Chapter 3. Using X-rays in the energy range between 350 and 950 eV, transitions from the 2p core level into the empty 3d valence levels are made. This process is dipole allowed and has a large absorption cross section. The 2p XAS spectrum consists of two clearly separated sets of absorptions, traditionally called the L_3 and L_2 edge. As an example, the NiO XAS spectrum is shown in Figure 4.5. The splitting is due to the SOC of the 2p core level from which the electrons are excited into the empty 3d level. The angular momentum j belonging to the split levels is 3/2 or 1/2 corresponding to the L_3 and L_2 edge, respectively. In an independent electron approach, the intensity ratio of the two sets of transitions should be two, because there are twice as many m_j sublevels for the $j = 3/2$ level, and the line shape of the two should be the same. In reality, the line shapes are quite different, and the intensity ratio is different from two due to strong correlation effects, not only between the 3d-electrons but also between the 2p core hole and the 3d electrons. To calculate a realistic absorption spectrum, we have to apply multiplet theory to the initial as well as the final state and calculate the transition probability between the two.

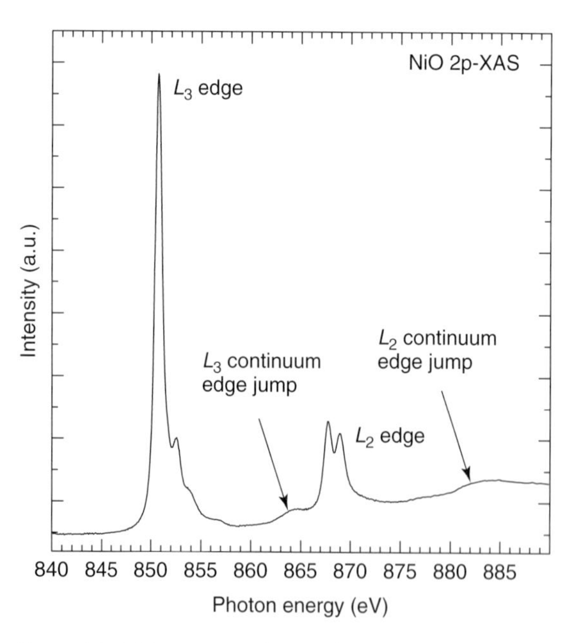

Figure 4.5 XAS spectrum of NiO, showing well-separated L_2 and L_3 edges as well as the corresponding continuum edge jumps. (Reprinted from [13].)

Not all of the multiplet states can be reached in the 2p XAS process. In the dipole approximation, the absorption intensity is given by

$$I_{\text{abs}} = |\langle i|\boldsymbol{\epsilon}.\mathbf{r}|f\rangle|^2 = |\langle i|P^q_\alpha|f\rangle|^2 \tag{4.1}$$

where the initial and final states are denoted by i and f, $\boldsymbol{\epsilon}$ is the polarization vector of the X-rays, and $\mathbf{r}$ the electron position vector. The second expression is written in terms of the dipole operator P^q_α , where q is the type of polarization of the radiation ($q = 0$ for linear and $q = \pm 1$ for circular polarization), and α the length of the position vector in the direction of the polarization vector in the coordinate system of the sample. The following selection rules can be derived

$$\begin{aligned} \Delta L &= 0, \pm 1 \\ \Delta S &= 0 \\ \Delta J &= 0, \pm 1 \\ \Delta l &= \pm 1 \\ \Delta j &= 0, \pm 1 \\ \Delta m_l &= q = 0, \pm 1 \\ \Delta m_s &= 0 \end{aligned} \tag{4.2}$$

The differences between the independent electron approximation and multiplet theory are nicely demonstrated in Figure 4.6, taken from a publication of Thole and van der Laan [22]. It shows calculated spectra for a ground-state d^8 configuration for an increasing ratio of the total electrostatic interaction U(p,d) and spin–orbit interaction ξ_p. The spectrum at the top corresponds to the independent electron result including 2p SOC. The bottom spectrum is the other extreme where the SOC can be neglected with respect to electron correlation. In the latter case, $L - S$

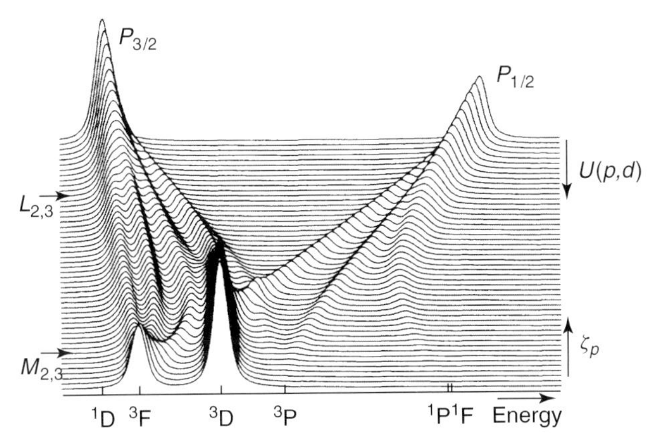

Figure 4.6 The calculated transition probability for the $2p^63d^8$ to $2p^53d^9$ excitation process as a function of the relative size of the 2p − 3d SOC $\xi\mathbf{L.S}$ and the Coulomb interaction $U(\text{p}, d)$. The top spectrum corresponds to pure SOC. The bottom spectrum to pure $L - S$ coupling. No LF effects were included, that is, spherical symmetry is assumed. (Reproduced from [22] with permission.)

coupling applies. The ground-state multiplet of a d^8 configuration is ^3F. The dipole selection rules $\Delta S = 0$ and $\Delta J = 0, \pm 1$ imply that only transitions to the excited ^3F and ^3D multiplets are allowed. The arrows at the left hand side of the picture point at the spectra most resembling the experimental $L_{2,3}$ and $M_{2,3}$ spectra for a d^8 configuration.

In addition to electron correlation, LF effects, 3d spin–orbit and exchange interactions should be taken into account. These interactions affect, as a general rule, both the initial and final states. LF splittings are energetically the most important ones and will strongly influence the lineshape. As a rule, the other interactions are only small perturbations and the lower split-off excited states of the ground state may be thermally accessible, making the spectra temperature dependent. The strict set of selection rules of Eq. (4.2) couples the initial state and close-by excited states to very specific selections of final states. Small changes in the interaction parameters will produce large changes in the spectrum. This makes XAS a very sensitive tool to study the detailed electronic and magnetic structure of TM compounds.

4.4.1 Magnetic Linear Dichroism

In general, the XAS spectra depend on the orientation of the sample and the polarization of the X-rays. They can be described by a complex conductivity tensor $\boldsymbol{\sigma}$, which has several nonzero components, depending on the symmetry of the system (see Table 4.1). The isotropic spectrum, which is equal to the trace of $\boldsymbol{\sigma}$, can

Table 4.1 Conductivity tensor in (i) cubic symmetry, (ii) tetragonal symmetry, (iii) orthorhombic symmetry, (iv) monoclinic symmetry, (v) triclinic symmetry, and (vi) spherical symmetry with a magnetization in the z direction. The case for nonspherical symmetry with a magnetization in an arbitrary direction is discussed in the appendix.

cubic

$$\begin{pmatrix} \sigma & 0 & 0 \\ 0 & \sigma & 0 \\ 0 & 0 & \sigma \end{pmatrix}$$

tetragonal

$$\begin{pmatrix} \sigma_{xx} & 0 & 0 \\ 0 & \sigma_{xx} & 0 \\ 0 & 0 & \sigma_{zz} \end{pmatrix}$$

orthorhombic

$$\begin{pmatrix} \sigma_{xx} & 0 & 0 \\ 0 & \sigma_{yy} & 0 \\ 0 & 0 & \sigma_{zz} \end{pmatrix}$$

monoclinic

$$\begin{pmatrix} \sigma_{xx} & \sigma_{yx} & 0 \\ \sigma_{yx} & \sigma_{yy} & 0 \\ 0 & 0 & \sigma_{zz} \end{pmatrix}$$

triclinic

$$\begin{pmatrix} \sigma_{xx} & \sigma_{yx} & \sigma_{zx} \\ \sigma_{yx} & \sigma_{yy} & \sigma_{zy} \\ \sigma_{zx} & \sigma_{zy} & \sigma_{zz} \end{pmatrix}$$

spherical+magnetic in z

$$\begin{pmatrix} \sigma^0 - \frac{1}{3}\sigma^2 & \sigma^1 & 0 \\ -\sigma^1 & \sigma^0 - \frac{1}{3}\sigma^2 & 0 \\ 0 & 0 & \sigma^0 + \frac{2}{3}\sigma^2 \end{pmatrix}$$

be obtained by measuring the spectra in three perpendicular orientations. In this paragraph, we limit the discussion to linear dichroism, which is just the symmetric part of $\boldsymbol{\sigma}$. We do not discuss circular dichroism, because it is absent in AF materials.

It is useful to distinguish two types of linear dichroism, that is, natural linear dichroism (NLD) and magnetic linear dichroism (MLD) [23]. For systems of lower than cubic symmetry, the natural linear dichroism can have two different causes: either the ground-state charge density is noncubic or the final-state orbital energies are nondegenerate. The first condition is fulfilled in CoO, where the open t_{2g} shell can give rise to a noncubic orbital occupation if the lattice is distorted. In MnO, having a d^5 configuration, on the other hand, the ground-state charge density stays close to spherical, even if a distortion is imposed. However, the degeneracy of the final state is lifted by the distortion, and the addition of an electron will cost a different amount of energy, depending on the orbital one adds the electron to. The magnetic linear dichroic effect is due to a spin alignment that breaks the symmetry. For absorption in the visible wavelength region, one is used to quite small magnetooptical effects. In the X-ray regime, these effects are much stronger due to the large core hole (2p) SOC constant.

A linear dichroism spectrum is often defined as the difference between two spectra measured at different sample orientations, for instance parallel to and at right angles to the main symmetry axis.

$$I_{\mathrm{XLD}} = I^{\parallel} - I^{\perp} \tag{4.3}$$

Using the optical conductivity tensor as defined in Table 4.1, one can relate the linear dichroism spectrum to $\boldsymbol{\sigma}$. In tetragonal symmetry, the relationship is $I_{XLD}^{D_{4h}} = \boldsymbol{\sigma}_{zz} - \boldsymbol{\sigma}_{xx}$. In orthorhombic symmetry, one can measure three different linear dichroism signals, depending on the crystallographic face one studies, that is, $I_{XLD}^{D_{2h}^{x}} = \boldsymbol{\sigma}_{yy} - \boldsymbol{\sigma}_{zz}$, $I_{XLD}^{D_{2h}^{y}} = \boldsymbol{\sigma}_{zz} - \boldsymbol{\sigma}_{xx}$, or $I_{XLD}^{D_{4h}^{z}} = \boldsymbol{\sigma}_{xx} - \boldsymbol{\sigma}_{yy}$. For a magnetic system in otherwise spherical symmetry, the linear dichroic signal is $\boldsymbol{\sigma}^2$.

The difference spectrum is zero for angle averaged situations, like the one encountered in random polycrystalline samples. Also in crystals having cubic symmetry, linear dichroism is expected to be absent. An example is shown in Figure 4.21. A rather thick bulklike relaxed layer of MnO on a Ag substrate shows no effect. The t_{2g} and e_g orbitals are both half filled, that is, the corresponding charge densities are spherically symmetric. The magnetic linear dichroism is zero as well, despite the fact that MnO is antiferromagnetically ordered at $T = 30$ K, because the linear dichroism contributions due to different magnetic domains cancel each other.

The fundamental reason for the occurrence of natural as well as magnetic linear dichroism is hidden in the dipole selection rules. For linear polarized light, one can write the polarization-dependent transition probability as

$$I_{\alpha}^{0} \propto |\langle i|\alpha|f\rangle|^2 \tag{4.4}$$

with α equal to x, y, or z depending on the polarization of the light. The initial and final states are also functions of the coordinates x, y, and z. In most cases, the

Table 4.2 Dipole selection rules between all possible 2p-3d transitions for different polarizations.

Polarization		d_{yz}	d_{xz}	d_{xy}	$d_{x^2-y^2}$	$d_{3z^2-r^2}$
	p_x	0	0	0	1/2	1/6
I_x	p_y	0	0	1/2	0	0
	p_z	0	1/2	0	0	0
	p_x	0	0	1/2	0	0
I_y	p_y	0	0	0	1/2	1/6
	p_z	1/2	0	0	0	0
	p_x	0	1/2	0	0	0
I_z	p_y	1/2	0	0	0	0
	p_z	0	0	0	0	2/3
$3I_0$		1	1	1	1	1
$I_z - I_x$		1/2	0	−1/2	−1/2	1/2

integral of equation 4.5 will be zero. Only if the product of all three components is even in all three coordinates the transition intensity can be nonzero. For later use we have collected all intensities for a transition between a p and d orbital in Table 4.2.

The isotropic intensity is the average value over the three polarization directions summed over all possible initial-state orbitals. In Table 4.2 the intensity has been normalized such that $3I_0 = 1$. We see that this intensity is independent of the final-state orbital one chooses. From Table 4.2, it is also seen that the linear dichroism is zero if the t_{2g} (e_g) orbitals are degenerate and equally occupied, but finite if they are nondegenerate.

Obviously, the above treatment in terms of noncorrelated electrons and without SOC is much too simple to explain the rich structure often encountered in XAS spectra. This flaw becomes even more evident if one wants to discuss *magnetic* linear dichroism. In that case one can obtain a dichroic effect even if the d orbitals are degenerate and equally occupied. The most important ingredients needed to explain magnetic linear dichroism are the 2p SOC, 2p-3d Coulomb interaction and a 3d magnetization of the initial-state wave function.

Let us start by including the 2p SOC in the independent electron approach. It is important for the linear dichroism, because the SOC couples the magnetic moments to the orbitals. In Table 4.3, we have summarized the results for the transition involving individual spin–orbit split $|j, m_j\rangle$-p-orbitals and d-orbitals. The numbers can easily be obtained by using Table 4.2 and the selection rule $\Delta m_s = 0$. It is seen from this table that no linear dichroism is obtained for equally occupied and degenerate t_g (e_g) orbitals, just like in the case without spin–orbit splitting discussed before. The reason is that the fully occupied $p_{3/2}$ and $p_{1/2}$ levels are also spherically symmetric. The only possible effect within an independent particle

Table 4.3 Transition intensities for the excitation of a spin-down electron from spin–orbit split 2p-states into individual 3d-states for x and z polarized light. The numbers have been multiplied with 36.

j_{2p}	m_j	$\|j, m_j\rangle$		$d^{\downarrow}_{yz}$	$d^{\downarrow}_{xz}$	$d^{\downarrow}_{xy}$	$d^{\downarrow}_{x^2-y^2}$	$d^{\downarrow}_{3z^2-r^2}$
3/2	3/2	$\frac{1}{\sqrt{2}}(-p_x^{\uparrow} - ip_y^{\uparrow})$	I_x	0	0	0	0	0
			I_z	0	0	0	0	0
	1/2	$\frac{1}{\sqrt{6}}(-p_x^{\downarrow} - ip_y^{\downarrow} + 2p_z^{\uparrow})$	I_x	0	0	3	3	1
			I_z	3	3	0	0	0
	−1/2	$\frac{1}{\sqrt{6}}(p_x^{\uparrow} - ip_y^{\uparrow} + 2p_z^{\downarrow})$	I_x	0	12	0	0	0
			I_z	0	0	0	0	16
	−3/2	$\frac{1}{\sqrt{2}}(p_x^{\downarrow} - ip_y^{\downarrow})$	I_x	0	0	9	9	3
			I_z	9	9	0	0	0
1/2	1/2	$\frac{1}{\sqrt{3}}(-p_x^{\downarrow} - ip_y^{\downarrow} - p_z^{\uparrow})$	I_x	0	0	6	6	2
			I_z	6	6	0	0	0
	−1/2	$\frac{1}{\sqrt{3}}(-p_x^{\uparrow} + ip_y^{\uparrow} + p_z^{\downarrow})$	I_x	0	6	0	0	0
			I_z	0	0	0	0	8

picture is due to the magnetic interaction acting on the p levels. But the resulting splitting is so small that the different m_j levels may be considered to be degenerate.

In fact, it is crucial to include the 2p-3d electron–electron interaction to simulate the magnetic linear dichroism, in reality even observed for equally occupied and degenerate t_g (e_g) orbitals. To demonstrate this, the NiO L_2 edge is taken as an example, because the observed magnetic linear dichroism is relatively easy to explain. The initial state has a $3d^8$ configuration. The 3d level is split by an octahedral LF in a t_{2g} level, which is completely filled, and an e_g level containing two unpaired electrons. The final state has a $2p^5 3d^9$ configuration, that is, one 2p core electron has been excited into the e_g level. Alternatively, one might say that one of the two holes in the e_g level has been transferred to either the $2p_{1/2}$ or $2p_{3/2}$ state.

We only consider the transitions to the $2p_{1/2}$ level (or the L_2 edge). At 0 K the initial state is fully spin polarized by the exchange field. In total, the basis for the final state consists of eight states. The 2p core hole is in the $j_{2p} = 1/2$ state with either $m_j = 1/2$ or $m_j = -1/2$. The remaining 3d hole will have either spin-up or spin-down and will be either in the $d_{x^2-y^2}$ orbital or the $d_{3z^2-r^2}$ orbital. The $2p_{j=1/2}$ state and the spin of the 3d-electron belong in cubic symmetry to the e'_2 and the e_g hole belongs to the e_g (e'_1) irreducible representation. The product of the three gives three possible symmetries for the final state, namely one doublet of E_1 symmetry and two triplets of T_1 and T_2 symmetry.

The energy difference between these three states is due to Coulomb repulsion and can be considered as the sum of two contributions, the energy difference between a

Table 4.4 L_2 hole final states for 2p-XAS of NiO. The states in the last row are the 2p core hole states with $j = \frac{1}{2}$ and $m_j = \pm\frac{1}{2}$. I_x and I_z are the intensities for x- and z-polarized light.

Initial-State wavefunction →		$\underline{d}^{\uparrow}_{x^2-y^2}\underline{d}^{\uparrow}_{3z^2-r^2}$
Final-State wavefunctions ↓	I_x	I_z
$\sqrt{\frac{1}{2}}\underline{\lvert\frac{1}{2},-\frac{1}{2}\rangle}\underline{d}^{\uparrow}_{x^2-y^2} - \sqrt{\frac{1}{2}}\underline{\lvert\frac{1}{2},\frac{1}{2}\rangle}\underline{d}^{\downarrow}_{x^2-y^2}$	1	0
$-\sqrt{\frac{1}{2}}\underline{\lvert\frac{1}{2},-\frac{1}{2}\rangle}\underline{d}^{\uparrow}_{3z^2-r^2} - \sqrt{\frac{1}{2}}\underline{\lvert\frac{1}{2},\frac{1}{2}\rangle}\underline{d}^{\downarrow}_{3z^2-r^2}$	3	0
$\sqrt{\frac{1}{2}}\underline{\lvert\frac{1}{2},-\frac{1}{2}\rangle}\underline{d}^{\uparrow}_{x^2-y^2} + \sqrt{\frac{1}{2}}\underline{\lvert\frac{1}{2},\frac{1}{2}\rangle}\underline{d}^{\downarrow}_{x^2-y^2}$	1	0
$-\sqrt{\frac{3}{4}}\underline{\lvert\frac{1}{2},\frac{1}{2}\rangle}\underline{d}^{\uparrow}_{3z^2-r^2} - \sqrt{\frac{1}{4}}\underline{\lvert\frac{1}{2},-\frac{1}{2}\rangle}\underline{d}^{\downarrow}_{x^2-y^2}$ 0	0	0
$\sqrt{\frac{1}{4}}\underline{\lvert\frac{1}{2},\frac{1}{2}\rangle}\underline{d}^{\uparrow}_{x^2-y^2} - \sqrt{\frac{3}{4}}\underline{\lvert\frac{1}{2},-\frac{1}{2}\rangle}\underline{d}^{\downarrow}_{x^2-y^2}$	0	2
$\sqrt{\frac{1}{2}}\underline{\lvert\frac{1}{2},-\frac{1}{2}\rangle}\underline{d}^{\uparrow}_{3z^2-r^2} + \sqrt{\frac{1}{2}}\underline{\lvert\frac{1}{2},\frac{1}{2}\rangle}\underline{d}^{\downarrow}_{3z^2-r^2}$	3	0
$\sqrt{\frac{1}{4}}\underline{\lvert\frac{1}{2},\frac{1}{2}\rangle}\underline{d}^{\uparrow}_{3z^2-r^2} + \sqrt{\frac{3}{4}}\underline{\lvert\frac{1}{2},-\frac{1}{2}\rangle}\underline{d}^{\downarrow}_{x^2-y^2}$	0	0
$\sqrt{\frac{3}{4}}\underline{\lvert\frac{1}{2},\frac{1}{2}\rangle}\underline{d}^{\uparrow}_{x^2-y^2} + \sqrt{\frac{1}{4}}\underline{\lvert\frac{1}{2},-\frac{1}{2}\rangle}\underline{d}^{\downarrow}_{x^2-y^2}$	0	6
$\mathrm{c}\underline{\lvert\frac{1}{2},\frac{1}{2}\rangle} = \sqrt{\frac{1}{3}}\left(-\underline{p}^{\uparrow}_{x} - i\underline{p}^{\uparrow}_{y} - \underline{p}^{\downarrow}_{z}\right)\underline{\lvert\frac{1}{2},-\frac{1}{2}\rangle} = \sqrt{\frac{1}{3}}\left(-\underline{p}^{\downarrow}_{x} + i\underline{p}^{\downarrow}_{y} + \underline{p}^{\uparrow}_{z}\right)$		

spin triplet and a spin singlet, and the difference in Coulomb repulsion between a p_z orbital and a $d_{3z^2-r^2}$ orbital and between a p_z and a $d_{x^2-y^2}$ orbital. Calculations, using atomic Hartree–Fock values for the Slater integrals, show that the first triplet is very close in energy to the doublet, the difference being only 140 meV, whereas the second triplet is 1.27 eV higher. This latter energy difference is the cause of the characteristic double peak structure observed in the experimental NiO L_2 edge (see for instance Figure 4.11). The expressions for the wave functions for the different multiplet components have been collected in Table 4.4. The transition intensity for x- and z-polarized light, calculated with the aid of Table 4.3, are given in the last two columns of the table. (Note that the spins for the hole state wave functions have been inverted with respect to the electron state wave functions of Table 4.3.) The intensity integrated over all states comprising the spectrum is seen to be the same for both polarizations. The first peak, consisting of the doublet and first triplet, however, exhibits a negative and the second peak a positive magnetic linear dichroism effect. This demonstrates that even if the overall integrated magnetic dichroism is zero, multiplets that are separately visible in the spectrum may show large dichroic effects.

To calculate the X-ray magnetic linear dichroism (XMLD) spectra at finite temperatures, one should realize that above we only calculated the spectra for a fully magnetized ground state. At finite temperatures, low-energetic excited states are populated according to Boltzman statistics. In spherical symmetry, it has been shown that the XMLD scales with $\langle M^2 \rangle$ [24]. Above the ordering temperature, the material is in the paramagnetic state and the magnetic linear dichroism vanishes. Below the ordering temperature, the temperature dependence of $\langle M^2 \rangle$ can be expressed within mean field theory as

$$\langle M^2 \rangle = S(S+1) + SB_S(S/t)\coth(1/2t) \tag{4.5}$$

where $B_S(S/t)$ is the Brillouin function and $t = k_B T/(g\mu_B H_{\text{exch}})$ the reduced temperature. We see in Section 4.6 that this characteristic temperature dependence is of great help to distinguish the magnetic and nonmagnetic contributions to the linear dichroism.

4.5 Strain

In the experimental work we are going to discuss, epitaxial strain plays a crucial role. It is used as a tool to make the d-electron charge and/or spin distribution anisotropic. In fact, it turns out to be possible to orient the spin in different directions by applying compressive or tensile strain.

The origin of the misfit strain was first explained properly by Frank and van der Merwe [25] in terms of the balance between the binding of the overlayer to the substrate and the elastic energy buildup in the overlayer. Within a Frenkel–Kontorowa model, they calculated the misfit limit for stable and metastable pseudomorphous growth of a single monolayer (ML) to be about 0.08 and 0.12, respectively. Above those limits, either empty or doubly occupied substrate sites, depending on whether the strain is tensile or compressive respectively, may, or will be, formed. For more than one ML, the value of the critical misfit decreases, because the strain energy increases with thickness, but the interface energy stays more or less constant. For a given misfit below the critical misfit for a single ML, fully strained epitaxial films can be grown up to a critical layer thickness. Beyond this limit, misfit dislocations appear and the strain is released slowly on further growth until finally the lattice is fully relaxed. To calculate the critical thickness, we employ the Matthews–Blakeslee model [26]. In our case, we cannot employ the original Matthews–Blakeslee expression, because it was specifically derived for semiconductor materials having the zincblende structure. Instead we have to go back to a more general expression for the equilibrium strain in a partly relaxed film of thickness h [27]

$$\varepsilon(h) = \frac{G_S b(1 - \nu\cos^2\alpha)}{4\pi h(G_S + G_O)(1+\nu)\cos\lambda}\left(\ln\left(\frac{h}{b}\right) + 1\right) \tag{4.6}$$

where G_S and G_O are the shear modulus of the substrate and overlayer, respectively, and ν is the Poisson ratio of the film. The angle α between the dislocation line and its Burgers vector $\mathbf{b}$ is 90° for a pure edge dislocation. The quantity $b\cos\lambda$ is the component of the Burgers vector responsible for the strain relief, that is, λ is the angle between the Burgers vector and the line in the interface plane normal to the

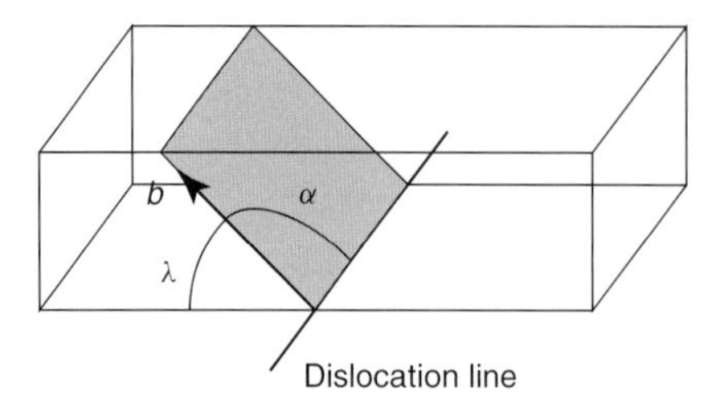

Figure 4.7 Misfit dislocation for a rocksalt type epitaxial layer grown in the [001] direction. The Burgers vector $\mathbf{b} = \frac{1}{2}[\bar{1}01]$ lies in a (101) slip plane and makes an angle λ of 45° with the surface and an angle α of 90° with the dislocation line.

dislocation line. The most important edge dislocation in the rocksalt structure has a Burgers vector $\mathbf{b} = \frac{1}{2}\langle\bar{1}01\rangle$, with {101} slip planes [28]. If the layer is grown in the [001] direction, as is true for all the TM oxide layers discussed in this chapter, there are four sets of edge dislocations contributing to the misfit relaxation, forming dislocation lines at the interface in the [100] and [010] direction (see Figure 4.7). Because we are dealing with a pure edge dislocation, the angle α is 90°, and the angle λ is equal to 45°. To get a single strain curve for all oxide layers independent of the substrate, we define a normalized strain parameter

$$\varepsilon_{\text{norm}}(h) = C\varepsilon(h) = \frac{3}{16\pi\sqrt{2}}\frac{b}{h}\left(\ln\left(\frac{h}{b}\right)+1\right) \tag{4.7}$$

where $C = 3(G_S + G_O)(1+\nu)/(8G_S)$ is the normalization constant. In case of a [001] growth direction, the best choice for the shear modulus is the so-called tetragonal shear modulus $\frac{1}{2}(C_{11} - C_{12})$. If the Poisson ratio is close to 1/3 and if the shear moduli of the substrate and overlayer are the same, the normalization factor is 1 and the normalized strain parameter is equal to the real strain. This is a good approximation if the AF monoxides are grown on oxide substrates, but not for growth on soft materials like silver or gold metal. Equation 4.7 can be used to calculate the critical thickness for coherent growth by making the strain equal to its maximum value, the lattice misfit f. In Table 4.5, we give the parameters needed to calculate the normalized misfit for the overlayer substrate combinations discussed in this chapter. The corresponding critical thickness is determined from the second point of intersection of the normalized misfit and the generalized strain curve Eq. (4.8) as shown in Figure 4.8. From this figure, we see that fully strained oxide layers can be grown on oxide substrates up to normalized misfits of about 4 percent. On silver, however, the critical misfit is at most 1 ML for NiO and CoO and zero for MnO.

The results are not very accurate for a number of reasons. One uncertainty is the size of the dislocation core in the logarithmic term of Eq. (4.8). Secondly, the elastic continuum theory breaks down at small layer thicknesses, that is, instead of turning down, the strain curve should continue to go up for decreasing thickness until the approximate value predicted for 1 monolayer by Frank and van der Merwe

Table 4.5 Parameters determining the critical misfit for different overlayer substrate combinations. a_O and a_S are the overlayer and substrate lattice constants, f is the misfit, G_O and G_S are the shear moduli of the overlayer and substrate, ν is the Poisson ratio, $C = 3(G_S + G_O)(1 + \nu)/(8G_S)$ is the normalization constant (see text), f_{norm} is the normalized misfit, and h_c is the critical thickness in monolayers.

	a_O/a_S (Å)	f (%)	G_O/G_S (10^{10}Pa)	ν	C	f_{norm} (%)	h_c (ML)
NiO/MgO	4.177/4.213	−0.85	7.25/10.0	0.31	0.85	−0.72	35
CoO/MgO	4.260/4.213	1.11	5.6/10.0	0.36	0.80	0.89	26
MnO/CoO	4.445/4.260	4.16	5.6/5.6	0.33	1.00	4.16	2
CoO/MnO	4.260/4.445	−4.16	5.6/5.6	0.36	1.02	−4.2	2
NiO/Ag	4.177/4.086	2.235	7.25/1.55	0.31	2.78	6.2	1
CoO/Ag	4.260/4.086	4.27	5.6/1.55	0.36	2.30	9.9	0
MnO/Ag	4.445/4.086	8.8	5.6/1.55	0.33	2.30	20	0

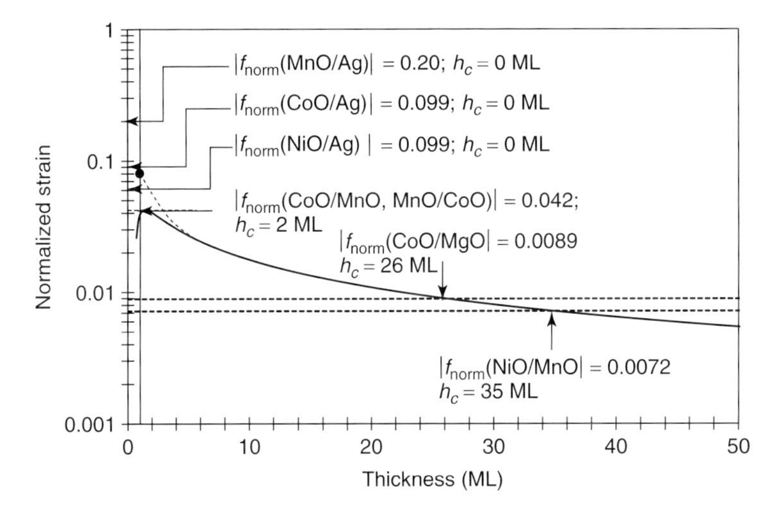

Figure 4.8 Equilibrium strain as a function of layer thickness for materials having the rocksalt structure based on the Matthews–Blakeslee formalism. A normalization constant $C = 3(G_S + G_O)(1 + \nu)/(8G_S)$ was applied to the misfit, containing the material dependent parameters, that is, the shear modulus of substrate and overlayer an the Poisson ratio, in order to get a universal curve for different overlayer materials. At low thicknesses, the elastic continuum theory breaks down. The Frank and van der Merwe value for the critical misfit of a single monolayer (ML) has been added as well. For a given normalized misfit, the critical thickness can be read from this curve. This has been done for all the overlayer substrate combinations discussed in this chapter.

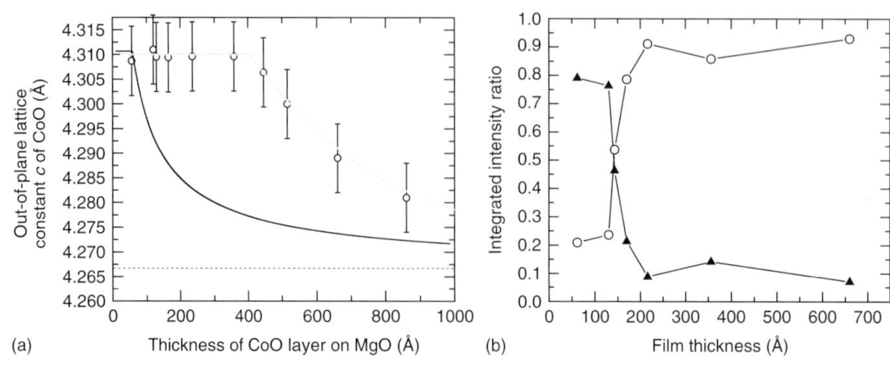

Figure 4.9 (a) Out-of-plane lattice constant of a CoO layer grown on a MgO substrate versus layer thickness determined from a radial $\theta - 2\theta$ X-ray Diffraction scan through the (002) diffraction peak. (b) The fractions of diffuse (open circles) and coherent (closed triangles) intensity near the (002) diffraction peak. As explained in the text, the drop in the coherent intensity is due to the appearance of misfit dislocations if the critical thickness is reached. The drop in the lattice constant, which is commonly (mis)taken as the critical thickness, occurs at a much higher value.(Reprinted from [29].)

is reached, as is indicated by the dotted line. Finally, there may be a kinetic barrier for relaxation keeping the coherent layer intact up to thicknesses exceeding the values predicted by equilibrium theory considerably. Consequently, the critical misfits listed in Table 4.5 should be considered as approximate lower limits to the experimental values. Many attempts have been made to introduce a realistic kinetic barrier by considering specific mechanisms of dislocation nucleation and motion. However, in most models, new and unknown parameters are introduced, making quantitative predictions questionable [27].

To determine the critical thickness experimentally, a method commonly employed is to measure the in-plane lattice constant as a function of thickness using X-ray or electron diffraction techniques. In Figure 4.9a, the relative change in the lattice constant versus thickness curve for CoO on MgO from X-ray diffraction is compared with the theoretical curve calculated using Eq. (4.8) [29]. As in the case of NiO on MgO [30], the critical thickness h_c appears to be much larger than the predicted value. However, one should realize that X-ray diffraction tends to overestimate the critical thickness. For thicknesses in excess of h_c, the diffraction signal consists of a coherent and diffuse part, as can be seen from Figure 4.10, the last one increasing at the expense of the first. As was pointed out by Kaganer [31], the maximum of the coherent signal is not a good measure of the average lattice constant. Only after the broad diffuse line has taken over completely, the line position is directly related to the average lattice constant and the average strain. This will happen if the average distance between the dislocations becomes smaller than the film thickness. A much better criterion for having reached the critical thickness is the appearance of diffuse scattering due to misfit dislocations. From Figure 4.9b,

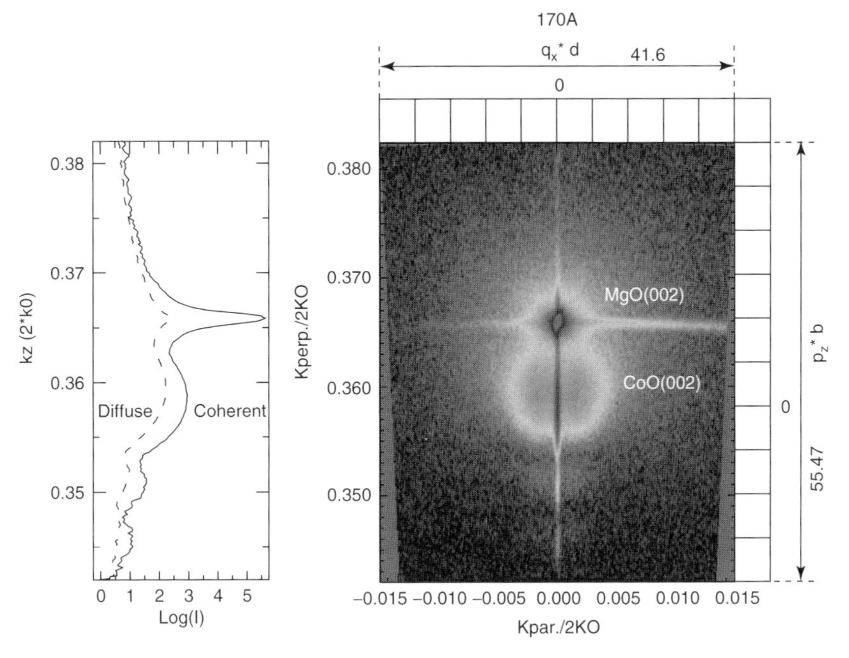

Figure 4.10 The X-ray scattering around the rocksalt (002) reflection for a 170 Å layer of CoO on a MgO substrate. The coherent scattering from the CoO is the streak down the middle of the picture at $q_x = 0$, showing fringes due to the finite thickness of the layer. The corresponding radial scan is the drawn curve in the graph to the left. A scan through the diffuse structure to the right and left of the coherent streak at a q_x value slightly different from zero is also shown. The diffuse scattering is due to misfit dislocations. Although the coherent intensity is in fact already much smaller than the total diffuse intensity, it is dominating the picture because of its narrow transverse structure. The maximum of the coherent peak is not related to the average lattice constant as is often assumed.(Reprinted from [29].)

one arrives at a value of 70 instead of 200 ML, which has to be compared with a value of 55 ML predicted by Eq. (4.8).

In conclusion, one may distinguish three ranges in the development of strain during growth: (a) the range below the critical thickness in which the film is uniformly and fully strained, (b) the range in which the average distance between dislocations is large with respect to the film thickness, that is, the dislocation density is low ($\rho h < 1$), and (c) the range in which the dislocation density is high ($\rho h > 1$). In range (b) the film is not uniformly strained, the deformation being large near the dislocation sites, but small in between. In range (c) the strain slowly decreases with thickness. Consequently, if one wants to use misfit strain as a tool to induce linear dichroism, the thickness and misfit should preferably be such that the film is fully strained (range a). In the partly relaxed range (b), the linearly polarized XAS spectrum is expected to be nonuniformly broadened.

4.6
Linear Dichroism Results for AF TM Monoxide Layers

4.6.1
XMLD of Epitaxial NiO(100)/MgO layers

In an octahedral environment the t_{2g} orbitals of the divalent Ni d^8-ion are completely filled with six paired electrons and the remaining two electrons are in the twofold degenerate e_g orbital with parallel spin. In terms of many-electron theory, this corresponds to a ${}^3A_{2g}$ triplet. The charge distribution has a cubic symmetry and the orbital moment is strongly quenched. Above T_N, in the paramagnetic phase, X-ray Linear dichroism due to an anisotropic charge distribution is expected to be zero, because the crystal structure is cubic. In the AF ordered phase at low temperatures, single domains should exhibit large XMLD effects, but we expect the individual domain contributions to be averaged out in a single crystal. By growing a strained epitaxial NiO layer on a suitable substrate like MgO, the symmetry lowering may induce magnetic as well as natural linear dichroism. We concentrate on the magnetic Linear Dichroism first [32, 33] and discuss the LF-induced effects later [34]. The lattice constant of MgO is 4.213 Å. The misfit for NiO is only −0.85%. Up to very thick layers, the growth is pseudomorphous and coherent, that is, the in-plane lattice constant of the NiO is equal to the MgO lattice constant. Therefore, the NiO is slightly expanded in-plane and contracted in the growth direction. The artificial contraction favors the preferential growth of the eight domains with spins pointing in $\langle 11\bar{2}\rangle$ directions closest to the film normal, that is, in $[\pm1\ \pm1\ \pm2]$ directions. For these domains, the average spin direction is perpendicular to the film. This preference can be understood by noting that in addition to the contraction along the trigonal axis, the S-domains in NiO are contracted slightly along the spin direction [17]. Consequently, the domains with more or less out-of-plane spins are favored if the strained film is grown on a substrate like MgO having a larger lattice constant.

The dependence of the linearly polarized spectrum on the angle θ between the polarization vector and the easy direction of the magnetic-moment can be written as [33]

$$I_{q=0}(\omega,\theta) = I_{\text{iso}}(\omega) + \delta I_{\text{iso}}^{\text{exch}}(\omega) + I_{LD}(\omega,\theta)(3\cos^2\theta - 1) \tag{4.8}$$

$I_{\text{iso}}(\omega)$ is the isotropic spectrum without exchange interaction, $I_{\text{iso}}^{\text{exch}}(\omega)$ is the exchange contribution due to the nn spin-spin correlation to the isotropic spectrum (an extensive account of this temperature-dependent contribution can be found in [35]), and $I_{LD}(\omega,\theta)$ is the anisotropic contribution due to the long range order part of the exchange interaction. Arenholz *et al.* [36] have recently shown that in general the anisotropic contribution is more complicated and is not properly represented by the last term of Eq. (4.9). However, as we demonstrate in the appendix, the above expression is correct for a set of equivalent domains with an average magnetic moment along the tetragonal axis.

It is convenient in practice to choose a characteristic feature of the spectrum to test the validity of Eq. (4.9). As was explained in Section 4.4.1, the double

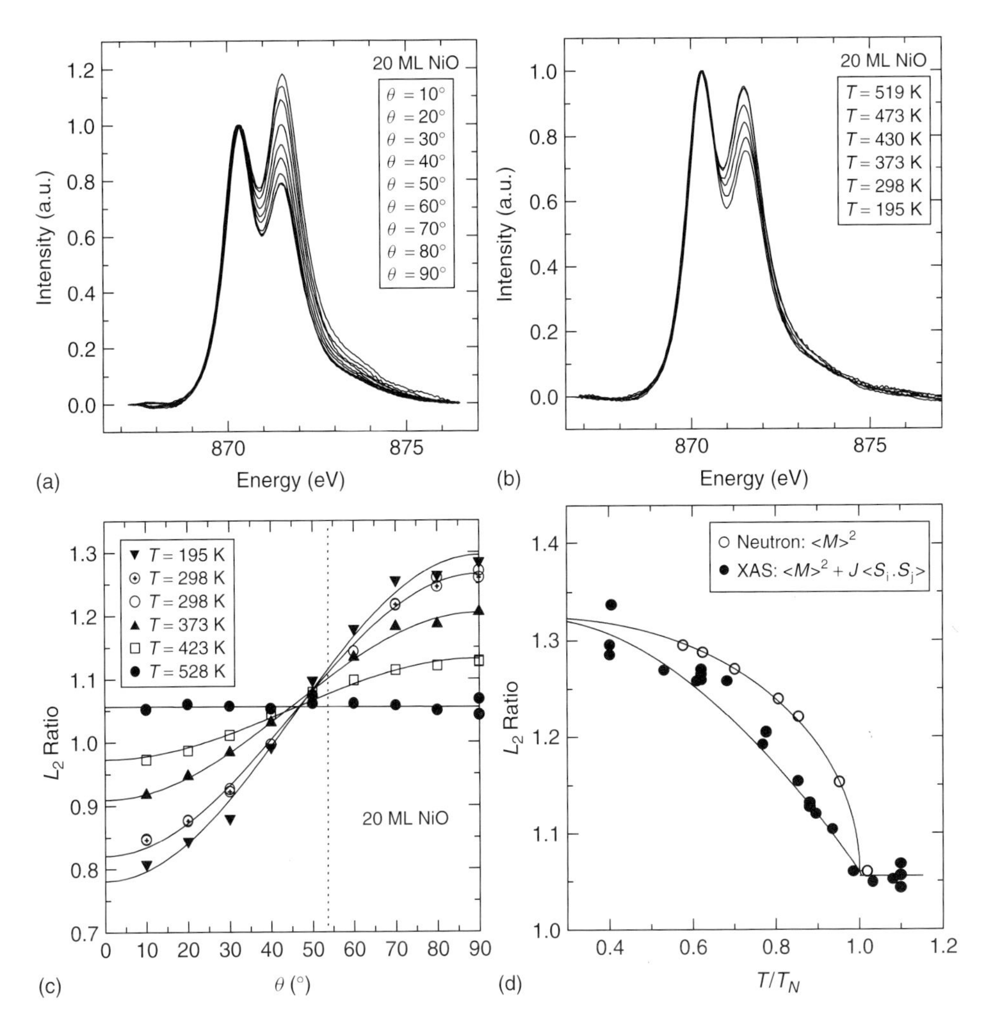

Figure 4.11 Ni L_2-XAS of a 20 ML NiO(001) film: (a) at different temperatures, (b) as a function of the angle θ between the polarization vector and the normal to the surface, (c) polarization and temperature dependance of the ratio of the two L_2 peaks, and (d) temperature dependence of the L_2 ratio. The data are compared to neutron scattering data. (Reprinted from [33].)

peak structure of the L_2 edge exhibits quite a strong contrast for perpendicular polarization directions. So, the ratio of the two peaks was taken as a measure of the exchange field. In Figure 4.11(a) the Ni L_2-XAS of a 20 ML NiO(100) film on a MgO substrate at 298 K is shown as a function of the angle θ between the polarization vector and the surface normal. The spectra are normalized to the first peak. In reality, the total intensity of the L_2 spectrum is constant. The temperature dependence of the effect is shown in Figure 4.11(b). Above 473 K the spectra do not change anymore, suggesting that the material is in the paramagnetic phase.

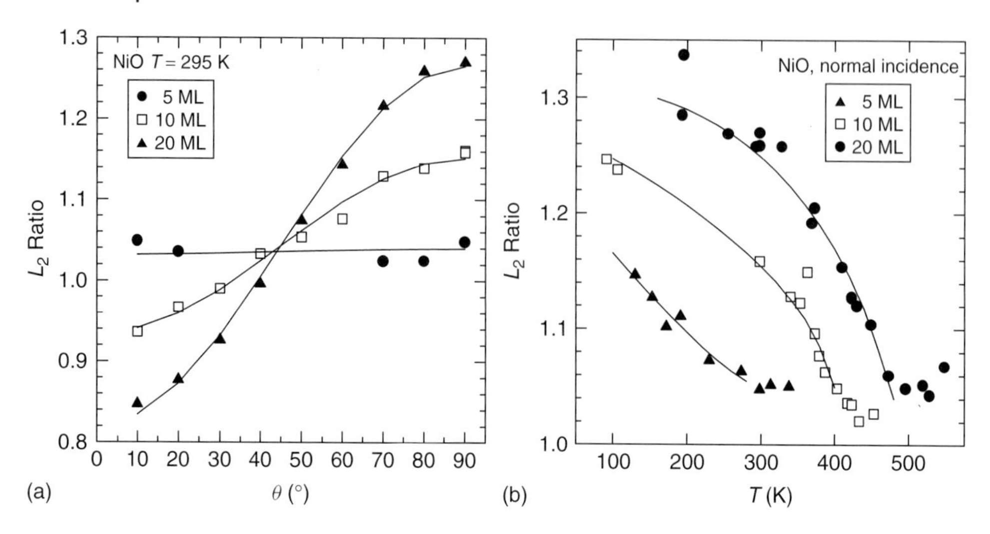

Figure 4.12 (a) polarization dependence for different NiO layer thicknesses at roomtemperature, and (b) temperature dependence of the L_2 ratio for different NiO layer thicknesses. (Reprinted from [33].)

In Figure 4.11(c), the ratio of the L_2 peaks of all the measured data is plotted as a function of the angle between the film normal and the polarization direction. The lines in Figure 4.11(c) are fits to Eq. (4.9). At 528 K , well above T_N, the L_2 peak ratio is independent of θ, that is, the linear dichroism has disappeared completely. This implies that the LF contribution to the linear dichroism is unimportant, although the symmetry of the coherent epitaxial layer is lowered to D_{4h}. Apparently the deformation is too small to cause a measurable effect. In the next paragraph, we show that crystal-field-induced effects are observed if NiO is grown on top of a substrate with a larger misfit like Ag.

The temperature dependence of the L_2 peak ratio at normal incidence is shown in Figure 4.11(d). For comparison also data for $\langle M \rangle^2$ obtained from neutron diffraction [20] is drawn in the same graph. The same experiments were repeated for 5- and 10-ML-thick samples (Figure 4.12). The effect decreases dramatically with decreasing layer thickness at room temperature. In fact, the linear dichroism has disappeared completely for the 5-ML sample. The temperature dependence of the L_2 peak ratio for the 5-, 10-, and 20-ML samples is shown in Figure 4.12b. T_N is seen to be very much dependent on sample thickness. Whereas T_N is 520 K for bulk samples, it is approximately 470, 430, and 300 K for the 20-, 10-, and 5-ML NiO samples, respectively. This was interpreted as a transition from 3D to 2D behavior.

4.6.2
Ligand-Field-Induced Linear Dichroism in Strained NiO/Ag(100) Layers

Fcc metals like Ag and Au can also be used as substrates for epitaxial growth of NiO. The misfit, however, is much larger than for MgO. For Ag it is −2.2% . From

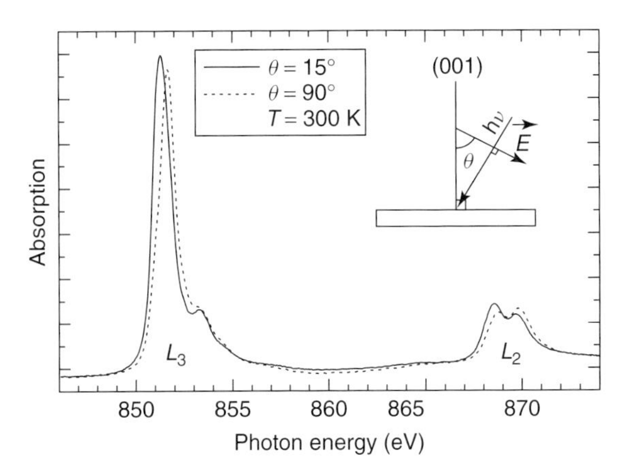

Figure 4.13 Experimental polarization-dependent Ni $L_{2,3}$ XAS of 1 ML of NiO(001) on Ag(001) covered with a caplayer of 10 ML of MgO. θ is the angle between the polarization vector and the surface normal. (Reprinted from [34].)

Section 4.5, the equilibrium critical thickness for coherent growth is estimated to be only about 1 ML. This small value is mainly caused by the much smaller shear modulus of silver metal. Haverkort *et al.* have studied the X-ray linear dichroism of a single monolayer of NiO grown on a Ag single crystal in the [001] direction [34]. The film was covered with a caplayer of MgO, which has a lattice constant close to that of NiO. It is unlikely that this layer, including the NiO monolayer, is coherent with respect to the substrate, but a considerable residual strain will still be present. In Figure 4.13, L_2 and L_3 edge spectra for two nearly perpendicular polarization directions are shown. The L_2 spectra are very similar to those observed for thick NiO films on MgO substrates. However, there are two characteristic differences. As can be seen from the close-ups of the L_2 region in Figure 4.14, the sign of the dichroism in the 1-ML NiO/Ag(001) spectrum is reversed as compared to the 20-ML NiO/MgO spectra, and the peaks are shifted by 0.35 eV. It is very unlikely that the dichroism in this case is of magnetic origin. On the basis of the data for NiO on MgO discussed in the previous section, T_N is predicted to be below room temperature. Also shown in Figure 4.14 are the data for 80 K, which appear to be identical to the room temperature data, suggesting that T_N is even lower than 80 K for a single NiO ML. A credible cause for the linear dichroism is the energy difference between the $d_{x^2-y^2}$ and $d_{3z^2-r^2}$ orbitals associated with the tetragonal deformation due to strained growth. The shifts in the peaks in both the L_2 and L_3 region are in favor of this explanation, since lower symmetry ligand fields split or alter initial- and final-state energy levels. MLFT calculations confirm this notion. The simulated spectra, also shown in Figure 4.14, are seen to reproduce the experimental spectra very well. The calculation was similar to that of Alders *et al.* [33], but now in the D_{4h} point-group. (Strictly speaking, the local symmetry is C_{4v} but for d-electrons the results are identical.) The shift of 0.35 eV, most clearly visible in the L_3 edge, can be understood in the independent electron approximation. The

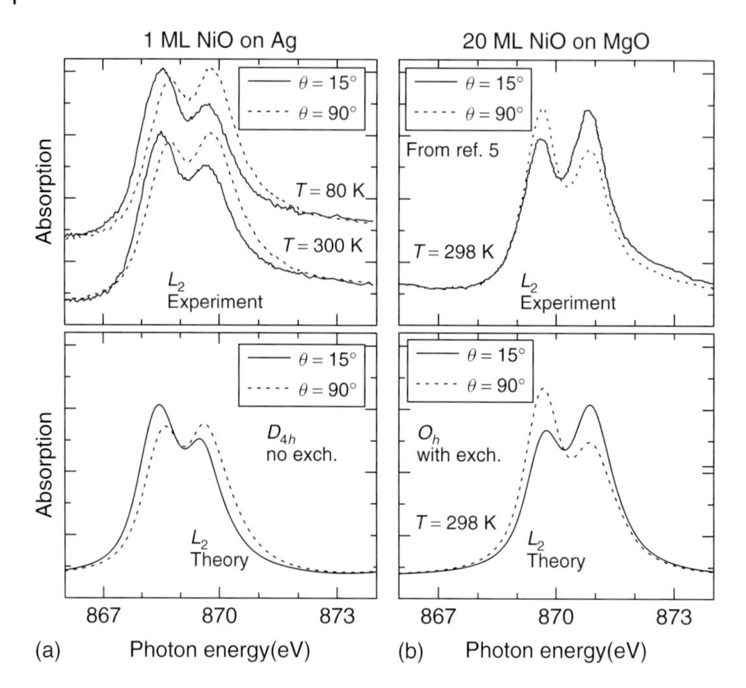

Figure 4.14 (b) The experimental and theoretical Ni L_2 XAS of 1 ML of NiO on Ag(001) capped with 10 ML of MgO. (a) Similar spectra for 20 Ml of NiO on MgO(001) taken from [34]. (Reprinted from [34].)

ground state has a hole in both of the split e_g levels, that is, one in the b_{1g} ($3d_{x^2-y^2}$) state and one in the a_{1g} ($3d_{3z^2-r^2}$) state. In the final state, one of the two holes will be occupied by the excited electron. Using Table 4.2, we see that with z-polarized light only the $3d_{3z^2-r^2}$ state can be reached, and with x-polarized light the probability to excite the electron into the $3d_{3z^2-r^2}$ is 1/4 and into the $3d_{x^2-y^2}$ is 3/4. The shift between the two polarizations is predicted to be approximately equal to the LF splitting of the e_g level. Another characteristic feature is the transfer of spectral weight between the two main peaks of the L_2 edge. As we have explained already in Section 4.4.1, in O_h symmetry the first peak is related to an E_1 and a T_2, and the second peak to a T_1 final state. In D_{4h} symmetry, these three final states split into the following pairs, A_1 and B_1, B_2 and E_1, and A_2 and E_1, respectively. The two E_1 states will mix and cause a change in the intensity ratio of the two peaks.

In the two cases discussed up to now, that is, the thicker layers of NiO on MgO substrates and the single ML of NiO on Ag, two extreme situations were encountered. In the first case, the charge anisotropy due to strain could be neglected, and in the second case the magnetic anisotropy is neglected. A general strategy to separate the two contributions is to take linear dichroism spectra below and above the ordering temperature. Above T_N, the magnetic contribution can be excluded and any linear dichroism must be due to charge anisotropy. In the case of NiO, we are helped by the fact that the ground state cannot be split by LF or spin–orbit interactions. The LF-induced linear dichroism, therefore, is independent of temperature, in contrast to the magnetic contribution below T_N which scales with $\langle M^2 \rangle$.

Altieri *et al.* [37] studied also thicker NiO layers grown on Ag single crystals, that is, 3 and 30 ML of NiO capped with 10 ML of MgO. In the 3 ML of NiO on Ag sample, the L_3 edge is shifted 0.1 eV in the high-temperature paramagnetic phase, indicating an LF contribution due to a tetragonal distortion, similar to the single ML case, but of smaller magnitude. The most surprising result is that the L_2 peak ratios at low temperatures, presumably well below the AF ordering temperature, are in the 3-ML NiO/Ag case inverted with respect to the 30-ML NiO/MgO case. Apparently, the average spin direction now has a main component parallel to the layer, implying that there is a preference for the formation of [±1 ±2 ±1] and [±2 ±1 ±1] domains in these thin strained layers. In the thicker, presumably fully relaxed NiO layers on Ag, the sign of the L_2 MLD is again the same as for thicker NiO layers on MgO. The same geometrical explanation for this preference given in the case of strained NiO layers on MgO can be applied, that is, compressive stress favors domains with in-plane spins, because the characteristic distortion of the S-domains is a contraction along the spin direction. A more sophisticated explanation was given by Finazzi and Altieri [38], assuming that dipole–dipole interaction is the main interaction responsible for the magnetic anisotropy in NiO. They prove that in bulklike samples, positive and negative strain favor in-plane and out-of-plane spin orientations, respectively, and that for thin layers the spins are strongly oriented in-plane.

4.6.3 Isotropic XAS of CoO

Compared to NiO, the (polarized) XAS of CoO is much more involved. The basic reason is that the t_{2g} levels of CoO are partly filled and degenerate, in contrast to NiO (and MnO). CoO has a high spin d^7 configuration. In an independent electron picture, the first five electrons go into a different 3d-orbital each with parallel spins, and the remaining two will form pairs in the lower t_{2g} shell. Consequently, there is one hole in the t_{2g} level, which may be in either the d_{xy} ,d_{xz}, or d_{yz} orbital. Because of this degeneracy, the hole carries a (pseudo) orbital moment of $\tilde{L} = 1$. The three linear orbital combinations with magnetic quantum numbers $m_L = -1$, 0, and 1 are $d_{-1} = \sqrt{1/2}(d_{xz} - id_{yz})$, $d_0 = d_{xy}$, and $d_1 = \sqrt{1/2}(-d_{xz} - id_{yz})$. This orbital moment couples to the total spin $S = \frac{3}{2}$, the total angular moment becoming $\tilde{J} = \frac{1}{2}$, $\tilde{J} = \frac{3}{2}$, or $\tilde{J} = \frac{5}{2}$. The corresponding energy levels are separated by SOC. The actual splittings between the $\tilde{J} = \frac{1}{2}$ ground state and the $\tilde{J} = \frac{3}{2}$ and $\tilde{J} = \frac{5}{2}$ excited states are 40 and 120 meV, respectively. So, the excited states will be occupied to some extent at higher temperatures. As a result, the XAS spectrum becomes temperature dependent. In Figure 4.15 the experimental isotropic XAS spectra of CoO grown as a polycrystaline film on Ag taken at different temperatures are shown [39]. The most pronounced effects are the reduction of the sharp peak at 777 eV and the shift of the L_2 edge toward lower energies with increasing temperature.

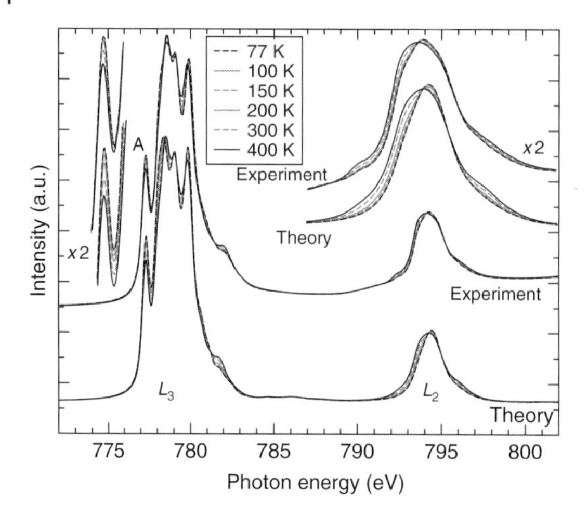

Figure 4.15 Temperature-dependent experimental and theoretical isotropic CoO $L_{2,3}$ spectra. (Reprinted from [39].)

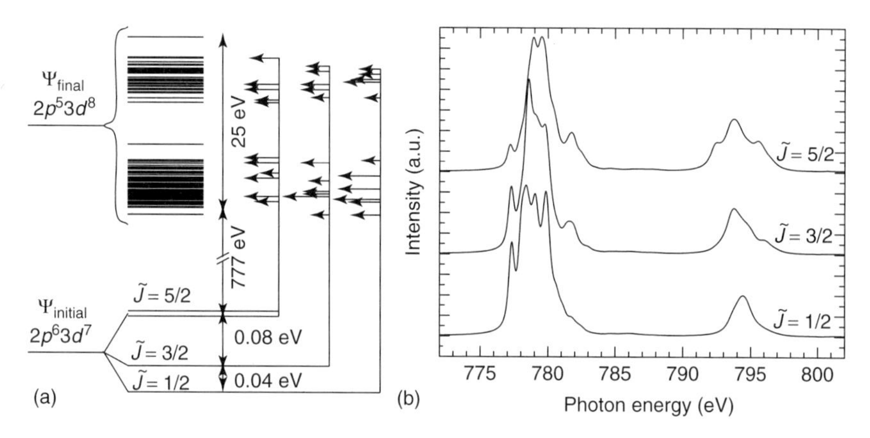

Figure 4.16 (a) Energy level diagram of CoO. The Initial state is split by SOC into a triplet, quintet, and septet. (b) The theoretical spectra for the triplet, quintet, and septet initial states. (Reprinted from [39].)

In Figure 4.16a, the energy level diagrams for the spin–orbit split initial-state and the many multiplet levels for the final-state $2p^5d^8$ configuration have been shown. The arrows point at the dipole allowed excited multiplet levels, the length of the arrows being proportional to the transition probability. In (b) the calculated XAS spectra for the initial-state levels with different $\tilde{J}$-values are shown. Using Boltzman statistics, the theoretical spectra can be calculated. For comparison, the results have been added to the experimental results in Figure 4.15. The observed features are reproduced very well.

It may be noted that the isotropic spectrum of NiO also displayed some temperature dependence, although the ground state is not split by spin–orbit or LF interactions. This effect was linked to the isotropic part of the exchange interaction. In principle, the same effect is active in CoO as well. However, the exchange field is much smaller in CoO and the temperature dependence of the isotropic spectrum is in fact dominated by the thermally accessible spin–orbit split excited states. Also, exchange striction plays a role [39].

4.6.4 Linear Dichroism of Strained CoO Layers

Like in NiO, epitaxial strain was used to induce an anisotropic distribution of charge, LF splittings, and magnetic moments in CoO, and polarized XAS was used to detect and quantify them [40]. The cubic lattice of CoO (a = 4.267 Å) can be expanded within the layer by choosing a cubic substrate with a larger lattice constant, like MnO (a = 4.445 Å), the misfit being −4.2% and the corresponding theoretical equilibrium critical thickness being a few ML (see Section 4.5). Instead of taking a single crystalline MnO substrate, a thick (100 Å) fully relaxed film of MnO was grown first on a silver substrate. The thickness of the CoO film was chosen to be 5 ML. From reflection high energy electron diffraction (RHEED), the in-plane lattice constant was determined to be 4.424 Å, which is still very close to the coherent value. To increase the critical thickness somewhat, the layer was capped with an additional 7 ML of MnO. Using the Poisson ratio 0.27 deduced from X-ray diffraction results, the tetragonal distortion is at least 6.2% for such a (14 Å)MnO/(10 Å)CoO/(100 Å)MnO/Ag(001) sample. A small compression was achieved by growing CoO directly on a Ag(001) substrate. The misfit between CoO and Ag is +4.3%. However, the oxide lattice will relax immediately at the first atomic layer due to the softness of silver metal. For the (90 Å)CoO/Ag(001) film, the in-plane and out-of-plane lattice constants were found to be 4.285 Å and 4.235 Å, respectively, indicating a residual strain of +0.75% and a tetragonal distortion of 1.1%.

To separate the LF and magnetic contributions to the XLD spectrum, the same strategy was employed as in the case of NiO. At temperatures exceeding the magnetic ordering temperature, any linear dichroism is related to the charge anisotropy and LF generated by the lowering of the cubic symmetry. The spectra for the samples under compressive and tensile stress taken at 400 K, which is well above T_N, are shown in Figure 4.17

The symmetry lowering due to the tetragonal distortion is accompanied by a splitting of the t_{2g} and e_g levels as shown inFigure 4.2. Depending on the sign of the distortion, the level splittings are inverted. In Figure 4.18, the theoretical linear polarized XAS spectra for different values of the splitting energies $\Delta_{t_{2g}}$ and Δ_{e_g} are shown. The spectra have been calculated at 400 K using the spin–orbit and $10Dq$ values deduced from the simulations of the isotropic spectrum. It is clear from these spectra that Δ_{e_g} does not produce much linear dichroism. The reason is simple. The charge distribution in the e_g shell is not anisotropic, because the e_g levels are filled with one electron each. The spectral changes are not completely zero, because the final state is affected by the splitting, just as in the case of NiO

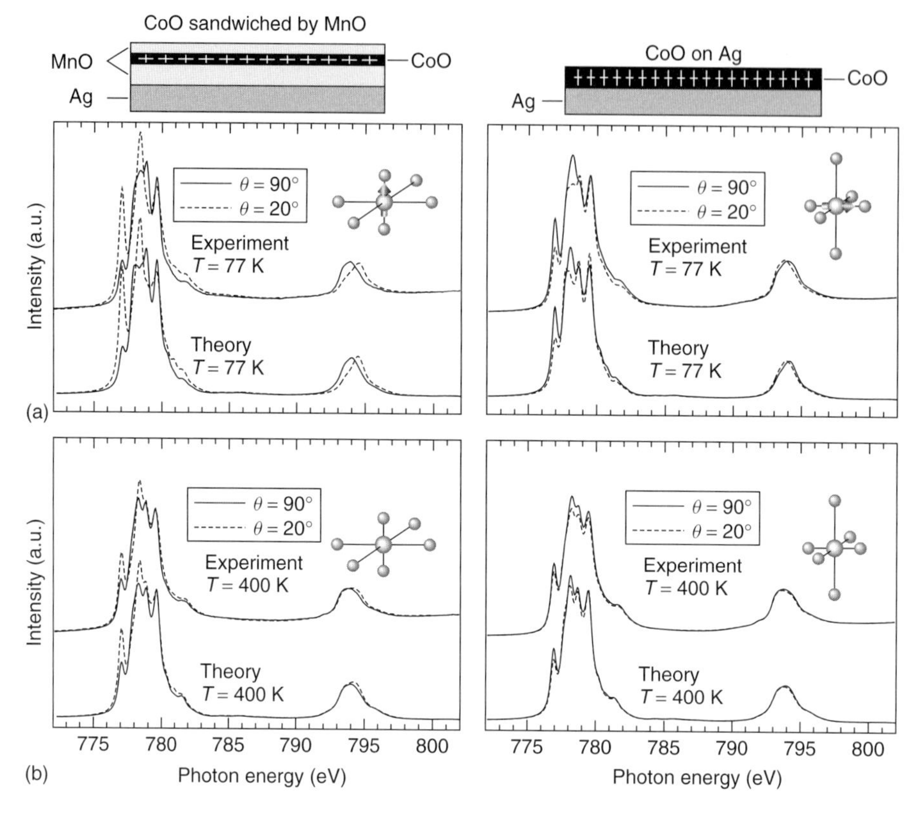

Figure 4.17 Experimental 2p-XAS spectra for an in-plane stretched CoO layer sandwiched between MnO layers and an in-plane compressed CoO layer grown on a silver substrate. The spectra in (b) were taken at 400 K in the paramagnetic phase and the spectra in (a) at 77 K in the magnetically ordered phase. The direction of the exchange field was determined from the best fit between theory and experiment to be perpendicular to the surface for the in-plane stretched film and parallel for the other. (Reprinted from [40].)

discussed earlier. On the other hand, the linear dichroism is seen to be very large if $\Delta_{t_{2g}}$ is varied. The t_{2g} shell is an open shell and the orbital occupation is anisotropic. In Figure 4.17, the simulated spectra have been added. For CoO sandwiched by MnO $\Delta_{t_{2g}} = -56$ meV and for CoO on Ag $\Delta_{t_{2g}} = 18$ meV gave the best fits. Because the spectra are not very sensitive to Δ_{e_g}, this parameter was not separately fitted, but taken to be $4\Delta_{t_{2g}}$. The size and signs of the splitting parameters is in qualitative agreement with the size and signs of the tetragonal distortion. For CoO sandwiched by MnO, the d_{xy} level is found to be the lowest and the d_{xz} and d_{yz} the highest levels, and for CoO on Ag the opposite holds. In addition, the absolute value of the splitting energy is largest for the most distorted layer.

From Figure 4.17, it is evident that the first peak of the L_3 edge is very distinct and has a high contrast. The origin of this peak is easily understood in an independent electron picture. The lowest possible excitation of a 2p electron is into the hole of

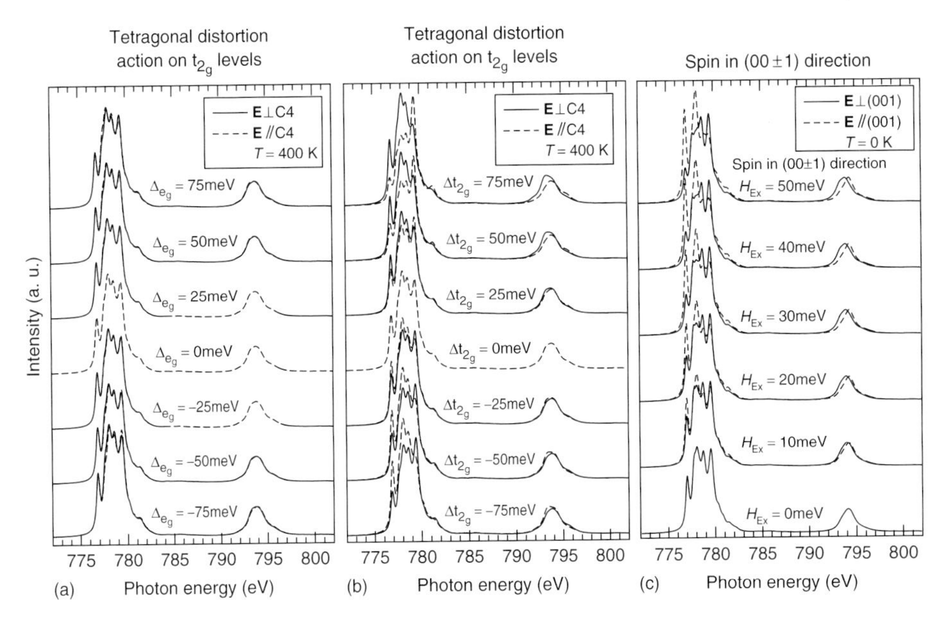

Figure 4.18 Theoretical 2p-XAS spectra for CoO as a function of the tetragonal LF parameters Δ_{eg} (a) and $\Delta_{t_{2g}}$ (b), and the exchange field (c). (Reprinted from [13].)

the t_{2g} level. The intensity of the transition is proportional to $|\langle p_i|\alpha|d_j\rangle|^2$, where i is x, y, or z, α is the polarization direction x, y, or z and j is xy, xz, or yz. From Table 4.2, we find that for every polarization direction q there is one t_{2g} orbital, which cannot be excited, that is, the d_{xy} level cannot be excited with z-polarized light, the d_{xz} level cannot be excited with y-polarized light, and the d_{yz} level cannot be excited with x-polarized light. So, if the hole is in the d_{xy} level there will be no intensity in the first peak for z-polarized light. In Figure 4.17, we notice that the peak at 777 eV is indeed reduced considerably for positive $\Delta_{t_{2g}}$ and $\mathbf{E}\|C_4$. However, the intensity decreases at a much slower pace with increasing LF parameter than might be anticipated on account of the above simple reasoning. After all, the intensity should be zero for $\Delta_{t_{2g}}$ exceeding $k_B T$. The explanation is that we have neglected SOC, which favors a spherical charge distribution and therefore delays the intensity decrease. Only when $\Delta_{t_{2g}} \gg \zeta$ the hole will be entirely in the d_{xy} orbital. For a tetragonal elongation, that is, a negative $\Delta_{t_{2g}}$, the hole is in the d_{xz} and d_{yz} levels.

Below T_N the magnetic moment distribution starts to become anisotropic due to collinear alignment and a magnetic component is expected to add to the linear dichroism. In Figure 4.18 also the calculated XMLD for cubic CoO at 0 K and an increasing exchange field is shown. The exchange field is taken to be parallel to the z-axis ([001]). Below 30 meV the XMLD scales with the size of the exchange field and above this value it starts to flatten. This behavior is radically different from NiO, for which the full effect is present at 0 K for an infinitesimally small exchange field. Again the spin–orbit interaction is at the basis of the far from abrupt changes in intensity, that is, it trims down the anisotropic charge distribution. The explanation is that SOC couples the $S = \frac{3}{2}$ state with the effective orbital momentum to a $\tilde{J} = \frac{1}{2}$ state. Doublets in cubic symmetry do not have an anisotropy for dipole transitions by symmetry constraints. The reason one can measure MLD at all is due to quantum mechanical mixing in of excited states through second-order perturbation in the exchange field. Therefore, the XMLD is zero for zero exchange field and grows only linearly with the size of the exchange field. In contrast to this, Ni^{2+} having a d^8 configuration shows its full XMLD effect as soon as the exchange field is larger than $k_B T$.

Also in the magnetic case the contrast for the first peak at 777 eV is very large. Like the tetrahedral LF, the exchange field generates a preferred orbital occupation. If the spins are for instance aligned in the z-direction, the spin–orbit interaction will align the orbital moment in the same direction. The orbital with a moment in the z-direction $d_1 = \sqrt{1/2}(-d_{xz} - id_{yz})$ will be coupled to the spin if it contains the hole. Consequently, the probability of exciting a 2p electron into this hole is reduced for $x-$ and $y-$ polarized light ($\mathbf{E} \perp [001]$), but not for z-polarized light ($\mathbf{E} \parallel [001]$). Similarly, if the spin is turned to an in-plane direction, for example [100] , the peak intensity will be reduced for y and z polarized light and not for x-polarized light, if we keep the coordinate system unchanged. The contrast of the first peak, therefore, is a measure of the spin direction. However, this contrast is also strongly affected by the LF and it is not easy to separate the two contributions. Comparing Figures 4.18 (b) and (c), a better criterion for the spin direction is the overall energy shift of the L_2 edge. For light polarized parallel to the spin direction, the energy position of the L_2 edge is lower than for light polarized perpendicular to the spin direction. The shifts caused by the LF are much smaller.

The low-temperature spectra of the CoO films are also shown in Figure 4.17. The temperature of 77 K is well below T_N of about 300 K and the spins are almost completely aligned. For the theoretical fits, the exchange field was taken to be equal to 12.6 meV, a value determined from neutron scattering experiments [41, 42] and the SOC and LF parameters were copied from the fits of the isotropic and high-temperature spectra, respectively. Hence, the only unknown parameter left is the orientation of the spin. The exchange field for CoO sandwiched between MnO, the layer under tensile stress, is found to be perpendicular to the layer and the exchange field for CoO on Ag, the slightly compressed layer, is in the plane of the layer. A tilt with respect to the film normal for CoO sandwiched between MnO or with respect to the plane of the film for CoO on Ag leads to a noticeable reduction of the linear dichroism. If preferential domain orientation would be the reason for the observed spin orientation, a tilt of 27° away from one of the $\langle 001 \rangle$ directions is expected. The nearly perfect perpendicular and parallel alignments suggest that a different explanation should be found.

In Figure 4.19 (a and b) the temperature dependence of the two characteristic features of the linearly polarized spectra, that is, the linear dichroism contrast of the first peak at 777 eV and the mean energy shift of the L_2 edge are shown. A clear kink is visible in the curves of the first quantity at about 300 K, which is T_N. Above this temperature the effect is different from zero, indicating an

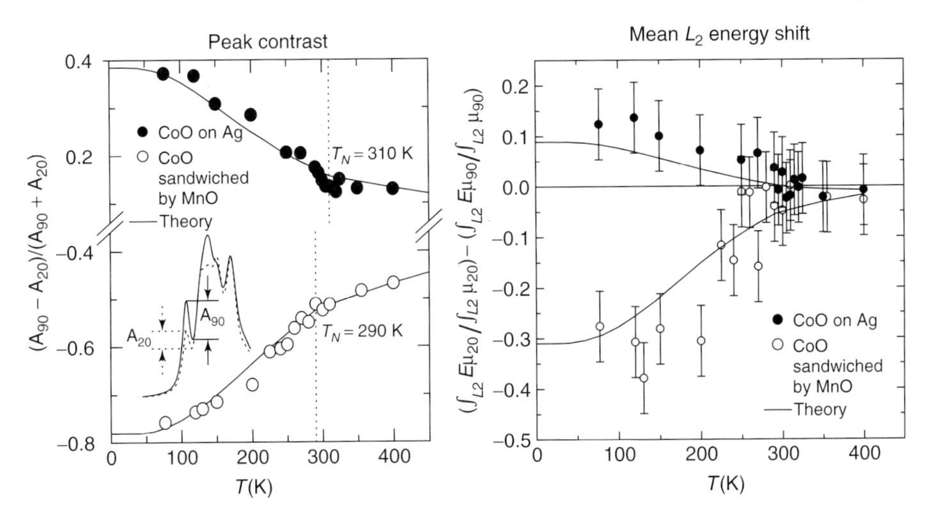

Figure 4.19 (a) Temperature dependence of the polarization peak contrast , which is the ratio of the difference and the sum of the the peak at 775 eV for polarization angles of $\theta = 20^\circ$ and $\theta = 90^\circ$. (b) Temperature dependence of the polarization-dependent energy shift of the L_2 edge. Filled and empty circles are data points and solid lines are theoretical simulations. (Reprinted from [40] and [13].)

LF-induced charge anisotropy, which is dependent on temperature because excited states split-off by the SOC are thermally accessible, as was already discussed for the isotropic spectrum. Below T_N the peak contrast increases beyond the LF value due to magnetic ordering. This figure nicely summarizes the analysis so far. As pointed out earlier, the mean energy shift of the L_2 edge (Figure 4.19) is mainly sensitive to the magnetic component of the anisotropy.

4.6.5
Spin Alignment in Strained CoO

The orientation of the spin was found to be in the plane of the CoO layer for compressive and perpendicular to the layer for tensile epitaxial strain. This can be understood qualitatively in the independent electron picture. Easiest to explain is the case of CoO sandwiched by MnO. The CoO distance is elongated in-plane and the d_{xy} level is lowered 56 meV in energy with respect to the d_{xz} and d_{yz} levels. The only hole in the t_{2g} level is in the twofold degenerate state, which was assigned an orbital momentum ± 1 in the z-direction by combining them into $d_1 = \sqrt{1/2}(-d_{xz} - id_{yz})$ and $d_{-1} = \sqrt{1/2}(d_{xz} - id_{yz})$. The electron spin wants to align with the orbital momentum in an antiparallel fashion. The spin, therefore, also wants to be in the z-direction, that is, perpendicular to the layer. Hence, the two spin-down electrons are in the d_{xy} and the d_1 orbitals and the hole is in the d_{-1} orbital.

The case of the compressively strained layer of CoO on Ag is more difficult to explain. The d_{xy} level now is higher in energy by 18 meV. If the spin–orbit interaction would have been small with respect to the LF splitting $\Delta_{t_{2g}}$, the hole would be in the d_{xy} level and the orbital momentum would have been completely quenched. However, in reality the spin–orbit interaction is considerably larger than the LF splitting. Under these circumstances, it is advantageous to choose a quantization axis which is in the plane of the sample, for instance the x-axis. A new orbital $d_{-1} = \sqrt{1/2}(d_{xz} - id_{yz})$ may be created with an orbital momentum of $-1\mu_B$. Putting the hole in this orbital, the LF stabilization does not acquire its maximum value, because there is only half a hole in the d_{xy} level instead of one, but this energy loss is more than compensated by the energy gain due to the SOC. In reality, one should realize that, as the SOC is more important than the noncubic distortions and the exchange fields, one has a ground-state hole density close to the one in cubic symmetry as depicted in Figure 4.3b, perturbed toward a hole carrying orbital momentum either in the z direction or in the plane of the thin film.

4.6.6
Electronic Stucture of Strained CoO

From the above-mentioned multiplet fits of the isotropic and linear polarized XAS spectra, a large number of parameters have been obtained and a detailed description of the electronic structure has become possible. The octahedral LF splitting between the t_{2g} and e_g levels is 1.1 eV. Covalency is of limited importance because there are 7.12 electrons on the cobalt ion, which is only slightly larger than the value of 7 for the nonbonded ion. In the independent electron (hole) picture, there is one hole in the t_{2g} level and two in the e_g level. The t_{2g} hole is threefold degenerate and the spin degeneracy is fourfold, because the total spin is $\frac{3}{2}$. Consequently, the ground state is overall 12-fold degenerate. This degeneracy is lifted by the 3d SOC, lower symmetry LF, and exchange fields. Because all three are of the same order of magnitude, perturbation theory cannot be applied. Nevertheless, we will start with the effect of SOC, which is in fact the largest interaction. In Figure 4.20a, the twelve spin–orbit split initial-state levels are shown as a function of the most important tetragonal LF splitting parameter $\Delta_{t_{2g}}$. In the middle, where this parameter is zero, we find the energy level diagram of cubic CoO. The spin–orbit interaction causes a splitting into a ground state with $\tilde{J} = \frac{1}{2}$, a first excited state with $\tilde{J} = \frac{3}{2}$, and a highest excited state with $\tilde{J} = \frac{5}{2}$. There is a small splitting of the latter state in a quartet and a doublet, which we will ignore. If we now go to the extreme limit of a large negative LF splitting, corresponding to tensile strain, the d_{xy} orbital is lowered in energy (see Figure 4.2) and the hole will be in the d_{xz} and d_{yz} orbitals. These can be combined into the d_1 and d_{-1} orbitals and the magnetic moment of these states will be parallel or antiparallel to the total spin moment of $\pm\frac{1}{2}$ or $\pm\frac{3}{2}$, giving four twofold degenerate orbitals going down with decreasing $\Delta_{t_{2g}}$. In the extreme case of $\Delta_{t_{2g}} \ll \zeta$ (far outside the graph), the hole would be in a Kramers doublet consisting of a d_1 orbital and magnetic spin momentum of $-\frac{3}{2}$ or a d_{-1} orbital and magnetic spin momentum of $\frac{3}{2}$. On the other extreme end of the graph, that is, large positive LF splitting, the hole is in the d_{xy} orbital and the orbital momentum is totally quenched. In Figure 4.20, the actual experimental values of the LF splitting

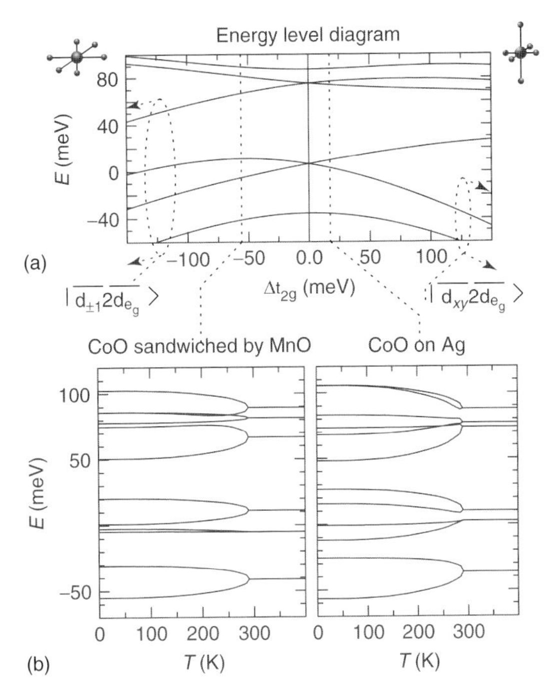

Figure 4.20 (a) Energy level diagram of the twelve lowest states of CoO as a function of the tetragonal distortion. (b) Energy level diagram as a function of the exchange field for the experimentally determined crystal fields for the in-plane stretched (left) and compressed (right) CoO layers. (Reprinted from [13].)

parameter for the strained samples discussed in the previous sections (−56 meV and +18 meV) have been indicated.

The effect of an exchange field on the energy level diagram is seen in Figure 4.20b. The energy is plotted as a function of temperature. The exchange field follows a $J = \frac{3}{2}$ Brillouin function.

4.6.7 Strain-Induced Linear Dichroism in MnO Layers

For the MnO studies [13, 29, 43], three types of samples were used, that is, (100 Å)MnO/Ag(001), (14 Å)MnO/(10 Å)CoO/(100 Å)MnO/Ag(001), and (22 Å) MnO/(90 Å)CoO/Ag(001). In fact, the second sample was the same as the one used in the CoO study, and the others were freshly made. The MnO and CoO measurements were all done during the same run.

The misfit of MnO on Ag is very large (8.8%) and the 100 Å thick layer of MnO on Ag is completely relaxed to its own lattice parameter. The XAS at two different orientations of the polarization vector are almost identical (Figure 4.21). Although the lattice is locally slightly distorted due to the AF ordering at low temperatures,

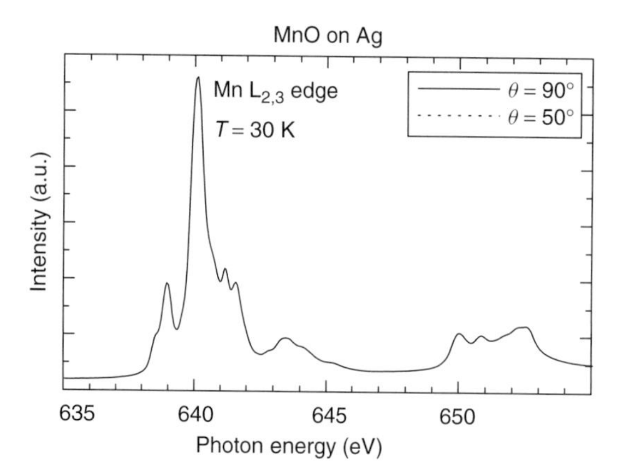

Figure 4.21 Mn $L_{2,3}$ XAS spectra of a 50-ML MnO film grown on Ag(001) and taken at 30 K. The spectra were taken with polarized light. θ is the angle between the electric vector and the surface normal. (Reprinted from [43].)

almost no linear dichroism can be detected, because the effects of the different domains present in the sample average out. Note that the sample is grown at a temperature exceeding T_N, where it still has the cubic structure. On cooling down, one gets an isotropic distribution of AF domains.

In the second sample, the strain is large (−4.2%) for the CoO layer and small but finite (about 0.5% as determined from the in-plane lattice parameter) for the MnO top layer. The strain in the MnO layer of the third sample, on the other hand, is expected to be much larger, of the order of 4% for coherent growth on top of an almost relaxed layer of CoO. The linear polarized XAS spectra for these samples are shown in Figure 4.22 for a temperature well above (200 K) and well below (77 K and 67 K) the magnetic ordering temperature of MnO. In both cases, a sizable linear dichroism is detected.

As in the previous sections, we first discuss the observed linear dichroism in the high-temperature paramagnetic state. MnO has a high spin d^5 configuration, that is, all d-levels will contain one electron and the total spin $S = \frac{5}{2}$. A tetragonal distortion cannot split the $^6A_1$ nondegenerate ground state, but as in the case of NiO, the final state will be affected. A 2p spin-down electron is excited into one of the half-filled t_{2g} or e_g orbitals. The corresponding peaks in the L_3 edge are the first two peaks at 639 and 640 eV. A tetragonal distortion will split these levels as shown in Figure 4.2. Because in both samples, the MnO is compressed in the plane of the layer, the highest t_{2g} level is d_{xy} and the highest e_g level is $d_{x^2-y^2}$. These levels can be accessed with x-polarized light ($\theta = 90^\circ$) but not with z-polarized light. The other levels can be accessed with z-polarized light, but the intensity is reduced for x-polarized light (see Table 4.2). This should result in an overall shift of the peaks to higher energy in going from z- to x-polarized light, which is indeed observed for the strongly distorted MnO on relaxed CoO sample. In the CoO sandwiched by MnO, the effect is small, in agreement with the fact that the MnO top layer is hardly distorted.

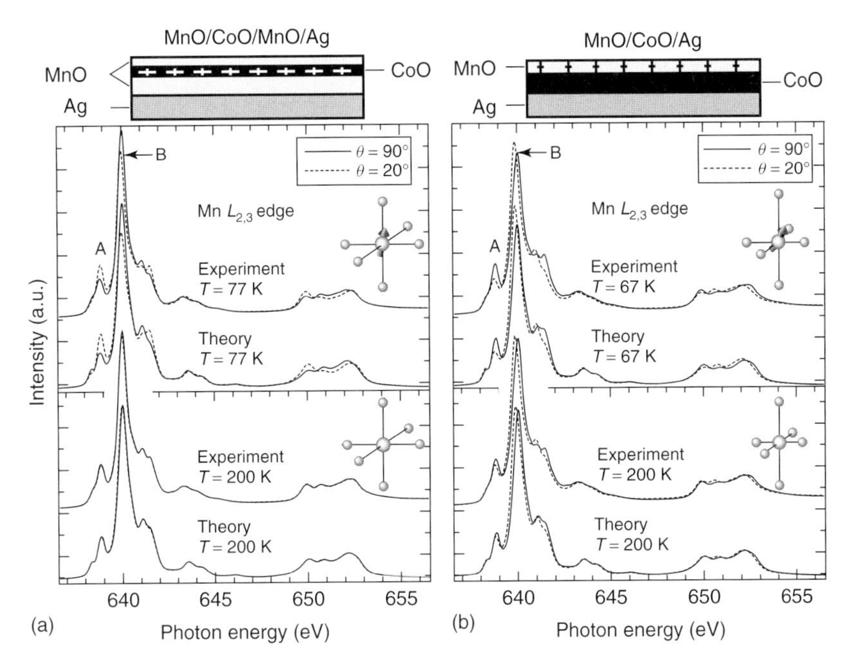

Figure 4.22 Experimental and calculated Mn $L_{2,3}$ XAS spectra of MnO in (a) (14 Å)MnO/(10 Å)CoO/(100 Å)MnO/Ag(001) and (b) (22 Å)MnO/(90 Å)CoO/Ag(001) for two different polarization directions, $\theta = 20^\circ$ and 90°. The spectra in the upper panels were taken below T_N, and the ones in the lower panels above T_N. (Reprinted from [43].)

To analyze the data more quantitatively, a multiplet LF or cluster calculation was performed similar to the one performed for NiO but now in the D_{4h} symmetry, thus including 2p − 3d and 3d − 3d electron correlation and O-2p Mn-3d LF effects and a Brillouin-type temperature-dependent exchange field. As far as possible, the parameters already known for bulk MnO [44] were used. The results of the calculations have been added to Figure 4.22. The LF parameters D_s and D_t giving the best fit are 9.3 and 2.6 meV for the CoO sandwiched between MnO layers and 48.6 and 11.1 meV for the MnO on a thick relaxed CoO layer. The LF parameter values confirm that the distortion for the two samples is an in-plane compression which is much stronger in the latter sample.

Below T_N, the (AF) magnetic ordering is an additional source of linear dichroism. In Figure 4.23 the polarization contrast, taken as the ratio of the difference and sum of the intensity of the first pronounced peak at 639 eV for $\theta = 20^\circ$ and 90°, for the strained and almost relaxed MnO layer is plotted as a function of temperature. The solid lines are the simulated curves, calculated under the assumption that the exchange field is parallel to the [112] direction for the relaxed MnO/CoO/MnO/Ag sample and in the [211] direction for the strained MnO/CoO/MnO sample. In NiO the reason for the occurrence of the effect was assumed to be the preferential growth of domains with a magnetic moment in a $\langle 11\bar{2}\rangle$ direction as close as possible to the surface normal, in order to adapt as much as possible to the tensile strain in

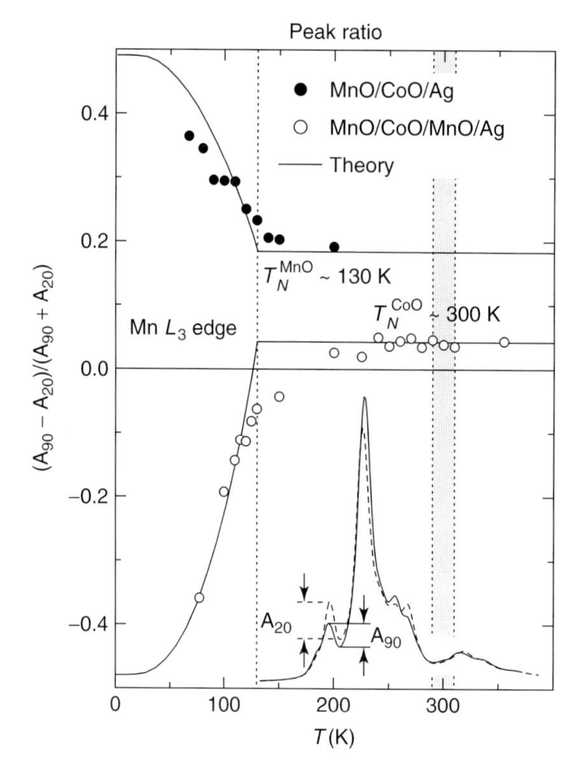

Figure 4.23 Temperature dependence of the polarization contrast of the L_3 peak at 639 eV. The solid lines are calculated curves, assuming that the Mn spins align with the Co spins of the underlying CoO layers. (Reprinted from [43].)

the sample. Applying the same reasoning to the two types of MnO samples, one would expect domains with the main component of the magnetic moment in the plane of the layer for both samples, because both MnO layers are compressed. This clearly is not observed. The low-temperature linear dichroism has the opposite sign for the two samples. A comparison with the simulated data suggests spin directions close to the [112] and [211] direction. The more or less out-of-plane [112] orientation of the spins for the weakly compressed MnO layer is not expected. Because the spins in the underlying CoO layer are oriented perpendicular to the layer surface, a likely explanation is that the spins are oriented by a strong interlayer exchange coupling. The evidence, therefore, is in favor of an AF exchange biasing effect of a CoO substrate. For the strongly compressed MnO layer, we cannot decide whether the close to in-plane orientation is due to the strain-induced preferential growth of domains having a close to in-plane spin direction or a strong interlayer exchange coupling, because both would lead to the observed in-plane spin direction.

4.7 Conclusions

The study of the magnetic structure of AF TM oxides has matured considerably since the introduction of polarized XAS techniques thanks to the development of reliable computational methods and programs to simulate the rather complicated highly resolved structures of the spectra. The first action one has to take experimentally is to make the magnetic effect visible to the XMLD technique. In bulk materials, the linear dichroism is usually averaged out. One possibility is to do the experiment on a single magnetically ordered domain using microscopic techniques. Another possibility, and that is the one we have chosen, is to deform the crystal by applying growth strain in order to get a nonuniform distribution of magnetically ordered domains. In simple cases, for instance for NiO grown on MgO, the measurement of a characteristic feature of the linear polarized XAS spectrum is sufficient to determine the magnitude and orientation of the exchange field. However, in the general case, one has to disentangle basically all components of the electronic strucure, such as the Coulomb interactions, LF, SOC, magnetic interactions, in order to single out the magnetic effects one is interested in. Even in NiO, despite its nondegenerate ground state, the linear dichroism in a strongly strained state is not only due to magnetic effects. A strategy to separate the natural and magnetic linear dichroism is to do temperature-dependent measurements. Above T_N, the magnetic linear dichroism must be absent, and below T_N the magnetic linear dichroism follows a Brillouin type temperature dependence. One of the findings in NiO is that the spins are oriented more or less in-plane or out-of-plane depending on whether the strain is tensile or compressive. Strain, therefore, can be used as a tool to orient the spins in AF films. In open shell systems like CoO, the finite orbital momentum of the d-electrons is an important additional factor. The dominating effect of strain now is that different orbitals are occupied depending on whether the strain is tensile or compressive. The orientation of the spin is determined by the coupling to the finite orbital momentum. Also in this case strain can be used to orient the spin, but only the mechanism is fundamentally different from the one in NiO and MnO. Finally, it was shown in the case of MnO grown on top of more or less strained CoO that exchange biasing can also happen for AF layers.

Appendix: Polarization and Spin Direction Dependence of the Linear Dichroism in Nonspherical Symmetry

It has recently been shown that the linear dichroism in the general case does not only depend on the direction of the polarization with respect to the spin but also on the direction of the spin with respect to the high symmetry local axes [36]. This can lead to quite different results and the sign of the MLD spectra can even be reversed. This is very important for single magnetic domain studies by means of microscopy techniques. In this chapter, we have studied only samples in which more than one domain was present and measurements were averaged over these domains.

This results in an average spin direction that is in a high symmetry direction. In these cases, the conductivity tensor ($\boldsymbol{\sigma}$), describing the absorption of the sample, becomes quite simple, as we show below.

In general $\boldsymbol{\sigma}$ of a magnetic system is a complex 3×3 non-Hermitian tensor which has energy-dependent nonorthogonal complex principle axes. In the wavelength region of visible light, one almost never encounters this general situation. Magnetooptical effects at these photon energies are generally quite small and symmetry distorting effects can be neglected or treated as a small perturbation. For X-ray absorption at the $L_{2,3}$ edge, one has a large 2p SOC and magnetooptical effects are generally quite strong. It then becomes important to treat the conductivity tensor in its full form.

In this appendix, we start from the general case and show how symmetries reduce the complexity of the tensor, especially in the case where several magnetic domains are present and only an average is measured. The general conductivity tensor can be written as

$$\boldsymbol{\sigma} = \begin{pmatrix} \sigma_{xx} & \sigma_{yx} & \sigma_{zx} \\ \sigma_{xy} & \sigma_{yy} & \sigma_{zy} \\ \sigma_{xz} & \sigma_{yz} & \sigma_{zz} \end{pmatrix} \tag{4.9}$$

In spherical symmetry with the spin in the (001) direction, this simplifies to

$$\boldsymbol{\sigma} = \begin{pmatrix} \boldsymbol{\sigma}^0 - \frac{1}{3}\boldsymbol{\sigma}^2 & \boldsymbol{\sigma}^1 & 0 \\ -\boldsymbol{\sigma}^1 & \boldsymbol{\sigma}^0 - \frac{1}{3}\boldsymbol{\sigma}^2 & 0 \\ 0 & 0 & \boldsymbol{\sigma}^0 + \frac{2}{3}\boldsymbol{\sigma}^2 \end{pmatrix} \tag{4.10}$$

where the isotropic spectrum is given by $\boldsymbol{\sigma}^0$, the circular dichroism by $\imath\boldsymbol{\sigma}^1$, and the linear dichroism by $\boldsymbol{\sigma}^2$. For a spin in an arbitrary direction given by θ and ϕ, the conductivity tensor should be rotated. With the aid of the rotation matrix

$$\mathbf{R}(\theta, \phi) = \begin{pmatrix} \cos(\theta)\cos(\phi) & -\sin(\phi) & \cos(\phi)\sin(\theta) \\ \cos(\theta)\sin(\phi) & \cos(\phi) & \sin(\theta)\sin(\phi) \\ -\sin(\theta) & 0 & \cos(\theta) \end{pmatrix} \tag{4.11}$$

the rotated conductivity tensor $\boldsymbol{\sigma}_{(x,y,z)} = \mathbf{R}.\boldsymbol{\sigma}.\mathbf{R}^{\mathrm{T}}$ becomes

$$\boldsymbol{\sigma}_{(x,y,z)} = \begin{pmatrix} \boldsymbol{\sigma}^0 + \boldsymbol{\sigma}^2(x^2 - \frac{1}{3}) & \boldsymbol{\sigma}^1 z + \boldsymbol{\sigma}^2 xy & -\boldsymbol{\sigma}^1 y + \boldsymbol{\sigma}^2 xz \\ -\boldsymbol{\sigma}^1 z + \boldsymbol{\sigma}^2 xy & \boldsymbol{\sigma}^0 + \boldsymbol{\sigma}^2(y^2 - \frac{1}{3}) & \boldsymbol{\sigma}^1 x + \boldsymbol{\sigma}^2 yz \\ \boldsymbol{\sigma}^1 y + \boldsymbol{\sigma}^2 xz & -\boldsymbol{\sigma}^1 x + \boldsymbol{\sigma}^2 yz & \boldsymbol{\sigma}^0 + \boldsymbol{\sigma}^2(z^2 - \frac{1}{3}) \end{pmatrix} \tag{4.12}$$

where x, y, and z are the spin directional cosines $\cos(\phi)\sin(\theta)$, $\sin(\phi)\sin(\theta)$, and $\cos(\theta)$.

To understand how this tensor behaves in lower symmetries, we expand the angular spin dependence of this tensor in terms of spherical harmonics

$$\boldsymbol{\sigma}(\theta, \phi) = \sum_{k=0}^{\infty} \sum_{m=-k}^{k} \begin{pmatrix} \boldsymbol{\sigma}_{xx}^{km} & \boldsymbol{\sigma}_{yx}^{km} & \boldsymbol{\sigma}_{zx}^{km} \\ \boldsymbol{\sigma}_{xy}^{km} & \boldsymbol{\sigma}_{yy}^{km} & \boldsymbol{\sigma}_{zy}^{km} \\ \boldsymbol{\sigma}_{xz}^{km} & \boldsymbol{\sigma}_{yz}^{km} & \boldsymbol{\sigma}_{zz}^{km} \end{pmatrix} Y_{km}(\theta, \phi) \tag{4.13}$$

As $\boldsymbol{\sigma}$ is an observable, it should be invariant under all symmetry operations of the local point-group. For a cubic symmetry, for example, a rotation of 90° around

z should bring $\boldsymbol{\sigma}$ back to itself. This will lead to relations between many expansion coefficients. Solving this set of equations (or using a reduction table that already contains these solutions) leads to the symmetry allowed tensor components. It may be noted that due to the local magnetic moment, one should in principle work with double or colored groups. We will normally label the symmetry as if the material is in the paramagnetic phase and then add a symmetry breaking magnetization.

In spherical symmetry, for example, there is no m-dependence and only expansion coefficents with $k = 0$, 1, or 2 remain. From formula 4.14, it then becomes clear why in spherical symmetry the labeling for the tensor elements has been chosen as $\boldsymbol{\sigma}^0$ for the isotropic spectra, $\boldsymbol{\sigma}^1$ for the circular dichroism, and $\boldsymbol{\sigma}^2$ for the linear dichroism. These are namely the expansion coefficients for $k = 0$, 1, and 2.

In cubic symmetry, where the x, y, and z are taken to be the C_4 axes, the conductivity tensor can be written as

$$\boldsymbol{\sigma}_{(x,y,z)} = \begin{pmatrix} \boldsymbol{\sigma}^0 + \boldsymbol{\sigma}^2_{eg}(x^2 - \frac{1}{3}) & \boldsymbol{\sigma}^1_{t1u} z + \boldsymbol{\sigma}^2_{t2g} xy & -\boldsymbol{\sigma}^1_{t1u} y + \boldsymbol{\sigma}^2_{t2g} xz \\ -\boldsymbol{\sigma}^1_{t1u} z + \boldsymbol{\sigma}^2_{t2g} xy & \boldsymbol{\sigma}^0 + \boldsymbol{\sigma}^2_{eg}(y^2 - \frac{1}{3}) & \boldsymbol{\sigma}^1_{t1u} x + \boldsymbol{\sigma}^2_{t2g} yz \\ \boldsymbol{\sigma}^1_{t1u} y + \boldsymbol{\sigma}^2_{t2g} xz & -\boldsymbol{\sigma}^1_{t1u} x + \boldsymbol{\sigma}^2_{t2g} yz & \boldsymbol{\sigma}^0 + \boldsymbol{\sigma}^2_{eg}(z^2 - \frac{1}{3}) \end{pmatrix} \tag{4.14}$$

The expansion in spherical harmonics was truncated at $k = 2$. In spherical symmetry, this truncation is exact, but in cubic symmetry it is not, although the deviation for several simple cases does not seem to be too big. One important observation is that the magnetocrystalline anisotropy that is, the dependence of the energy on the spin direction, only shows up at higher expansion orders in k. It can be measured with XAS, but in order to describe it one needs a notation that is a bit more involved than the one presented here.

In cubic symmetry, $\boldsymbol{\sigma}^2$ becomes different for the off diagonal part of the tensor ($\boldsymbol{\sigma}^2_{t2g}$) and for the diagonal part ($\boldsymbol{\sigma}^2_{eg}$). This leads to the effect that the angle of the spin with respect to the crystal axes determines if one measures $\boldsymbol{\sigma}^2_{eg}$ (spin parallel to a C_4 axes) or $\boldsymbol{\sigma}^2_{t2g}$ (spin parallel to the C_3 axes).

In tetragonal symmetry with the C_4 axis along the z axis and a C_2 axis along the x and y axis, the conductivity tensor becomes

$$\boldsymbol{\sigma}_{(x,y,z)} = \begin{pmatrix} \begin{matrix}\boldsymbol{\sigma}^0 + \frac{1}{2}\boldsymbol{\sigma}^2_{b_{1g}}(x^2 - y^2) \\ -\frac{1}{2}\boldsymbol{\sigma}^2_{a_{1g}}(z^2 - \frac{1}{3})\end{matrix} & \boldsymbol{\sigma}^1_{a_{2u}} z + \boldsymbol{\sigma}^2_{b_{2g}} xy & -\boldsymbol{\sigma}^1_{eu} y + \boldsymbol{\sigma}^2_{eg} xz \\ -\boldsymbol{\sigma}^1_{a_{2u}} z + \boldsymbol{\sigma}^2_{b_{2g}} xy & \begin{matrix}\boldsymbol{\sigma}^0 - \frac{1}{2}\boldsymbol{\sigma}^2_{b_{1g}}(x^2 - y^2) \\ -\frac{1}{2}\boldsymbol{\sigma}^2_{a_{1g}}(z^2 - \frac{1}{3})\end{matrix} & \boldsymbol{\sigma}^1_{eu} x + \boldsymbol{\sigma}^2_{eg} yz \\ \boldsymbol{\sigma}^1_{eu} y + \boldsymbol{\sigma}^2_{eg} xz & -\boldsymbol{\sigma}^1_{eu} x + \boldsymbol{\sigma}^2_{eg} yz & \boldsymbol{\sigma}^0_{zz} + \boldsymbol{\sigma}^2_{a_{1g}}(z^2 - \frac{1}{3}) \end{pmatrix} \tag{4.15}$$

This equation shows that in tetragonal symmetry also the circular dichroism becomes direction dependent.

For the measurements presented in the chapter, the general formalism is valid, but not needed to its full extent. All measurements have been done on layered multidomain samples, having an inequivalent out-of-film direction (z) and in-plane direction. If, for example, a single domain exists having its spin direction in the

$[xyz]$ direction, than domains with the spins oriented in the $[\pm x \pm y \pm z]$ direction and domains with the spin in the $[\pm y \pm x \pm z]$ direction are also present in equal quantities. The spectrum of a multidomain sample is the same as the spectrum of a sample with a conductivity tensor averaged over all the above-mentioned spin directions, which is

$$\boldsymbol{\sigma}^{D_{4h}}_{(x,y,z)} = \begin{pmatrix} \sigma^0 - \frac{1}{2}\sigma^2_{a_{1g}}(z^2 - \frac{1}{3}) & 0 & 0 \\ 0 & \sigma^0 - \frac{1}{2}\sigma^2_{a_{1g}}(z^2 - \frac{1}{3}) & 0 \\ 0 & 0 & \sigma^0_{zz} + \sigma^2_{a_{1g}}(z^2 - \frac{1}{3}) \end{pmatrix} \tag{4.16}$$

A measurement with the polarization given by $\varepsilon = (\varepsilon_x, \varepsilon_y, \varepsilon_z)$ then gives the intensity

$$\varepsilon^* \cdot \boldsymbol{\sigma}^{D_{4h}}_{(x,y,z)} \cdot \varepsilon = \sigma^0(\varepsilon_x^2 + \varepsilon_y^2) + \sigma^0_{zz}\varepsilon_z^2 + \sigma^2_{a_{1g}}\left(z^2 - \frac{1}{3}\right)\left(\varepsilon_z^2 - \frac{1}{3}\right) \tag{4.17}$$

which is the formula used in the current chapter to determine the spin direction in the measured films.

References

1. Finazzi, M., Duò, L., and Ciccacci, F. (2009) Magnetic properties of interfaces and multilayers based on thin antiferromagnetic oxide films. *Surf. Sci. Rep.*, **64**, 139.
2. Mott, N.F. (1949) The basis of the electron theory of metals, with special reference to the transition metals. *Proc. Phys. Soc. A*, **62**, 416.
3. Hubbard, J. (1963) Electron correlations in narrow energy bands. *Proc. Roy. Soc. London Ser. A*, **276**, 238.
4. Zaanen, J., Sawatzky, G.A., and Allen, J.W. (1985) Band gaps and electronic structure of transition-metal compounds. *Phys. Rev. Lett.*, **55**, 418.
5. Rogojanu, O.C. (2002) Stabilizing CrO by epitaxial growth. PhD thesis, University of Groningen, Groningen.
6. Bethe, H.A. (1929) Term seperation in crystals. *Ann. Phys.*, **3**, 133.
7. van Vleck, J.H. (1935) The group relation between the Mulliken and Slater-Pauling theories of valence. *J. Chem. Phys.*, **3**, 803.
8. Ballhausen, C.J. (1962) *Introduction to Ligand Field Theory*, McGraw-Hill, New York.
9. Cowan, R.D. (1981) *The Theory of Atomic Structure and Spectra*, University of California Press, Berkeley.
10. de Groot, F. and van der Laan, G. (1997) Collected works of Theo Thole: the spectroscopy papers. *J. Electron Spectrosc. and Relat. Phenom.*, **86**, 25.
11. Kanamori, J. (1963) Electron correlation and ferromagnetism of transition metals. *Prog. Theo. Phys.*, **30**, 275.
12. van der Marel, D. (1985) The electronic structure of embedded transition metal atoms. PhD thesis, Rijksuniversiteit te groningen, Groningen.
13. Haverkort, M.W. (2005) Spin and orbital degrees of freedom in transition metal oxides and oxide thin films studied by soft x-ray absorption spectroscopy. PhD thesis, Universität zu Köln, Cologne.
14. Bruno, P. (1989) Tight-binding approach to the orbital magnetic-moment and magnetocrystalline anisotropy of transition-metal monolayers. *Phys. Rev. B*, **39**, 865.
15. Abragam, A. and Bleaney, B. (1970) *Electron Paramagentic Resonance of Transition Ions*, Oxford University Press, Oxford.
16. Shull, C.G., Stauser, W.A., and Wollan, E.O. (1951) Neutron diffraction by

paramagnetic and antiferromagnetic substances. *Phys. Rev.*, **83**, 333.

17. Sänger, I., Pavlov, V.V., Bayer, M., and Fiebig, M. (2006) Distribution of antiferromagnetic spin and twin domains in NiO. *Phys. Rev. B*, **74**, 144401.
18. Goodwin, A.L., Tucker, M.G., Dove, M.T., and Keen, A. (2006) Magnetic structure of MnO at 10k from total neutron scattering data. *Phys. Rev. Letters*, **96**, 47209.
19. Jauch, W., Reehuis, M., Bleif, H.J., and Kubanek, F. (2001) Crystallographic symmetry and magnetic structure of CoO. *Phys. Rev. B*, **64**, 052102.
20. Roth, W.L. (1958) Multispin axis structures for antiferromagnets. *Phys. Rev.*, **111**, 772.
21. Nagamiya, T. and Motizuki, K. (1958) Theory of the magnetic scattering of neutrons by CoO. *Rev. Mod. Phys.*, **30**, 89.
22. van der Laan, G. and Thole, B.T. (1988) Local probe for spin-orbit interaction. *Phys. Rev. Letters*, **60**, 1977.
23. Siegmann, H.C. and Stöhr, J. (2006) *Magnetism*, Springer, Berlin, Heidelberg, New York.
24. Thole, B.T., van der Laan, G., and Sawatzky, G.A. (1985) Strong magnetic dichroism predicted in the $M_{4,5}$ x-ray absorption spectra of magnetic rare-earth materials. *Phys. Rev. Lett.*, **55**, 2086.
25. Frank, F.C. and van der Merwe, J.H. (1949) One dimensional dislocations I static theory and II misfitting monolayers and oriented overgrowth. *Proc. Royal Soc. (London)*, **A198**, 205, 216.
26. Matthews, J.W. and Blakeslee, A.E. (1974) Defects in epitaxial multilayers. *J. Cryst. Growth*, **27**, 118.
27. Fitzgerald, E.A. (1991) Dislocations in strained-layer epitaxy: theory, experiment, and applications. *Materials Science Reports*, **7**, 87.
28. Hull, D. and Bacon, D.J. (1984) *Introduction to Dislocations*, Butterworth and Heinemann, Oxford.
29. Csiszar, S.I. (2005) X-ray diffraction and X-ray absorption of strained CoO and MnO thin films. PhD thesis, University of Groningen, Groningen.
30. James, M.A. and Hibma, T. (1999) Thickness-dependent relaxation of NiO(001) overlayers om MgO(001) studied by x-ray diffraction. *Surf. Sci.*, **433-435**, 718.
31. Kaganer, V.M., Kohler, R., Schmidbauer, M., and Opitz, R. (1997) X-ray diffraction peaks due to misfit dislocations in heteroepitaxial structures. *Phys. Rev. B*, **55**, 1793.
32. Alders, D., Vogel, J., Levelut, C., Peacor, S.D., Sacchi, M., Hibma, T., Tjeng, L.H., Chen, C.T., van der Laan, G., Thole, B.T., and Sawatzky, G.A. (1995) Magnetic x-ray dichroism study of the nearest-neighbor spin-spin correlation function and long-range magnetic order parameter in antiferromagnetic NiO. *Europhys. Lett.*, **32**, 259.
33. Alders, D., Tjeng, L.H., Voogt, F.C., Hibma, T., Sawatzky, G.A., Chen, C.T., Vogel, J., Sacchi, M., and Iacobucci, S. (1998) Temperature and thickness dependence of magnetic moments in NiO epitaxial films. *Phys. Rev. B*, **57**, 11623.
34. Haverkort, M.W., Csiszar, S.I., Hu, Z., Altieri, S., Tanaka, A., Hsieh, H.H., Lin, H.-J., Chen, C.T., Hibma, T., and Tjeng, L.H. (2004) Magnetic versus crystal-field linear dichroism in NiO thin films. *Phys. Rev. B*, **69**, 020408.
35. van Veenendaal, M.A., Alders, D., and Sawatzky, G.A. (1995) Influence of superexchange on NI $2p$ x-ray absorption spectroscopy. *Phys. Rev. B*, **51**, 13966.
36. Arenholz, E., van der Laan, G., Chopdekar, R.V., and Suzuki, Y. (2007) Angle-dependent NI^{2+} x-ray magnetic linear dichroism: interfacial couplin revisited. *Phys. Rev. Letters*, **98**, 197201.
37. Altieri, S., Finazzi, M., Hsieh, H.H., Lin, H.-J., Chen, C.T., Hibma, T., Valeri, S., and Sawatzky, G.A. (2003) Magnetic dichroism and spin structure of antiferromagnetic NiO(001) films. *Phys. Rev. Lett.*, **91**, 137201.
38. Finazzi, M. and Altieri, S. (2003) Magnetic dipolar anisotropy in strained antiferromagnetic films. *Phys.Rev. B*, **68**, 54420.
39. Haverkort, M.W., Tanaka, A., Hu, Z., Hsieh, H.H., Lin, H.-J., Chen, C.T., and Tjeng, L.H. (2008) Low energy excitations in CoO studied by temperature

dependent x-ray absorption spectroscopy cond-mat 0806.3736.

40. Csiszar, S.I., Haverkort, M.W., Hu, Z., Tanaka, A., Hsieh, H.H., Lin, H.-J., Chen, C.T., Hibma, T., and Tjeng, L.H. (2005) Controling orbital occupation and spin orientation in CoO layers by strain. *Phys. Rev. Lett.*, **95**, 187205.

41. Tanaka, A. and Jo, T. (1992) Temperature dependence of 2*p*-core x-ray absorbtion spectra in 3*d* transition-metal compounds. *J. Phys. Soc. Japan*, **61**, 2040.

42. Rechtin, M.D. and Averbach, B.L. (1972) Short-range magnetic order in CoO. *Phys. Rev. B*, **5**, 2693.

43. Csiszar S.I., Haverkort, M.W., Hu, Z., Burnus, T., Tanaka, A., Hsieh, H.H., Lin, H.J., Chen, C.T., Cezar, J.C., Brookes, N.B., Hibma, T., and Tjeng, L.H. Aligning spins in antiferromagnetic films using antiferromagnets. *Phys. Rev. B*, cond-mat 0504520.

44. Tanaka, A. and Jo, T. (1994) Resonant 3*d*, 3*p* and 3*s* photoemission in transition metal oxides predicted at 2*p* threshold. *J. Phys. Soc. Jpn.*, **63**, 2788.

5
Exchange Bias by Antiferromagnetic Oxides

Marian Fecioru-Morariu, Ulrich Nowak, and Gernot Güntherodt

5.1
Introduction

For magnetic bilayers consisting of a ferromagnet (FM) in contact with an antiferromagnet (AFM), a shift of the hysteresis loop along the magnetic field axis can occur, which is called exchange bias (EB). This shift is observed after cooling the entire system in an external magnetic field below the Néel temperature, T_N, of the AFM in the case of a saturated FM (for a review see [1–4]). The role of the AFM is to provide at the interface a net magnetization which is stable or irreversible during reversal of the FM magnetization, consequently shifting the hysteresis loop. The key to understanding EB is to identify and explain this interface net magnetization and its stability. The exchange coupling at the interface between an FM and an AFM should obviously be very strong for periodic, uncompensated magnetic moments at a perfectly flat surface of the AFM. This ideal interface model by Meiklejohn and Bean [5] was confronted with the puzzling experimental facts that similar EB effects were observed for periodically uncompensated as well as compensated moments at the surface of the AFM and that the EB field was about a factor of 100 less than that calculated by the ideal interface model.

For a long time, the bulk AFMs were considered as systems with zero magnetization. In the search for the microscopic origin of EB a major advancement was made by the "random field model" proposed by Malozemoff [6–8]. Assuming compensated AFM spins and roughness at the interface, the FM–AFM exchange interaction is random in sign, leading to a random exchange field. As an outgrowth of this model, magnetic domains in the AFM were considered whose variable size due to exchange and anisotropy constants was assumed to account for the different strengths of the EB field.

In the approach of Malozemoff [6–8], a small net magnetization occurs at the FM/AFM interface due to its roughness, giving rise to domain walls in the AFM perpendicular to the FM/AFM interface during cooling in the presence of the saturated FM. However, the formation and the energetics of such domain walls as well as the stability of the interface magnetization have never been proven. In addition, the formation of domain walls perpendicular to the FM/AFM interface

Magnetic Properties of Antiferromagnetic Oxide Materials.
Edited by Lamberto Duò, Marco Finazzi, and Franco Ciccacci

ISBN: 978-3-527-40881-8

exclusively due to interface roughness is energetically unfavorable and therefore unlikely to occur. Therefore, other approaches have been developed, where a domain wall forms in the AFM parallel to the interface while the magnetization of the FM rotates [9, 10]. However, it was shown by Schulthess and Butler [11] that in this model, EB vanishes if the motion of the spins in the AFM is not restricted to a plane parallel to the film as was done in the work of Koon [10]. To obtain EB, Schulthess and Butler [11] assumed uncompensated AFM spins at the interface, but their occurrence and stability during a hysteresis loop are not explained, neither in their model nor in other similar models [12, 13].

In a recent experiment Miltényi *et al.* [14] showed that it is possible to strongly influence EB in Co/CoO bilayers by diluting the CoO AFM layer, that is, by inserting nonmagnetic substitutions ($Co_{1-x}Mg_xO$) or defects ($Co_{1-y}O$) not at the FM/AFM interface, but rather throughout the volume part of the AFM. In the same work it was shown that a corresponding theoretical model, the so-called "domain state (DS) model," based on Monte Carlo simulations shows a behavior very similar to the experimental results. According to these findings the observed EB has its origin in a DS in the whole AFM, which carries a net magnetization in the volume and also at the interface of the AFM. This DS occurs during cooling in an external magnetic field and is stabilized by nonmagnetic defects. Later on, it was shown that a variety of experimental facts associated with EB can be explained within this DS model [15–17]. The importance of defects for the EB effect is also confirmed by recent experiments on $Fe_{1-x}Zn_xF_2$/Co bilayers [18] and by investigations which showed that it is possible to modify EB by irradiating an FeNi/FeMn system using He ions in the presence of a magnetic field [19–21]. On the other hand, models which assume the formation of a domain wall parallel to the interface during field cooling cannot explain these findings. A domain wall parallel to the interface stores the energy for redirecting the FM magnetization like a spring that is wound up, and nonmagnetic defects will rather suppress this spring effect leading to a decrease of EB with increasing defect concentration. Further support for the relevance of domains in EB systems is given by a direct spectroscopic observation of AFM domains [22, 23].

In this chapter, we will present our work mainly on the model EB system Co/CoO, for which we have developed on experimental and theoretical grounds the "DS-model" – based on an imbalanced or noncompensating sublattice magnetization of the AFM. The motivation or advantage of using an antiferromagnetic oxide for EB, such as CoO, is twofold: on the one hand, from an experimental point of view the Néel temperature of CoO ($T_N = 293$ K) is slightly below the room temperature so that the field cooling is easily feasible. Also, the EB effect can be switched on and off easily with temperature. Moreover, nonmagnetic defects as pinning sites of AFM domain walls can be introduced in a controlled way into the fcc lattice by either substitution or overoxidation. On the other hand, from a theoretical point of view, the very high anisotropy of CoO allowed for a straightforward modeling in the framework of the Ising model. More generally, CoO and other AFM oxides may play an important role in future all-oxide spintronics devices. In fact, CoO was the material which led to the discovery of EB more than 50 years ago [5].

A recent review of the magnetic properties of antiferromagnetic oxide films at interfaces and in multilayers has been given by Finazzi *et al.* [24].

We want to demonstrate that this model can account for most of the salient experimental features. The crucial role is played by nonmagnetic volume defects in the AFM, giving rise to the formation of domain walls throughout the volume. The latter in turn are responsible for the occurrence of uncompensated moments in the AFM, which after field cooling give rise to the EB effect. The role of interface roughness for realizing EB [5–7] is revisited [25]. For a smooth interface, it will be shown that having defects only at the interface is not sufficient to yield EB. However, the interdiffusion of FM spins into the AFM will be shown to enhance the interface coupling. The latter increases both the exchange field provided by the FM during cooling and the action of the AFM upon the FM during hysteresis. In the limit of vanishing bulk AFM dilution our model approaches the assumptions of Malozemoff for his model. We show that our assumption of an Ising spin model for CoO is not a crucial constraint of the model [15]. In fact we show that the EB exhibits a maximum for intermediate anisotropies.

5.2 Theoretical Background

5.2.1 Diluted Antiferromagnets in a Magnetic Field

Considerable interest has been focused in recent years on the understanding of magnetic systems with quenched randomness. One prominent example is the diluted Ising antiferromagnet in an external magnetic field (DAFF), a model for antiferromagnetic materials with high anisotropy so that the Ising limit is justified, with part of the magnetic ions randomly substituted by nonmagnetic ones. Among the best experimental realizations is $Fe_{1-p}Zn_pF_2$, where FeF_2 is an AFM with very strong uniaxial anisotropy and the Zn atoms are the nonmagnetic defects. The DAFF is an ideal system to study typical properties of structurally disordered systems, as there are domains, metastability, and slow dynamics (for reviews on DAFF see [26, 27]). In addition, many of the findings of the DAFF are also relevant for the random field Ising model (RFIM), which has been shown to be in the same universality class [28–30].

The Hamiltonian of the DAFF can be written as

$$H = -J_{\mathrm{AFM}} \sum_{<i,j>} \varepsilon_i \varepsilon_j \sigma_i \sigma_j - B \sum_i \varepsilon_i \sigma_i \qquad (5.1)$$

with the AFM nearest-neighbor exchange constant $J_{\mathrm{AFM}} < 0$ and the Zeeman energy B describing the interaction with the magnetic field. The $\sigma_i = \pm 1$ are normalized Ising spin variables representing spins with atomic moment μ. A fraction p of sites is left without a magnetic moment ($\varepsilon_i = 0$, representing a defect) while the other sites carry a moment ($\varepsilon_i = 1$). The defect distribution is assumed to be random and quenched in time.

Let us first focus on the phase diagram of the three-dimensional DAFF, that is, on equilibrium properties. In zero field, the system undergoes a phase transition from the disordered, paramagnetic (PM) phase to the long-range ordered AFM phase at the dilution-dependent Néel temperature T_N as long as the dilution p is small enough, so that the lattice of occupied sites is above the percolation threshold. In the low-temperature region, for small magnetic fields, $B \ll J_{\mathrm{AFM}}$, the long-range ordered phase remains stable in three dimensions [31, 32] while for higher fields the DAFF develops a DS [33–35] with a spin-glass-like behavior. The reason for the domain formation was originally investigated by Imry and Ma for the RFIM [36]. Transferring the so-called Imry–Ma argument to the DAFF, the driving force for the domain formation is a statistical imbalance of the number of impurities of the two AFM sublattices within any *finite* region of the DAFF. This imbalance leads to a net magnetization within that region, which couples to the external field. A spin reversal of the region, that is, the creation of a domain, can hence lower the energy of the system. The necessary energy increase due to the formation of a domain wall can be minimized if the domain wall passes preferentially through nonmagnetic defects at a minimum cost of exchange energy. Hence, these domains have nontrivial shapes following from an energy optimization. They have been shown to have a fractal structure with a broad distribution of domain sizes and with scaling laws quantitatively deviating from the original Imry–Ma assumptions [37, 38].

A two-dimensional schematic spin configuration illustrating the Imry–Ma argument is shown in Figure 5.1. The black dots denote defects (nonmagnetic ions or vacancies) and the solid line surrounds a domain in which the staggered magnetization is reversed with respect to the background staggered magnetization outside this domain. The number of uncompensated spins of the domain is 3 and the number of broken bonds at the domain boundary is 5. Therefore, for $B > 5/3|J_{\mathrm{AFM}}|$ the shown spin configuration is stabilized by the field.

In small fields, the equilibrium phase of the three-dimensional DAFF is long-range ordered. However, if cooled in a field B below a certain (irreversibility)

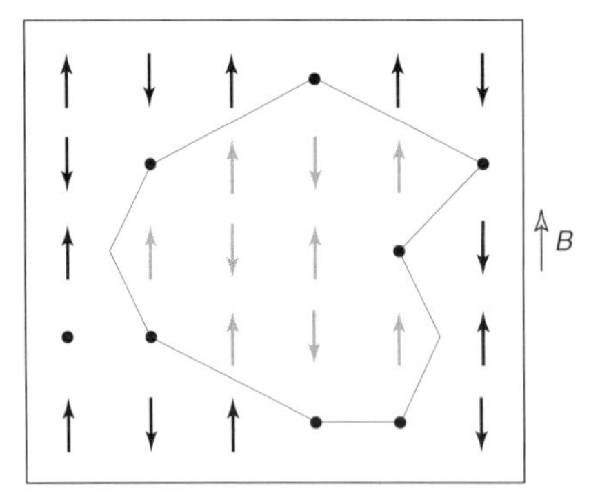

Figure 5.1 Schematic illustration of the Imry–Ma argument for domain formation.

temperature $T_i(B)$, the system usually develops metastable domains [39, 40]. The reason for this metastability is a strong pinning which hinders domain wall motion. These pinning effects are due to the dilution (random-bond pinning) as well as to the fact that a rough domain wall also carries magnetization in a DAFF (following again the Imry–Ma argument) which couples to the external field and hinders domain wall motion (random-field pinning) [41]. Consequently, after cooling the system from the PM phase within an external field, a DAFF freezes in a metastable DS which survives even after switching off the field, then leading to a remanent magnetization which decays extremely slow [42–44].

The origin of domain wall pinning is also illustrated in Figure 5.1. Consider the case that the field is lowered so that B changes from above to below $5/3|J_{\mathrm{AFM}}|$. Then, it is energetically favorable to turn the whole domain. But the corresponding dynamical process will be extremely slow, since the domain wall is pinned at the defects as well as between pairs of spins which are aligned with the field. Hence, during a movement of the domain-wall energy, barriers have to be overcome by thermal activation. This explains why a large domain in general will stay in a metastable state on exponentially long time scales. As a consequence, irreversibilities can be observed in a DAFF for temperatures lower than a critical value $T_i(B)$ [34, 45]. During field cooling from the PM state, the DAFF develops a DS with a certain surplus magnetization as compared to the long-range ordered state. The long-range ordered antiferromagnetic state develops during zero field cooling, and it remains stable even in a field for temperatures $T < T_c(B)$. This ordered state carries also a finite magnetization due to the response of the system to the field. The difference between these two magnetization curves is the irreversible surplus magnetization stemming from the DS of the DAFF.

In Figure 5.2, the considerations above are gathered in a schematic phase diagram of the three-dimensional DAFF (see also [33, 34, 37]). Shown are the equilibrium phases – long-range ordered (AFM) and PM – as well as the so-called irreversibility line $T_i(B)$. During field cooling below this line the system develops a frozen DS. Note that both the critical temperature $T_c(B)$ and the characteristic (irreversibility) temperature T_i are field dependent and that it is $T_i(B) > T_c(B)$, where both these temperatures approach the Néel temperature for a small magnetic field. The critical temperature is a decreasing function of the dilution p, so that the region of the

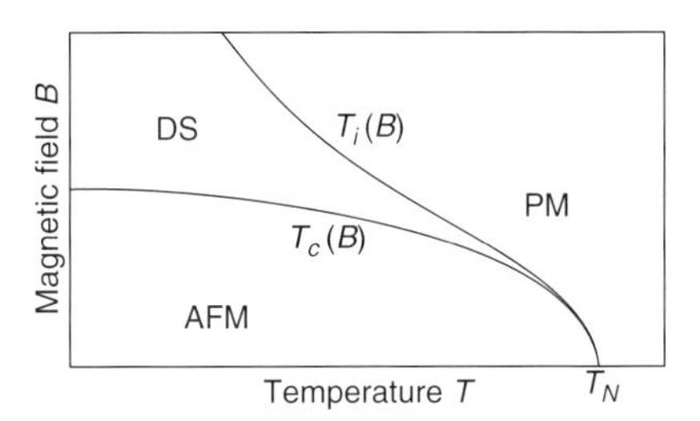

Figure 5.2 Schematic phase diagram of a three-dimensional DAFF in the temperature-field plane.

phase diagram where AFM long-range order can occur shrinks with increasing dilution. Below the percolation threshold no long-range order can occur.

In the following, we will argue that these well-established properties of the DAFF are the key to understanding EB. During preparation of an EB system, the AFM is usually cooled in an external magnetic field and additionally under the influence of an effective interface exchange field stemming from the FM magnetized to saturation. Hence, the AFM will develop a DS with an irreversible surplus magnetization similar to that of a DAFF after field cooling. This irreversible surplus magnetization then controls the EB.

5.2.2 Domain-State Model for Exchange Bias

The DS model for EB [14] consists of t_{FM} monolayers of FM and t_{AFM} monolayers of diluted AFM. The FM is exchange coupled to the topmost layer of the AFM. The geometry of the model is sketched in Figure 5.3. In order to account for finite rather than infinite crystalline anisotropies, both FM and AFM are described by a classical spin model with uniaxial anisotropy and Heisenberg exchange for nearest neighbors on a simple cubic lattice (exchange constants J_{FM} and J_{AFM} for the FM and the AFM, respectively, while J_{INT} stands for the exchange constant between FM and AFM). For simplicity we assume that the values of the magnetic moments of FM and AFM are identical (and included in the magnetic field energy $\mathbf{B}$). The Hamiltonian of the system is then

$$\begin{aligned} H = & -J_{\mathrm{FM}} \sum_{<i,j>} \mathbf{S}_i \cdot \mathbf{S}_j - \sum_i \left(d_z S_{iz}^2 + d_x S_{ix}^2 + \mathbf{S}_i \cdot \mathbf{B} \right) \\ & -J_{\mathrm{AFM}} \sum_{<i,j>} \varepsilon_i \varepsilon_j \boldsymbol{\sigma}_i \cdot \boldsymbol{\sigma}_j - \sum_i \varepsilon_i \left(k_z \sigma_{iz}^2 + \boldsymbol{\sigma}_i \cdot \mathbf{B} \right) \\ & -J_{\mathrm{INT}} \sum_{<i,j>} \varepsilon_j \mathbf{S}_i \cdot \boldsymbol{\sigma}_j \end{aligned} \tag{5.2}$$

where $\mathbf{S}_i$ denote normalized, classical spins (i.e., unit vectors) at sites of the FM layer and $\boldsymbol{\sigma}_i$ denote normalized spins at sites of the AFM. The first line of the

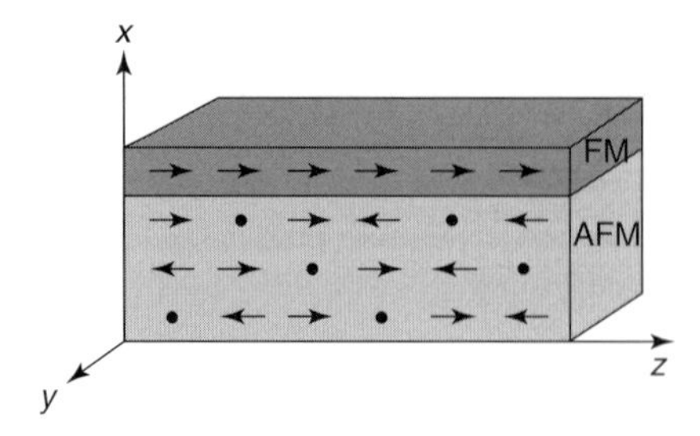

Figure 5.3 Sketch of the DS model with one FM layer and three diluted AFM layers. The dots mark defects. The easy axis of both FM and AFM is the z-axis.

Hamiltonian describes the energy of the FM with uniaxial anisotropy and the z-axis as its easy axis (anisotropy constant $d_z > 0$). The dipolar interaction is approximated in the model by an additional anisotropy term (anisotropy constant $d_x = -0.1 J_{\mathrm{FM}}$ in the present case) which includes the shape anisotropy, leading to a magnetization which is preferentially in the y–z-plane. The second line is the contribution from the AFM, once again with uniaxial anisotropy and its easy axis along the z-direction (anisotropy constant $k_z > 0$). The AFM is diluted, that is, a fraction p of sites is left without a magnetic moment ($\varepsilon_i = 0$) while the other sites carry a moment ($\varepsilon_i = 1$). The last term describes the interaction of the FM with the interface AFM monolayer.

Equation (5.2) suggests a simple ground-state argument for the strength of the EB field. Assuming that all spins in the FM remain parallel during field reversal and that some net magnetization of the interface layer of the AFM remains constant during the reversal of the FM, a simple calculation gives the usual estimate for the EB field B_{EB},

$$t_{\mathrm{FM}} B_{\mathrm{EB}} = J_{\mathrm{INT}} m_{\mathrm{INT}} \tag{5.3}$$

where m_{INT} is the stable part of the interface magnetization of the AFM (per spin) which is responsible for the EB and B_{EB} is the EB field (energy). For an ideal uncompensated and totally stable interface one would expect $m_{\mathrm{INT}} = 1$. As is well known, this estimate leads to a much too high bias field, while for an ideal compensated interface, on the other hand, one would expect $m_{\mathrm{INT}} = 0$ and hence, $B_{\mathrm{EB}} = 0$. Experimentally, however, often there is on the one hand no big difference between compensated and uncompensated interfaces and on the other hand, it is found that B_{EB} is much smaller than $J_{\mathrm{INT}}/t_{\mathrm{FM}}$, that is, about a few percent of it. The solution of this puzzle is that m_{INT} is neither constant during field reversal nor is it a simple known quantity [16, 17], and we will discuss this quantity and its origin later on.

5.2.3
Mean-Field Solution of the Domain-State Model

In the present subsection, we focus on the results obtained with a mean-field-type calculation for an EB system consisting of a FM film in contact with an AFM film having a large anisotropy like CoO or FeF_2, so that the AFM layer can be modeled as Ising spin system [46]. The AFM layer is magnetically diluted and different degrees of dilution are considered. Both the EB field B_{EB} and the coercive field B_c are calculated. The results are in qualitative agreement with those obtained previously with Monte Carlo simulations. For the coercivity we find, in agreement with experiments, a significant maximum of B_c around the blocking temperature at which B_{EB} vanishes. This maximum is analyzed in detail as a function of temperature for an ideal AFM layer as well as for different degrees of dilution.

Most of the results are obtained under the simplified assumption that the exchange interaction in the FM layer is very large. Under this assumption reversal of the FM layer is, to a good approximation, by coherent rotation, making it possible to obtain a number of analytic results for the fields B_c and B_{EB} from which more insight into the mechanisms of EB and coercivity is gained. From these analytic results we conclude, in agreement with earlier findings [47], that the appearance of EB and the enhancement of the coercivity in EB systems are of different origin, both can be present, and they are not correlated in contrast to what was found in [48]. The results are supplemented by a calculation in which a finite exchange interaction is assumed in the FM layer. In this case, the mean-field equations for the AFM layer have to be complemented by the corresponding equations for the FM spins. These equations are solved numerically for a certain set of parameters, showing that a finite FM exchange does not change the results significantly at low temperatures.

The Hamiltonian of the system is the sum of three terms, $H = H_{\text{FM}} + H_{\text{AFM}} + H_{\text{INT}}$, with

$$H_{\text{FM}} = -J_{\text{FM}} \sum_{<i,j>} \mathbf{S}_i \cdot \mathbf{S}_j - \sum_i \left(D S_{ix}^2 + S_{ix} B \right) \tag{5.4}$$

denoting the Heisenberg-Hamiltonian of the FM layer,

$$H_{\text{AFM}} = -J_{\text{AFM}} \sum_{<i,j>} \varepsilon_i \varepsilon_j \sigma_i \sigma_j - \sum_i (\varepsilon_i \sigma_i B) \tag{5.5}$$

denoting the Ising Hamilton of the AFM layer, and

$$H_{\text{INT}} = -J_{\text{INT}} \sum_{i \varepsilon (\text{INT})} \varepsilon_i S_{ix} \sigma_i \tag{5.6}$$

denoting the interaction energy of both, where the label i runs over all spins in the AFM interface layer, and we use the same index to enumerate the adjacent spin in the FM layer. Three-dimensional unit vectors $\mathbf{S}_i$ and Ising variables $\sigma_i = \pm 1$ denote spins in the FM and AFM layers, respectively. The magnetic field (with field energy B) is applied along the x-direction, which is parallel to the easy axis of the FM layer (anisotropy constant $D > 0$), whilst the z-direction is normal to the layers. We consider an AFM layer with quenched disorder, $\varepsilon_i = 0, 1$ with probabilities p and $1 - p$, respectively, also having its easy axis along the x-axis. Furthermore, we consider the nearest-neighbor interactions on a simple cubic lattice with exchange constants $J_{\text{FM}} > 0$ and $J_{\text{AFM}} < 0$ for the FM and the AFM, respectively, while J_{INT} stands for the exchange constant between FM and AFM.

Providing that the Curie temperature is very large as compared to all other relevant energies, reversal of the FM is, to a good approximation, by coherent rotation (see also [49]) and a temperature dependence of the magnetization in the FM can be neglected if we restrict ourselves to relatively low temperatures. This means that the magnetization of the FM layer acts as an external parameter on the AFM layer similar to the applied magnetic field. Thus, within this approximation, the free energy appears to be a function of the external field B and of the magnetization direction of the FM layer. For an Ising-type AFM it only depends on its x-component, S_x, and it is given by

$$F(S_x) = -NlDS_x^2 - NlBS_x - k_B T \text{Tr}\, e^{-\beta(H_{\text{AFM}} + H_{\text{INT}})} \tag{5.7}$$

where l is the number of FM monolayers, N the number of spins in an FM monolayer, and $\beta = k_B T$. In the following, we mainly concentrate on this approximation which enables us to derive a number of results analytically, resulting in a deeper understanding of the complex behavior of these exchange-coupled FM/AFM layers.

The free energy given in Eq. 5.7 is a function of S_x and the external field. It expresses the free energy of the AFM spin system keeping these parameters fixed. As a function of S_x, it may have different local minima for fixed external field B. These local minima give rise to (metastable) branches in the hysteresis loops. Indeed, starting with a sufficiently large field applied in the x-direction and $S_x = 1$ the effective field acting on the AFM interface layer is $B + J_{\text{INT}} S_x > 0$. Decreasing the external field, we follow the descending branch of the hysteresis loop. For small enough external fields, the free energy for $S_x = 1$ becomes metastable, that is, it will get larger than the free energy for $S_x = -1$. The system stays in the metastable minimum during further decrease of the external field until this minimum vanishes at a certain field B_-, at which the AFM switches from $S_x = 1$ to $S_x = -1$. This procedure is very similar to the Stoner–Wohlfarth scenario for switching of a magnetic particle. The only difference is that in the present case the FM magnetization is coupled to an AFM layer which may or may not be in thermal equilibrium depending on its dilution or disorder. A similar discussion holds for the ascending branch of the hysteresis loop. We now elucidate this scenario in more detail, restricting ourselves first to an ideal (undiluted) AFM layer.

The first derivative of the free energy with respect to S_x in the undiluted case ($\varepsilon_i = 1$ for all i) can be expressed in terms of the interface magnetization of the AFM layer,

$$F'(S_x) = -2NlDS_x - NlB - J_{\text{INT}} \sum_{i \in \text{INT}} \langle \sigma_i \rangle \tag{5.8}$$

and the second derivative of F is given by

$$F''(S_x) = -2NlD - \beta J_{\text{INT}}^2 \left\langle \left(\sum_{i \in \text{INT}} (\sigma_i - \langle \sigma_i \rangle) \right)^2 \right\rangle \tag{5.9}$$

where $\langle \sigma_i \rangle$ denotes a thermal average. This second derivative is negative showing that F' is decreasing monotonously. Thus in the interval $-1 < S_x < 1$, the first derivative of the free energy has one zero or no zeros. In the first case, F has two local minima at the end points, $S_x = -1, +1$ and one maximum in between, and in the second case only one global minimum at one of these end points, a situation similar to a simple Stoner–Wohlfarth scenario. Consequently, the fields B_- and B_+ at which the magnetization of the FM switches can be obtained in the same way, i. e. from $F' = 0$ at $S_x = 1$ and $S_x = -1$, respectively, and we obtain

$$\begin{aligned} B_- &= -2D - J_{\text{INT}} m_{\text{INT}} (B_-, S_x = 1)/l, \text{ and} \\ B_+ &= 2D - J_{\text{INT}} m_{\text{INT}} (B_+, S_x = -1)/l \end{aligned} \tag{5.10}$$

with $m_{\text{INT}} = \frac{1}{N} \sum_{i \in \text{INT}} \varepsilon_i \langle \sigma_i \rangle$ being the AFM interface magnetization. Note that these Equations are exact as far as the AFM layer is concerned, but fluctuations in the FM layer are neglected. In Eq. 5.10 the magnetizations of the AFM interface layer enter. These magnetizations have to be calculated with the fixed field applied and for both $S_x = 1$ and -1, respectively. For an ideal AFM layer, the induced magnetization in the AFM interface is completely reversible, and it is an odd function of the effective field. Therefore, the coercive field is given by $B_c = B_+ = -B_-$ and there is no EB.

If the coupling of the AFM layer to the FM layer is weak, the AFM interface magnetization can be linearized. Under this condition, explicit expressions for the coercive fields can be obtained from which a deeper insight into the complex behavior of these coupled systems can be gained. The linearized induced interface magnetization contains two parts, one which is proportional to the sum of the external field and the exchange field from the FM layer and the other which is proportional to the external field only. This second term arises indirectly from an exchange coupling of the AFM interface layer to its neighboring AFM-monolayer which only sees the applied field resulting in a term linear in the external field. However, its contribution to the interface magnetization is very small (and absent for an AFM monolayer) and will be shown to be negligible for moderate values of J_{INT}, resulting in the following approximate expression for m_{INT} in the linear regime,

$$m_{\mathrm{INT}} = \chi_{\mathrm{AF}}^{(1)} \left(B + J_{\mathrm{INT}} S_x\right) \tag{5.11}$$

Thus, the AFM interface magnetization as a function of B contains two branches during a hysteresis cycle, in the case of $J_{\mathrm{INT}} > 0$: an upper branch for $S_x = 1$ when reducing the external field and a lower branch after switching of the FM layer from $S_x = 1$ to -1 when increasing the field again. For $J_{\mathrm{INT}} < 0$ the behavior is reversed.

Within this linear approximation an explicit Equation for the coercive field B_c can be obtained from Eqs. 5.10 and 5.11,

$$B_c = \frac{2D + J_{\mathrm{INT}}^2 \chi_{\mathrm{AF}}^{(1)}/l}{1 + J_{\mathrm{INT}} \chi_{\mathrm{AF}}^{(1)}/l} \tag{5.12}$$

where the susceptibility entering this equation is obtained from the usual mean-field equations. Neglecting the influence of the homogeneous magnetization of the neighboring AFM layer on the AFM interface layer, this susceptibility is given by

$$\chi_{\mathrm{AF}}^{(1)} = \frac{\beta}{\cos h^2 \left(\beta J_{\mathrm{AF}} \left(z m_s^{(1)} + m_s^{(2)}\right)\right) + \beta J_{\mathrm{AF}} z} \tag{5.13}$$

In these equations the staggered magnetizations for $B = 0, m^{(k)}{}_s$, enter where k labels the AFM layers ($k = 1$ being the AFM interface layer). These quantities have to be obtained numerically from the self-consistency equations. Equation 5.12 nicely illustrates that the coercive field depends on two contributions, one coming from the FM itself (2D) and one from the interaction with the AFM. The denominator depends on the sign of the interface coupling.

It is possible to generalize the present approach to systems in which the FM does not have an infinite exchange interaction, as it has been assumed up to now, but a finite one. In this case one has to consider, in addition to the mean-field equations for the AFM, mean-field equations for the spins in the FM layer which in [46] were solved numerically for a single FM monolayer ($l = 1$). In previous Monte Carlo simulations [13, 15], it was shown that in exchange-coupled FM/AFM multilayer, magnetic dilution leads to a stabilization of domains in the AFM that carry a net magnetization. This magnetization is frozen at low temperatures leading to a frozen exchange field in the FM layer and thus to EB. These frozen domains depend on the history, which means that one has to specify the way the low-temperature state in which the hysteresis loop is calculated (or measured in experiments) is reached. Thus, in the diluted case it is necessary to follow exactly the same procedure as it is done in experiments: one starts at a temperature well above T_N with a fully magnetized FM layer (sometimes also with an applied field parallel to this magnetization) and cools the system slowly down to temperatures below T_N. This cooling process has previously been carried out with Monte Carlo simulations which mimic a dynamical process. However, the details of the dynamical aspect are less important. Rather these calculations intend to find local free energy minima. Therefore, this cooling procedure can alternatively also be carried out within local mean-field theory, in which the local mean-field equations, which in the present case are given by

$$m_i = \varepsilon_i \tan h \left[\beta \left(-J_{\mathrm{AFM}} \sum_j \varepsilon_j m_j + J_{\mathrm{INT}} S_x + B \right) \right] \tag{5.14}$$

are iterated at a fixed temperature until a (metastable) self-consistent solution is obtained. Then, the temperature is lowered by a small amount and a new iteration process is started with the previously obtained values of the magnetization as initial conditions. This procedure can be continued until the final temperature at which the hysteresis curve is going to be calculated is reached. This approach has been applied successfully to random field systems in connection with irreversibilities and frozen states [50].

It is important to note that the corresponding mean-field energy also satisfies Eq. 5.8 where at the right-hand side the local magnetization has to be replaced by the local magnetization in the mean-field approximation, that is,

$$F_{\mathrm{MF}}{}' (S_x) = -2NlDS_\chi - NlB - J_{\mathrm{INT}} \sum_{i \in \mathrm{INT}} \varepsilon_i \langle \sigma_i \rangle \tag{5.15}$$

Cooling the system as described above means that one stays in local minima of the free energy.

In [46], hysteresis loops were calculated in a similar way by changing the external field in small steps, solving the mean-field equations numerically in each step by iteration. The AFM interface magnetization $m_{\mathrm{INT}}(B)$ was recorded for each value of B for the descending branch $S_x = 1$ and the ascending branch $S_x = -1$, respectively, and the fields B_- and B_+ at which the magnetization of the FM switches were then obtained as before, that is, from $F' = 0$ at $S_x = 1$ and -1, respectively, that is from the implicit Equation

$$B_\pm = \pm 2D - J_{\mathrm{INT}} m_{\mathrm{INT}} \left(B_\pm, S_x = \mp 1 \right) / l \tag{5.16}$$

In these calculations, a system of size 96 × 96 lattice sites per monolayer was investigated with periodic boundary conditions in the plane. The fields $B_{\pm}$ depend on the disorder configuration. Therefore, for each degree of dilution, 16 different realizations of the disorder were generated and the fields obtained for each configuration were averaged. The AFM interface magnetization m_{INT} obtained in this way can be decomposed into a sum of two terms, an irreversible (metastable) part m_{irr}, which does not change when going through the hysteresis loop at low temperatures, and a part, which follows the field, m_{rev}, having two branches for $S_x = 1$ and -1, respectively. Because of this frozen interface magnetization the fields $B_{\pm}$ are no longer equal with the consequence that the system shows EB.

Before we discuss these numerical results further, we can go, similar to the undiluted case, one step further noting that, in the limit of the small effective fields, m_{rev} can be linearized,

$$m_{\text{rev}} = \chi_{\text{AF}}^{(1)} J_{\text{INT}} S_x + \chi_{\text{AF}}^{(2)} B \qquad (5.17)$$

The first term corresponds to the response to the exchange field, while the second term is the response to the applied field. Only for the case of an undiluted AFM monolayer the induced AFM magnetization is strictly proportional to the effective field $J_{\text{INT}}\, S_x + B$, in which case both susceptibilities in Eq. 5.17 are equal.

Thus, within linear approximation, the AFM interface magnetization, which is shifted by m_{irr}, contains two branches during a hysteresis cycle, in the case of $J_{\text{INT}} > 0$: an upper branch for $S_x = 1$ when reducing the external field and a lower branch after switching of the FM layer from $S_x = 1$ to -1 when increasing the field again. For $J_{\text{INT}} < 0$ a reversed behavior is observed. This scenario is exactly what was also obtained in the earlier Monte Carlo simulations, see Figures 5.4 and 5.5 in [16]. Within this linear approximation explicit Equations for the fields B_+ and B_- can be obtained from Eq. 5.16,

$$B_{\pm} = \frac{\pm 2D - J_{\text{INT}} m_{\text{irr}}/l \pm J_{\text{INT}}^2 \chi_{\text{AF}}^{(1)}/l}{1 + J_{\text{INT}} \chi_{\text{AF}}^{(2)}/l} \qquad (5.18)$$

resulting in the following expressions for the coercive field and the bias field, respectively,

$$B_{\text{EB}} = \frac{1}{2}(B_+ + B_-) = \frac{-J_{\text{INT}} m_{\text{irr}}/l}{1 + J_{\text{INT}} \chi_{\text{AF}}^{(2)}/l} \qquad (5.19)$$

$$B_{\text{c}} = \frac{1}{2}(B_+ - B_-) = \frac{2D + J_{\text{INT}}^2 \chi_{\text{AF}}^{(1)}/l}{1 + J_{\text{INT}} \chi_{\text{AF}}^{(2)}/l} \qquad (5.20)$$

From Eq. 5.19, it can be concluded that EB only occurs if the interface magnetization contains a part which is frozen during field reversal. Even though this is rather obvious, it is seen here most clearly. Note that the second term in the denominator of Eq. 5.19 is missing in the usual estimate for the bias field (see Eq. 5.3). This is consistent with the current approach, since in the usual estimate a part of an AFM interface magnetization which follows the external field is generally not considered.

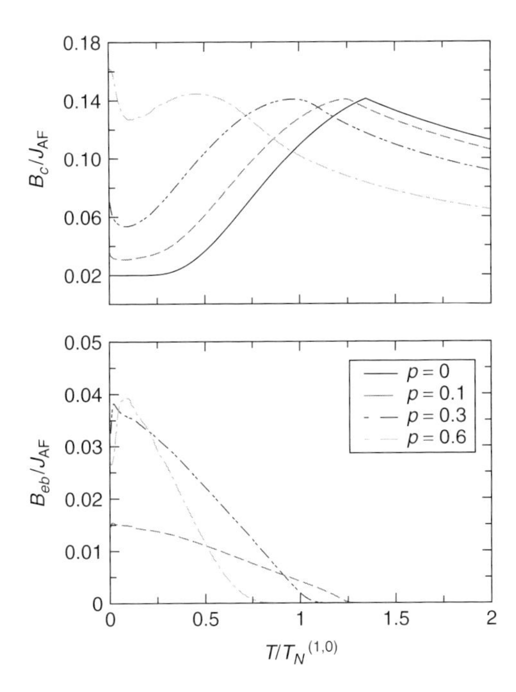

Figure 5.4 Coercivity and bias field for different dilutions p of the AFM as a function of reduced temperature. $L = 3$ and $J_{INT} = -J_{AF}$. Figure taken from [46].

However, the linear approximation is expected to be valid for not-too-large interface coupling just as in the undiluted case.

These analytic results obtained in the linear approximation contribute to a deeper understanding of bias and coercivity. For a numerical calculation of these quantities, however, it is more convenient to start with the calculation of the interface magnetization m_{INT} during a hysteresis cycle in the way described above. From this quantity the fields $B_{\pm}$ can be obtained directly using Eq. 5.16, without invoking the linear approximation. Corresponding results for the bias field and the coercive field for different dilutions of the AFM layer are shown in Figure 5.4 for negative interface coupling and in Figure 5.5 for positive coupling.

For zero dilution the coercive fields have a cusp at the onset of the AFM order in the AFM layer. For small number of AFM layers, L, this is also the maximum of the coercivity while for larger L (not shown) the maximum is slightly shifted to smaller temperatures [46]. In the diluted cases the cusp is smeared out. The bias field shows a nearly linear decrease as a function of temperature, in agreement with previous investigations both experimentally and theoretically. It vanishes roughly at that temperature at which the coercive field has its maximum. Note, however, that for stronger dilution there is a pronounced shift of this maximum to temperatures lower than the temperature at which the bias vanishes.

The coercive fields show an interesting behavior at low temperatures, where in the diluted case a dip occurs for FM interface coupling and an upward turn

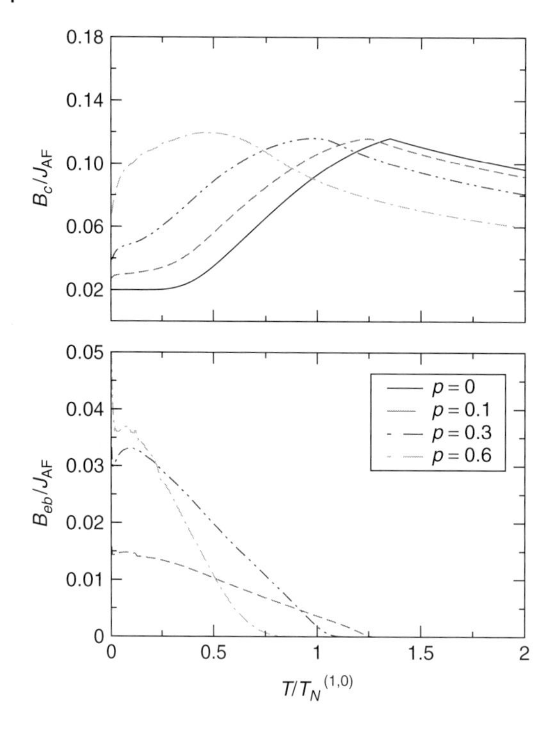

Figure 5.5 Coercivity and bias field for different dilutions p of the AFM as a function of reduced temperature. $L = 3$ and $J_{INT} = J_{AF}$. Figure taken from [46].

for the antiferromagnetically coupled FM/AFM layers. This feature has its origin in nearly loose spins in the AFM interface layer which contribute significantly to the susceptibility $\chi_{AF}^{(2)}$ at low temperatures, leading to an enhancement or a depression of the coercive field due to the changing sign of J_{INT} in the denominator in Eq. 5.20. This sensitivity of the coercive field with respect to the sign of J_{INT} has an important consequence for the possibility of an experimental determination of the sign of the exchange interaction. In [51], the measured coercive fields of IrMn multilayers show an upward shift very similar to our findings. Thus, it is tempting to conclude that the exchange interaction at the FM/AFM interface in these layers is antiferromagnetic.

5.3 Experiments

5.3.1 Antiferromagnetic Oxides: CoO, $Co_{1-y}O$, $Co_{1-x}Mg_xO$

Most of the reported investigations concerning EB with AFM oxides involve the monoxides CoO, NiO, and FeO. Since the bulk T_N of CoO is about room temperature, and its magnetocrystalline anisotropy is high, the AFM CoO has become one of the most investigated model systems for EB since comprehensive magnetic characterization can be conveniently performed. In the following, we will describe the EB properties of the AFM CoO.

The AFM CoO has an fcc structure in the PM phase above T_N, with a slight tetragonal contraction along [100] below T_N [52]. Bulk CoO has a $T_N = 293$ K. The AFM ordering below T_N is characterized by ferromagnetically aligned moments in (111) planes with the moments in adjacent (111) planes being aligned antiparallel, that is, the spins of next nearest neighbors along [100] are antiparallel. The AFM alignment results from the superexchange coupling between the cobalt cations via the intervening p orbitals of the oxygen ions.

The EB samples that we will discuss were prepared by molecular beam epitaxy (MBE). (0001)-oriented single crystalline sapphire (Al_2O_3) was used as the substrate for Co/CoO film deposition. On the cleaned sapphire substrate, a 6-nm-thick Co layer was first deposited by electron-beam evaporation at a temperature of $T_{Co} = 575$ K and rate of 0.2 nm min^{-1}. The Co layer was subsequently annealed at a temperature of $T = 775$ K for 10 minutes. The pressure during evaporation was better than 5×10^{-9} mbar. For all samples, a 0.4-nm-thick CoO film was grown on top of the Co layer at a temperature of $T = 350$ K and at a rate of 0.3 nm min^{-1} by evaporating Co in an oxygen pressure of $p(O_2) = 3.3 \times 10^{-7}$ mbar. This ensures that all samples have an identical FM/AFM interface at the lowest possible defect concentration, independent of the dilution of the subsequently deposited AFM layer. The layered sample structure is schematically illustrated in the inset of Figure 5.8a.

The AFM CoO allows for two different ways to introduce nonmagnetic defects. The first type of defect can be formed with nonmagnetic MgO since it has the same crystal structure as CoO and only a 1.1% lattice mismatch. The substitution of Co atoms by nonmagnetic Mg atoms results in the diluted AFM $Co_{1-x}Mg_xO$. Second, Co can oxidize in any stoichiometry ($Co_{1-y}O$) ranging from CoO to Co_3O_4 ($y = 0.25$) [53]. Co_3O_4 has a $T_N = 33$ K. This means that because of Co deficiency or, in other words, because of overoxidation of CoO, defects can be created in the magnetic structure.

On top of the CoO interface layer two different sets of diluted AFM layers were deposited. In a first set of samples, CoO was diluted by nonmagnetic MgO forming $Co_{1-x}Mg_xO$. Co and MgO were co-evaporated in an oxygen atmosphere of $p(O_2) = 3.3 \times 10^{-7}$ mbar at a substrate temperature of $T_{oxide} = 350$ K and a deposition rate of 0.3 nm min^{-1}. The Mg concentration was varied between $x = 0$ and 1, while the diluted AFM layer thickness was kept constant at 20 nm. For a second set of samples, nonmagnetic defects were introduced by overoxidation of CoO, yielding Co-deficient $Co_{1-y}O$. The Co deficiency y was controlled by varying the partial pressure of oxygen during evaporation between $p(O_2) = 3.3 \times 10^{-7}$ mbar and 1.0×10^{-5} mbar, while growth temperature, growth rate, and AFM layer thickness were all identical to those of the first set of samples.

All samples were characterized *in situ* by reflection high energy electron diffraction (RHEED) and low energy electron diffraction (LEED). The RHEED patterns of the individual layers which the samples consist of are depicted in Figure 5.6a–c.

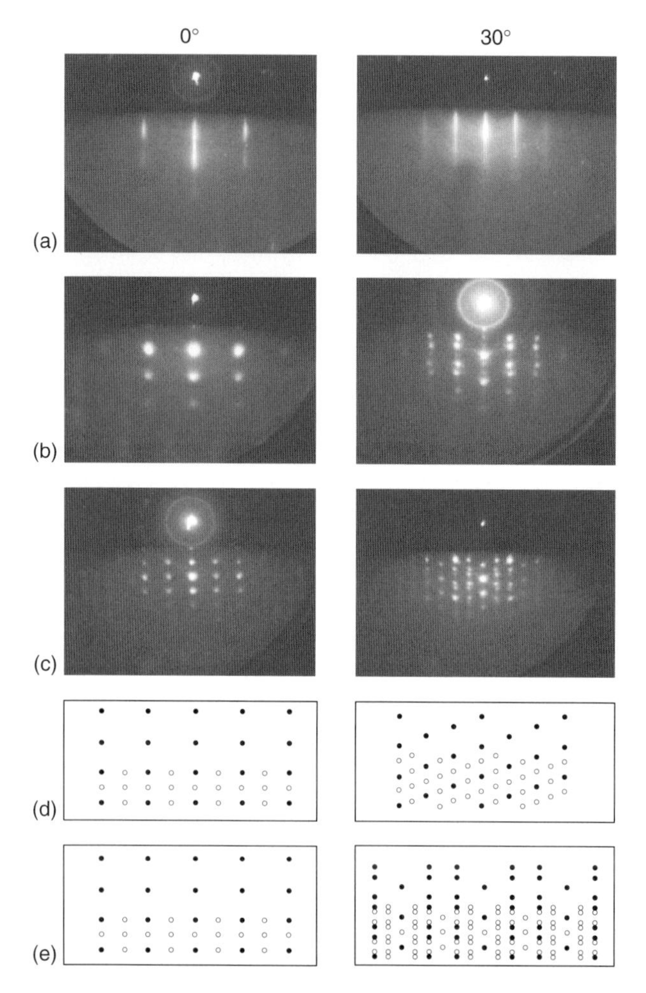

Figure 5.6 RHEED images of the (a) 6-nm Co layer on sapphire, and two 20-nm $Co_{1-y}O$ layers prepared at (b) $p(O_2) = 3.3 \times 10^{-7}$ mbar and at (c) $p(O_2) = 1.0 \times 10^{-5}$ mbar on a 6-nm-thick Co layer. (d) and (e): Simulated reflections of the diffraction patterns of $Co_{1-y}O$ (111); (d) without twins and (e) with 60° in-plane twins. Only solid dots fulfill the diffraction condition for the CoO fcc lattice. The two vertical panels show the patterns for 0° and 30° in-plane orientation of the incident electron beam relative to the sapphire $[\overline{11}\overline{2}0]$ axis. Figure taken from [17].

The left panels show the diffraction patterns for the electron beam incident parallel to the $[\overline{1}\overline{1}20]$ direction (0°) of the (0001)-oriented sapphire substrate and the right panels, for the beam parallel to the $[\overline{1}010]$ direction (30°). For the 6-nm-thick Co layer on the sapphire substrate in Figure 5.6a, we observe diffraction from a two-dimensional surface. Hence, Co grows epitaxially in either fcc (111) or in hcp (0001) orientation, which differ only in their so-called ABAB or ABCABC stacking order, respectively, along the surface normal [54]. Additional LEED investigations (not shown) clearly reveal the corresponding sixfold symmetry.

RHEED images of the 20-nm-thick $Co_{1-y}O$ layers are shown in Figure 5.6b,c, grown at oxygen pressures of $p(O_2) = 3.3 \times 10^{-7}$ mbar and 1.0×10^{-5} mbar, respectively. All diffraction patterns from the AFM layers show a transmission image, that is, diffraction from a rough surface with islands [55]. In order to explain the observed RHEED patterns, a (111) orientation of fcc $Co_{1-y}O$ is assumed. The calculated diffraction patterns are shown in Figure 5.6d. The filled circles represent the reciprocal lattice points of the undiluted CoO fcc lattice. In the 0° direction, the calculated pattern of filled circles (Figure 5.6d) fits to the RHEED image of the $Co_{1-y}O$ layer prepared at low oxygen pressure (Figure 5.6b). For the samples prepared at higher oxygen pressures, additional diffraction spots (open circles in Figure 5.6d,e) appear at half the distance between the filled circles, showing an additional structure with approximately twice the lattice constant in real space. We believe that these additional spots are due to the formation of Co_3O_4 upon dilution, which is also consistent with the results from X-ray diffraction (see below). We conclude that the almost defect-free CoO is deposited at low oxygen pressure ($p(O_2) = 3.3 \times 10^{-7}$ mbar), while for higher oxygen pressures Co-deficient $Co_{1-y}O$ is formed. To further investigate the formation of the additional phase with increasing oxygen pressure, we analyze the evolution of the RHEED intensity of the (0, 1/2) spot (open circles in Figure 5.6d,e) as a function of oxygen pressure.

Figure 5.7a shows a line scan through the RHEED diffraction pattern of $Co_{1-y}O$ prepared at a high oxygen pressure ($p(O_2) = 1 \times 10^{-5}$ mbar). The (0, 1) peak corresponds to the undiluted fcc lattice of CoO, while the (0, 1/2) peak appears for Co-deficient $Co_{1-y}O$. The relative intensity of the (0, 1/2) and (0, 1) peaks strongly increases as a function of oxygen pressure as seen in Figure 5.7b. This supports the notion that the number of volume defects in the AFM layer can be controlled by the oxygen pressure during deposition, that is, the number of defects continuously increase with increasing oxygen pressure.

We now discuss the RHEED patterns along the 30° direction in Figure 5.6, right panels. The calculated RHEED pattern in Figure 5.6d does not reproduce the double spot structures as observed for all oxygen concentrations (see Figure 5.6b,c, right panels). In order to explain these diffraction patterns, we furthermore have to assume that $Co_{1-y}O$ grows in a twinned structure where crystallites are oriented at 60° relative to each other (compare Figure 5.6e with 5.6b,c, right panels). Similar to the 0° direction the undiluted sample (Figure 5.6b) only shows reflections from a CoO fcc lattice, while the diluted samples show reflections from the defect phase also.

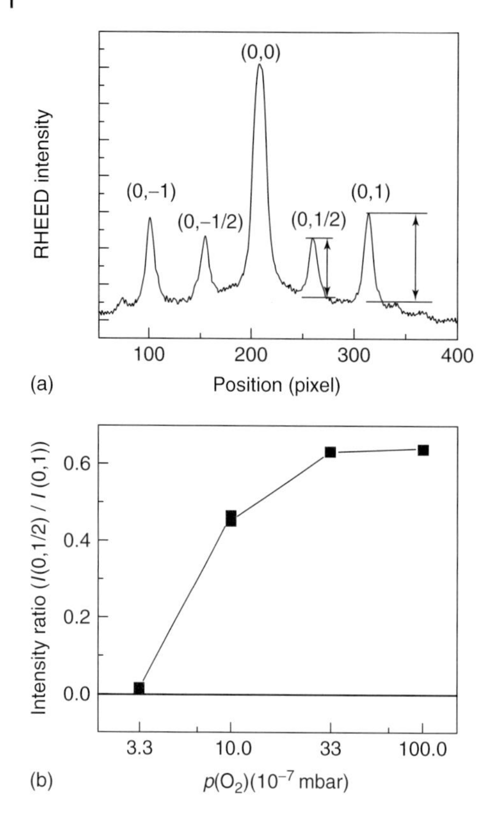

Figure 5.7 (a) Line scan of the RHEED image shown in Figure 5.6c, left panel, for a $Co_{1-y}O$ layer prepared at $p(O_2) = 1.0 \times 10^{-5}$ mbar. (b) Dilution dependence of intensity ratio between the additional (0, 1/2) spot for overoxidized $Co_{1-y}O$ and the (0, 1) reflections for the CoO fcc structure. Figure taken from [17].

In conclusion, the main experimental features of the RHEED investigation of the $Co_{1-y}O$ layers are (i) the number of defects in $Co_{1-y}O$ continuously increases with increasing oxygen pressure during evaporation and (ii) the layers grow with 60° twins. The same qualitative findings were observed in Mg-diluted CoO layers (not shown). In the following discussion, we will refer to the samples with the lowest defect concentration ($p(O_2) = 3.3 \times 10^{-7}$ mbar and $x(\text{Mg}) = 0.0$) as unintentionally diluted.

Additional structural characterization was carried out by *ex situ* X-ray diffraction using Cu K_α radiation ($\lambda = 0.15418$ nm). A high-angle $\theta - 2\theta$ scan of a $Co/Co_{1-y}O$ bilayer with the AFM prepared at $p(O_2) = 3 \times 10^{-6}$ mbar is shown in Figure 5.8a. Besides prominent [001] Al_2O_3 substrate peaks, only [111] reflections and those of higher order are seen for both Co and CoO, which is consistent with the RHEED results. In addition, [111]-oriented Co_3O_4 is observed. To further investigate the surface of the $Co_{1-y}O$ layers, *ex situ* atomic force microscopy images were taken.

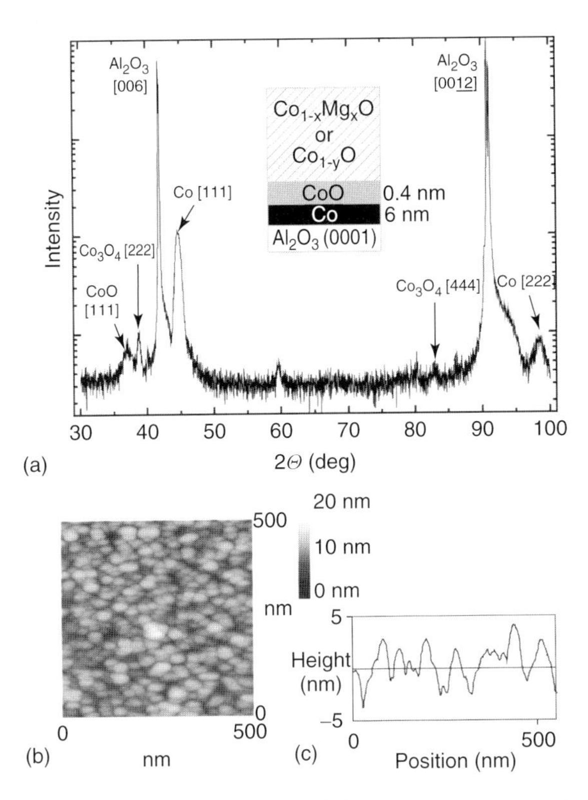

Figure 5.8 (a) High-angle X-ray diffraction ($\lambda = 0.15418$ nm) for Co/$Co_{1-y}O$ with $Co_{1-y}O$ prepared at $p(O_2) = 3 \times 10^{-6}$ mbar. The sample structure is schematically depicted in the inset. (b) Atomic force microscopy image of a 20-nm-thick $Co_{1-y}O$ layer prepared at $p(O_2) = 3 \times 10^{-6}$ mbar on a 6-nm-thick Co layer. Crystallite sizes range between 25 and 35 nm. (c) Atomic force microscopy line scan of image in (b). Figure taken from [17].

Figure 5.8b shows an atomic force microscope image for a 20-nm-thick $Co_{1-y}O$ layer prepared at $p(O_2) = 3 \times 10^{-6}$ mbar on top of a 6-nm Co layer. As expected from the RHEED investigations, a rough surface is found. The crystallite size ranges between 25 and 35 nm. As seen from the line scan in Figure 5.8c, the surface has a peak-to-peak height variation of approximately 6 nm.

5.3.2 Exchange Bias between the Ferromagnet Co and the Diluted Antiferromagnet CoO

Magnetic characterization of the Co/CoO samples was performed using a superconducting quantum interference device (SQUID) magnetometer. The samples were cooled from 320 K, that is, from above T_N of CoO, to 5 K in the presence of an external magnetic field $+B_{FC}$, oriented parallel to the plane of the film. Subsequently, the temperature was gradually increased up to 320 K through a sequence of intermediate temperatures. At each of these temperatures, the applied field was

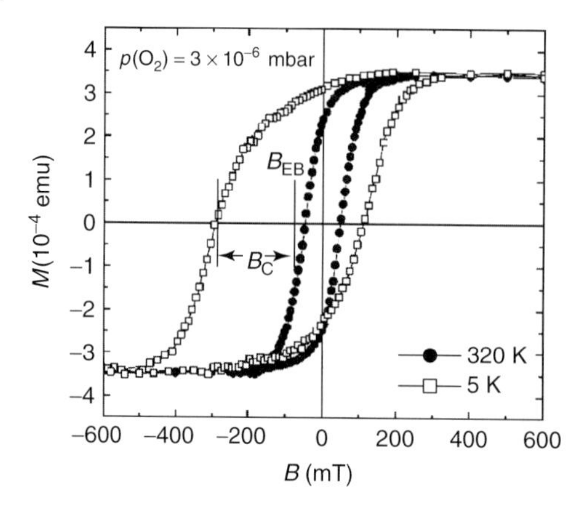

Figure 5.9 Hysteresis loops of $Co_{1-y}O/Co/Al_2O_3$ (0001) at $T = 5$ and 320 K with the $Co_{1-y}O$ prepared at $p(O_2) = 3 \times 10^{-6}$ mbar. The EB field B_{EB} and the coercive field B_c are indicated in the figure. Figure taken from [17].

cycled from $+B_{FC}$ to $-B_{FC}$ and back to $+B_{FC}$ in order to measure a hysteresis cycle. Except for the explicit studies of the cooling-field dependence, a cooling field of $B_{FC} = 5$ T was chosen for all measurements presented. This value is larger than the saturation field of the FM Co layer. The switching fields of the hysteresis cycles, B_1 and B_2, were used to determine the EB field B_{EB} and the coercive field B_c according to $B_{EB} = (B_1 + B_2)/2$ and $B_c = (B_1 - B_2)/2$, respectively.

Figure 5.9 shows typical hysteresis loops above ($T = 320$ K) and below ($T = 5$ K) T_N of a sample with the $Co_{1-y}O$ layer grown at $p(O_2) = 3 \times 10^{-6}$ mbar. The loop at low temperature exhibits a large B_{EB} toward negative magnetic fields, which is opposite to the cooling-field direction. In addition, a strong increase in the B_c is observed at 5 K compared to 320 K, leading to a significant broadening in the width of the loops.

5.3.2.1 Nonmagnetic Dilution of the Antiferromagnet

The dilution dependence of both the $|B_{EB}|$ and B_c for Mg-diluted $Co_{1-x}Mg_xO$ samples is shown in Figure 5.10a,b, while in Figure 5.10c,d analogous results of Co-deficient $Co/Co_{1-y}O$ samples are presented. The EB is enhanced by a factor of 3–4 for both types of defects in the AFM layer. Maximum enhancement is obtained for $x(\text{Mg}) = 0.1$ and $p(O_2) = 5 \times 10^{-6}$ mbar. Within the DS model, the observed increase in the EB shift with an increasing number of defects can be related to the formation of volume domain walls, which preferentially pass through the nonmagnetic defects at no cost of exchange energy. This leads to an excess magnetization of the AFM. For large dilutions ($x(\text{Mg}) > 0.25, p(O_2) > 5 \times 10^{-6}$ mbar), the EB again decreases as the AFM order is increasingly suppressed and eventually the connectivity in the AFM lattice is lost. Residual EB at high dilutions ($x(\text{Mg}) = 1.0$) has to be attributed to the 0.4-nm CoO interface layer and the underlying oxidized layer.

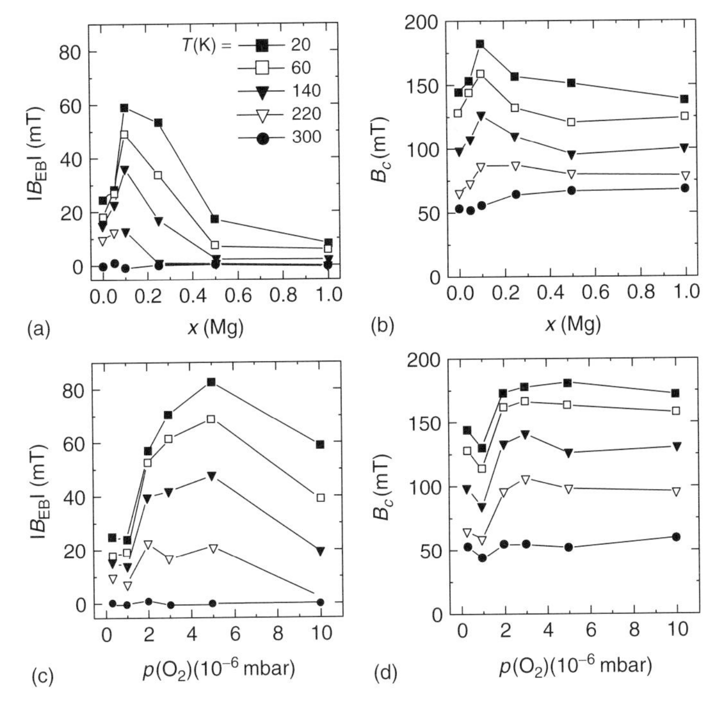

Figure 5.10 (a) EB field and (b) coercive field versus Mg concentration x in the $Co_{1-x}Mg_xO$ layer for various temperatures. (c) EB field and (d) coercive field versus oxygen pressure during deposition of the $Co_{1-y}O$ layer at the same temperatures. Note that only absolute EB values are plotted. All lines are guides to the eye. Figure taken from [17].

Monte Carlo (MC) simulations [16] qualitatively reproduce both the initial increase in EB with increasing dilution and its decrease at larger dilutions. However, the concentrations for optimally diluted samples differ usually between the experiment (x(Mg) ~ 0.15) and theory ($p \sim 0.6$). A possible origin of this difference is the presence of grain boundaries in the twinned AFM layer which reduce the domain-wall energy, thus leading to a finite EB without intentional dilution of the AFM layer as observed for the unintentionally diluted samples. This conclusion is consistent with the rather small EB found in untwinned and single crystalline AFM [56]. As seen in Figure 5.10a,c, the defect concentration for maximum EB depends on the temperature, that is, it shifts toward smaller values at elevated temperatures, which is also qualitatively observed in MC simulations [16]. Furthermore, it should be noted that the maximum of the EB field is obtained at lower values of the defect concentration when an additional roughness of the interface is assumed (see Section 5.4.3 and [23]).

The dilution dependence of B_c is shown in Figure 5.10b,d. Similar to the EB field, B_c changes nonmonotonically with dilution and shows maximum values at the same defect concentrations as the EB, although its relative changes are smaller than those observed in the EB field. Like the EB, the coercivity also strongly decreases with increasing temperature, which is common for many EB systems. The change in coercivity with temperature is caused by the change in the coupling of the FM (Co) layer with the AFM (CoO) layer.

5.3.2.2 **Hysteresis Curves and Uncompensated Magnetic Moments (Vertical Shift)**

In Figure 5.9 we have shown typical hysteresis loops of $Co_{1-y}O/Co$ exhibiting EB. Until now we have concentrated only on the horizontal shift of the hysteresis loop, that is, the EB. However, if the magnetization of the AFM (due to the uncompensated magnetic moments in the AFM) is irreversible under field reversal, it can be identified as a vertical shift of the measured FM/AFM hysteresis loop. Such an AFM magnetization was first observed in hysteresis loops of Fe/FeF_2 and Fe/MnF_2 [57] and recently discussed in detail for the EB system NiFe/FeMn [58]. Both positive and negative vertical shifts were found and attributed to positive (FM) and negative (AFM) FM/AFM interface coupling, respectively.

Like most conventional magnetization probes, SQUID magnetometry is not layer- or element-specific but rather measures the whole FM/AFM bilayer magnetization. In addition to the magnetization of the FM layer, both interface and volume magnetizations of the diluted AFM layer contribute to the total magnetization.

To investigate the change of the AFM magnetization with the number of introduced volume defects in the AFM layer, we performed high-accuracy magnetization measurements of the vertical magnetization shift for both unintentionally diluted and oxygen-diluted samples grown at $p(O_2) = 3.3 \times 10^{-7}$ mbar and $p(O_2) = 3 \times 10^{-6}$ mbar, respectively. By cooling the FM/AFM samples from above T_N to lower temperatures in the presence of a magnetic field, it is possible to set the AFM magnetization direction and to measure its irreversible part as being the vertical shift of the low-temperature hysteresis loop. The vertical shift was determined at $T = 20$ K and is given by $M_{shift} = |M(B_+)| - |M(B_-)|$, where B_+ and B_- are chosen so that the FM layer is fully saturated with $|B_+| = |B_-|$. The data shown in Figure 5.11 were taken at $B_\pm = \pm 0.8T$. As is seen in Figure 5.11, at large cooling fields M_{shift} is positive and overall increases with dilution of the AFM layer at all cooling fields. This increase can directly be linked to the creation of additional volume defects in the AFM layer as shown by the above RHEED analysis. It further supports that a DS is developed in the AFM after field cooling, carrying a surplus magnetization, which increases with dilution.

It is important to note that in our experiments we measure the total AFM surplus magnetization as was also investigated by the similar thermoremanent magnetization (TRM) probes of CoO/MgO multilayers. The temperature dependence of the TRM exhibits a strong similarity to that of $Ni_{81}Fe_{19}/CoO$ bilayers [59]. Although this surplus magnetization does not equal the irreversible DS magnetization m_{irr} of the AFM interface layer (defined in Section 5.2.3) which is at the origin of EB, we find striking qualitative agreement that the EB field indeed is proportional to the measured AFM magnetization (see Figures 5.11 and 5.19).

In the MC simulations (Section 5.4.1), contributions from both interface and bulk magnetization of the AFM layer can be separated. After zero-field cooling, the bulk

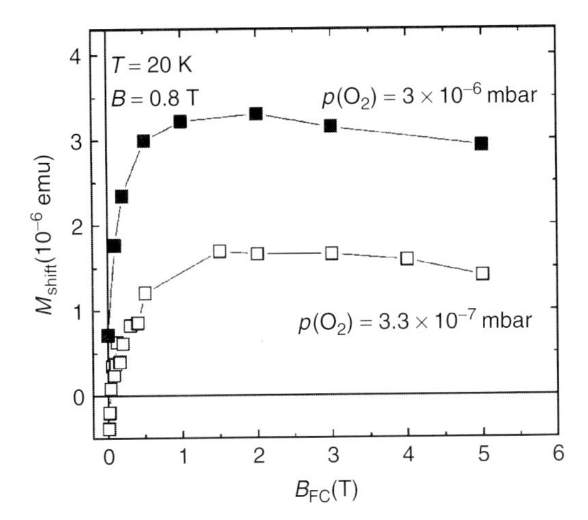

Figure 5.11 Vertical magnetization shift M_{shift} versus cooling field B_{FC} for $Co/Co_{1-y}O$ samples with $Co_{1-y}O$ prepared at different oxygen pressures. Data are taken at $B = \pm 0.8\,T$ and $T = 20\,K$ and are extracted as described in the text. Figure taken from [17].

of the AFM layer does not carry surplus magnetization, while it should dominate the total surplus magnetization for larger cooling fields. Indeed, we observe a strong reduction in M_{shift} for cooling fields below 1 T as is seen in Figure 5.11. Thus, it is suggestive that the low-field AFM magnetization primarily originates from the AFM spins close to the interface. While for diluted samples, M_{shift} remains finite and positive for zero cooling field, it changes sign for unintentionally diluted samples. The former case is consistent with positive (FM) FM/AFM interface coupling in Co/CoO. This conclusion is also consistent with the cooling-field dependence of the EB shift, which will be discussed later in the Section 5.3.2.6.

5.3.2.3 Substitutional versus Structural Defects

In terms of the DS model, we expect for undiluted $Co_{1-x}Mg_xO$ ($x \to 0$) and $Co_{1-y}O$ ($y \to 0$) a vanishing B_{EB} ($B_{EB} \to 0$). However, experimentally we still observe a small but finite B_{EB} at low temperatures in undiluted, twinned films (see Figure 5.10 above). This is attributed to disorder other than dilution, possibly grain boundaries in the twinned CoO layer. In the following discussion, we present experimental evidence of the direct influence of structural defects on EB, such as twinning and surface morphology, in addition to dilution. These additional defects have to be taken into account for the complete assessment of the EB.

For this study, epitaxial CoO(111)/Co(111) bilayers grown on MgO(111) substrates have been used. The substitutional defects were introduced by changing the partial pressure of oxygen, $p(O_2)$, as explained in the Section 5.3.1. A complex RHEED analysis [17, 60] has revealed that while the Co layer grows as smoothened 3D layer on MgO (111), the $Co_{1-y}O$ layer grown on top of the Co layer exhibits a twinned structure with crystallites rotated by 60° relative to each other. On the

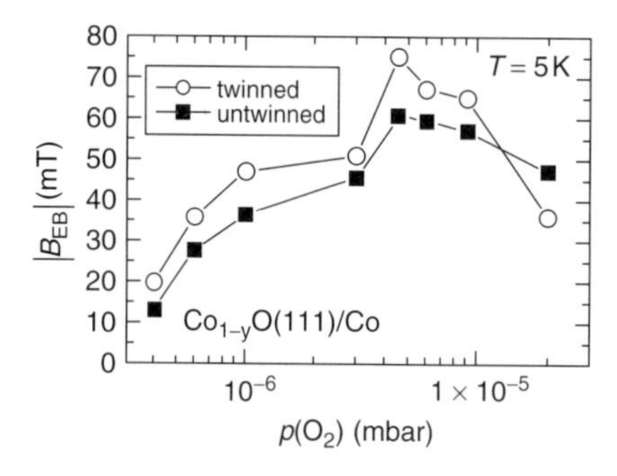

Figure 5.12 EB field at 5 K as a function of the oxygen pressure during deposition of the $Co_{1-y}O$ layer for twinned and untwinned AFM in $Co_{1-y}O(111)/Co(111)$ on MgO(111). Figure taken from [60].

other hand, the growth of the $Co_{1-y}O$ as the first layer on the MgO(111) substrate, followed by the Co layer on top, leads to untwinned $Co_{1-y}O$ and to a typical 3D-growth.

Figure 5.12 shows the dependence of B_{EB} at 5 K on the partial pressure of oxygen, $p(O_2)$, during deposition of twinned as well as untwinned $Co_{1-y}O$ in the $Co_{1-y}O(111)/Co$ system. We observe a strong dependence of B_{EB} on $p(O_2)$, that is, on the substitutional defects (Co deficiencies), similar to that shown in Figure 5.10c. Moreover, B_{EB} for the twinned $Co_{1-y}O$ is higher than that for untwinned samples except for the highest pressure in Figure 5.12. The vertical shift of the dilution dependence of B_{EB} for twinned $Co_{1-y}O$ compared to untwinned $Co_{1-y}O$ indicates that besides dilution other nonmagnetic defects in $Co_{1-y}O$, such as twin boundaries as structural defects, can enhance B_{EB}. In other words, the twin boundaries in twinned $Co_{1-y}O$, similar to dilution of magnetic sites, also reduce the domain-wall energy favoring the formation of domains in the AFM and hence the increase of m_{irr} and B_{EB}.

We already mentioned that the Co layer grows as smoothened 3D-layer on MgO (with twinned $Co_{1-y}O$ growing on top) and that the (untwinned) $Co_{1-y}O$ on MgO as well as the Co layer on top shows a 3D-growth. This means that although the interface roughness in the case of twinned samples is less than that in untwinned samples, the values of B_{EB} of twinned samples are larger (see Figure 5.12). This evidences the dominant role of the twin boundaries as defects for EB compared to interface roughness. The experimental evidence of the dominance of nonmagnetic defects in the bulk of the AFM over interface roughness is very much consistent with the DS model. The untwinned–undiluted sample prepared at 4×10^{-7} mbar shows a B_{EB} of about 12 mT at 5 K. The origin of this residual, nonzero B_{EB} can be explained by the presence of other imperfections in the CoO, such as 3D surface morphology, as observed by RHEED [60].

In conclusion disorder in the volume part of the AFM, such as substitutional defects (dilution) or structural defects (twinning and surface morphology), significantly enhances EB. These defects support the formation of volume domains in the AFM, which are responsible for EB.

5.3.2.4 Temperature Dependence

The temperature dependence of the $|B_{EB}|$ for the Mg-diluted $Co/Co_{1-x}Mg_xO$ samples is shown in Figure 5.13a–e at the same Mg concentrations as in Figure 5.10a. For all samples, the $|B_{EB}|$ monotonically increases with decreasing temperature. It varies almost linearly with temperature near optimum dilution (x(Mg) $= 0.1$ and x(Mg) $= 0.25$), while it saturates at low temperatures and low dilutions (Figure 5.13a,b). The former dependence agrees well with the temperature behavior, as can be seen from the mean-field calculations for near-optimum dilutions (Figures 5.4 and 5.5).

The results are described by the DS model as follows: A metastable DS is frozen at low temperatures after field cooling, which inhibits domain-wall motion. Thermally activated domain-wall motion becomes more favorable at elevated temperatures, which leads to a reduction of m_{irr} and thus to a decrease of the EB field as observed experimentally.

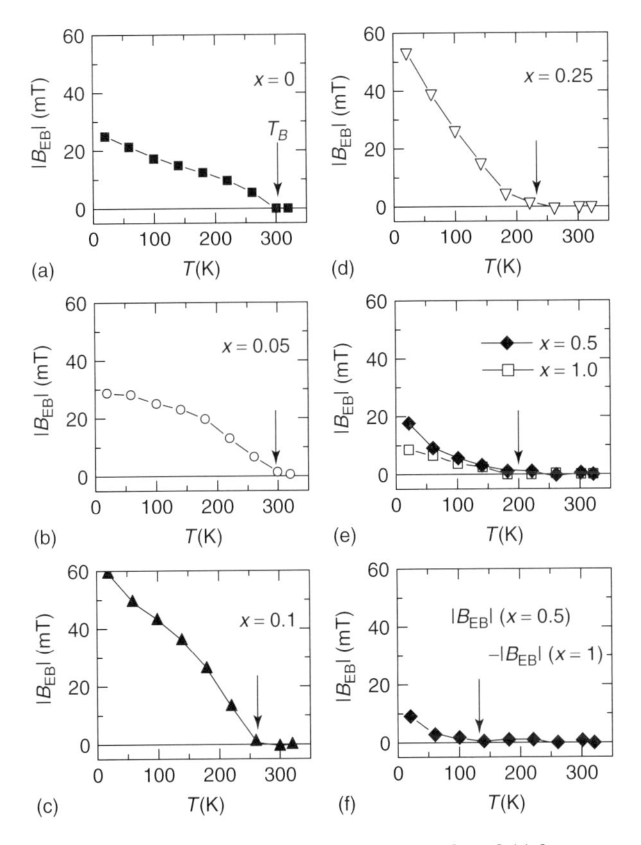

Figure 5.13 (a–e) Temperature dependence of EB field for the Mg-diluted $Co/Co_{1-x}Mg_xO$ samples with $x = 0$, 0.05, 0.1, 0.25, 0.5, and 1.0. (f) Difference in EB fields for samples with $x = 0.5$ and 1.0. The blocking temperature is marked by an arrow in each Figure. Figure taken from [17].

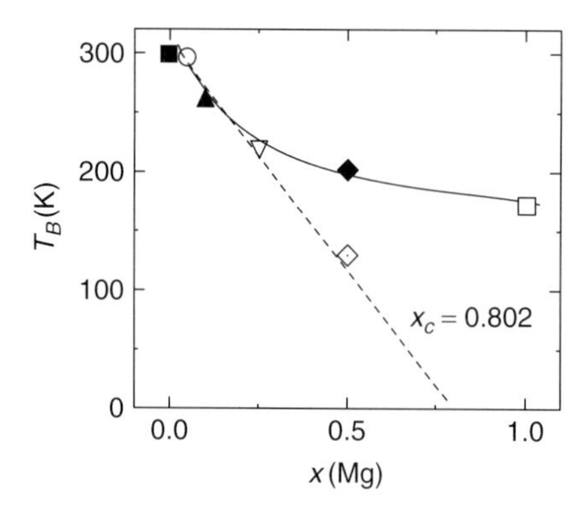

Figure 5.14 Dilution dependence of blocking temperature T_B for the Mg-diluted $Co/Co_{1-x}Mg_xO$ samples as shown in Figure 5.13. For $x = 0.5$ the lower T_B (dot-centered diamond) on the dashed line is taken from Figure 5.13f. All lines are guides to the eye. The percolation threshold x_c for a three-dimensional fcc lattice with nearest neighbor interaction is included. Figure taken from [17].

The EB vanishes above the so-called blocking temperature, T_B, which can significantly differ from T_N [16]. Its monotonic decrease with increasing Mg dilution in $Co/Co_{1-x}Mg_xO$ samples can be seen by the arrows in Figure 5.13a–e and is also plotted in Figure 5.14. We observed a similar but weaker decrease of T_B for the $Co/Co_{1-y}O$ samples (not shown). Note that the blocking temperature is not completely diminished for the fully diluted sample with $x(\text{Mg}) = 1.0$. As already discussed above, we attribute this remaining EB coupling to the CoO interface layer and the underlying oxidized layer. This residual EB coupling is also likely to be relevant for the temperature dependence of the EB field for the sample with $x(\text{Mg}) = 0.5$, as can be seen in Figure 5.13e. To further separate the contributions of the undiluted interface layer to the EB coupling from that of the diluted AFM volume layer, we subtract the EB fields for the sample with $x(\text{Mg}) = 1.0$ from that for the sample with $x(\text{Mg}) = 0.5$, which is depicted in Figure 5.13f. The resulting blocking temperature, which is strongly reduced, is also included in Figure 5.14 at $x(\text{Mg}) = 0.5$. Note that all other samples do not show any significant change in blocking temperature when performing a similar subtraction (not shown).

As seen in Figure 5.14, the modified blocking temperatures linearly decrease with increasing Mg dilution of the AFM volume layer (dashed line). In particular, the extrapolated dilution, above which EB coupling vanishes completely ($T_B \to 0$), is close to the percolation threshold ($x_c = 0.802$) for AFM in a three-dimensional fcc lattice with nearest-neighbor interaction [61]. This supports the notion that no global EB coupling remains once the connectivity of the AFM spin lattice is lost.

According to the DS model, the decrease of B_{EB} with temperature is due to the reduction of m_{irr} as a consequence of thermally activated domain-wall motions. In the following discussion, we prove this statement for another group of $Co_{1-y}O/Co$ samples grown on MgO(100). In Figure 5.15a, we show the temperature dependence of $|B_{EB}|$ of an undiluted sample (grown at $p(O_2) = 4 \times 10^{-7}$ mbar) together with an optimally diluted sample (grown at $p(O_2) = 5 \times 10^{-6}$ mbar).

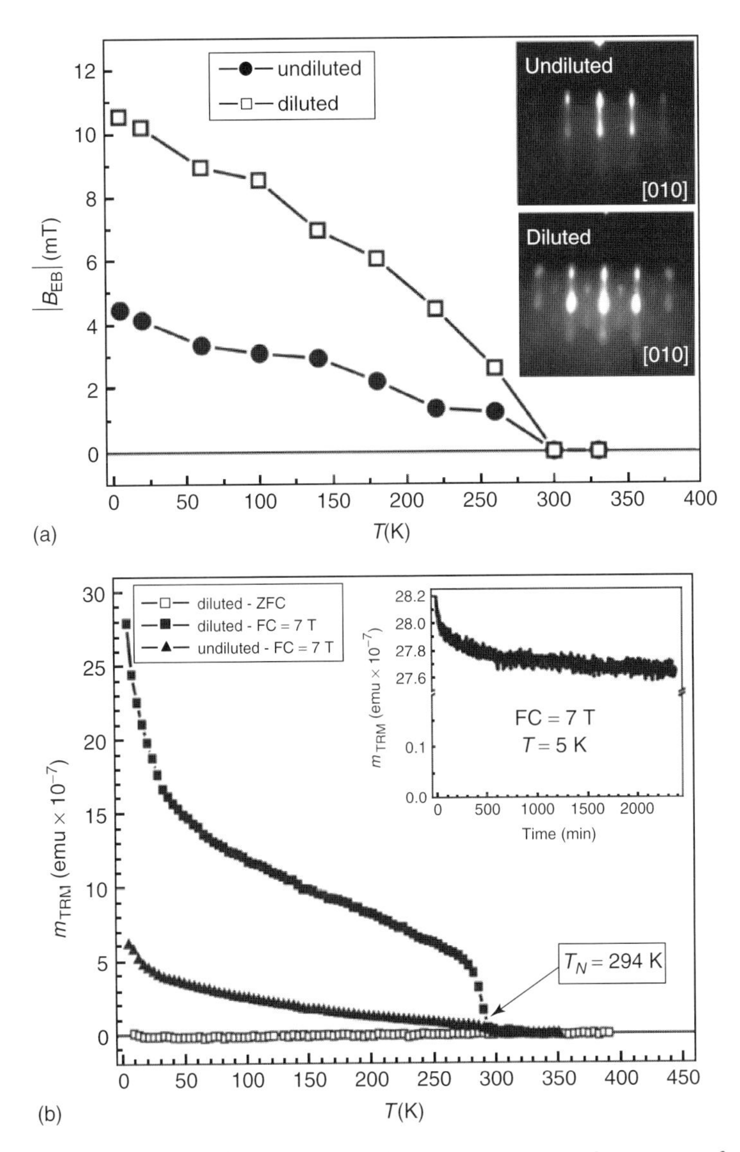

Figure 5.15 (a) EB field of MgO(100)/$Co_{1-y}O$(100)/Co(11$\bar{2}$0) as a function of temperature for undiluted and diluted samples. The inset shows RHEED patterns of the undiluted and diluted $Co_{1-y}O$ layer. (b) Thermoremanent magnetization (m_{TRM}) of the field-cooled MgO(100)/$Co_{1-y}O$(100) samples as a function of temperature for undiluted and diluted samples. The zero-field-cooled (ZFC) curve of the diluted sample is also shown for comparison. The inset shows that the relaxation of m_{TRM} in time is very weak. Figure taken from [62].

The RHEED patterns corresponding to the diluted and undiluted samples are also shown as inset in Figure 5.15a. For both samples, $|B_{EB}|$ decreases with temperature and shows an enhancement upon dilution in the temperature range between 5 and 294 K. The temperature dependence and dilution dependence of $|B_{EB}|$ of the $Co_{1-y}O/Co$ samples grown on MgO(100) are in fact very similar to those of the $Co_{1-y}O/Co$ samples grown on (0001) sapphire substrates (shown in Figure 5.13) as well as on MgO (111) substrates [59]. This indicates that the crystal orientation of CoO has no significant effect on the overall trend of $|B_{EB}|(T)$ curves for different defect concentrations.

For comparison, we used two samples containing only the AFM $Co_{1-y}O$ grown directly on top of the MgO(100) substrates (without the Co layer). The deposition was performed with the same deposition parameters and $p(O_2)$ values as in the case of the $Co_{1-y}O/Co$ samples. These two AFM-only samples were cooled from above T_N of CoO to 5 K in a magnetic field of +7 T, and subsequently their TRM m_{TRM} was recorded by SQUID magnetometry during heating up in the absence of any external magnetic field. Figure 5.15b shows the temperature dependence of m_{TRM} for both diluted and undiluted AFM-only samples. A strong enhancement of m_{TRM} (> 300%) is observed for the diluted sample in comparison with the undiluted one, which reflects the $|B_{EB}|$ enhancement upon dilution of the $Co_{1-y}O/Co$ bilayers. For T > 50 K, the m_{TRM} of the AFM-only samples decreases with temperature in a similar way as the corresponding $|B_{EB}|$ of EB bilayers. However, at low temperatures (T < 50 K), m_{TRM} shows an abrupt decrease with increasing temperature as opposed to a monotonic decrease of $|B_{EB}|$ for the entire temperature range between 5 K and T_N. This behavior is consistent with the results of [59] for the polycrystalline NiFe/CoO system.

In conclusion, although the m_{TRM} of the AFM-only samples (measured by SQUID) does not precisely represent the m_{irr} of the AFM interface layer, we can observe a strikingly good agreement between the temperature dependence and dilution dependence of $|B_{EB}|$ and of m_{TRM} at temperatures above 50 K. This is in strong agreement with the DS model. The differences between the $m_{TRM}(T)$ and $|B_{EB}|(T)$ curves below 50 K are mainly due to the low-anisotropy uncompensated spins, which give rise to the low-temperature enhancement of m_{TRM} but which are not capable of pinning the Co layer (and hence of producing EB) in the $Co_{1-y}O/Co$ samples [59, 63].

5.3.2.5 **Thermoremanent Magnetization and Training Effect**

Most thin-film EB systems show a reduction of the EB shift upon subsequent magnetization reversals of the FM layer, which is the so-called training effect. For all Mg-diluted $Co/Co_{1-x}Mg_xO$ samples, we measured the training effect at $T = 5$ K after field cooling in $B_{FC} = 5$ T. Typical magnetization reversals corresponding to the first and 51st hysteresis loops are shown in Figure 5.16 for the sample with $x(Mg) = 0.5$. Besides a clear, but rather small, reduction of the EB shift, a decrease in the coercive field is observed.

The training effect implies that during magnetization reversal, the FM layer reverses neither homogeneously nor reversibly. According to the DS model, the

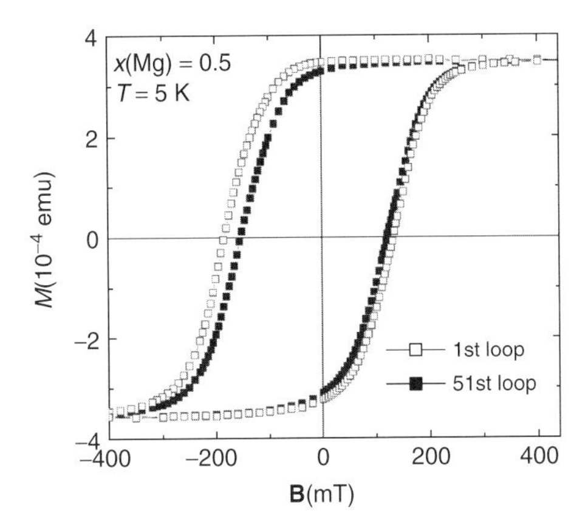

Figure 5.16 Training effect of $Co/Co_{1-x}Mg_xO$ sample with $x = 0.5$. Plotted are the first and the 51st hysteresis loops at $T = 5$ K after field cooling. Figure taken from [17].

training effect is due to a rearrangement of the AFM domain structure, which results in a partial loss of the irreversible DS magnetization m_{irr} of the AFM interface layer during field cycling [16]. This magnetization loss leads to a reduction of the EB shift.

In order to prove the above statement, we have measured by SQUID magnetometry the irreversible part of m_{TRM} of the above-discussed $Co_{1-y}O$ AFM-only samples. This will be denoted in the following discussion as $m_{\text{TRM}}^{\text{irr}}$ of the whole AFM layer. $m_{\text{TRM}}^{\text{irr}}$ was measured at 5 K after cooling in a magnetic field of +7 T by cycling the magnetic field between −7 and +7 T and recording the difference between the m_{TRM} values each time when passing through 0 T (scanning along the ascending and descending field brances), for each field cycle n. $m_{\text{TRM}}^{\text{irr}}$ is at the origin of the vertical shift of the FM/AFM hysteresis loop, as shown in Figure 5.11. The dependence of $m_{\text{TRM}}^{\text{irr}}$ on the field cycle n is plotted together with that of $|B_{\text{EB}}|$ for $Co_{1-y}O/Co$ bilayers in the same graph in Figure 5.17. One can see that both $|B_{\text{EB}}|$ and $m_{\text{TRM}}^{\text{irr}}$ show a decrease with n, that is, they exhibit a training effect. The maximum decrease in both the $m_{\text{TRM}}^{\text{irr}}$ and the $|B_{\text{EB}}|$ occurs between the first and second field cycles, and they asymptotically approach almost constant values for the remaining cycles. This observed correlation between the training effect of the $m_{\text{TRM}}^{\text{irr}}$ and that of $|\mathbf{B}_{\text{EB}}|$ clearly suggests that it is the loss of $m_{\text{TRM}}^{\text{irr}}$ due to irreversible changes in the AFM domain structure during the field cycles that causes a reduction of $|B_{\text{EB}}|$ [62]. It is important to realize here that the training effect of EB is believed to be due to a reduction in $m_{\text{TRM}}^{\text{irr}}$ of the AFM surface next to the interface with the FM [16]. However, by SQUID magnetometry the $m_{\text{TRM}}^{\text{irr}}$ of the whole AFM layer is measured, which includes both volume and surface contributions. This could be one reason that an exact quantitative agreement between the training effect of $m_{\text{TRM}}^{\text{irr}}$ and of $|B_{\text{EB}}|$ was not observed. Another reason for this may arise from the observed difference between the behavior of $m_{\text{TRM}}(T)$ and $|B_{\text{EB}}|(T)$ for $T < 50$ K (Figure 5.15).

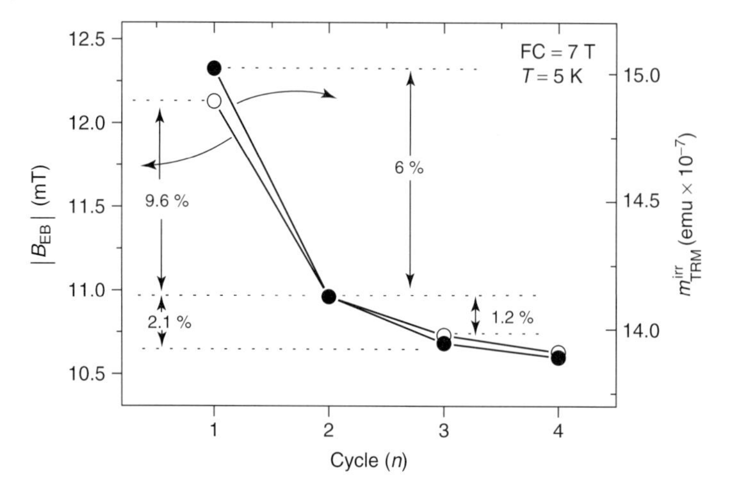

Figure 5.17 EB field (○) for a diluted CoO/Co sample and the irreversible thermoremanent magnetization $m_{\mathrm{TRM}}^{\mathrm{irr}}$ (•) of a diluted AFM-only (CoO) sample as a function of the number of field cycles (n). Figure taken from [62].

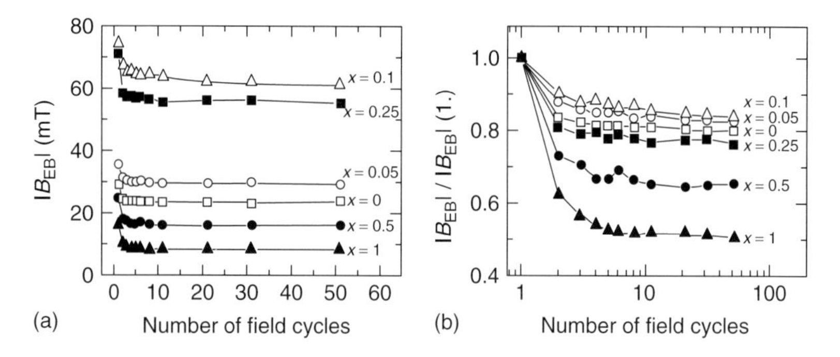

Figure 5.18 Dilution dependence of the training effect for $Co/Co_{1-x}Mg_xO$ samples. (a) EB shift as a function of subsequent hysteresis cycles for different Mg concentrations x at $T = 5$ K. (b) Normalized EB values from (a) plotted on a semilog scale. Note that there is almost no training effect after ten field cycles. Figure taken from [17].

To further investigate the relevance of the AFM domain structure for the training effect, we focus on the dilution dependence of the Mg-diluted samples, which is depicted in Figure 5.18 at $T = 5$ K. Independent of the dilution, the EB shift is strongly reduced only during the first field cycles and then remains almost constant. This behavior is in striking qualitative agreement with MC simulations [16].

In Figure 5.18b, we plot the relative decrease of the EB shift by normalizing the data from Figure 5.18a by their initial value at each dilution. We can see that the magnitude of this relative training effect behaves nonmonotonically with dilution, with the smallest effect observed for samples near optimum dilution (x(Mg) $= 0.1$).

In other words, at optimum dilution the EB is the strongest with the smallest training effect, while at high and low dilutions the EB decreases while the training effect is increased. This supports the notion that the formation of volume domains in the AFM layer plays a crucial role in the EB interaction at the FM/AFM interface.

The observed dilution dependence of the training effect can be interpreted within the DS model as follows. AFM domain walls are most strongly pinned near optimum dilution. This pinning results in large energy barriers and prohibits domain-wall motion upon FM magnetization reversal leading to a small training effect. At larger dilutions the size and the connectivity of the AFM spin lattice get reduced. This results in a decrease of the AFM domain-wall barrier and a decrease of EB. At the same time AFM domain-wall motion and relaxation become easier upon FM magnetization reversal, leading to an enhancement of the training effect.

5.3.2.6 **Cooling-Field Dependence**

The EB effect reveals a striking dependence on the magnitude of the cooling field, B_{FC}. At large cooling fields, the EB (i) is either constant [64] or moderately reduced in most EB systems, or (ii) changes its sign (positive EB) in systems such as Fe/FeF_2 and Fe/MnF_2 [65]. These results were explained by AFM interface coupling between FM and AFM. While the former case (i) is obtained for positive (ferromagnetic) FM/AFM interface coupling, a negative (AFM) interface coupling yields the latter case (ii).

To investigate the sign of the interface coupling in Co/CoO and its dependence on the AFM volume dilution, we studied the cooling-field dependence of the same $Co/Co_{1-y}O$ samples as shown in Figure 5.10c at both low ($p(O_2) = 3.3 \times 10^{-7}$ mbar) and optimum ($p(O_2) = 5 \times 10^{-6}$ mbar) defect concentrations. For all cooling fields, the FM layer was first magnetized at a field of 5 T and a temperature of $T = 320$ K. Then, it was cooled to $T = 20$ K for cooling fields B_{FC} between 0 and 5 T. As shown in Figure 5.19, the $|B_{EB}|$ increases at low cooling fields while it slightly decreases

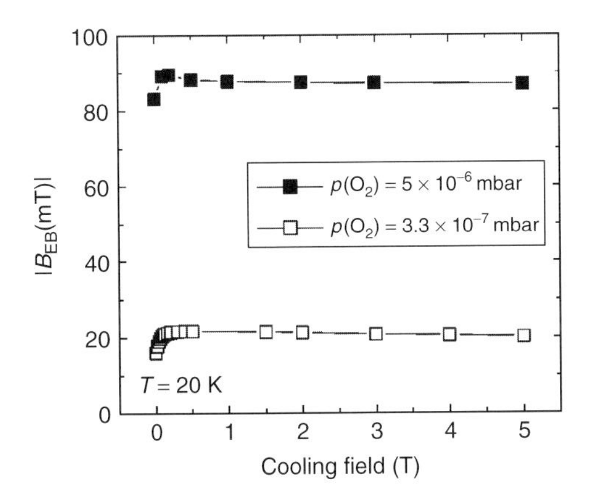

Figure 5.19 EB field versus cooling field for $Co/Co_{1-y}O$ samples at $T = 20$ K with the $Co_{1-y}O$ layer prepared at different oxygen pressures. Figure taken from [17].

at larger fields for both samples. We primarily attribute the initial increase to the magnetization of the FM layer, which is not fully saturated during field cooling at low cooling fields. Since the magnetization of the FM layer determines the global EB [66], its decrease toward small fields will reduce the EB coupling.

Since there is only a slight decrease of the $|B_{EB}|$ at high cooling fields and no observation of positive EB, we conclude that we have positive (FM) interface coupling between Co and CoO. This is consistent with no change in the sign of B_{EB} in NiFe/CoO up to $B_{FC} = 5\,T$ [64]. This conclusion is further confirmed by the observation of a positive vertical shift of the hysteresis loops (see Figure 5.11).

5.3.2.7 Antiferromagnetic Thickness Dependence

Most of the reports in the literature generally agree that there must be a minimum AFM thickness in order to yield EB. For larger thicknesses the observations can be classified into two characteristic types of dependence. (i) For thicknesses larger than a minimum value, the EB remains constant as a function of the AFM layer thickness. (ii) With increasing AFM layer thickness the EB field goes through a maximum and then continuously decreases.

We therefore investigated the AFM layer thickness dependence of the EB field for both unintentionally diluted ($p(O_2) = 3.3 \times 10^{-7}$ mbar) and optimally diluted ($p(O_2) = 5 \times 10^{-6}$ mbar) $Co_{1-y}O$ layers, which is depicted in Figure 5.20 at $T = 5$ K. For optimally diluted samples, the EB field strongly increases with increasing AFM layer thickness and saturates above 20 nm, which roughly corresponds to the size of the AFM crystallites. The EB field for unintentionally diluted samples, however, decreases with increasing AFM layer thickness and levels off at large thickness. We note that just by varying the defect density in the volume of the AFM layer we observe both types of thickness dependence as reported in other EB systems.

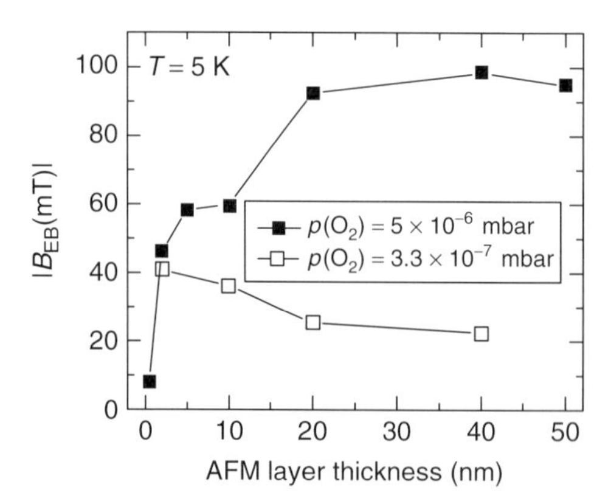

Figure 5.20 EB field as a function of the AFM layer thickness in $Co/Co_{1-y}O$ samples for both unintentionally diluted and optimally diluted $Co_{1-y}O$ layers prepared at oxygen pressures of $p(O_2) = 3.3 \times 10^{-7}$ mbar and $p(O_2) = 5.0 \times 10^{-6}$ mbar, respectively. Figure taken from [17].

Within the DS model, the different dependence is described in the following way. At small levels of disorder (unintentionally diluted samples), AFM domains can only be created at the cost of high energy. Since it is energetically unfavorable to close the domains parallel to the layers, the domain walls will extend through the whole thickness of the AFM layer perpendicular to the FM/AFM interface. For that case, the domain-wall energy increases proportionally to the AFM layer thickness [7, 8]. Thus, the formation of a domain wall in the AFM layer becomes less favorable with increasing AFM layer thickness. This results in a reduction of both the number of domain walls and the irreversible DS magnetization m_{irr}, thus leading to a drop in the EB field with increasing AFM layer thickness. For large AFM layer thicknesses, the low defect density in the volume of the AFM layer prohibits the formation of domain walls even if there is disorder at the interface. For very thin AFM layers, on the other hand, the disorder from the interface dominates and domain wall formation is energetically favorable, which then leads to a large EB field.

For the optimally diluted samples, domain walls can be created in an external magnetic field at a low cost of energy due to the nonmagnetic defects. The irreversible DS magnetization m_{irr} at the interface, which is responsible for the EB, is stabilized by the AFM volume domain structure. This is consistent with the results of the MC simulations [16]. Hence, the domain structure becomes more and more stable with increasing AFM layer thickness, leading to an increase of the EB. Assuming that grain boundaries strongly lower the AFM coupling strength, the grains may act in a magnetically independent manner, leading to a constant EB for AFM thicknesses larger than the grain size. This is consistent with our experimental observations in Figure 5.20.

5.3.2.8 **Blocking Temperature Distribution**

The thermal stability of the AFM is an important issue in EB systems. The blocking temperature distribution of an EB system accounts for the thermal stability of the AFM, and it reveals the temperature range where the thermally activated reversals take place in the AFM. An initial and important contribution to the understanding of the blocking temperature distributions belongs to Soeya *et al.* [67]. Using a special measurement procedure, they proved the existence of a variety of "exchange paths" in exchange-coupled NiFe/NiO bilayers. Each "exchange path" produces its own local unidirectional anisotropy and a different local blocking temperature. The measured exchange coupling was considered as the sum of the contributions of individual "exchange paths," each with its own local blocking temperature. The authors did not associate the notion of "exchange paths" to AFM domains or grains, but they suggested that the variety of "exchange paths" was caused by interface disorder and fluctuations in the atomic arrangement at the FM/AFM interface. More recent work including thermal energy barriers for magnetization reversal can be found in [68].

According to the DS model, a DAFF develops in a DS after cooling in an external magnetic field at a temperature below T_N. Each AFM domain carries a local DS magnetization m_{DS}, which originates from the uncompensated moments due to the formation of domain walls. Therefore, the total DS magnetization (M_{DS}) results from the sum of individual m_{DS}. The irreversible DS magnetization at the interface to the FM layer, m_{irr}, gives rise to the EB. In other words, each AFM domain has its own local unidirectional anisotropy (EB) and its own blocking temperature T_B, which are domain-size dependent. In this context, the notion of "exchange path," introduced by Soeya *et al.* [67], can be easily linked to an AFM domain.

For the measurement of the blocking temperature distribution, we present below the results for two samples of Co_{1-y}/Co grown on MgO(111), namely, unintentionally diluted ($p(O_2) = 3.3 \times 10^{-7}$ mbar) and optimally diluted ($p(O_2) = 5 \times 10^{-6}$ mbar). The measurement procedure implies several steps, as described in the following:

- First, the samples are cooled in the presence of a saturating magnetic field from above T_N of CoO to 5 K. As a consequence, the AFM enters a DS, that is, the AFM CoO is decomposed into AFM domains of different sizes which can be treated as independent Ising-type domains, with their own T_B. All AFM domains which are formed during the field cooling will contribute to B_{EB} at 5 K.
- At 5 K, the magnetic field is reversed and the temperature is increased to certain values denoted as reversal temperatures T_{rev} in the following discussion. During heating at T_{rev}, the AFM domains with T_B smaller than T_{rev} will enter their PM phase. As mentioned above, T_B is domain-size dependent, being larger for larger domains.
- The temperature is decreased again from T_{rev} to 5 K in the presence of the same reversed magnetic field. The AFM domains which entered the PM state during heating to T_{rev} will re-enter their AFM phase during field cooling from T_{rev} to 5 K. However, because of the reversed cooling-field direction, the uncompensated moments belonging to the AFM domains will gain a reversed orientation as compared to their initial orientation. A hysteresis curve measured after this step will exhibit a weaker EB due to those AFM domains which reversed their orientation.
- The procedure described above is repeated for different values of T_{rev}, yielding a successive domain reversal within the AFM.

With increasing T_{rev}, there will be a decrease in the magnitude of B_{EB} due to an increasing fraction of AFM domains which undergo thermally activated reversals. For a certain value of T_{rev} (called median blocking temperature [69]),the same fraction of AFM domains will have an opposite orientation, which will result in a zero EB.

The dependence of B_{EB} on T_{rev} of the two samples measured at 5 K is shown in Figure 5.21a. Starting from 5 K, B_{EB} decreases in magnitude with T_{rev}, goes through zero at the median blocking temperature, and then changes sign. For temperatures close to T_N, the absolute values of B_{EB} are equal to those measured after the initial field cooling at 5 K. Overall, the magnitude of B_{EB} is larger for the diluted sample as compared to the undiluted one, which is in agreement with our results shown in Figure 5.15a.

The first derivative of B_{EB} with respect to reversal temperature T_{rev}, that is, dB_{EB}/dT_{rev}, is proportional to the blocking temperature distribution $f(T_{rev})$ of

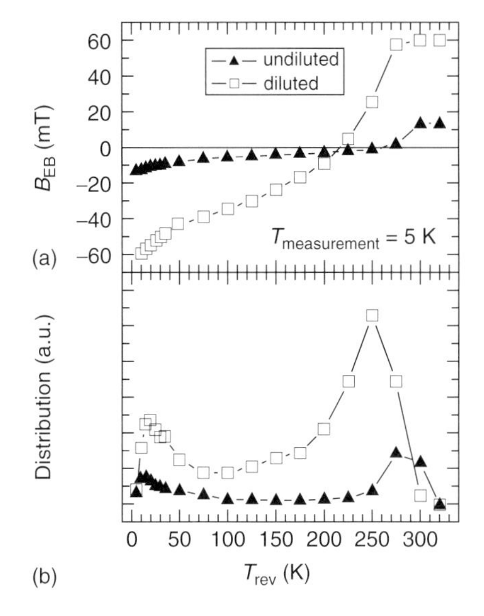

Figure 5.21 (a) B_{EB} and (b) distribution profiles of the blocking temperature as a function of T_{rev} for undiluted and diluted CoO/Co samples. At the respective T_{rev}, a certain fraction of AFM domains enters the paramagnetic state and it is subsequently reversed upon cooling to 5 K in the reversed field. Figure taken from [69].

the AFM domains [67, 70]. Similarly, when discussing this quantity in terms of domain size, dB_{EB}/dT_{rev} is proportional to the distribution of domain size $f(d)$. It is important to emphasize that d accounts for the size of AFM domains as well as for the low-temperature-isolated uncompensated moments [71].

Figure 5.21b shows the blocking temperature distribution of the AFM domains of the undiluted and diluted Co_{1-y}/Co samples. Both samples show a two-maxima distribution, with one maximum at low temperatures and the other one at temperatures close to T_N. For undiluted CoO, a plateau is observed in the temperature range from 100 to about 230 K, indicating that no significant reversals of the AFM domains take place in this temperature range. At higher temperatures, close to T_N, a narrow distribution can be observed for the undiluted sample compared to the diluted one. This points out that the undiluted CoO presents a DS with a narrow blocking temperature distribution $f(T_{rev})$, that is, a narrow domain size distribution $f(d)$. In contrast, $f(T_B)$ of the diluted sample shows a broad maximum at high temperature which extends over a much wider temperature range. In addition, both the maximum of $f(T_{rev})$ and the median blocking temperature shift to lower temperatures for the diluted sample. Hence, the AFM domains in the diluted AFM undergo thermally activated reversals over a wide temperature range. In conclusion, by (intentionally) diluting the AFM CoO, a broadening of the blocking temperature distribution $f(T_{rev})$ and of the domain size distribution $f(d)$ is obtained. In terms of thermal stability, this means that dilution results in a degradation of the thermal stability of the diluted AFM.

The low-temperature peak of $f(T_{rev})$ of the diluted sample is more pronounced than the corresponding one of the undiluted sample, but both show the maximum at roughly the same temperature $T_{rev} \approx 20$ K. We believe that this is due to the contribution of isolated AFM spin clusters, the small CoO grains, and some defects within these grains, all these effects being more pronounced in the diluted sample [69]. Moreover, our XRD studies (Figure 5.8a) reveal the appearance of the Co_3O_4 spinel phase within CoO for the diluted sample. Co_3O_4 is also an AFM with $T_N \approx 33$ K and may contribute to the low-temperature peak of $f(T_{rev})$ of the diluted sample (Figure 5.21b). Moreover, the exchange coupling between the Co_3O_4 phase and the host CoO can result in an increase in T_N of Co_3O_4 up to 80 K [72], which can explain the broadening of the low-temperature peak of $f(T_{rev})$ to higher temperatures.

5.4 Model Calculations

5.4.1 Modeling of Experimental Data

Apart from the mean-field work by Scholten *et al.* [46], mainly Monte Carlo methods were used to numerically investigate the DS model. Some of them focused on the Ising limit for the AFM [16, 25, 49, 73, 74], while others used the full Heisenberg Hamiltonian of Eq. 5.2 [15, 21]. In the latter case a heat-bath algorithm with single-spin flip dynamics was used, where the trial step of the spin update consisted of two steps: firstly, a small variation within a cone around the former spin direction, followed, secondly, by a total spin flip. This twofold spin update is ergodic and symmetric and can take care of a broad range of anisotropies, from very soft spins up to the high anisotropy (Ising) limit [75]. To observe the domain structure of the AFM, one has to guarantee that typical length scales of the domain structure fit into the system and that typical system sizes are of a lateral extension of 128×128 and a thickness of $t_{FM} = 1$ and t_{AFM} ranging from 3 to 9. Periodical boundary conditions were used within the film plane and open-boundary conditions perpendicular to it.

The main quantities which were monitored were the thermal averages of the z-component of the magnetic moment for each individual monolayer normalized to the magnetic moment of the saturated monolayer. In simulations, the system is first cooled from above to below the ordering temperature of the AFM. During cooling, the FM which is initially magnetized along the easy z axis provides a nearly constant exchange field at the interface with the AFM monolayer. Also, the system is cooled in the presence of an external magnetic field, the cooling field. In addition to the exchange field from the ordered FM, the cooling field also acts on the AFM. When the desired final temperature is reached, a magnetic field is applied along the easy axis and reduced in small steps down to a certain minimum value and afterwards raised again up to the initial value. This corresponds to one cycle of the

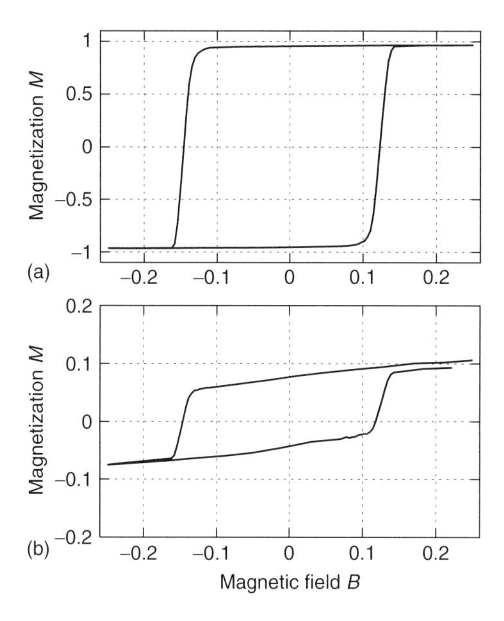

Figure 5.22 Simulated hysteresis loops of the DS model as explained in the text (see Eq. 5.2). Dilution $p = 0.4$, $k_B T = 0.1 J_{FM}$, positive interface coupling, $J_{INT} = |J_{AFM}|$. AFM anisotropy $k_z = J_{FM}/2$. The cooling field was $B_{cool} = 0.25\ J_{INT}$. Shown are (a) the magnetic moment of the FM and (b) the interface monolayer of the AFM (normalized to its saturation value.) Taken from [14].

hysteresis loop. A simulated hysteresis loop obtained as described above is depicted in Figure 5.22. Shown are the results for the magnetization of (a) the FM and (b) the AFM interface monolayer. EB is clearly observed.

An analysis of the magnetization curve of the interface layer gives an interesting insight into the nature of EB. After the field-cooling procedure, the AFM interface carries a magnetization. A part of this AFM interface magnetization is stable during sweeping the hysteresis loop and leads to the fact that the magnetization curve of the interface layer of the AFM is shifted upwards. This irreversible part of the interface magnetization, m_{irr}, of the AFM acts as an additional effective field on the FM, resulting in EB. Note that the interface magnetization of the AFM also displays hysteresis as a result of the exchange coupling to the FM. This means that the whole interface magnetization of the AFM consists of a reversible part leading to an enhanced coercivity and an irreversible part leading to EB. In experiments, usually the magnetization of the whole FM/AFM bilayer is measured. The corresponding sample magnetization loop might not only be shifted horizontally but also vertically. The vertical shift contains contributions from the volume part of the AFM as well as from its interface (for a comparison with experiment see Section 5.3.2.2). The volume magnetization of the AFM is induced by the cooling field, and hence it is not shifted when the cooling field is zero and shifted upwards when it is finite. The interface contribution depends on the sign of the interface coupling and may be positive, as in our calculation, or even negative for negative interface coupling [16, 17, 56].

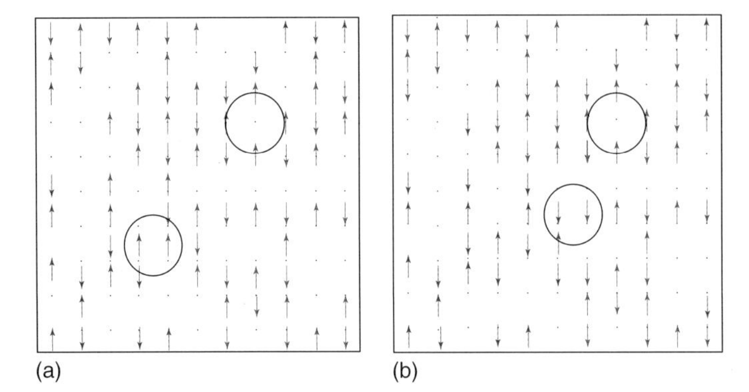

Figure 5.23 Snapshots of spin configurations in a small portion of the interface monolayer of the AFM after field cooling with the external field and FM magnetization, both pointing upward (a) and after reversal of the FM (b). The interface coupling is assumed to be positive. The colors distinguish different AFM domains. The circles mark sources of magnetization, wall magnetization as well as volume magnetization.

With the two sketches (Figure 5.23a,b), we wish to illustrate the origin of interface magnetization of the AFM on a more microscopic basis, including its partitioning in reversible and irreversible parts. Figure 5.23a shows spin configurations in a small portion of the interface monolayer of the AFM after field cooling. The simulated system size is $64 \times 64 \times 10$ with only one FM monolayer. For simplicity, this simulation was performed in the Ising limit for the AFM ($k_z \to \infty$). The dilution p of the AFM is 50%; nevertheless the spins are much more connected via the third dimension than it appears from the sketch.

Obviously, the AFM is in a DS, where a domain is defined as a region of undisturbed AFM order. The reason for the domain formation and, consequently, for the lack of long-range order is the interface magnetization which couples to the exchange field coming from the FM and the external field (both pointing up), lowering the energy of the system. The interface magnetization follows from two contributions. Examples for both are indicated by circles. One contribution comes from parallel spin pairs in the domain walls (domain-wall magnetization), all pointing up in our example (Figure 5.23a), that is, in the direction of the exchange field of the FM and the external field. A second contribution comes from an imbalance of the number of defects of the two AFM sub-lattices (volume magnetization). The imbalance in the number of defects of the two AFM sub-lattices also leads to a net magnetization within a domain, which couples to the exchange field of the FM and the external field. The reason for the imbalance is that the domain structure is not random. Rather, it is an optimized structure arising during the initial cooling procedure with as much magnetization as possible coupling to the exchange field of the FM and the external field, following the energy minimization principle.

However, an AFM-interface magnetization alone cannot lead to EB. Only the irreversible part of it (during hysteresis) may lead to EB. Figure 5.23b shows for comparison spin configurations in the same portion of the interface monolayer of the AFM after reversal of the FM. Clearly, the major part of the domain structure did not change during reversal of the FM. However, there are rearrangements on smaller length scales, leading mainly to the fact that the domain-wall magnetization changes its sign. In Figure 5.23b, all of the spin pairs within domain walls now point down following the reversed FM and the external field while the volume magnetization coming from the defects remains frozen. The stability of the domain structure stems from the fact that the domain walls are pinned at defect sites as well as between pairs of spins which are aligned with the field. Hence, during the movement of the domain-wall energy, barriers have to be overcome by thermal activation. This explains why a large domain in general remains in a metastable state on exponentially long time scales, while rearrangements on a shorter length scale are possible, of course depending on the waiting time, the temperature, and the material parameters of the AFM.

Important features of EB systems found experimentally [17] (some of them described in Section 5.3) have their counterpart in simulations of the DS model [16], such as the order of magnitude of EB fields, the shape of hysteresis curves, the dilution dependence of EB, its temperature dependence, the training effect, and the occurrence of positive EB. Other properties of EB systems which were successfully investigated within the framework of the DS model are the dependence of EB on thickness of the AFM [51, 73], the influence of ion irradiation [21], asymmetric reversal modes [49], properties of the AFM domain structures [76], the enhanced coercivity [46], and the cooling-field dependence [48]. In the following subsections we will focus on the influence of the anisotropy of the AFM [14] and on newer results on the influence of interface roughness [25].

5.4.2 Anisotropy Dependence

As already argued in the section before, during the initial cooling procedure the AFM becomes frozen in a DS, the structure of which depends on the system parameters. Some of these model parameters can be estimated from measurements, for example, the exchange interactions from critical temperatures. The anisotropy of the AFM is probably the most unknown quantity since it is hard to measure and only few *ab initio* calculations exist (see, e.g., [77]). In the following discussion, we will treat this anisotropy as an unknown parameter and discuss its influence on the EB.

Typical staggered domain configurations of the bulk AFM are shown for three different values of the AFM anisotropy in Figure 5.24. For low anisotropies, $k_z < J_{AFM}$, AFM domain walls have a width of the order of $(J_{AFM}/k_z)^{1/2}$. Even for the lowest anisotropy shown, the width is only of the order of a few lattice constants, which can hardly be detected in the figure. Also, because of the dilution domain walls tend to follow the holes so that the wall width is further reduced at those places. Interestingly, the domain structure itself depends on k_z. The system has the smallest domains for an intermediate value of $k_z = J_{FM}$ and not for the Ising case corresponding to the high anisotropy limit $k_z = 30J_{FM}$ as one might expect.

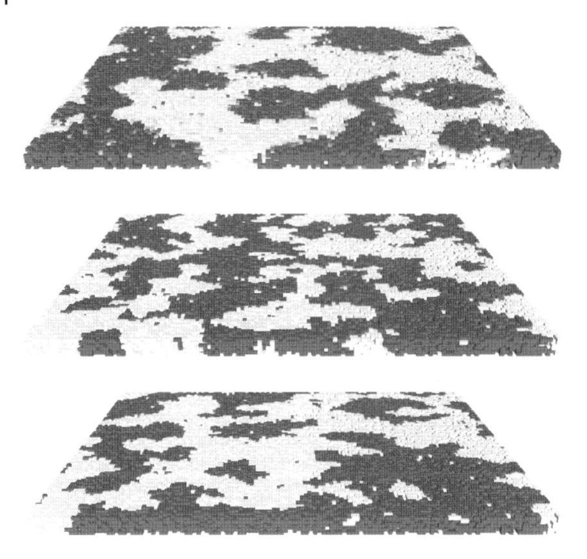

Figure 5.24 Frozen DS of a 40% diluted AFM consisting of six monolayers for different values of the AFM anisotropy, $k_z = 0.1J_{FM}$, $1.0J_{FM}$, $30J_{FM}$ (from top). The shading marks the z-component of the staggered magnetization. Taken from [14].

We will discuss the results following from this behavior later in connection with the anisotropy dependence of the EB.

In the absence of any anisotropy in the FM and for very low dilution of the AFM, we observe a perpendicular coupling across the interface between FM and AFM. The magnetization reversal in the FM is here by coherent rotation. The picture changes with increasing uniaxial anisotropy in the FM and upon increasing dilution of the AFM. The magnetization reversal in the FM is now by domain-wall motion and the perpendicular coupling becomes less significant. This is because uniaxial anisotropies in both the FM and the AFM having the same axis no longer lead to an energy minimum for a perpendicular coupling across the interface. Moreover, AFM spins with missing AFM neighbors can lower their energy by rotating parallel to their FM neighbors. Therefore, in the framework of our calculations a spin-flop coupling is not an essential mechanism for EB.

In [14] the EB field was calculated for a wide range of values of k_z, starting from very soft spins to rigid, Ising-like spins. Figure 5.25 shows the results for different thicknesses of the AFM and for a dilution of $p = 0.4$. Interestingly, one finds a (negative) peak in the EB field for an intermediate value of k_z for a sufficiently thick AFM, while at lower thicknesses the EB field increases with the anisotropy and saturates in the Ising limit.

The key to the understanding of EB is the understanding of AFM domain configurations and domain walls. AFM domains are required to carry a surplus magnetization at the interface, which must be stable during hysteresis in order to produce any EB. In general, one might expect that the most stable domain

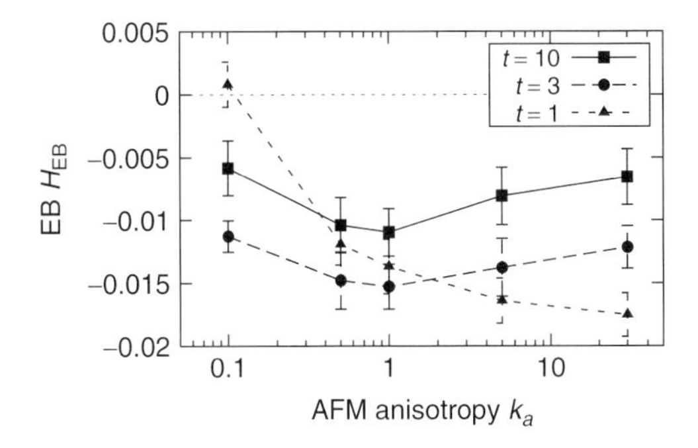

Figure 5.25 EB field versus anisotropy of the AFM for different AFM thicknesses t (numbers of AFM layers). Field and anisotropy are in units of J_{FM}. Taken from [14].

configurations are obtained for the Ising limit. But Figure 5.25 and also Figure 5.24 suggest that the behavior of domains is more complex. Let us start considering the Ising limit where some domain wall is formed upon field cooling. This domain wall preferentially passes through defects thereby minimizing the exchange energy, and at the same time it gathers magnetization thereby lowering the Zeeman energy. When the anisotropy k_z is decreased, the energy to create a domain wall will decrease. Thus, the system will respond by roughening the domain boundaries (see Figure 5.24, where the domain configuration in the middle is more complex than the lower one which represents the Ising limit). This roughening enhances the possibility for the domains to carry any surplus magnetization, and hence the EB will increase. However, there exists a counter-effect. With decreasing k_z the width of the domain wall increases, so that less energy can be saved through the dilution. Hence, for still lower anisotropy the domain walls will smoothen thereby lowering again the exchange energy (see once again Figure 5.24, now comparing the domain configuration in the middle with the upper one, which has much smoother domain walls because of still lower anisotropy). Since flat walls carry less remanent magnetization, the EB will decrease again with decreasing anisotropy. The compromise between these two opposite effects is achieved at some intermediate value of k_z, where the bias shows a maximum.

However, the maximum disappears at lower values of t. This happens since for only one monolayer of AFM we are close to the percolation threshold where the domain walls pass nearly exclusively through the defects at very little or no cost of energy. Therefore, the first mechanism discussed above is less important and the EB increases with k_z till it saturates in the Ising limit.

5.4.3 Structural Dependence

The DS model correctly predicts that some dilution of the AFM bulk – which is conducive to the formation of DS in the AFM – leads to an increased EB. However, prior simulations [16] have predicted an optimal dilution of the order of 50%, while experiments [14, 17] have found it to be of the order of 10%. In the following discussion, the aim is to modify the DS model to include nonplanar structure at the interface. As will be shown, this leads to a new estimate of the optimal dilution, which is much closer to that seen in the experiment and to new insight regarding the dependence of EB on the details of the interface structure [25]. The model comprises an Ising AFM of thickness t_{AFM}, coupled to a Heisenberg FM which is only one monolayer thick. The overall system is hence a cuboid of size $(t_{AFM} + 1) \times L \times L$ where L is set to 128 for all the results presented here. Periodic boundary conditions are applied in-plane and open boundary condition out-of-plane. Spins are arranged on a simple cubic lattice. Energies are calculated using the appropriate Ising or Heisenberg Hamiltonians, considering only the nearest neighbor exchange interactions and disregarding the dipole interaction.

As explained before the bulk of the AFM is randomly diluted with nonmagnetic defects which replace a proportion p of the spins. As a new aspect in the model, we now introduce some intermixing of the FM and AFM sites in one monolayer at the interface. This intermixing represents a minimum amount of roughness which in a real system will always exist. This mixing is described by a mixing coefficient R. In the layer adjacent to the FM monolayer, R is the proportion of spins which are part of the FM rather than part of the AFM. The dilution p continues to act in this layer, such that this layer consists of a proportion $(1 - p)\ R$ of FM sites, and $(1 - p)(1 - R)$ of AFM sites. Note, that for $R \to 0$ and $R \to 1$ the interface is perfectly flat and compensated while the roughness is maximal for $R = 0.5$. The model presented here is clearly a vast simplification of a real interface, which usually will have larger roughness with more than one monolayer involved, but as we will show in the following discussion the results it produces are nevertheless intriguing.

The effect of the interface mixing R upon the EB field is shown in Figure 5.26. We see that systems with low or no dilution feature a strong response in EB to the introduction of roughness, while systems with high dilution that previously [16] gave maximal EB are less affected. Comparing these results to the previous work [16], we note two most important features: firstly, that EB is exhibited even for systems with no dilution whatsoever ($p = 0$) and, secondly, that the rough interface

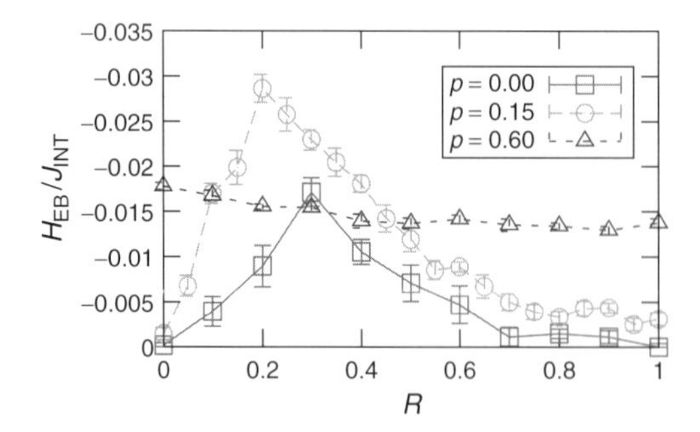

Figure 5.26 Variation of the EB field with the interface mixing R for different degrees of AFM dilution p. The thickness of the AFM is four monolayers. Taken from [25].

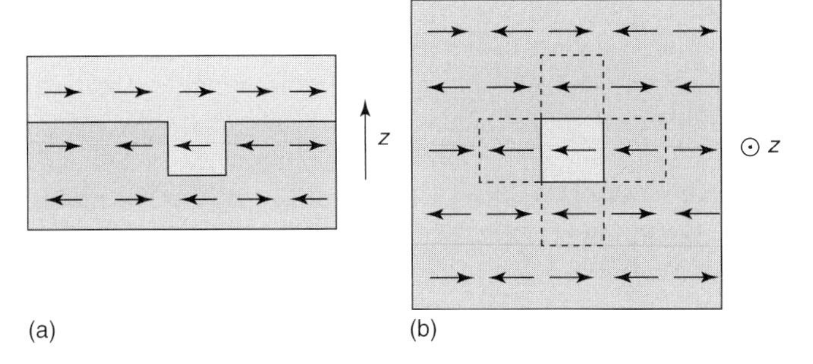

Figure 5.27 Isolated FM spin in the interface layer of the AFM as it is typical for low FM/AFM interface spin mixing coefficient R. The sketch (a) is a cross section while (b) shows the top view of the interface. On the (a) the FM is reversed while the lone FM spin embedded in the AFM is pinned by the interaction with its AFM neighbors.

renders a greater EB field ($\approx 0.027 J_{\mathrm{INT}}$) than the greatest previous result from the DS model ($\approx 0.02 J_{\mathrm{INT}}$). Interestingly, the maximal EB is achieved for $R \approx 0.2$, that is, for systems with a small amount of FM spins in the interface layer of the AFM.

The reason for the enhanced EB lies in the optimized interface coupling associated with low values of R. For a given DS, EB is greater in slightly rough systems because it effectively enhances the FM–AFM coupling. To understand this, consider an FM spin embedded in the AFM (as sketched in Figure 5.27), having five AFM neighbors and one FM neighbor. This lone FM spin will be pinned by the AFM, since $5|J_{\mathrm{AFM}}| > |J_{\mathrm{FM}}|$. For positive interface exchange interaction, it acts like a reversed AFM spin, except that unlike a normal AFM spin it has a FM interaction with its FM neighbor, thus increasing the net exchange coupling between FM and AFM since in usual EB systems the FM interactions are the strongest interactions. Note that the above argument works only in the limit of small R where one finds mainly isolated FM spins embedded in the AFM, instead of larger groups of FM spins or even lone AFM spins in an FM environment. This fact explains the asymmetry of the curves in Figure 5.26, which one might naively not expect. The largest EB field is achieved for a rather small amount of FM spins in the mixed interface layer. However, it is not obvious how the interface mixing could be controlled microscopically in an experiment.

The peak of the EB field with respect to R results from the tendency of excessive R to destroy the enhanced interface coupling described above: if there are too many FM spins in the interface layer, then they form groups which are no longer frozen into the AFM. For example, for $R = 0.5$ an FM interface-layer spin has on average three FM neighbors and three AFM neighbors, so with $J_{\mathrm{FM}} > J_{\mathrm{AFM}}$ it follows that the FM during hysteresis is reversible, rather than being pinned by the AFM. Furthermore, the peak in Figure 5.26 moves left with dilution because replacing some AFM sites with defects reduces the R threshold required to allow FM spins to be mixed with the AFM to move with the FM rather than being pinned to the AFM.

The presence of DS-fostering EB even at dilution $p = 0$ may be explained by the defect-like domain-wall-pinning action of FM spins in the AFM interface layer. For a smooth interface, it was found that having defects only at the interface was not sufficient to yield an EB, but the use of FM spins conveys an advantage over defects since they enhance rather than diminish the interface coupling. This strengthened interface coupling increases both the exchange field provided by the FM during cooling and the action of the AFM upon the FM during hysteresis. Note that in the limit of vanishing bulk dilution our model coincides with the model of Malozemoff [6–8] for minimal roughness, so that our findings qualitatively support his earlier ideas. To the best of our knowledge, the simulations in [25] are the only numerical investigations of Malozemoff's model.

For highly diluted ($p \approx 0.6$) systems, R appears to have only little effect. To understand this, consider the case $p = 0.5$. Here, FM interface spins have on average 1 FM neighbor, 2.5 defect neighbors, and 2.5 AFM neighbors. They are no longer firmly tied to the AFM, which annuls the enhanced interface coupling. For such high dilutions, roughness does not change the interface structure so much. Furthermore, because the DS is already well formed due to dilution, the roughness does not improve the AFM domain structure either. Overall, for $p = 0.6$ we see a slight decrease in EB with R as can be explained by the increasing thickness of the FM. The values of the EB field for $R = 1$ are in general about the same as those for $R = 0$. This is quite unexpected since the EB should decrease with increasing thickness of the FM, that is, it should be proportional to $1/t_{FM}$. However, within the error bars it does not decrease at all. Probably, this is an artifact of the small AFM thicknesses used ($t_{AFM} = 4$) since the net decrease in the thickness of the AFM with R increases its magnetization due to the lower energy cost of domain formation.

Maximal EB is found in systems with a small dilution ($p \approx 0.1$), as well as some mixing ($R \approx 0.2$) at the interface. This is an important result since it agrees with the experimental fact that the largest EB is found in systems with dilutions of the order of 10% (see Section 5.3.2.1 and [14, 17]), while previous simulations of the DS model massively overestimated the optimal dilution [14, 16].

5.5 Conclusions

Characteristic features of EB found experimentally in Co/CoO-based systems can be described theoretically by the DS model. Among these are the order of magnitude of EB fields, the shape of hysteresis curves, the dilution dependence of EB, its temperature dependence, the training effect, and the occurrence of positive EB. Other typical features of EB systems which were successfully investigated within the framework of the DS model are the dependence of EB on the thickness of the AFM, the dependence on the anisotropy of the AFM, the influence of ion irradiation, asymmetric reversal modes, properties of the AFM domain structures, the enhanced coercivity, the cooling-field dependence, and the influence of interface roughness.

However, one should note that most of the AFMs used in EB systems have a polycrystalline structure. Some model calculations for the metallic AFMs rest on a granular model for the AFM [12, 78, 79]. So far, the grainy structure of a realistic AFM was taken into account within the DS model only in the case of numerical simulations for $Ni_{80}Fe_{20}/(Fe_{50}Mn_{50})_{1-x}Cu_x$ [80]. In the metallic AFM, diluted by Cu of concentration x, the exchange coupling between the grains was omitted. A Heisenberg model was used, which explicitly accounted for the energy barriers upon reversal of the AFM spins and for the thermal relaxation of the magnetization according to the Arrhenius–Néel law. The role of the AFM domains in $\mathbf{B}_{EB}$ was tested by comparing samples for increasing dilution x with those in which for $x = 0$ the grain size was reduced. The finding was that decreasing the grain size and lowering the defect concentration x have the same effect on $\mathbf{B}_{EB}$ as increasing x for larger grains. Work following these lines is still missing for AFM oxides and would certainly contribute to the further understanding of EB.

References

1. Nogués, J. and Schuller, I.K. (1999) Exchange bias. *J. Magn. Magn. Mater.*, **192**, 203.
2. Berkowitz, A.E. and Takano, K. (1999) Exchange anisotropy – a review. *J. Magn. Magn. Mater.*, **200**, 552.
3. Stamps, R.L. (2000) Mechanism for exchange bias. *J. Phys. D: Appl. Phys.*, **33**, R247.
4. Radu, F. and Zabel, H. (2007) in *Springer Tracts in Modern Physics*, vol. 227 (eds H. Zabel and S.D. Bader), Springer-Verlag, Berlin Heidelberg, pp. 97–184.
5. (a) Meiklejohn, W.H. and Bean, C.P. (1956) New magnetic anisotropy. *Phys. Rev*, **102**, 1413; (b) Meiklejohn, W.H. and Bean, C.P. (1957) New magnetic anisotropy. *Phys. Rev.* **105**, 904.
6. Malozemoff, A.P. (1987) Random-field model of exchange anisotropy at rough ferromagnetic-antiferromagnetic interfaces. *Phys. Rev. B*, **35**, 3679.
7. Malozemoff, A.P. (1988) Heisenberg-to-Ising crossover in a random-field model with uniaxial anisotropy. *Phys. Rev. B*, **37**, 7673.
8. Malozemoff, A.P. (1988) Mechanism of exchange anisotropy. *J. Appl. Phys.*, **63**, 3874.
9. Mauri, D., Siegmann, H.C., Bagus, P.S., and Kay, E. (1987) Simple model for thin ferromagnetic films exchange coupled to an antiferromagnetic substrate. *J. Appl. Phys.*, **62**, 3047.
10. Koon, N.C. (1997) Calculations of exchange bias in thin films with ferromagnetic/antiferromagnetic interfaces. *Phys. Rev. Lett.*, **78**, 4865.
11. Schulthess, T.C. and Butler, W.H. (1998) Consequences of spin-flop coupling in exchange biased films. *Phys. Rev. Lett.*, **81**, 4516.
12. Stiles, M.D. and McMichael, R.D. (1999) Model for exchange bias in polycrystalline ferromagnet-antiferromagnet bilayers. *Phys. Rev. B*, **59**, 3722.
13. Kiwi, M., Mejía-López, J., Portugal, R.D., and Ramírez, R. (1999) Exchange bias model for Fe/FeF_2: role of domains in the ferromagnet. *Europhys. Lett.*, **48**, 573.
14. Miltényi, P., Gierlings, M., Keller, J., Beschoten, B., Güntherodt, G., Nowak, U., and Usadel, K.D. (2000) Diluted antiferromagnets in exchange bias: proof of domain state model. *Phys. Rev. Lett.*, **84**, 4224.
15. Nowak, U., Misra, A., and Usadel, K.D. (2002a) Modeling exchange bias microscopically. *J. Magn. Magn. Mater.*, **240**, 243.
16. Nowak, U., Usadel, K.D., Miltényi, P., Keller, J., Beschoten, B., and Güntherodt, G. (2002b) The domain

state model for exchange bias I: theorie. *Phys. Rev. B*, **66**, 14430.
17. Keller, J., Miltényi, P., Beschoten, B., Güntherodt, G., Nowak, U., and Usadel, K.D. (2002) The domain state model for exchange bias II: experiment. *Phys. Rev. B*, **66**, 14431.
18. Shi, H.T., Lederman, D., and Fullerton, E.E.C. (2002) Exchange bias in $Fe_xZn_{1-x}F_2$/Co bilayers. *J. Appl. Phys.*, **91**, 7763.
19. Mewes, T., Lopusnik, R., Fassbender, J., and Hillebrands, B. (2000) Suppression of exchange bias by ion irradiation. *Appl. Phys. Lett.*, **76**, 1057.
20. Mougin, A., Mewes, T., Jung, M., Engel, D., Ehresmann, A., Schmoranzer, H., Fassbender, J., and Hillebrands, B. (2001) Local manipulation and reversal of the exchange bias field by ion irradiation in FeNi/FeMn double layers. *Phys. Rev. B*, **63**, 060409.
21. Misra, A., Nowak, U., and Usadel, K.D. (2003) Control of exchange bias by diluting the antiferromagnetic layer. *J. Appl. Phys.*, **93**, 6593.
22. Nolting, F., Scholl, A., Stöhr, J., Seo, J.W., Fompeyrine, J., Siegwart, H., Locquet, J.-P., Anders, S., Lüning, J., Fullerton, E.E., Toney, M.F., Scheinfein, M.R., and Padmor, H.A. (2000) Direct observation of the alignment of ferromagnetic spins by antiferromagnetic spins. *Nature (London)*, **405**, 767.
23. Ohldag, H., Scholl, A., Nolting, F., Anders, S., Hillebrecht, F.U., and Stöhr, J. (2001) Spin eorientation at the antiferromagnetic NiO(001) surface in response to an adjacent ferromagnet. *Phys. Rev. Lett.*, **86**, 2878.
24. Finazzi, M., Duò, L., and Cicacci, F. (2009) Magnetic properties of interfaces and multilayers based on thin antiferromagnetic oxide films. *Surf. Sci. Rep.*, **64**, 139.
25. Spray, J. and Nowak, U. (2006) Exchange bias in ferromagnetic / antiferromagnetic bilayers with imperfect interfaces. *J. Phys. D*, **39**, 4536.
26. Kleemann, W. (1993) Random-field induced antiferromagnetic, ferroelectric and structural domains states. *Int. J. Mod. Phys. B*, **7**, 2469.
27. Belanger, D.P. (1998) Experiments on the random field Ising model, in *Spin Glasses and Random Fields* (ed. A.P.Young), World Scientific, Singapore, pp. 251–276.
28. Fishman, S. and Aharony, A. (1979) Random field effects in disordered anisotropic antiferromagnets. *J. Phys. C*, **12**, L729.
29. Cardy, J.L. (1984) Random-field effects in site-disordered Ising antiferromagnets. *Phys. Rev. B*, **29**, 505.
30. Hartmann, A. and Nowak, U. (1999) Universality in three dimensional random-field ground states. *Eur. Phys. J. B*, **7**, 105.
31. Imbrie, J.Z. (1984) Lower critical dimension of the random-field Ising model. *Phys. Rev. Lett.*, **53**, 1747.
32. Bricmont, J. and Kupiainen, A. (1987) Lower critical dimension for the random-field Ising model. *Phys. Rev. Lett.*, **59**, 1829.
33. Montenegro, F.C., King, A.R., Jaccarino, V., Han, S.-J., and Belanger, D.P. (1991) Random-field-induced spin-glass-like behaviour in the diluted Ising antiferromagnet $Fe_{0.31}Zn_{0.69}F_2$. *Phys. Rev. B*, **44**, 2255.
34. Nowak, U. and Usadel, K.D. (1991) Diluted antiferromagnets in a magnetic field: a fractal domain state with spin-glass behaviour. *Phys. Rev. B*, **44**, 7426.
35. Usadel, K.D. and Nowak, U. (1992) Diluted antiferromagnets in a magnetic field: evidence for a spin-glass phase. *J. Magn. Magn. Mater.*, **104–107**, 179.
36. Imry, Y. and Ma, S. (1975) Random-field instability of the ordered state of continuous symmetry. *Phys. Rev. Lett.*, **35**, 1399.
37. Nowak, U. and Usadel, K.D. (1992) Structure of domains in random Ising magnets. *Phys. Rev. B*, **46**, 8329.
38. Esser, J., Nowak, U., and Usadel, K.D. (1997) Exact ground state properties of disordered Ising systems. *Phys. Rev. B*, **55**, 5866.
39. Birgeneau, R.J., Cowley, R.A., Shirane, G., and Yoshizawa, H. (1984) Metastability in DAFF after field cooling. *J. Stat. Phys.*, **34**, 817.

40. Belanger, D.P., Rezende, M., King, A.R., and Jaccarino, V. (1985) Hysteresis, metastability, and time dependence in d=2 and d=3 random-field Ising systems. *J. Appl. Phys.*, **57**, 3294.
41. Villain, J. (1984) Nonequilibrium "critical" exponents in the random-field Ising model. *Phys. Rev. Lett.*, **52**, 1543.
42. Han, S.-J., Belanger, D.P., Kleemann, W., and Nowak, U. (1992) Relaxation of the excess magnetization of random-field induced metastable domains $Fe_{0.47}Zn_{0.53}F_2$. *Phys. Rev. B*, **45**, 9728.
43. Nowak, U., Esser, J., and Usadel, K.D. (1996) Dynamics of domains in diluted antiferromagnets. *Phys. A*, **232**, 40.
44. Staats, M., Nowak, U., and Usadel, K.D. (1998) Non-exponential relaxation in diluted antiferromagnets. *Phase Transitions*, **65**, 159.
45. Pollak, P., Kleemann, W., and Belanger, D.P. (1988) Metastability of the uniform magnetization in three-dimensional random-field Ising model systems. II. $Fe_{0.47}Zn_{0.53}F_2$. *Phys. Rev. B*, **38**, 4773.
46. Scholten, G., Usadel, K.D., and Nowak, U. (2005) Coercivity and exchange bias of ferromagnetic/antiferromagnetic multilayers. *Phys. Rev. B*, **71**, 64413.
47. Schulthess, T.C. and Butler, W.H. (1999) Coupling mechanisms in exchange biased films. *J. Appl. Phys.*, **85**, 5510.
48. Wee, L., Stamps, R.L., and Camley, R.E. (2001) Temperature dependence of domain wall bias and coercivity. *J. Appl. Phys.*, **89**, 6913.
49. Beckmann, B., Nowak, U., and Usadel, K.D. (2003) Asymmetric reversal modes in ferromagnetic/antiferromagnetic multilayers. *Phys. Rev. Lett.*, **91**, 187201.
50. Grest, G.S., Soukoulis, C.M., and Levin, K. (1986) Comparative Monte Carlo and mean-field studies of random-field Ising systems. *Phys. Rev. B*, **33**, 7659.
51. Ali, M., Marrows, C.H., Al-Jawad, M., Hickey, B.J., Misra, A., Nowak, U., and Usadel, K.D. (2003) Antiferromagnetic layer thickness dependence of the IrMn/Co exchange bias system. *Phys. Rev. B*, **68**, 214420.
52. Battle, P.D., Cheetham, A.K., and Gehring, G.A. (1979) A neutron diffraction study of the structure of the antiferromagnet $Co_pNi_{1-p}O$. *J. Appl. Phys.*, **50**, 7578.
53. Dieckmann, R. (1977) Cobaltous oxide point defect structure and nonstoichiometry, electrical conductivity, cobalt tracer diffusion. *Z. Phys. Chem.*, **107**, 189.
54. Kittel, C. (1971) *Introduction to Solid State Physics*, John Wiley & Sons, Ltd, Chichester.
55. Lagally, M.G., Savage, D.E., and Tringides, M.C. (1987) in *in Reflection High-Energy Electron Diffraction and Reflection Electron Imaging of Surfaces* (eds P.K.Larsen and P.J. Dobson), Plenum Press, New York, p. 139.
56. Nogués, J., Moran, T.J., Lederman, D., Schuller, I.K., and Rao, K.V. (1999) Role of interrfacial structure on exchange-biased FeF_2-Fe. *Phys. Rev. B*, **59**, 6984.
57. Nogués, J., Leighton, C., and Schuller, I.K. (2000) Correlation between antiferromagnetic interface coupling and positive exchange bias. *Phys. Rev. B*, **61**, 1315.
58. Fecioru-Morariu, M., Wrona, J., Papusoi, C., and Güntherodt, G. (2008) Training and temperature effects of epitaxial and polycrystalline $Ni_{80}Fe_{20}/Fe_{50}Mn_{50}$ exchange biased bilayers. *Phys. Rev. B*, **77**, 054441.
59. Takano, K., Kodama, R.H., Berkowitz, A.E., Cao, W., and Thomas, G. (1997) Interfacial uncompensated antiferromagnetic spins: role in unidirectional anisotropy in polycrystalline NiFe/CoO bilayers. *Phys. Rev. Lett.*, **79**, 1130.
60. Ghadimi, M.R., Beschoten, B., and Güntherodt, G. (2005) Role of structural defects on exchange bias in the epitaxial CoO/Co system. *Appl. Phys. Lett.*, **87**, 261903.
61. Stauffer, D. and Aharoni, A. (1991) Percolation network of polypyrrole in conducting polymer composites, in *Synthetic Metals*. Taylor & Francis, London.
62. Ghadimi, M.R., Ali, S.R., Fecioru-Morariu, M., and Güntherodt, G. (2010) to be published.
63. Fecioru-Morariu, M., Ali, S.R., Papusoi, C., Sperlich, M., and Güntherodt, G. (2007) Effects of Cu dilution in IrMn

on the exchange bias of CoFe/IrMn bilayers. *Phys. Rev. Lett.*, **99**, 097206.

64. Ambrose, T. and Chien, C.L. (1998) Dependence of exchange field and coercivity on cooling field in NiFe/CoO bilayers. *J. Appl. Phys.*, **83**, 7222.
65. Leighton, C., Nogués, J., Suhl, H., and Schuller, I.K. (1999) Competing interfacial exchange and Zeeman energies in exchange biased bilayers. *Phys. Rev. B*, **60**, 12837.
66. Miltényi, P., Gierlings, M., Bamming, M., May, U., Güntherodt, G., Nogués, J., Gruyters, M., Leighton, C., and Schuller, I.K. (1999) Tuning exchange bias. *Appl. Phys. Lett.*, **75**, 2304.
67. Soeya, S., Imagawa, T., Mitsuoka, K., and Narishige, S. (1994) Distribution of blocking temperature in bilayered $Ni_{81}Fe_{19}$/NiO Films. *J. Appl. Phys.*, **76**, 5356.
68. Vallejo-Fernandez, G., Fernandez-Outon, L.E., and O'Grady, K. (2008) Thermal activation of bulk and interfacial order in exchange biased systems. *J. Appl. Phys.*, **103**, 07C101.
69. Ghadimi, M.R., Fecioru-Morariu, M., Beschoten B., and Güntherodt, G. (2010) to be published.
70. Fecioru-Morariu, M. (2008) Exchange bias of metallic ferro-/antiferromagnetic bilayers: effects of structure, dilution, anisotropy and temperature. PhD Thesis, RWTH Aachen.
71. Papusoi, C., Sousa, R., Herault, J., Prejbeanu, I.L., and Dieny, B. (2008) Probing fast heating in magnetic tunnel junction structures with exchange bias. *New J. Phys.*, **10**, 103006.
72. Borchers, J.A., Carey, M.J., Erwin, R.W., Majkrzak, C.F., and Berkowitz, A.E. (1993) Spatially modulated antiferromagnetic order in CoO/NiO superlattices. *Phys. Rev. Lett.*, **70**, 1878.
73. Nowak, U., Misra, A., and Usadel, K.D. (2001) Domain state model for exchange bias. *J. Appl. Phys.*, **89**, 7269.
74. Beckmann, B., Usadel, K.D., and Nowak, U. (2006) Cooling field dependence of asymmetric reversal modes for ferromagnetic/antiferromagnetic multilayers. *Phys. Rev. B*, **74**, 054431.
75. Nowak, U. (2001) Thermally activated reversal in magnetic nanostructures. *Annu. Rev. Comput. Phys.*, **9**, 105.
76. Misra, A., Nowak, U., and Usadel, K.D. (2004) Structure of domains in an exchange bias model. *J. Appl. Phys.*, **95**, 1357.
77. Szunyogh, L., Lazarovits, B., Udvardi, L., Jackson, J., and Nowak, U. (2009) Giant magnetic anisotropy of the bulk antiferromagnets IrMn and IrMn3 from first principles. *Phys. Rev. B*, **79**, 020403(R).
78. Suess, D., Kirschner, M., Schrefl, T., Fidler, J., Stamps, R.L., and Kim, J.-V. (2003) Exchange bias of polycrystalline antiferromagnets with perfectly compensated interfaces. *Phys. Rev. B*, **67**, 54419.
79. Craig, B., Lamberton, R., Johnston, A., Nowak, U., Chantrell, R.W., and O'Grady, K. (2008) A model of the temperature dependence of exchange bias in coupled ferromagnetic/antiferromagnetic bilayers. *J. Appl. Phys.*, **103**, 07C102.
80. Papusoi, C., Hauch, J., Fecioru-Morariu, M., and Güntherodt, G. (2006) Tuning the exchange bias of soft metallic antiferromagnets by inserting nonmagnetic defects. *J. Appl. Phys.*, **99**, 123902.

6
Theory of Ferromagnetic–Antiferromagnetic Interface Coupling

Alexander I. Morosov and Alexander S. Sigov

6.1
Introduction

This chapter describes spin frustration that may occur on the boundary between the layers in the problem of spin-exchange interaction between ferromagnetic and antiferromagnetic layers, where we consider the cases of rough (for uncompensated surface of the antiferromagnet) and atomically smooth (for compensated surface of the antiferromagnet) interface between the layers. We study the distribution of magnetic order parameters that arises in the near-interface region, and obtain a magnetic phase diagram in the parameter space defined by the layer thickness and by the surface roughness for the two-layer ("ferromagnet–antiferromagnet") and spin-valve three-layer ("ferromagnet–antiferromagnet–ferromagnet") cases. By incorporating the single-ion anisotropy energy, our model improves upon the results of the exchange approximation.

Multilayered systems composed from alternating ferromagnetic and antiferromagnetic layers are widely used in magnetoresistive detectors based on the giant magnetoresistance (GMR) effect, in the reading heads of hard drives, and in magnetoresistive memory chips [1]. As layer thickness is usually on the order of nanometers or fractions of a nanometer, the interlayer boundary and its morphology have a substantial and often decisive influence on physical properties of multilayered magnetic structures.

Frustration of spin-exchange interaction between spins belonging to different layers on the interlayer boundary can give rise to a fairly complex magnetic phase diagram, which should be studied in detail in order to achieve optimization of element properties and devices in magnetoelectronic applications. Such an optimization also requires manufacturers to adopt production procedures ("technological paths") that would guarantee the necessary structure of interface boundaries.

This chapter is organized as follows. Section 6.2 discusses the concept of frustration on the ferromagnetic–antiferromagnetic interface. Section 6.3 lays out the mathematical model based on which a spatial distribution of ferro- and antiferromagnetic order parameters has been computed for a multilayered magnetic system. Section 6.4 considers distortion of magnetic order parameters

Magnetic Properties of Antiferromagnetic Oxide Materials.
Edited by Lamberto Duò, Marco Finazzi, and Franco Ciccacci

ISBN: 978-3-527-40881-8

near the interface between sufficiently thick ferromagnetic and antiferromagnetic layers, as well as the resulting effective interaction between order parameters in the bulk of the layers. It should be mentioned here that single-ion anisotropy energy is taken into proper consideration in this section, which is an improvement upon the exchange approximation.

The magnetic phase diagram of a thin ferromagnetic layer on a thick antiferromagnetic substrate is discussed at length in Section 6.5, and the phase diagram of a spin-valve system of the ferromagnetic–antiferromagnetic–ferromagnetic type in Section 6.6. Additionally, these sections provide a description of a new type of domain walls that arise due to frustrations of spin-exchange interaction.

Conclusions and practical recommendations are given in the last section, which is followed by the list of references.

6.2
Frustrations on the Ferromagnet–Antiferromagnet Interface

In this chapter, we study the case when lattice constants of the ferro- and antiferromagnet are closely matched, and the crystal lattice of one layer can be thought of as a continuation of the crystal lattice of the other layer. In addition, we shall assume a collinear antiferromagnet with two sublattices.

Our analysis reveals an essential distinction between the two following cases:

1) Atomic planes of the antiferromagnet that are parallel to the interface are uncompensated, in other words, in every such plane all spins belong to one sublattice of the antiferromagnet, whereby the spins of neighboring planes belong to different sublattices.
2) Atomic planes of the antiferromagnet that are parallel to the interface are compensated, in other words, they contain an equal number of spins belonging to each sublattice, and thus have a zero magnetic moment in the absence of external fields.

Let us consider the concept of frustration for each of these two cases, limiting our analysis to the model of Heisenberg exchange interaction between localized quasi-classic spins in the nearest-neighbor interaction approximation:

$$\hat{H}_H = -J_{ij}(\hat{S}_i, \hat{S}_j) \tag{6.1}$$

where J_{ij} equals $J_f \geq 0$ if both spins belong to the ferromagnet, $J_{ij} = J_{af} < 0$ if both spins belong to the antiferromagnet, and $J_{ij} = J_{f,af}$ if the spins belong to different layers. In the discussion below we assume, for definiteness, that $J_{f,af} > 0$.

6.2.1
Uncompensated Surface of the Antiferromagnet

Consider the case of layers having (100) orientation, and the crystal lattice having tetragonal volume-centered structure, with easy axis of magnetization [001] lying in the plane of the layers.

In the case of an ideal atomically smooth interface, it is easy to find the orientation of ferro- and antiferromagnetic order parameters (which are equal, respectively, to the magnetization of the ferromagnet and to the magnetization difference between sublattices of the antiferromagnet) that corresponds to the energy minimum of exchange interaction: inside the layers, order parameters are uniformly distributed and their mutual orientation is such that spins of the ferromagnet are parallel to spins of the atomic plane of the antiferromagnet that is closest to the interface since $J_{f,af} > 0$ (Figure 6.1a). In the case of $J_{f,af} < 0$, the mutual orientation of the spins would be antiparallel.

However, a real interface always contains atomic steps whose height is equal to one or several interatomic distances. In the following discussion we consider, for the sake of simplicity, only those steps whose height is equal to one interplane distance. However, one can easily show that steps whose height is equal to an odd

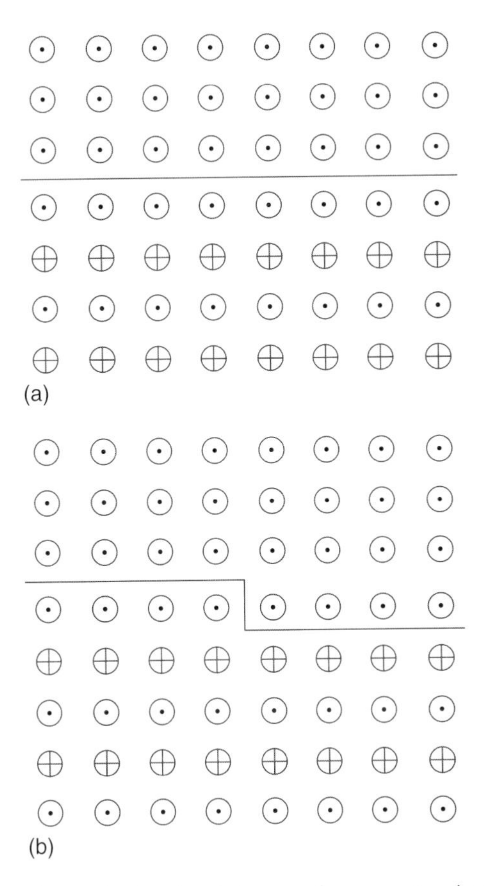

Figure 6.1 Interface between a ferromagnet and an uncompensated antiferromagnet: (a) perfectly smooth planar interface and (b) interface containing an atomic step.

number of interplane distances are fully equivalent to the ones we consider as far as frustrations are concerned, whereas those steps whose height is equal to an even number of interplane distances do not cause any frustrations and can therefore be excluded from consideration. It is thus sufficient to limit our analysis to steps of the former type, that is, step height is equal to an odd number of interplane distances.

On the opposite sides of a step of the first type, spins of the ferromagnet interact with those of the antiferromagnet that belong to different sublattices. It is impossible to identify a uniform distribution of magnetic order parameters that would minimize the energy of exchange interaction.

Indeed, if on the one side of the step the mutual orientation of the ferro- and antiferromagnetic order parameters is associated with the minimum energy of exchange interaction between the layers, then the same energy is maximum on the other side (Figure 6.1b). One thus observes a frustration that is due to the presence of an atomic step at the interface. A description of nonuniform distributions of the order parameters that arise due to frustrations is the subject of the key sections of this chapter.

6.2.2
Compensated Surface of the Antiferromagnet

Consider, for instance, the (110) orientation of layers having a tetragonal volume-centered crystal lattice, the easy axis of magnetization [001] lying in the plane of the layers.

In this case, spin frustration takes place even when the interface is atomically smooth. Indeed, exchange interaction between spins of the ferromagnet tends to align them parallel to each other, while exchange interaction between spins of the ferro- and antiferromagnet also tends to orient spins in the upper layer of the antiferromagnet parallel to the spins of the ferromagnet and thereby parallel to each other. On the other hand, exchange interaction between spins of the antiferromagnet tends to make neighboring spins of the antiferromagnet (including those in the upper layer) antiparallel to each other. The end result is a frustration of exchange interaction: one cannot find a uniform distribution of magnetic order parameters that would minimize the energies of all three spin–exchange interactions.

As spin frustration takes place even at atomically smooth interfaces, the possible presence of atomic steps on the physical interface is not nearly as critically important as it was for the uncompensated surface of the antiferromagnet. Except some special orientations of step edges that are outside the scope of this chapter, roughness of the interface does not cause any substantial change in the spin distribution. Thus when the surface of the antiferromagnet is compensated, it will be enough to consider only the case of an atomically smooth interface between the ferromagnetic and antiferromagnetic layers.

6.3 Mathematical Model

The majority of results discussed in this chapter have been obtained by computer simulation based on the following discrete model.

The temperature range is $T \ll T_C, T_N$ where T_C is the Curie temperature of the ferromagnet and T_N is the Neel temperature of the antiferromagnet. The magnetic moments of atoms are assumed to be constant in magnitude. A lattice of localized quasi-classical spins is studied in the approximation of Heisenberg exchange interaction between nearest neighbors. The direction of the localized spin is given by a unit vector $\mathbf{s}_i$ and its magnitude enters the corresponding interaction constant J_{ij}. The exchange interaction energy (Eq. 6.1) then takes the form

$$W_{ij}^{\text{ex}} = -J_{ij}\left(\mathbf{s}_i, \mathbf{s}_j\right) \tag{6.2}$$

We examine the cuts of a tetragonal body-centered spin lattice whose ratio a/c is such that the atom located at the center of the unit cell has atoms in its vertices as the atom nearest neighbors. In this case, the (100) and (110) cuts correspond, respectively, to the uncompensated and the compensated surface of the antiferromagnet.

An orthogonal Cartesian system is introduced, whose x-axis is lying in the plane of the layers and is pointing along the easy axis of magnetization [001], while the z-axis is perpendicular to the plane of the layers.

With this notation, the energy of the single-ion anisotropy has the form

$$W_{\text{an}} = K_{\perp} \sum_{i \in f} \left(s_i^{(z)}\right)^2 - K_{\|} \sum_{i} \left(s_i^{(x)}\right)^2 \tag{6.3}$$

where summation is over the localized spins, $K_{\|} > 0$ is the uniaxial anisotropy constant, and $K_{\perp} > 0$ is the surface anisotropy constant of the ferromagnet that implies the energy disadvantage of states having a nonzero z-component (i.e., component normal to the layer plane) of magnetization. An attempt to take dipole–dipole spin interaction into account explicitly makes the problem dramatically more complicated and leads to longer computation times.

In addition to tetragonal lattices, cubic crystal lattices with layer orientation (100) have been studied as well. In a body-centered cubic lattice, the (100) cut is uncompensated, and the difference from a tetragonal lattice is that two mutually perpendicular easy axes are present in the layer plane. The second term in Eq. (6.3) should be replaced by the expression

$$-K_{\|} \sum_{i} \left[\left(s_i^{(x)}\right)^4 + \left(s_i^{(y)}\right)^4\right] \tag{6.4}$$

A compensated surface (100) of the antiferromagnet is observed in the case of a simple cubic lattice with antiferromagnetic ordering of type C, when the neighboring spins belong to two different sublattices. In this case, we again observe two mutually perpendicular easy axes in the plane of the layer, and the expression (6.4) should be used to obtain the energy of single-ion anisotropy.

The equilibrium distribution of localized spins was obtained by modeling their dynamics on the basis of the system of Landau–Lifshitz–Gilbert equations:

$$\hbar \frac{d\mathbf{s}_i}{dt} = [\mathbf{s}_i \mathbf{H}_{\text{eff}}] + \mu \mathbf{H}_{\text{eff}} \tag{6.5}$$

where μ is the damping constant, and the effective field H_{eff} is

$$H^p_{\text{eff}} = -\frac{\partial W}{\partial s^p_i} \tag{6.6}$$

where $p = x, y, z$, and W is the sum of the exchange energy and the anisotropy energy.

The solution of the system (Eq. 6.5) was found by the "classical" fourth-order Runge–Kutta method. A transition to the equilibrium was identified by looking at the temporal behavior of the overall energy of the spin system. As a rule, the initial state was chosen in such a way as to be almost identical to the state that would occur in the absence of spin frustration. A weak perturbation is necessary to prevent the initial state from being an extremal state. In this case, there is no time evolution of spins.

The simulation was carried out for the case when atomic step edges on the boundaries between the layers are parallel to each other and to the y-axis of a Cartesian system of coordinates; in other words, the problem was a two-dimensional one with periodic (with respect to x) boundary conditions. This last assumption does not cause any substantial qualitative change to the magnetic phase diagram.

6.4 The Interface between Thick Ferromagnet–Antiferromagnet Layers

The term *thick* should not be taken too literally with regard to the layers. What is assumed in this section is that the thickness of layers is far larger than the characteristic distance R between atomic step edges at the boundary separating the layers.

In addition, in the case of layers with single-ion anisotropy of the "easy-axis" type, that is, when there is just one easy axis lying in the plane of the layers, it is necessary that layer thickness must substantially exceed the thickness of Bloch, Neel, and hybrid walls in the respective layers. As the thicknesses $\Delta_{f(af)}$ and surface energies $\varepsilon_{f(af)}$ of these walls (where indexes f and af relate to the ferro- and antiferromagnet, respectively) have the same order of magnitude [2]

$$\Delta_f \sim b \left(J_f / K_{||}\right)^{1/2} \tag{6.7}$$

where b is the lattice constant,

$$\varepsilon_f \sim \left(J_f K_{||}\right)^{1/2} / b^2 \tag{6.8}$$

we refer to them as *traditional* domain walls.

When these conditions are satisfied, all inhomogeneities of magnetic order parameters that arise due to spin frustration will be concentrated in the near-interface region.

6.4.1
Uncompensated Surface of the Antiferromagnet

In this case, one should consider two possible limiting-case formulas relating the above-mentioned parameters $\Delta_{f(af)}$ and R.

6.4.1.1 $R \ll (\Delta_f, \Delta_{af})$

Under this condition, a system of static spin vortices arises near the interface (Figure 6.2). At the interface, their boundaries coincide with atomic step edges. The location of vortices relative to the interface depends on the ratio of exchange stiffness constants of the layers. When the exchange interaction constant of one layer is much greater in magnitude than that of the other layer, the vortex is almost completely localized in the layer having a smaller exchange integral (Figure 6.2a). On the other hand, when the constants have equal magnitude, the vortex will be positioned symmetrically with respect to the interface (Figure 6.2b).

The dimensions of the vortex in the direction perpendicular to the interface are of the same order of magnitude as the smallest size of the vortex in the interface plane.

The emergence of static spin vortices due to frustration has been predicted theoretically in papers [3, 4] (see also the review paper [5]). As of this writing, we are not aware of any studies in which this effect has been observed experimentally. This lack of experimental data is due to the fundamental difficulty in observing

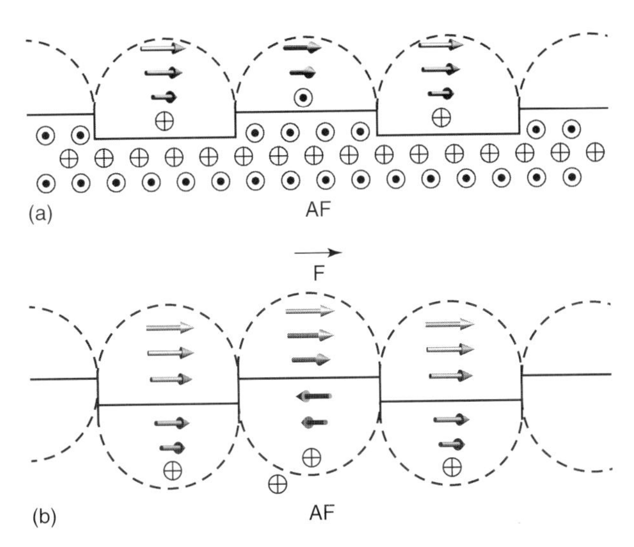

Figure 6.2 (a) Spin orientations in the ferromagnetic and antiferromagnetic layers in the vicinity of the interface for the case $|J_{af}| \ll J_f$ in the single-domain phase; (b) orientations of ferromagnetic and antiferromagnetic order parameters near the interface for the case $|J_{af}| \approx J_f$ in the single-domain phase. The arrows indicate the direction of the corresponding order parameter. The rotation of the order parameters takes place in the layer plane.

distributions of magnetic order parameters on a nanometer scale in the bulk of the magnetic layer.

Atomic steps divide the interface into regions of two types. In regions of the first type, the parallel orientation of ferromagnetic and antiferromagnetic order parameters is energetically favorable, whereas in regions of the second type, the antiparallel orientation is energetically favorable.

Since shape anisotropy makes energetically favorable those states whose magnetization vector (and thereby, due to interaction between layers, the antiferromagnetism vector as well) is lying in the layer plane, the direction of order parameters can be uniquely characterized by the angle $\theta_{f(af)}$ formed by the corresponding order parameter vector and the easy axis in the layer plane. As mentioned above, magnitudes of the order parameters are assumed to be invariable.

Spin vortices can also be subdivided into two types depending on the type of the region in which the vortex arises. One can describe the vortex structure using the exchange approximation. Since the characteristic size R of the vortex is much smaller than the traditional domain wall thickness, the vortex structure will remain essentially unchanged even when the anisotropy energy is taken into consideration. In the exchange approximation, magnetic order parameters inside the layers remain homogeneous at $|z| \gg R$, that is, at large distances from the interface, which is located at $z = 0$. Suppose the directions of the order parameters form an angle ψ. Then in the vortices of the first type, the net order parameter rotation angle is equal to ψ and the rotation partially takes place in one layer and partially in the other. In vortices of the second type, a rotation by the total angle $\pi - \psi$ takes place.

The vortex structure near an atomic step edge is of special interest. In the continuum approximation, the quantity $\nabla\theta_{f(af)}$ in the $r < R$ region (where r is the distance from the step edge) is inversely proportional to r: $\left|\nabla\theta_{f(af)}\right| \propto r^{-1}$. It is this region that makes the greatest contribution to the vortex energy. If exchange interaction between the layers is characterized by the same (to an order of magnitude) energy as the exchange interaction in the layer having the smaller exchange interaction constant, then the dependence just mentioned remains valid even when r is on the order of atomic distances. The region near the step edge will be discussed more extensively in Section 6.5.1.

The surface energy density averaged over the vortex can be estimated as [4, 5]

$$w_{\text{vor}} \sim \frac{\min\left(J_f, \left|J_{af}\right|\right)\alpha^2}{bR} \ln\frac{R}{b} \tag{6.9}$$

where α is the angle of rotation of the order parameter within the vortex.

At atomic distances from the step edge, the volume energy density of distortions is high. This may cause changes not just in the direction but also in the magnitude of the order parameter. On the other hand, at distances from the step edge exceeding several interatomic distances, the volume energy density of distortions, which is proportional to $\left|\nabla\theta_{f(af)}\right|^2$, apparently is no longer enough to cause a considerable change in the magnitude of the order parameter.

By using Eq. (6.9), one can determine the mutual orientation of the ferro- and antiferromagnetic order parameters outside the vortex region that will correspond to the energy minimum.

This problem is completely analogous to the "magnetic-proximity" model put forth by Slonczewski [6]. The total energy of the system of vortices can be written in the form

$$W = C_1\psi^2 + C_2(\pi - \psi)^2 \tag{6.10}$$

where

$$C_i = \frac{\min\left(J_f, |J_{af}|\right)\sigma_i}{Rb}\ln\frac{R}{b} \tag{6.11}$$

and σ_i is the total interface area occupied by the regions of type i (where $i = 1, 2$).

It is easy to see that when regions of both types are represented with equal probabilities ($\sigma_1 = \sigma_2$), the energy minimum (in the exchange approximation) in the absence of external magnetic fields is achieved if magnetic order parameters are mutually perpendicular at sufficiently large distances from the interface.

The question that arises is, how will our conclusions change if we go beyond the scope of the exchange approximation by taking into account the single-ion anisotropy energy?

The answer depends on the number and orientation of easy axes in the layer plane. In the case of two mutually perpendicular easy axes in the layer plane, there will be no change in the configuration suggested by the exchange approximation. Each order parameter will orient itself parallel to a corresponding easy axis, so they will still remain perpendicular to one another.

However, if there is only one easy axis in the layer plane, then a mutually perpendicular orientation of order parameters far from the interface does not yield a minimum of the single-ion anisotropy energy. Instead, this energy reaches its minimum when magnetic order parameters are collinear.

As a result of "competition" between the energy of vortices and the energy of single-ion anisotropy, a 90° exchange spring arises where the mutual orientation of the order parameters changes from perpendicular to collinear [7].

We prefer to use the term *exchange spring* rather than *domain wall*, because a domain wall separates domains having different orientations of the order parameter that has the same physical nature. In contrast, an exchange spring appears near the interface between layers whose order parameters differ by their physical nature. The structure of the exchange spring in each layer is completely analogous to the structure of a Bloch domain wall in the corresponding material. The width of the space region occupied by the exchange spring is of the same order of magnitude as the thickness $\Delta_{f(af)}$ of the corresponding domain wall and is much larger than the characteristic size R of a vortex (we recall that $R \ll \Delta_{f(af)}$). The energy of the exchange spring is smaller than the overall energy of the vortices by approximately the same factor.

For the most part, the exchange spring is located in the layer having a lower surface energy of the domain wall (Figure 6.3). The structure of the exchange spring is discussed in more detail in Section 6.4.2.

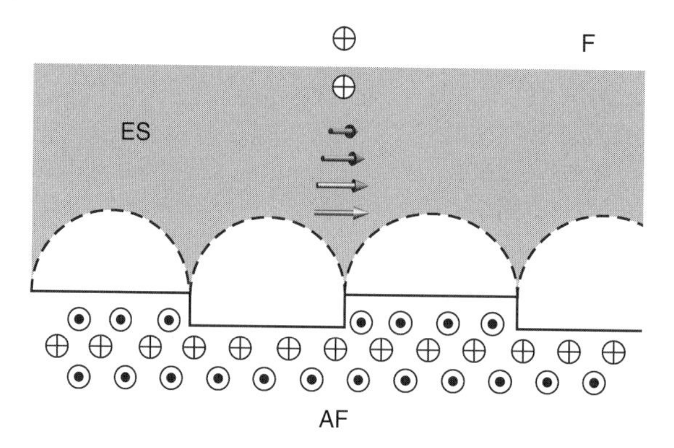

Figure 6.3 Exchange spring in the ferromagnet. The arrows indicate the direction of the corresponding order parameter. The rotation of the order parameters takes place in the layer plane.

6.4.1.2 $R \gg (\Delta_f, \Delta_{af})$

It is for this relation between parameters that the first attempts to come up with a qualitative picture of spin distribution in a frustrated system have been made [8, 9]. In these studies, domain walls were assumed to be atomically sharp. A computer simulation making use of the hypothesis of collinear spin orientation [10] has confirmed the existence of domain walls in one type of regions near the interface (out of the possible two types). The domain walls appeared in those layers where their surface energy was smaller, and they formed at the step edges bounding the given area (Figure 6.4a).

However, the hypothesis of collinear spin orientation is only valid when the single-ion anisotropy energy is considerably higher than the exchange interaction energy. However, we are discussing the opposite limiting case, where the spin distribution appears to be essentially noncollinear.

What happens to the spin distribution shown in Figure 6.3 as we increase the distance R between step edges at the interface? Let us assume for the sake of simplicity that the anisotropy constants of the layers are equal. Then the energy of a traditional domain wall will be lower in the layer having the lower exchange interaction energy, and both the vortices and the exchange spring will be mostly confined to this layer.

When the value of R approaches the thickness of a traditional domain wall, the transverse (i.e., in the direction perpendicular to the interface) size of the vortex stops growing and approaches the value of $\min(\Delta_f, \Delta_{(af)})$. The vortex changes its shape from a semicircle to a flattened semicircle.

If there exist two mutually perpendicular easy axes in the layer plane, the vortices sufficiently far away from the step edges transform into 90° exchange springs whose structure (as mentioned above) is similar to that of the Bloch domain wall in the corresponding layer. There is no qualitative change in the order parameter distribution (Figure 6.4b).

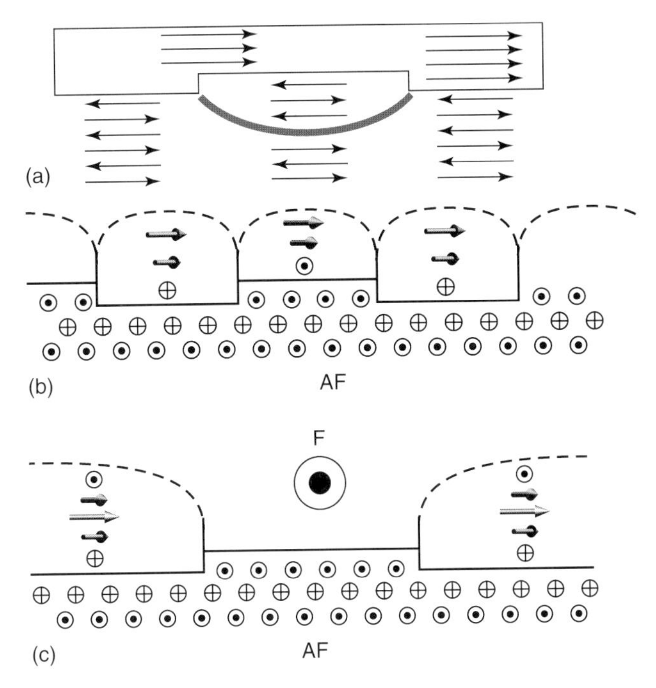

Figure 6.4 (a) Domain wall in the case of collinear orientation of spins in the layer plane; (b) exchange springs in the case of two mutually perpendicular easy axes; and (c) the case of a single easy axis. The arrows indicate the direction of the corresponding order parameter. The rotation of the order parameters takes place in the layer plane.

If there exists a single easy axis in the layer plane, an "annihilation" of the vortex and the 90° exchange spring takes place in the regions of the same (e.g., first) type. The resulting distribution of order parameters in the regions of this type becomes uniform.

In the regions of the other type, distortions of the vortex and of the spring add up, resulting in 180° exchange springs whose structure is analogous to that of 90° exchange springs in a system with two mutually perpendicular easy axes (Figure 6.4c). It is easy to see that unlike in the case shown in Figure 6.4a, exchange springs in the layers start from the interface.

6.4.2
Compensated Surface of the Antiferromagnet

The spin configuration arising due to frustrations near the smooth interface between ferromagnetic and compensated antiferromagnetic layers was first proposed by Koon [11]. It was found that in the exchange approximation, the magnetization vector of the ferromagnet orientates itself perpendicular to the antiferromagnetism vector in the absence of an external magnetic field.

The exchange interaction between ferromagnet and antiferromagnet spins leads to a canting of magnetizations of antiferromagnet sublattices, which is analogous to the canting of magnetizations of antiferromagnet sublattices that takes place in an external magnetic field (spin-flop orientation) (Figure 6.5). The only distinction is that in the nearest-neighbor approximation, the exchange field of the ferromagnet acts just on a single atomic layer of antiferromagnet spins. Therefore the canting angle θ in the antiferromagnet layer decreases rapidly at atomic distances from the interface.

Along with this canting, the spins in the ferromagnet undergo angular displacement and an antiferromagnetic order parameter vector is then induced in the ferromagnet. Just like in the antiferromagnet layer, the value of angular displacement φ falls off at atomic distances from the interface. As shown in [12], the decrease of θ and φ is described by a characteristic length that depends on the crystal lattice type and on the type of the cut.

In particular, one must mention the important role of the ratio of numbers z_1 and z_2, where z_1 is the number of nearest (with respect to the given spin) neighbors in the nearest atomic plane parallel to the interface, and z_2 is the number of nearest neighbors in the atomic plane (also parallel to the interface) to which the given spin belongs. Then the quantity $2z_1 + z_2 = z$ is the total number of nearest neighbors for the given spin.

In the region of small angles $|\varphi_j| \ll 1$, where j is the index of the atomic plane of the ferromagnet ($j = 1$ at the interface), the following equation holds [12]:

$$\varphi_j = \kappa \varphi_{j-1} \tag{6.12}$$

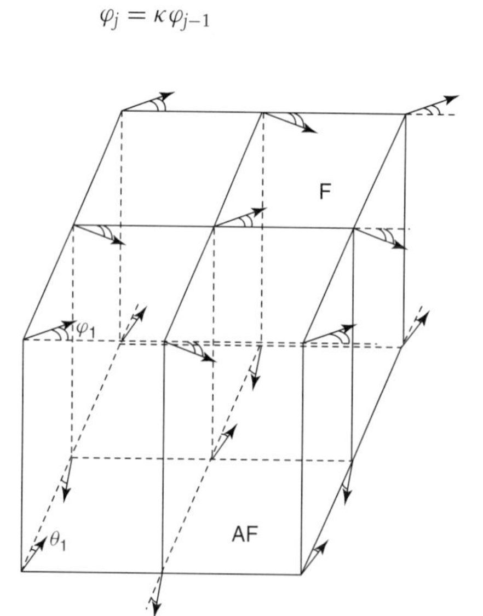

Figure 6.5 Orientation of spins in the bottom atomic layer of the ferromagnet and in the top atomic layer of the antiferromagnet for the (100) cut of a simple cubic lattice.

Table 6.1 The value of Z_1, Z_2, and K for several cuts.

Type of the lattice	Cut	z_1	z_2	κ
SC	(100)	1	4	0.101
SC	(110)	2	2	0.268
BC	(110)	2	4	0.172

where

$$\kappa = \frac{z_2}{z_1} + 1 - \sqrt{\frac{z_2}{z_1}\left(\frac{z_2}{z_1} + 2\right)} \tag{6.13}$$

Similarly, for the antiferromagnet layer,

$$\theta_j = -\kappa\theta_{j-1} \tag{6.14}$$

where κ has the same meaning as above.

Thus the antiferromagnet sublattices are canted in the opposite directions in each pair of neighboring atomic planes parallel to the interface. The values of z_1, z_2, and κ for several possible cuts are given in Table 6.1.

We must emphasize that it is precisely the presence of induced "improper" order parameters in each magnetic layer that is responsible for the magnetic interaction between the layers.

Far from the interface (at distances greater than the typical atomic distance), magnetic order parameters remain homogeneous in the framework of the exchange approximation.

What happens to the system when we go beyond the limits of the exchange approximation?

When two mutually perpendicular easy axes are present in the layer plane, the single-ion anisotropy energy can be taken into account without causing any changes in the above-described picture.

However, if there is only one easy axis in the plane of each layer, the spin-flop orientation causes frustration of single-ion anisotropy: it is impossible to realize such an orientation and achieve a minimum of single-ion anisotropy energy at the same time. This type of frustrated system is described in [13].

As a result of competition between the exchange energy and the energy of single-ion anisotropy, an exchange spring analogous to the one discussed in Section 4.1.1 arises near the interface. In the near-interface region, the magnetization vector in the ferromagnet layer and the antiferromagnetism vector in the antiferromagnet layer orientate themselves perpendicular to one another, deflecting from the easy axis. As the distance from the interface increases, the vectors turn in such a way as to become collinear to the easy axis (Figure 6.6). The characteristic length at which this change takes place is the thickness of the traditional domain wall in the corresponding layer.

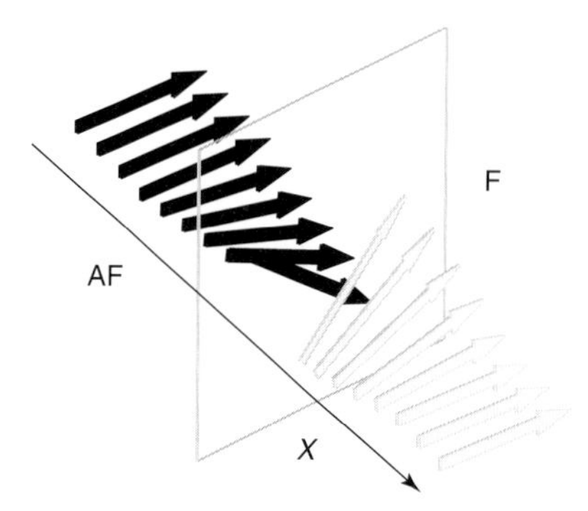

Figure 6.6 Order parameter distribution near the interface between ferromagnet and antiferromagnet.

The exchange spring arising in the process of remagnetization of a two-layer "hard ferromagnet–soft ferromagnet" system has been the subject of many studies (see, e.g., [14–16]). In the system under consideration, as opposed to a system of two ferromagnetic layers, the exchange spring arises in the absence of external magnetic fields and corresponds to an equilibrium, rather than metastable, state of the system.

If the thicknesses of traditional domain walls inside the layers are much larger than interatomic distances, an analytical solution of the given problem can be found easily by using the continuum approximation. In this approximation, one can neglect the induced "improper" order parameters that originate in each magnetic layer at atomic distances near the interface.

The energy $\tilde{W}$ of the exchange spring, which is obtained in the continuum approximation as the difference between the total energy of the layers and the energy of their uniform ground state, has the form

$$\begin{aligned}\tilde{W} &= NJ_f z_1 \left\{ \gamma_{af} \int_{-\infty}^{0} \left[(\theta'_{af}(x))^2 + \alpha_{af}(1 - \cos 2\theta_{af}(x)) \right] dx + \right. \\ &\left. + \int_0^{\infty} \left[(\theta'_f(x))^2 + \alpha_f(1 - \cos 2\theta_f(x)) + 2\beta(1 - \cos \theta_f(x)) \right] dx \right\} \qquad (6.15)\end{aligned}$$

Here we assume that the ferromagnet and the antiferromagnet are occupying space regions $x > 0$ and $x < 0$, respectively; N is the number of spins in the atomic plane parallel to the interface; distance is measured in the interplane distance units, and the dimensionless quantities γ_{af}, $\alpha_{f(af)}$, and β are correspondingly equal to

$$\gamma_{af} = \left| J_{af} \right| / J_f \qquad (6.16)$$

$$\alpha_{f(af)} = 2K_{\|} / z_1 \left| J_{f(af)} \right| \ll 1 \qquad (6.17)$$

$$\beta = \mu_{\text{at}} B_0 / z_1 J_f \qquad (6.18)$$

where μ_{at} is the atomic magnetic moment and B_0 is the strength of the external magnetic field.

Since the exchange energy is considerably higher than the anisotropy energy, we have

$$\theta_f(0) - \theta_{af}(0) = \frac{\pi}{2} \tag{6.19}$$

Without restricting generality, we choose the numeration of the two sublattices of the collinear antiferromagnet in such a way as to ensure that at sufficiently large distances from the interface the ferromagnet's magnetization vector should be parallel rather than antiparallel to the antiferromagnetism vector. Then the two vectors deflect from the easy axis of magnetization in the opposite directions. Let the corresponding deflection angles be $\theta_f > 0$ and $\theta_{af} < 0$.

By varying the energy given by Eq. (6.15), one obtains the following Euler equations:

$$\theta''_{af} = \alpha_{af} \sin 2\theta_{af} \tag{6.20}$$

and

$$\theta''_f = \alpha_f \sin 2\theta_f + \beta \sin \theta_f \tag{6.21}$$

The solution of the first Euler equation is of the form [17]

$$\cos \theta_{af}(x) = \mathrm{th}\left[-(2\alpha_{af})^{1/2}(x + x_1)\right] \tag{6.22}$$

while the solution of the second one is

$$\sin \theta_f(x) = \frac{2\left[\beta(2\alpha_f + \beta)\right]^{1/2} \mathrm{sh}\left[(2\alpha_f + \beta)^{1/2}(x + x_2)\right]}{2\alpha_f + \beta \mathrm{ch}^2\left[(2\alpha_f + \beta)^{1/2}(x + x_2)\right]} \tag{6.23}$$

The constants x_1 and x_2 are expressed through $\theta_f(0)$ and $\theta_{af}(0)$. Making use of Eq. (6.17) and minimizing the combined energy of the spring with respect to the last remaining parameter, we obtain

$$\cos \theta_f = \frac{\varepsilon_f \left\{\left[\varepsilon_f^2 \beta^2 + 4\alpha_f(\alpha_f + \beta)(\varepsilon_f^2 + \varepsilon_{af}^2)\right]^{1/2} - \varepsilon_f \beta\right\}}{2\alpha_f(\varepsilon_f^2 + \varepsilon_{af}^2)} \tag{6.24}$$

where $\varepsilon_f \propto \alpha_f^{1/2}$ and $\varepsilon_{af} \propto \gamma_{af}\alpha_{af}^{1/2}$ are the surface energy densities of domain walls in the corresponding layers in the absence of an external magnetic field.

If $B_0 = 0$ one has

$$\mathrm{tg}\theta_f(0) = \frac{\varepsilon_{af}}{\varepsilon_f} \tag{6.25}$$

and thus the exchange spring is mostly confined to the layer having a lower surface energy of the domain wall.

As the external magnetic field is increased, the exchange spring is forced out into the antiferromagnetic layer. For large values of β ($\beta \gg \alpha_f$) one gets

$$\theta_f(0) = \frac{\varepsilon_{af}}{\varepsilon_f}\sqrt{\frac{2\alpha_f}{\beta}} \propto \beta^{-1/2} \tag{6.26}$$

In the derivation of this result, we have neglected the canting of antiferromagnet sublattices by assuming the external magnetic field to be much weaker than the characteristic spin-flop field in the antiferromagnet.

6.5 A Thin Ferromagnetic (Antiferromagnetic) Layer on a Thick Antiferromagnetic (Ferromagnetic) Substrate

6.5.1 Uncompensated Surface of the Antiferromagnet

It should be noted that in the absence of an external magnetic field the problem of constructing the phase diagram of a thin layer of ferromagnet on a thick antiferromagnetic substrate is completely equivalent to the similar problem about the phase diagram of a thin layer of antiferromagnet on a thick ferromagnetic substrate. For the purpose of definiteness, we shall assume the thin layer to be ferromagnetic.

If the layer is thin enough, frustration-induced distortions of magnetic order parameters may spread over the whole volume of the layer and largely determine its magnetic phase diagram. In this section, we calculate the phase diagram for the two variables: thickness of the thin layer and characteristic distance between atomic step edges at the interface. Along with traditional domain wall thicknesses for the two layers, these two geometrical parameters are the most important characteristics of the system under consideration.

This problem has been analyzed by Levchenko, Morosov, A. Sigov, and Yu. Sigov [3, 4] (see also the review paper [5]) within the framework of the exchange approximation, when

$$(a, R) \lll (\Delta_f, \Delta_{af}) \tag{6.27}$$

where a is the thin layer thickness. The single-ion anisotropy has been taken into account by Morosov [7], who obtained a significantly more complex phase diagram for the case when the inequality (Eq. 6.27) is no longer valid.

We start our analysis with the region in the phase diagram in which all applicability conditions for the exchange approximation are satisfied. Then we are left with two geometrical parameters (a and R) and two dimensionless parameters: γ_{af} given by Eq. (6.16) and

$$\gamma_{f,af} = \frac{|J_{f,af}|}{J_f} \tag{6.28}$$

which completely determine the behavior of a two-layer system in the framework of the continuum approximation.

6.5.1.1 The Case of $\gamma_{af} \gg 1$

If $\gamma_{af} \gg 1$, then the substrate remains homogeneous for all practical purposes, and all distortions of the magnetic order parameter are located in the thin layer. In this

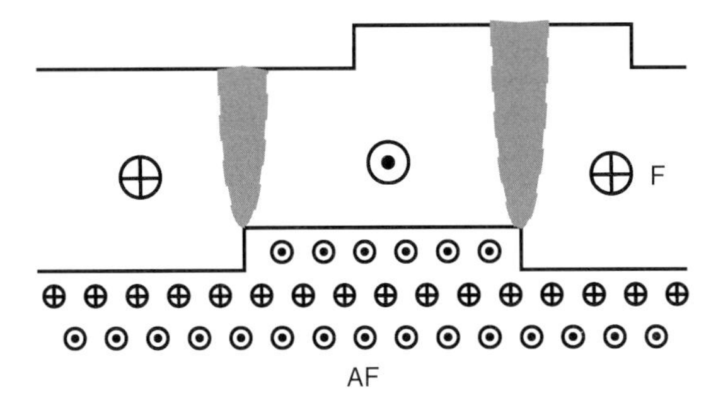

Figure 6.7 Nanodomain phase of a bilayer system in the case $\gamma_{af} \gg 1$.

case, we have only three parameters to consider: a, R, and $\gamma_{f,af}$. When using these parameters, we assume a and R to be measured in the units of the corresponding lattice constants.

Let us discuss various relations between these two geometrical parameters. If $R \gg a$, the thin ferromagnetic layer breaks up into nanodomains with domain walls perpendicular to the layer plane. The domain boundaries coincide with atomic step edges on the interface [18, 19], meaning that they coincide with the boundaries of regions of two types, where (respectively) parallel or antiparallel relative orientation of the ferromagnetic and antiferromagnetic order parameters is energetically favorable (Figure 6.7). The relative orientation of magnetic order parameters in each region corresponds to the minimum of the interface energy. Therefore the domain wall arising in such a case appears to be a 180° wall.

However, the papers [18, 19] offer no discussion of the structure of the resulting domain wall. As demonstrated in our papers [3, 5], the structure of such a wall is substantially different from the structure of traditional domain walls (Bloch, Neel cross-tie, etc.) – a fact that justifies our use of the label "traditional" to emphasize this distinction.

Let us now consider the structure of this nontraditional wall that appears due to frustration of exchange interaction at the interface.

The main distinguishing features of this wall are as follows:

1) small thickness determined by the competition of in-layer and interlayer exchange interactions, rather than by the competition between the exchange energy and the anisotropy energy (as is the case for traditional domain walls);
2) increase of domain wall thickness δ in proportion to the distance from the atomic step edge at the interface.

Figure 6.8 shows a typical dependence $\delta(z)$ of domain wall thickness on the z-coordinate (the z-axis is perpendicular to the layer planes and the interface is positioned at $z = 0$) calculated for the case $\gamma_{f,af} a \gg 1$ [3]. One can see that the function $\delta(z)$ is linear near the substrate and has a zero derivative at the free boundary surface of the layer.

On the basis of simple energy considerations, one can evaluate the domain wall thickness $\delta_0 \equiv \delta(z = 0)$ at the interface as well as the quantity ζ that is given by the

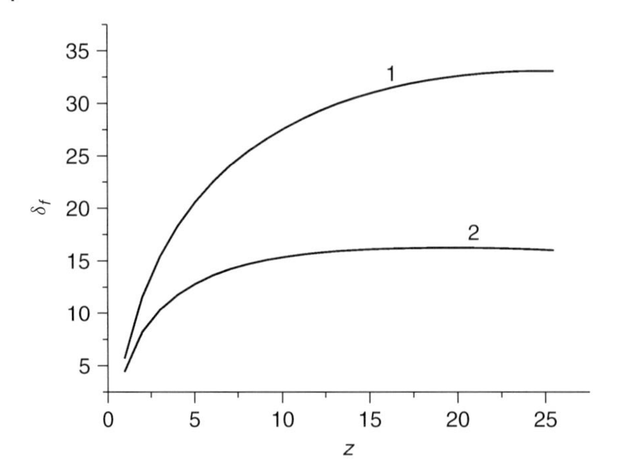

Figure 6.8 Dependence of the domain wall thickness on the distance from the interface calculated for (1) $\Delta_f > a$ and (2) $\Delta_f < a$. All distances are in units of the appropriate lattice parameters.

derivative function $\delta\prime(z)$ averaged over the layer thickness. Let us approximate the function $\theta(x, z)$ by the following expression:

$$\theta(x, z) = \begin{cases} \pi, \ x \geq \delta(z) \\ \dfrac{\pi}{2}\left(1 + x/\delta(z)\right), \ -\delta(z) < x < \delta(z) \\ 0, \ x \leq -\delta(z) \end{cases} \tag{6.29}$$

where

$$\delta(z) = \delta_0 + \zeta z, 0 < z < a \tag{6.30}$$

The inhomogeneity of the order parameter in the domain wall results in the following contribution to the energy of the exchange interaction in the thin layer (here energy is per unit length along the y-axis):

$$w_f = \frac{J_f}{2b}\int_0^a dz \int_{-\infty}^{\infty} dx\left[(\theta_x')^2 + (\theta_z')^2\right] \sim \frac{\pi^2 J_f}{4b}\left(\frac{1}{\zeta} + \frac{\zeta}{3}\right)\ln\frac{\zeta a + \delta_0}{\delta_0} \tag{6.31}$$

The increase in the energy of interlayer exchange interaction equals

$$w_{f,af} = \frac{2J_{f,af}}{b}\int_0^{\infty} dx[1 - \cos\theta(x, 0)] \sim \frac{2J_{f,af}}{b}\delta_0 \tag{6.32}$$

where the center of the domain wall is assumed to be located at $x = 0$. By minimizing the overall energy $w = w_f + w_{f,af}$ with respect to parameters ζ and δ_0, one immediately obtains estimated values for these parameters.

For $\gamma_{f,af}a \ll 1$ one has

$$\zeta \sim \sqrt{\gamma_{f,af}a} \ll 1 \tag{6.33}$$

so it is possible to neglect the increase of domain wall thickness. The initial wall thickness is

$$\delta_0 \sim \sqrt{a/\gamma_{f,af}} \tag{6.34}$$

If, on the other hand, $\gamma_{f,af}a \gg 1$, the minimization yields

$$\zeta \sim 1 \tag{6.35}$$

$$\delta_0 \sim \max(1, \gamma_{f,af}^{-1}) \tag{6.36}$$

The characteristic thickness of a nontraditional domain wall induced by the frustration is given by

$$\delta_f \sim \begin{cases} \delta_0, \text{ for } \gamma_{f,af}a \ll 1 \\ a, \text{ for } \gamma_{f,af}a \gg 1 \end{cases} \tag{6.37}$$

We see that when the thickness of magnetic layers is on a nanometer scale, the value of δ_f is much smaller than the thickness of a traditional domain wall. Equation 6.37 remains valid until domain wall thickness becomes comparable to the thickness Δ_f of a traditional domain wall. Thereafter, domain wall thickness approaches a limiting value due to the contribution from the anisotropy energy (Figure 6.8).

The energy per unit length of a domain wall of the new type is approximately (to an order of magnitude) equal to

$$w \sim \begin{cases} \dfrac{J_f}{b}\sqrt{\gamma_{f,af}a}, \ \gamma_{f,af}a \ll 1 \\ \dfrac{J_f}{b}\ln(\gamma_{f,af}a), \ \gamma_{f,af}a \gg 1 \end{cases} \tag{6.38}$$

The increase of domain wall thickness that takes place when $\gamma_{f,af}a \gg 1$ leads to only a logarithmic increase of energy with thin layer thickness.

It is worth noting that since the exchange interaction between spins belonging to different layers is generally of the same order of magnitude as the interaction between spins within a layer, fulfillment of the condition $\gamma_{f,af}a \ll 1$ (where a is on the order of 10 atomic distances) seems to be highly improbable. Therefore the following discussion focuses on the case $\gamma_{f,af}a \gg 1$.

Domain walls of the new type predicted theoretically in [3] have been observed experimentally in thin antiferromagnetic layers of manganese on an iron substrate by using spin-polarized scanning tunneling microscopy [20, 21]. As mentioned earlier, the problem of a thin antiferromagnetic layer on a ferromagnetic substrate is completely equivalent to the one just discussed. The positions of domain walls observed in the experiment perfectly coincided with the edges of atomic steps on the substrate surface, and their thickness increased linearly with thickness of the manganese layer, the proportionality coefficient being of the order of unity.

We now return to our theoretical analysis and try to predict what will happen to the nanodomain structure in a bilayer system as the characteristic distance R between atomic step edges becomes smaller.

When the distance R becomes comparable to the characteristic thickness of the nontraditional domain wall δ_f given by Eq. (6.37), the domain walls begin to overlap.

In the hypothetical case $\gamma_{f,af}a \ll 1$, in the $R \ll \delta_f$ limit, the thin layer goes into a single-domain state, and only slight distortions of the magnetic order parameter can take place within this layer. Thus the transition from a polydomain to a single-domain state in the thin layer occurs at $R \sim \delta_f$, and not at $R \sim a$.

In the case when $\gamma_{f,afa} \gg 1$ these conditions are equivalent. However, for $\delta_0 \ll R \ll \delta_f$, there remain large distortions of the ferromagnetic order parameter in the thin layer. All these distortions are concentrated near the interface, while in the bulk of the layer magnetization remains uniform. The distortions of the order parameter have the form of static spin vortices discussed earlier using thick layers as an example. If $\gamma_{af} \gg 1$, the vortices are located in the thin ferromagnetic layer and have the shape of a semicircle.

If $\gamma_{f,af} \ll 1$, then it is possible, in principle, that as the surface roughness and the value of δ_0 increase (see Eq. 6.36), the distance R becomes much shorter than δ_0. Then static spin vortices disappear and only weak distortions of the order parameter are preserved in the thin layer.

In the region where vortices are formed (as well as in the region where weak distortions take place) the minimum exchange energy corresponds to a mutually perpendicular orientation of ferromagnetic and antiferromagnetic order parameters outside of the region in question, whose size in the direction perpendicular to the layer planes is on the order of R. The reasons that cause this orientation of order parameters in the exchange approximation were discussed at length in Section 6.4. For such an orientation, magnetization rotates through the angle $\pi/2$ in each vortex, the rotation angles being opposite in the regions of different types.

The transition from a polydomain to a single-domain phase that takes place as we decrease parameter R or increase thin layer thickness a (assuming the applicability conditions for the exchange approximation are satisfied) occurs continuously and, strictly speaking, is not a phase transition. Such a transition can be realized in the process of atomic layer-by-layer deposition of the thin magnetic layer. As layer thickness increases, the condition $R \gg a$ will eventually fail and the system will go into a single-domain state.

Let us see how the picture described above will change if we go beyond the limits of the exchange approximation and take into account the single-ion anisotropy.

The phase diagram in Figure 6.9 "characteristic distance R between step edges–thin layer thickness ". There is no change in the region of the diagram where both parameters (R and a) satisfy the inequality $(a, R) \ll \Delta_f$, where Δ_f is the thickness of a traditional domain wall in the ferromagnet. However, outside this region, the real physical situation cannot be described adequately unless single-ion anisotropy is taken into account. In this case, just as in the case of thick layers, the outcome will depend on the number of easy axes (one easy axis or two mutually perpendicular easy axes) in the layer plane.

Let us start with the case when $R \ll \Delta_f$, and the ferromagnetic layer thickness a is gradually increased until it equals the quantity Δ_f and then exceeds it. In this

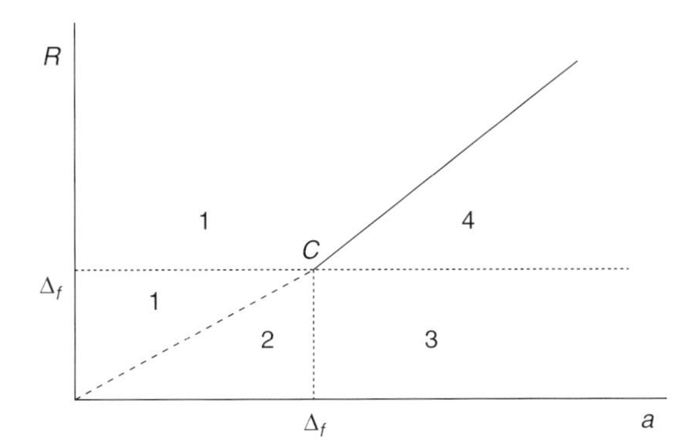

Figure 6.9 Phase diagram for the case $\gamma_{af} \gg 1$: (1) the polydomain phase; (2) the phase with semicircular vortices and with the film magnetization perpendicular to the easy axis; (3) the phase with vortices and a 90° exchange spring; and (4) the phase with 180° exchange springs parallel to the interface in one type of regions. The solid line is a first-order phase transition line, and C is the critical point.

process, the system makes a continuous transition to the thick layer case that was discussed in Section 6.4.

If there exist two mutually perpendicular axes of easy magnetization in the layer plane, the situation is similar to that considered in the exchange approximation for the case $R \ll a$.

If only one easy axis exists in the layer plane, then a (almost) 90° exchange spring arises in the thin layer near the interface, so that near the interface the ferromagnetic order parameter is perpendicular to the antiferromagnetic one, and far from the interface the magnetization in the ferromagnet deflects from the easy axis by the angle φ_f, which can be found as follows. The rotation angle of magnetization in the exchange spring is $\pi/2 - \varphi_f$ and the spring surface energy for $\varphi_f \ll 1$ is

$$\varepsilon_f = \varepsilon_f^{(0)} \left[1 - \frac{2\varphi_f}{\pi} \right] \tag{6.39}$$

where $\varepsilon_f^{(0)}$ is the surface energy of a 90° exchange spring. The anisotropy-energy loss (per unit surface area of the thin layer) due to the deflection of magnetization vector from the easy axis is

$$K_{\parallel} a (1 - \cos\varphi_f)/S_0 \tag{6.40}$$

where S_0 is the unit cell area in the thin layer plane.

One should keep in mind that the quantity a is expressed in the units of interatomic distance in the direction perpendicular to the layer plane.

By requesting that the sum of Eqs. (6.39) and (6.40) should be minimized, one finds for $a \gg \Delta_f$

$$\varphi_f \sim \frac{\Delta_f}{a} \ll 1 \tag{6.41}$$

Here it should be emphasized once again that for $a \ll \Delta_f$ the angle φ_f is close to $\pi/2$.

Suppose we have conditions corresponding to region 3 of the phase diagram (see Figure 6.9), that is, $a \gg \Delta_f$. If we now increase the value of R so that it exceeds Δ_f (but is still smaller than a), we will find the system in the state related to region 4 of the phase diagram that was described in Section 6.4.1. Under these conditions, the thin layer near the interface contains either 90° exchange springs of two types (for the two mutually perpendicular easy axes in the layer plane) or 180° exchange springs in the regions of one type (for one easy axis in the layer plane). The rest of the thin layer remains uniformly magnetized.

It remains for us to discuss the transition from phase 4 to the polydomain phase 1. As we mentioned already, as long as the exchange approximation applies, the transition process is taking place continuously and is therefore not a true phase transition.

The situation is quite different for $(a, R) \gg \Delta_f$. In this range of parameter values, domain walls do not overlap as the distance R decreases. The transition from phase 1 to 4 is therefore a jumplike phase transition of the first order. Indeed, for a certain critical value of parameter $R = R_c$, domain walls perpendicular to the layer plane get replaced by exchange springs parallel to the interface.

The critical value R_c is easily estimated by the following simple reasoning. A part of the wall positioned near the atomic step edge makes a small contribution to the wall energy at $a \gg \Delta_f$, whereas in the remaining part of the layer the domain wall can be considered as a traditional one. The energy (per unit interface area) of domain walls perpendicular to the layer plane can be estimated as $\nu\varepsilon_f a/R$, where $\nu \sim 1$ ($\nu = 1$ for parallel step edges and $\nu = 2$ when the steps form a square grid), and ε_f is the surface energy of a traditional 180° domain wall.

The energy (per unit interface area) of exchange springs parallel to the interface in the single-domain phase equals $\mu\varepsilon_f$. In the case of one easy axis, there appears a factor $\mu = 0.5$, because exchange springs appear only in regions of the same type. In the case of two crossing easy axes in the layer plane, this factor is $\mu < 1$ because the energy of a 90° domain wall is lower than the energy of a 180° wall.

By equating the energies of domain walls and exchange springs, one finds the critical value $R = R_c$ corresponding to the phase transition:

$$R_c \sim a \tag{6.42}$$

At $R > R_c$, the thin layer is in a polydomain state (phase 1) and at $R < R_c$ the thin layer is in a single-domain state, with exchange springs located near the interface.

For real samples, the phase transition gets "smeared out" (diffuse phase transition) because instead of having one constant width of atomic steps at the interface, we actually have to deal with some range of possible values.

Since for $R < \Delta_f$ the transition between phases 1 and 2 (see Figure 6.9) is a continuous process, while for $R \gg \Delta_f$ there takes place a first-order phase transition from phase 1 to 4, there should exist, on the phase diagram "characteristic distance between atomic step edges–thin layer thickness", a critical point C at $R \sim a \sim \Delta_f$. At this point C (Figure 6.9), the first-order phase transition line ends.

6.5.1.2 A Thin Layer with a Much Higher Exchange Rigidity

We now switch to the case of $\gamma_{af} \ll 1$, where the exchange interaction energy (exchange rigidity) in the thin layer is much higher than in the substrate. In such a case, magnetic order parameter distortions can no longer be considered as localized solely in the thin layer because frustration also causes fairly significant distortions of the antiferromagnetic order parameter in the substrate.

We start our consideration with the case of an isolated domain wall perpendicular to the layer plane, which corresponds to the polydomain phase of the thin layer.

If the interlayer exchange interaction is of the same order of magnitude as in the substrate, then numerical calculations show that in the exchange approximation, the domain wall has the shape displayed in Figure 6.10 [4, 5]. It can be described by two characteristic lengths. The first one is the domain wall thickness in the thin ferromagnetic layer δ_f, and the second is the size δ_0 of that area on the interface in which the difference $\theta_f - \theta_{af}$ differs from its optimum value which equals zero on one side of the atomic step and π on its another side. This optimum value corresponds to the minimum of interlayer interaction energy.

When estimating the energy of distortions of the ferromagnetic order parameter in the thin layer, one can neglect the widening of the domain wall in the thin layer. Then one has $\nabla\theta \sim \delta_f^{-1}$ and the energy of the domain wall in the thin layer (per unit length of the wall) is equal to

$$w_f \sim J_f a/b\delta_f \tag{6.43}$$

by the order of magnitude.

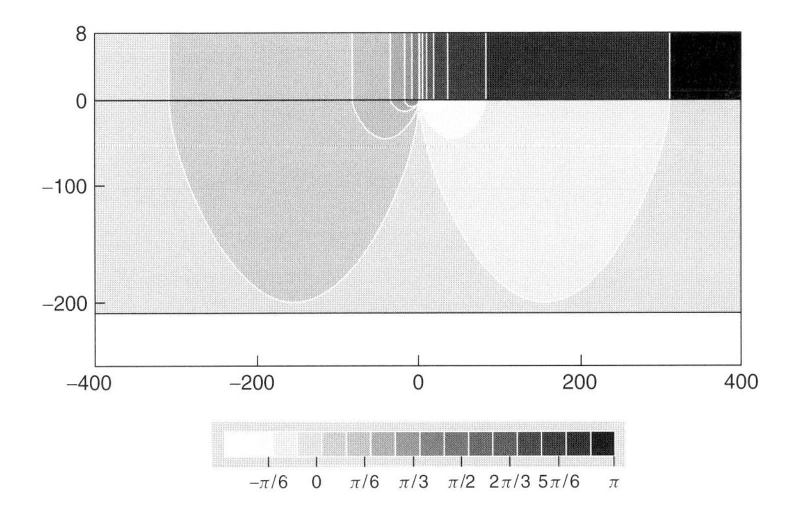

Figure 6.10 Distributions of the order parameters over the domain wall. The ordinate is equal to zero at the interface. All distances are in units of the appropriate lattice parameters. The correspondence between the hatching and the value of $\theta_{f(af)}$ (measured in radians) is shown in the inset.

The energy of distortions of the antiferromagnetic order parameter in the substrate can be easily estimated as well, because the structure of distortions of the order parameter near the step edge is analogous to that in the thin layer given the condition $\gamma_{af} \gg 1$. The only difference is that the widening of the distortion region is limited from above by the quantity δ_f rather than by the thin layer thickness. The energy of distortions of the order parameter in the substrate (per unit length) is estimated as

$$w_{af} \sim \frac{J_{af}}{b} \ln \frac{\delta_f}{\delta_0} \tag{6.44}$$

Minimizing the sum $w_f + w_{af} + w_{f,af}$ ($w_{f,af}$ is given by Eq. (6.32) with respect to the parameters δ_0 and δ_f, we find

$$\delta_0 \sim \max\left(1, \frac{\gamma_{af}}{\gamma_{f,af}}\right) \tag{6.45}$$

$$\delta_f \sim a/\gamma_{af} \tag{6.46}$$

Since $\delta_f \gg a$, it is possible to neglect the widening of the domain wall in the ferromagnetic layer.

It can be easily seen that the main contribution to the domain wall energy is made by the antiferromagnetic order parameter distortions in the substrate.

All these estimates are true if the condition $\delta_0 \ll \delta_f$ is valid, which means that

$$\frac{\gamma_{f,af}\, a}{\gamma_{af}^2} \gg 1 \tag{6.47}$$

Otherwise (though such a situation seems to be only hypothetical), order parameter distortions in the substrate are small enough, and domain wall parameters are analogous to the ones obtained for the case $\gamma_{af} \gg 1$ when $\gamma_{f,af}\, a \ll 1$.

If the substrate thickness is less than δ_f, the domain wall penetrates through the entire substrate, that is, the antiferromagnetic layer breaks up into domains, while the ferromagnetic layer remains, nearly uniform. In other words, we arrive at the case $\gamma_{af} \gg 1$ where the roles of the layers have been inverted.

We have thus obtained the critical thickness of the substrate. If the substrate thickness exceeds this critical value, the substrate can be considered as sufficiently thick. For large distances between atomic steps on the interface, the critical substrate thickness equals δ_f.

Let us see what changes should be introduced to the picture displayed in Figure 6.10 if one goes beyond the limits of the exchange approximation.

For the sake of simplicity, we assume that anisotropy constants in the layers are equal, so the ratio of thicknesses of traditional domain walls is defined by the ratio of exchange energies in the layers:

$$\frac{\Delta_f}{\Delta_{af}} = \gamma_{af}^{-1/2} \tag{6.48}$$

Since $\gamma_{af} \ll 1$ in the case under consideration, we have the condition $\Delta_f \gg \Delta_{af}$. Therefore, as the thickness a of the thin ferromagnetic layer increases, the thickness of the domain wall δ_f induced by frustration first reaches the value of Δ_{af}. After that, for $\Delta_f \gg \delta_f \gg \Delta_{af}$ the shape of the region of magnetic order parameter distortions in the substrate changes from semicircle to a flattened semicircle, elongated along the interface. When the distance from the step edge exceeds Δ_{af}, distortions of the antiferromagnetic order parameter assume the shape of an exchange spring parallel to the interface.

Let us estimate domain wall thickness δ_f in the ferromagnet assuming that $\delta_f \ll \Delta_f$. The energy of distortions of the antiferromagnetic order parameter in the substrate w_{af} at $\Delta_{af} \ll \delta_f \ll \Delta_f$ is given by the expression

$$w_{af} \sim \frac{J_{af}}{b} \ln \frac{\Delta_{af}}{\delta_0} + (\delta_f - \Delta_{af}) b \varepsilon_{af} \tag{6.49}$$

instead of Eq. (6.44). Here again all lengths are given in the units of the corresponding interatomic distances.

Minimizing the sum $w_f + w_{af} + w_{f,af}$ with respect to the parameter δ_f, we obtain [7]

$$\delta_f \sim \left(\frac{J_f a}{\varepsilon_{af} b}\right)^{1/2} \sim \left(\frac{a \Delta_{af}}{\gamma_{af}}\right)^{1/2} \propto a^{1/2} \tag{6.50}$$

Thus we see that for the given range of thin ferromagnetic layer thicknesses, the linear increase of thickness δ_f is replaced by a square root dependence on a.

Once thickness δ_f (which is increasing with a) becomes equal to the traditional domain wall thickness Δ_f, it stops increasing with increase of thin ferromagnetic layer thickness, and we have $\delta_f \approx \Delta_f$. Under our simplifying assumptions, the condition $\delta_f = \Delta_f$ is equivalent to the condition $\Delta_{af} = a$.

We may now consider the "thin layer thickness a–characteristic distance between step edges R" phase diagram (Figure 6.11). If parameter R exceeds the domain wall thickness δ_f in the range $R < \Delta_f$, then the thin layer gets subdivided into domains by domain walls of a new type, which are perpendicular to the layer plane and coincide in position with the atomic step edges on the interface (phase 1 in Figure 6.11).

As parameter R decreases, the domain walls begin to overlap and the system makes a continuous transition to a state in which the thin layer is actually a single domain and significant distortions of the antiferromagnetic order parameter occur in the substrate near the interface. When $R \ll \Delta_{af}$, those distortions are just semicircular static spin vortices (phase 2). If $\Delta_{af} \ll R \ll \delta_f \ll \Delta_f$, the vortices become flattened (phase 3) and their size along a normal to the layer plane becomes close to Δ_{af} and is less than their size R in the layer plane.

The energy of the vortices is minimum when the ferromagnetic and antiferromagnetic order parameters are mutually perpendicular, just like in the case $\gamma_{af} \gg 1$. Therefore in the case of a single easy axis in the layer plane, magnetization of the thin layer for $a < \Delta_{af}$ is directed along the axis of hard magnetization.

For the range of values $a < \Delta_{af}$, the transitions between phases 1, 2, and 3 are continuous, and therefore they are not "true" phase transitions.

As we increase the thickness of the thin layer, the system at $a > \Delta_{af}$ makes a transition from phase 2 to 4 that was discussed in Section 6.4. The difference from

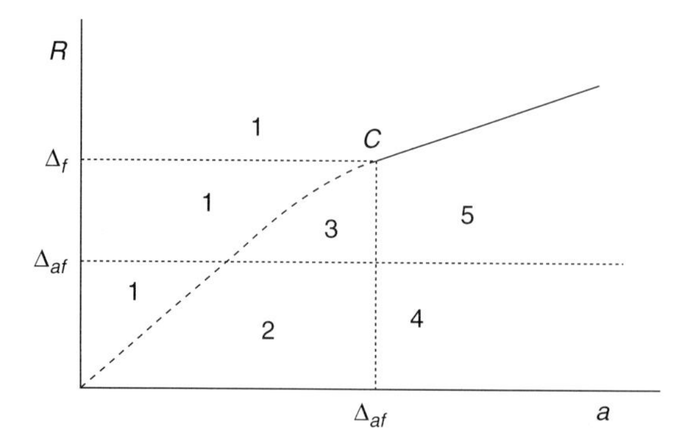

Figure 6.11 Phase diagram for the case $\gamma_{af} \ll 1$: (1) the polydomain phase; (2) the phase with semicircular vortices and with the layer magnetization perpendicular to the easy axis; (3) the phase with flattened vortices and with the layer magnetization perpendicular to the easy axis; (4) the phase with semicircular vortices and a 90° exchange spring; and (5) the phase with 180° exchange springs parallel to the interface in one-type regions. The dashed line describes the $\delta_f(a)$ dependence, the solid line is a first-order phase transition line, and C is the critical point.

the case $\gamma_{af} \gg 1$ is that, first, the distortions of the magnetic order parameter near the interface (including the 90° exchange spring arising in the system with a single easy axis in the layer plane) are "shifted" from the thin layer into the substrate, and, second, in such a system the deflection of thin-layer magnetization from the easy axis at $a \gg \Delta_{af}$ is given by Eq. (6.41), where Δ_{af} is substituted for Δ_f.

If we start out from phase 4 (see Figure 6.11) and increase the characteristic distance R between atomic steps, so that $R \gg \Delta_{af}$ (but R should remain smaller than its critical value R_c that corresponds to the transition of the thin layer into a polydomain phase), then we get to phase 5. This phase is characterized by 90° exchange springs in the substrate in the case of mutually perpendicular easy axes in the layer plane; or, in the case of one easy axis in the layer plane, by 180° exchange springs in the regions of one type in the substrate plus a uniform distribution of magnetic order parameters in the regions of the other type. The situation is analogous to the one that arises when $\gamma_{af} \gg 1$, but now the exchange springs are located in the substrate, rather than in the thin layer.

The transition between the single-domain phase 5 and the polydomain phase 1 of the thin layer for $R > \Delta_f$ is a smeared (diffused) phase transition of the first order, just like in the case $\gamma_{af} \gg 1$. Therefore when $R \sim \Delta_f$ there exists a critical point C on the phase diagram of Figure 6.11.

The critical value R_c corresponding to the phase transition point can be found in the same manner as in the case $\gamma_{af} \gg 1$. The only difference is that in the expression for the energy of exchange springs parallel to the interface inside the antiferromagnet, the quantity ε_f should be replaced with ε_{af}. This energy (per unit area of the interface) is on the order of $\mu\varepsilon_{af}$. Then the value R_c can be estimated as

$$R_c \sim a/\gamma_{af}^{1/2} \tag{6.51}$$

In the intermediate case $\gamma_{af} \sim 1$, the distortions of magnetic order parameters localized near the interface populate the boundary regions of both layers – the thin ferromagnetic layer and the substrate. The estimations for the phase boundaries adduced above are valid (by the order of magnitude) for this case ($\gamma_{af} \sim 1$) as well.

It is possible, of course, to increase the number of physical parameters, for example, by introducing two substantially different anisotropy constants for different layers. However, this would tend to make the phase diagrams far more complex and less self-intuitive. We therefore leave it for the reader to generalize the results given in this subsection for a large number of parameters, using the information and methods given here. At the same time, we feel that such a project should only be undertaken when there is a compelling need to explain or predict some specific experimental results.

It is beyond the scope of this chapter to discuss the behavior of the above-described bilayer systems under the influence of an external magnetic field, though we are well aware that the phenomenon of hysteresis loop shift (the "exchange bias") in a ferromagnet due to the interaction of ferromagnetic and antiferromagnetic order parameters plays an important part in technical applications where "ferromagnet–antiferromagnet" multilayer structures are involved. This topic is the subject of a chapter 7 in this book.

The reader who would like to find the exchange bias for each of the phases that form in the system, and to learn how the character of the exchange bias depends on the number of easy magnetization axes in the layer plane, can refer to a recently published book which contains a whole chapter written by us that is specifically dedicated to these questions [22].

The question then arises about the best way to achieve phase transitions between different phases shown in the diagrams in this subsection in a practical experimental setup. If $\gamma_{af} \gg 1$, this can only be done by increasing the thickness of the thin layer in the process of deposition. In the case of $\gamma_{af} \ll 1$, when the Curie temperature of the ferromagnet is higher than the Neel temperature of the antiferromagnet, such a transition can be initiated by heating the sample. When we approach the Neel temperature, the parameter $\delta_f \propto \gamma_{af}^{-1} \propto T_N/(T_N - T)$ increases indefinitely, so one can achieve a transition from a polydomain to a single-domain state in the process of sample heating.

The situation considered in Section 6.5.1.2 is in good agreement with the results given in [23], which describes an experimental study of the "thickness–vicinal angle" phase diagram for an iron film on a chromium (001) substrate. This would be the case of $\gamma_{af} \ll 1$ in our model. For vicinal angles χ close to zero, the polydomain phase was observed at film thicknesses $a < a_c = 3.5\,\text{nm}$. In a film with critical thickness a_c, the characteristic distance between the edges of randomly positioned atomic steps corresponds to the value δ_f given by Eq. (6.46). For larger values of a, a single-domain phase was observed with magnetization oriented perpendicular to atomic step edges. According to the theory developed in this subsection, the antiferromagnetism vector must be parallel to the step edges.

As the vicinal angle gets larger, we observe some regularly aligned parallel steps in addition to the ones positioned at random. When the concentration of regular steps becomes predominate ($\chi \geq 1^{\circ}$), the value of a_c begins to decrease. According to the theory developed here, one can write

$$a_c \propto ctg\chi \sim \chi^{-1} \tag{6.52}$$

6.5.2
Compensated Surface of an Antiferromagnetic Substrate

As shown in Section 6.4, in the region near an atomically smooth interface between a ferromagnet and an antiferromagnet whose spin-compensated atomic planes are parallel to the interface, in the case of one easy axis in the layer plane, there arises a 90° exchange spring that is mostly located in the layer having a lower traditional domain wall energy.

Suppose now that one of the layers – say, for the sake of definiteness, the ferromagnetic one – is of thickness $a \ll \Delta_f$. Then its contribution to the system energy is insignificant. As the result, the antiferromagnetic order parameter in the substrate is parallel to the easy axis, whereas the ferromagnet magnetization vector is perpendicular to it: $\theta_{af}(0) = 0$, $\theta_f(0) = \pi/2$.

It follows from numerical calculations [13] that as the thin layer thickness increases (though we still require that $a \ll \Delta_f$), the value of $\theta_{af}(0)$ increases proportionally:

$$\theta_{af}(0) = \theta_{af}^{\text{bulk}}(0)a/\Delta_f \tag{6.53}$$

where $\theta_{af}^{\text{bulk}}(0)$ is the angle formed at the interface by the antiferromagnetism vector and the easy axis in the case of thick layers.

It is important to keep in mind that $\theta_f(0) = \pi/2 - \theta_{af}(0)$.

What happens to the deflection angle of the ferromagnet's magnetization on the free surface of the thin layer? The quantity $\theta_f(a)$ decreases with increase of thin layer thickness a. There are two major reasons for this: first, because the thin layer makes a larger contribution to the system energy, which leads to a corresponding decrease of the quantity $\theta_f(0)$; and second, because the magnetization vector can rotate through a larger angle inside the layer since at larger values of a, a larger part of the exchange spring can fit within the layer.

The dependence of the quantities $\theta_f(0)$ and $\theta_f(a)$ on the number of atomic planes in the thin layer obtained by numerical calculation is presented in Figure 6.12.

In the case of a small number of atomic planes, magnetization of the ferromagnetic layer remains uniform and $\theta_f(0) = \theta_f(a)$ (point $P = 5$ in Figure 6.12).

As the number of atomic planes gets larger, one has $\theta_f(a) \rightarrow 0$, while $\theta_f(0)$ approaches a constant value given by Eq. (6.25).

The above-described behavior of the quantity $\theta_f(a)$ is in qualitative agreement with experimental observations [24] for the system consisting of a thin NiO layer on a Fe (001) surface (as mentioned earlier, the problem about the phase diagram of a thin antiferromagnetic layer on a thick ferromagnetic substrate is completely equivalent

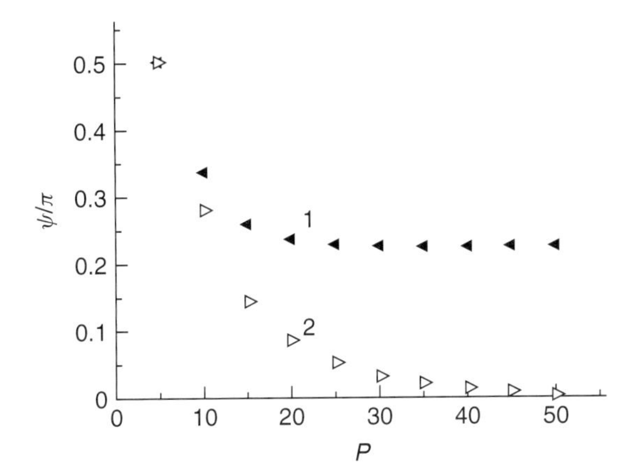

Figure 6.12 Rotation angle of the magnetization vector versus the number of atomic planes in the ferromagnet, calculated for $\alpha_f = \alpha_{af} = 0.004$ and $\gamma_{af} = 1$: (1) at the interface between the ferromagnetic and antiferromagnetic layers and (2) at the free surface of the ferromagnet.

to the previously considered problem that deals with a thin ferromagnetic layer on an antiferromagnetic substrate).

6.6
Spin-Valve Ferromagnet–Antiferromagnet–Ferromagnet System

In this section, the phase diagram of a spin-valve system comprising two ferromagnetic layers separated by an antiferromagnetic interlayer is discussed.

If the atomic planes of a collinear two-sublattice antiferromagnet that are parallel to the interface are uncompensated, then the presence of atomic steps at the interface causes frustration of the exchange interaction between the ferromagnetic layers (see, e.g., review article [5]). Indeed, on one side of the atomic step edge, the number of atomic planes is even, while on the other side it is odd. If the number of uncompensated planes of the antiferromagnet is odd, spins of the ferromagnetic layers interact with the nearest neighboring spins of the antiferromagnet belonging to one and the same sublattice (Figure 6.13a). Regardless of the sign of the exchange integral $J_{f,af}$ for neighboring spins located in different layers, parallel orientation of magnetization vectors of the ferromagnetic layers is energetically favorable. If the number of atomic planes in the antiferromagnetic layer is even, then spins of the ferromagnetic layers interact with the nearest neighboring spins of the antiferromagnet belonging to different sublattices, and antiparallel orientation of magnetization vectors of ferromagnetic layers is energetically favorable (Figure 6.13b). We thus observe frustration caused by the atomic step.

Thus atomic steps on both interfaces break up the plane parallel to the layer planes into regions of two types. For the regions of the first type, parallel orientation of magnetizations of the ferromagnetic layers is energetically favorable; in contrast,

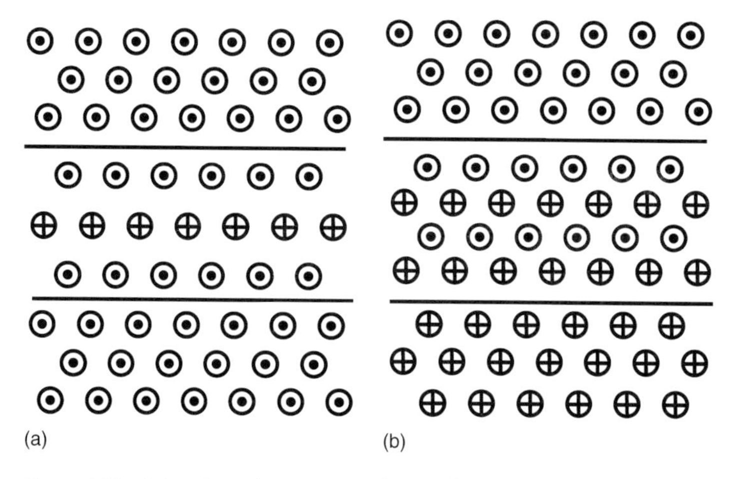

Figure 6.13 Spin orientation corresponding to the energy minimum: (a) odd number of atomic planes in the antiferromagnetic interlayer and (b) even number of atomic planes in the antiferromagnetic interlayer.

for the regions of the second type, antiparallel orientation of magnetizations of the ferromagnetic layers is favorable.

The three-layered "ferromagnet–antiferromagnet–ferromagnet" frustrated system was studied by our group several years ago using the exchange approximation [25, 26]. A "layer thickness–roughness" phase diagram was predicted for such a system. The exchange approximation is valid if the atomic step width R and the layer thickness are much smaller than the thicknesses of traditional domain walls in the ferromagnet and the antiferromagnet, which are determined by the ratio of the exchange and anisotropy energies. The achievements of modern technology make it possible to produce films and multilayer structures with large atomic step widths at the interface. Therefore it becomes important to go beyond the limits of the exchange approximation and to take into account the single-ion anisotropy. This task was performed in [27].

In this section, we restrict ourselves to the case $\gamma_{af} \ll 1$. Otherwise, order parameter distortions in the antiferromagnetic spacer are practically absent, and the problem reduces to that of order parameter distortions in a ferromagnetic film on a rigid antiferromagnetic substrate. Such a problem was discussed in the previous section. In addition, in order to reduce the number of parameters, we assume that the layers are of equal thickness and that their single-ion anisotropy constants are equal as well.

6.6.1 Domain Walls in a Three-Layer System

The most important geometrical parameters of the system are the layer thickness a and the characteristic distance R between step edges on the layer interface. If the

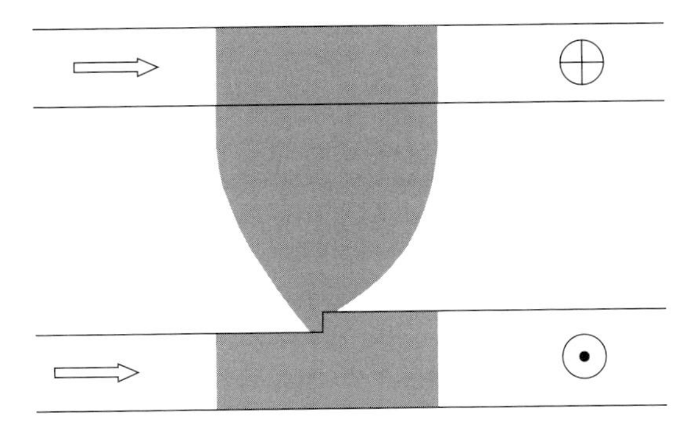

Figure 6.14 Distribution of order parameters near a frustration-induced domain wall (calculated in the exchange approximation).

distance R exceeds a certain critical value (which is specified later on), then all layers break up into domains with parallel or antiparallel orientation of magnetizations of ferromagnetic layers. The domain wall of the new type that arises due to frustration penetrates all three layers; its coordinates in the layer plane coincide with those of the atomic step edge at each of the two interfaces. In the exchange approximation, magnetization vectors of the ferromagnetic layers in the domain wall rotate through the angle of 90° in opposite directions. The antiferromagnetic order parameter rotates together with magnetization of that ferromagnetic layer whose interface has no atomic step at the given location [5, 25, 26] (Figure 6.14).

The structure and the energy of the domain wall depend on the parameter $\gamma_{f,af}a/\gamma_{af}$.

6.6.1.1 $\gamma_{f,af}a/\gamma_{af} \ll 1$

In this case, one can neglect the dependence $\theta_{f,af}(z)$, (i.e., domain wall widening is negligible), so one deals with a one-dimensional problem. The domain wall takes the shape that is shown in Figure 6.15a. It is characterized by two thicknesses δ_f and δ_{af} corresponding to the ferromagnet and antiferromagnet layers, respectively. The energy (per unit length of the domain wall) of ferromagnetic order parameter inhomogeneity is given by Eq. (6.43). By analogy, the energy of antiferromagnetic order parameter inhomogeneity can be estimated as

$$w_{af} = \left|J_{af}\right| a/b\delta_{af} \tag{6.54}$$

Let the atomic step be located at the first interface between the layers and have the coordinate $x = 0$. The interlayer interaction energy (per unit length of the domain wall) at this interface is

$$\begin{aligned} w_{f1,af} &= \frac{J_{f,af}}{b^2}\left\{\int_{-\infty}^{0}\left[1-\cos(\theta_{f,1}(x)-\theta_{af}(x))\right]dx+\right. \\ &\left.+\int_{0}^{+\infty}\left[1+\cos(\theta_{f,1}(x)-\theta_{af}(x))\right]dx\right\} \end{aligned} \tag{6.55}$$

At the second interface, where there is no atomic step at the given location, the analogous energy contribution is

$$w_{f2,af} = \frac{J_{f,af}}{b^2} \int_{-\infty}^{\infty} \left[1 - \cos(\theta_{f,2}(x) - \theta_{af}(x))\right] dx \qquad (6.56)$$

To estimate the thicknesses δ_f and δ_{af}, one can make use of the wall symmetry: $\theta_{f,1}(0) = -\pi/4$ and $\theta_{af}(0) = \theta_{f,2}(0) = \pi/4$ (to the accuracy of corrections of the order of γ_{af}). In addition, one can assume that $\delta_{af} \ll \delta_f$, as shown later on. Then the region of integration over the x coordinate can be divided into the intervals $|x| \leq \delta_{af}$, in which the function $\theta_{f,i}(x)$ can be considered equal to $\theta_{f,i}(0) (i = 1, 2)$, and $\delta_{af} \leq |x| \leq \delta_f$, in which the quantity θ_{af} can be considered equal to 0 and $\pi/2$ to the left and the right of the step edge, respectively. with account of this one can write down

$$w_{f2,af} + w_{f1,af} \sim \frac{J_{f,af}}{b}(c_1\delta_f + c_2\delta_{af}) \qquad (6.57)$$

where the coefficients c_1 and c_2 are positive, and their magnitudes are of the order of unity.

Minimization of the net energy

$$w = w_{f,1} + w_{af} + w_{f,2} + w_{f1,af} + w_{f2,af}$$

yields

$$\delta_f \approx (a/\gamma_{f,af})^{1/2} \qquad (6.58)$$

$$\delta_{af} \approx \left(\frac{a\gamma_{af}}{\gamma_{f,af}}\right)^{1/2} = \delta_f \gamma_{af}^{1/2} \ll \delta_f \qquad (6.59)$$

which is accurate to the order of magnitude.

These results justify the assumption made above. The total energy of the domain wall per unit length in the layer plane is (to the order of magnitude)

$$w \sim \frac{J_f}{b}(a\gamma_{f,af})^{1/2} \qquad (6.60)$$

Computer simulation for a wide range of values for all parameters confirms the estimations obtained above.

6.6.1.2 $\gamma_{f,af}a/\gamma_{af} \gg 1$

In this case, which is of greater practical interest than the previous one, the domain wall thickness in the antiferromagnetic interlayer increases dramatically with the distance from the interface containing the atomic step, just like in the case $\gamma_{af} \ll 1$ for the two-layer system (Figure 6.15b). The minimum value δ_0 of domain wall thickness is found in exactly the same way and given by Eq. (6.45).

The thickness δ_f of the domain wall in ferromagnetic layers can also be found by making use of simple energy considerations similar to those in the preceding subsection.

Let us suppose that $\delta_f \gg a$ and the atomic step is located at the point $x = 0$. Then in the region $a \ll |x| \ll \delta_f$, isoclines of θ_{af} are essentially parallel to the

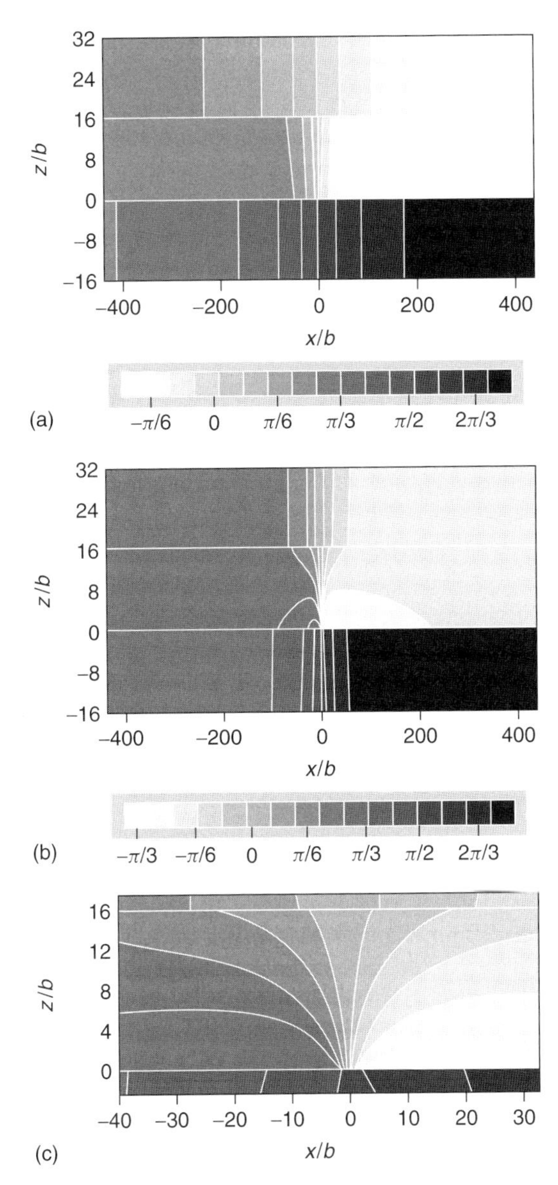

Figure 6.15 Domain wall in a spin-valve system for the case of (a) $\gamma_{faf}a/\gamma_{af} \ll 1$ and (b) $\gamma_{f,af}a/\gamma_{af} \gg 1$. The correspondence between the hatching and the value of θ (measured in radians) is shown in the inset. Diagram (c) shows the central part of diagram (b). Coordinates $z = 0$ and $z = 16$ correspond to the interfaces. The step is located at the point $x = z = 0$.

interface (Figure 6.15c). In this region, we have $|\nabla\theta_{af}| \approx a^{-1}$. The energy of antiferromagnetic order parameter distortions per unit length of the domain wall is

$$w_{af} = \frac{|J_{af}|}{b}\left(\frac{\delta_f}{a} + \ln\frac{a}{\delta_0}\right) \tag{6.61}$$

and the quantity w_f is given by Eq. (6.54).

By minimizing the net energy of the domain wall one obtains

$$\delta_f \approx a/\gamma_{af}^{1/2} \gg a \tag{6.62}$$

which justifies our assumption. Therefore one can neglect domain wall widening in the ferromagnetic layer.

Let us see how domain wall structure will change when the single-ion anisotropy energy is taken into account. The condition of applicability of the exchange approximation (Eq. 6.27) can be violated in two ways: by increasing the step size in the case of thin layers ($a \ll \Delta_{af}$), and by increasing both a and R simultaneously.

In the case of two mutually perpendicular easy axes in the layer plane, there will not be any changes in the patterning of thin layers. However, in the presence of only one easy axis in the layer plane, the breaking up of the thin layers into 90° domains ceases to be energetically favorable at $R \gg \Delta_f$. Indeed, let the magnetizations be directed along the easy axis in the domains of one type. Then they must be perpendicular to the easy axis in the domains of the other type. As a result, the anisotropy energy increases and the total energy is minimum when magnetization is collinear with the easy axis both in domains with parallel orientation of magnetizations of the ferromagnetic layers and in domains with antiparallel orientation of those magnetizations. In this case, however, the structure of the domain walls that are perpendicular to the layer planes and separate these domains changes radically: the domain walls become three-layered ones.

Let us consider a planar domain wall that is perpendicular to the layer plane and separates a domain with parallel orientation of magnetizations of ferromagnetic layers ($\xi < 0$, with ξ being the coordinate in units of the lattice parameter along an axis perpendicular to the step edge from which the domain wall nucleates) from a domain with antiparallel orientation of magnetizations ($\xi > 0$). In the first region of the domain wall, both magnetizations rotate in the layer plane through an angle of 45° from the easy axis. The thickness of this region is of the same order as the traditional domain wall thickness Δ_f, and magnetizations of the ferromagnetic layers remain parallel to each other within this region.

The second region is similar to the frustration-induced domain wall described above. The width of this region is given by Eq. (6.62) or (6.58). Here the magnetization of one ferromagnetic layer rotates through 90° in the same direction as in the first region, while the magnetization of the other layer rotates through 90° in the opposite direction. As a result, the magnetizations rotate through the angles of 135° and -45°, respectively, from their initial direction (Figure 6.16), and their mutual orientation becomes antiparallel.

Finally, in the third region, both magnetizations remain antiparallel and rotate through 45° in the same direction as in the first region. As they exit the domain wall,

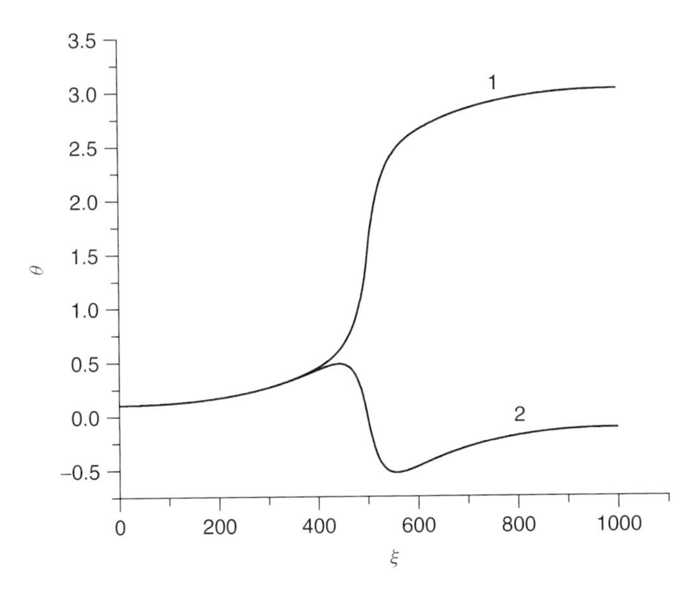

Figure 6.16 Angles of rotation of the magnetization vectors of the ferromagnetic layers in a three-layered domain wall: (1) an atomic step is present at the interface and (2) the interface is smooth. The atomic step is located at the point $\xi = 500$.

these magnetizations form the angles of 180° and 0° with the easy axis, respectively. The antiferromagnetic order parameter rotates together with the magnetization of the second ferromagnetic layer whose interface does not contain an atomic step at the given location.

As the layer thickness increases, the estimation (Eq. 6.62) remains valid until the thickness δ_f of the domain wall induced by the frustration of exchange interaction becomes of the same order of magnitude as the traditional domain wall thickness. Thereafter, due to the contribution from the anisotropy energy, domain wall thickness remains unchanged and equal to Δ_f.

Under the assumptions we have made, the condition of applicability of the exchange approximation, $a/\gamma_{af}^{1/2} \ll \Delta_f$, is equivalent to the condition $a \ll \Delta_{af}$, where $\Delta_{af} \sim b\left(J_{af}/K_{||}\right)^{1/2}$ is the thickness of a traditional domain wall in the antiferromagnet.

As layer thicknesses increase and reach $a \sim \Delta_{af}$, three-layered domain walls normal to the layer planes undergo a transformation. The central region of the domain wall becomes comparable in width to the other regions, and the domain wall in the second ferromagnetic layer disappears. The domain wall in the first ferromagnetic layer (whose interface has an atomic step at the given location) transforms into a traditional 180° domain wall. The characteristic size of the region around the atomic step edge where one can observe antiferromagnetic order parameter distortions is of the order of Δ_f in the direction parallel to the layer

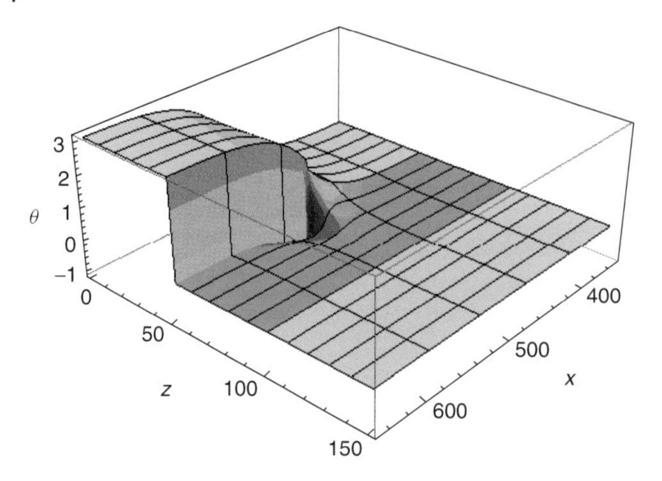

Figure 6.17 Domain wall in one layer for the case of $\gamma_{af} = 0.1$, $a = 50$, and $\Delta_f = 14$. An atomic step is located at $x = 500$ and $z = 50$.

plane and perpendicular to the step edge, and is of the order of Δ_{af} in the direction normal to the layer plane. Therefore, the central part of the antiferromagnetic interlayer is in a single-domain state. The structure of the isolated domain wall induced by the frustration of exchange interaction is shown in Figure 6.17 for $a \gg \Delta_{af}$ (see the color inset).

6.6.2
Phase Diagram

In the exchange approximation, the polydomain phase exists at $R > \delta_f$. At $a \sim 1$ nm and $\gamma_{af} \sim 0.1$, this condition is satisfied even for domains with size of the order of 10 nm. One can thus speak of a nanodomain state whose investigation requires extremely sensitive techniques.

The domain structure of a Fe/Cr multilayer system has been studied experimentally [28]. In this system, the average thickness of the antiferromagnetic layers corresponded to the antiparallel orientation of magnetizations of the neighboring ferromagnetic layers. It was found that as the roughness of the interfaces increases, the percentage of regions with parallel orientation of magnetizations of the neighboring ferromagnetic layers increases as well and finally reaches 50%.

The question that arises is, what happens to the domains as we decrease the size R of the atomic steps? Within the range of applicability of the exchange approximation, the domain walls begin to overlap and at some critical value $R \sim \delta_f$ the system makes a continuous transition to a state in which ferromagnetic layers are almost homogeneous; only weak distortions of magnetization can take place in these layers.

If $\gamma_{f,af} a/\gamma_{af} \ll 1$ and step sizes belong to the range $\delta_{af} \ll R \ll \delta_f$, the antiferromagnetic interlayer gets broken up into 90° domains. In the domains

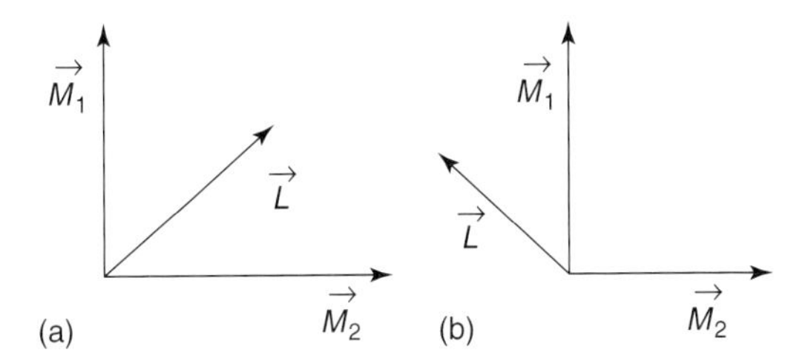

Figure 6.18 Orientations of the order parameters within domains.

of the first type, the antiferromagnetism vector is parallel or antiparallel to the bisector of the angle formed by the magnetizations (Figure 6.18a), whereas in the domain of the second type, the antiferromagnetism vector is orthogonal to this bisector (Figure 6.18b) and can have two opposite directions. The domain type is determined by the type of the region, that is, by the number (even or odd) of atomic planes in the antiferromagnetic interlayer.

As shown in [25], the additional (as compared to a state with no frustration) energy is related to the interlayer interaction energy and has the form

$$W = -\frac{2J_{f,af}}{b^2}\left(\sigma_1 \cos\frac{\psi}{2} + \sigma_2 \cos\frac{\pi - \psi}{2}\right) \tag{6.63}$$

where σ_1, σ_2 are the areas of the regions of the first and second types on the surface of the antiferromagnetic layer, and ψ is the angle between magnetizations of the ferromagnetic layers. If $\sigma_1 = \sigma_2$, then the additional energy reaches its minimum at $\psi = \pi/2$; in other words, in the absence of an external magnetic field, magnetization vectors of the ferromagnetic layers are perpendicular to each other.

If, on the other hand, $a \ll R \ll \delta_{af}$, then the system is in a state where distortions of magnetic order parameters are small and the orientation of order parameters corresponds to the one shown in Figure 6.18a.

If $\gamma_{f,af}a/\gamma_{af} \gg 1$ and step sizes are in the range $a \ll R \ll \delta_f$, the additional (as compared to a state with no frustration) energy is related to the energy of the exchange springs of two types arising in the antiferromagnetic interlayer. Springs of either the first or the second type appear in the corresponding type of regions. Such a state can be described within the framework of the Slonczewski "magnetic-proximity" model [6]. Specifically, the dependence of the system energy on the angle ψ between magnetization vectors of the ferromagnetic layers can be described by Eq. (6.10), where ψ and $\pi - \psi$ are the rotation angles of the antiferromagnetic order parameter in the exchange springs of the first and the second type, respectively. The constants C_1 and C_2 were estimated [25, 26] to be equal to

$$C_{1,2} \approx \frac{J_{af}}{a} \cdot \frac{\sigma_{1,2}}{b^2} \tag{6.64}$$

For $\sigma_1 = \sigma_2$, the energy reaches a minimum at $\psi = \pi/2$; in other words, in the absence of an external magnetic field, magnetization vectors of the ferromagnetic layers are mutually perpendicular. The condition that the energy of exchange springs should be much higher than the anisotropy energy yields the well-known inequality $a \ll \Delta_{af}$. In order to decrease the anisotropy energy, magnetization vectors of the ferromagnetic layers are arranged in such a way that the easy axis becomes the bisector of the right angle between them.

The noncollinear orientation of ferromagnetic layer magnetizations in multilayered structures with an antiferromagnetic interlayer has been discussed extensively (see, e.g., [29, 30]). It should be noted, however, that unless the experiment provides a sufficiently high spatial resolution, one has to rely on data averaged over a large part of the surface of the layer, and as a consequence, it becomes hard to distinguish a polydomain state of ferromagnetic layers with nanosized domains from a noncollinear state of homogeneous ferromagnetic layers. In fact, in the majority of experimental studies, the possibility of a polydomain state is simply ignored.

Let us now switch to the range of values $R \ll a$. In this case, all distortions of the order parameters are concentrated near the interfaces, interaction between the ferromagnetic layers becomes weak, and the energy of interaction between the adjacent layers discussed in Section 6.4 plays the dominant role.

As a result, at $\sigma_1 = \sigma_2$, the antiferromagnetic order parameter is oriented perpendicular to the magnetizations of the ferromagnetic layers, which thus become collinear vectors.

The transition from a phase with mutually perpendicular magnetizations of the ferromagnetic layers to a phase with collinear orientation of the magnetizations is a first-order phase transition [26]. Both states coexist for a certain range of the values of R, and the energies become equal at a certain value $R^* \sim a$. The value of the quantity R^* does not depend on temperature, so one cannot observe this phase transition by raising or lowering the temperature of the system.

Figure 6.19 shows a phase diagram "layer thickness–characteristic distance between atomic step edges" for a spin-valve system for the case when $J_{f,af} \sim J_{af}$, that is, when there are no weak distortions of the order parameter.

Let us discuss how phase 3 will change when we take the range of layer thicknesses $a > \Delta_{af}$, for which the single-ion anisotropy energy becomes important. We consider the case of a single easy magnetization axis in the layer plane. The exchange springs in the antiferromagnetic interlayer transform into 180° domain walls of the following shape: at distances of the order of Δ_{af} from atomic step edges, which serve as boundaries of the domain wall, the thickness of this wall increases from the interatomic distance to a value of the order of Δ_{af}, while in the remaining area, the domain wall is still a traditional one. Therefore, only the antiferromagnetic interlayer will be in the polydomain state (phase 4 in Figure 6.19). The magnetizations of the ferromagnetic layers are collinear, and their transition from parallel orientation to antiparallel one should be accompanied by collective "switching" of the domain walls in the antiferromagnetic interlayer (Figure 6.20). Since atomic step edges are arranged randomly, the energies of the initial and the final ("switched") state are equal.

At this range of layer thicknesses, the transition from a single-domain state of the ferromagnetic layers to a polydomain state (phase 2 in Figure 6.19) is a first-order

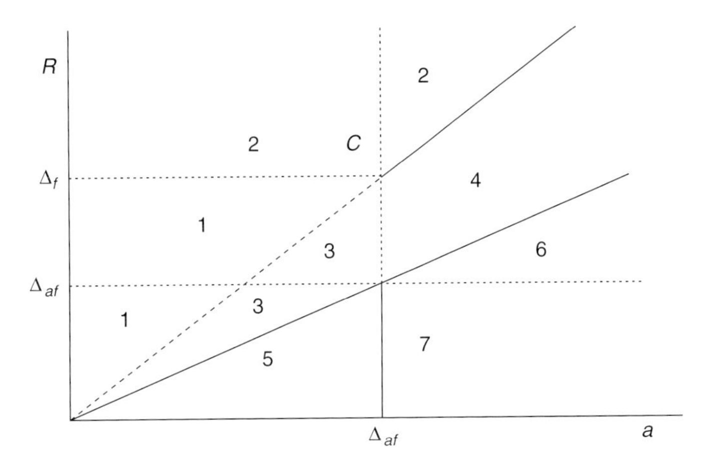

Figure 6.19 Phase diagram of a spin-valve system: (1) the phase with 90° domains in the ferromagnetic layers; (2) the phase with 180° domains in the ferromagnetic layers; (3) the phase in which the ferromagnetic layers are in a single-domain state and the antiferromagnetic interlayer contains exchange springs (Slonczewski phase); (4) the phase with 180° domains in the antiferromagnetic layer; (5) the phase with static 90° spin vortices and with the antiferromagnetic order parameter directed perpendicular to the magnetizations of the ferromagnetic layers; (6) the phase with 180° exchange springs parallel to the interfaces in the antiferromagnetic layer; and (7) the phase with spin vortices and two 90° exchange springs parallel to the interfaces in the antiferromagnetic layer. Solid lines are first-order phase transition lines and C is the critical point.

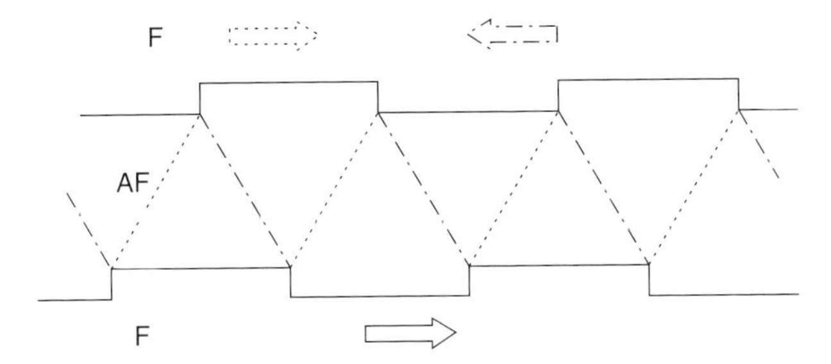

Figure 6.20 "Switching" of domain walls in the interlayer. Dot-and-dash lines and dotted lines show the positions of domain walls in the antiferromagnetic interlayer before and after the switching, respectively. The dot-and-dash arrow and the dotted arrow indicate the direction of magnetization of the upper layer before and after the switching, respectively. The solid arrow indicates the direction of magnetization of the bottom layer.

phase transition. Since this transition is continuous at small layer thicknesses, the phase diagram has a critical point at $a \sim \Delta_{af}$.

Far enough from the critical point ($a \gg \Delta_{af}$), the contribution to the domain wall energy from the transition region near the step edge is small and the parameter R_c corresponding to the phase transition point at a given layer thickness can be found from the following considerations. In the polydomain phase, the energy (per unit layer area) of transverse domain walls is given by

$$w_p \sim 2\nu\varepsilon_f a/R \tag{6.65}$$

where $\nu \sim 1$ ($\nu = 1$ when atomic step edges are parallel, and $\nu = 2$ when they form a square grid).

The energy (per unit square area of the layer) of the domain walls in the antiferromagnetic interlayer in the single-domain phase is equal to

$$w_m \sim \varepsilon_{af} L/R \tag{6.66}$$

where ε_{af} is the surface energy density of a traditional domain wall in the antiferromagnet, and L is the characteristic distance between the nearest-neighbor steps on the opposite interfaces (Figure 6.20). By equating the energies given by Eqs. (6.65) and (6.66) and estimating L as $(a^2 + R^2/4)^{1/2}$, we find the critical value

$$R_c \approx 4\nu a/\gamma_{af}^{1/2} \tag{6.67}$$

At $R > R_c$, the ferromagnetic layers are in a polydomain state, and at $R < R_c$, they are in a single-domain state. In a physical spin-valve system, the phase transition will be a diffused transition due to the spread in the widths of atomic steps at the interfaces.

It can be easily seen from Figure 6.20 that in the antiferromagnetic layer, a domain wall connecting steps on the opposite interfaces is energetically favorable only if parameter R exceeds a certain critical value R^*. For a periodic arrangement of steps shown in Figure 6.20, this value can be estimated as

$$R^* = (4/3)^{1/2}\, a \tag{6.68}$$

For $R < R^*$, distortions of the antiferromagnetic order parameter are concentrated near the interfaces, and the bulk of the antiferromagnetic layer is in a single-domain state (phase 6 in Figure 6.19). The interaction between the ferromagnetic layers is weak in this case, and the interaction between adjacent layers becomes dominant. The transition of the antiferromagnetic layer from a polydomain state to a single-domain state is also a first-order phase transition diffused due to the spread in the widths of atomic steps on the interface.

For the range of values $\Delta_{af} \ll R < R*$, distortions of the antiferromagnetic order parameter have the form of 180° exchange springs that are parallel to the interface and exist in those regions on the interface where the spins of the ferromagnetic layer are the nearest neighbors of the spins of a certain (let us say, the second – for the sake of definiteness) antiferromagnet sublattice. In those regions where the spins of the ferromagnetic layer are the nearest neighbors of the spins of the first sublattice of the antiferromagnet, the antiferromagnetic order parameter remains homogeneous (Figure 6.21).

As parameter R decreases within the range $R \ll \Delta_{af} \ll a$, static spin vortices arise in both types of regions in the interlayer. The edges of these vortices at the interfaces coincide with the step edges, and their size along the normal to the interface is equal to their minimum size in the interface plane. In order to decrease the vortex energy, the antiferromagnetic order parameter near the interfaces orients

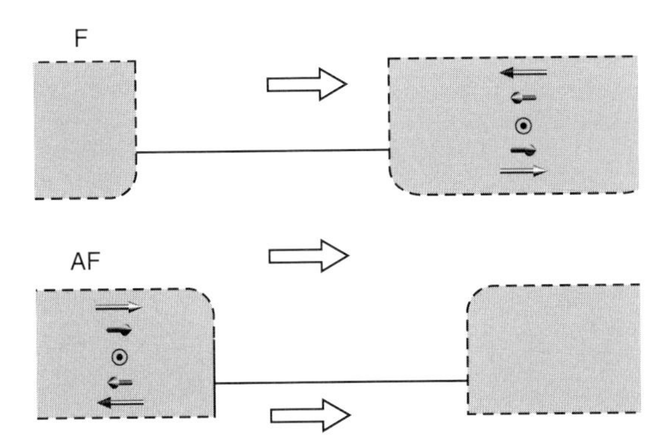

Figure 6.21 Exchange springs parallel to the interface in the antiferromagnetic interlayer. The arrows indicate the direction of the corresponding order parameter. The rotation of the order parameters takes place in the layer plane.

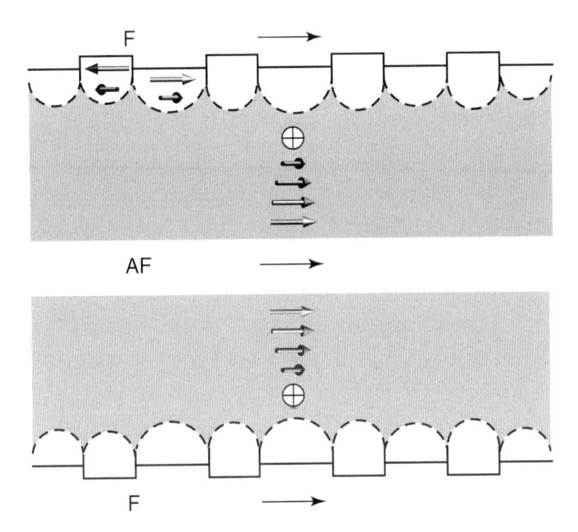

Figure 6.22 Spin vortices and two 90° exchange springs in the antiferromagnetic interlayer. The arrows indicate the direction of the corresponding order parameters. The rotation of the order parameters takes place in the layer plane.

itself perpendicular to the easy axis. In each vortex, the antiferromagnetism vector rotates through the angle of 90°, the directions of rotation being opposite in the regions of different types. In addition to static 90° vortices, 90° exchange springs are formed near the two interfaces within the interlayer, so that the order parameter in the center of the antiferromagnetic layer is parallel to the easy axis (Figure 6.22). This state is denoted as phase 7 in Figure 6.19.

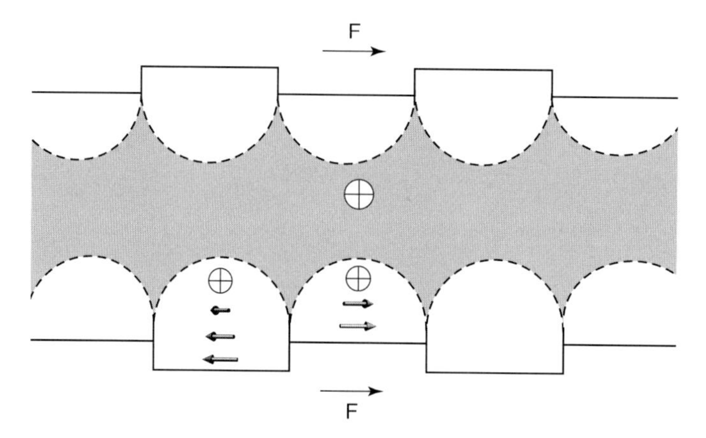

Figure 6.23 The phase in which there are spin vortices and the antiferromagnetic order parameter is perpendicular to the magnetizations of the ferromagnetic layers. The arrows indicate the direction of the corresponding order parameter. The rotation of the order parameters takes place in the layer plane.

As the layer thickness a decreases and reaches the value $a \sim \Delta_{af}$, these domain walls disappear abruptly ("jumpwise") and the spin-valve system undergoes a phase transition to the state with static 90° spin vortices, the antiferromagnetic order parameter being perpendicular to the magnetizations of the ferromagnetic layers and to the easy axis (Figure 6.23). This state is denoted by the number 5 in Figure 6.19.

The phase diagram displayed in Figure 6.19 will change in the case when we have two mutually perpendicular easy axes in the layer plane. The three-layered domain walls will no longer occur and the whole region of the polydomain state will correspond to phase 1 with 90° domains. In this case, phase 7 is absent altogether and the region where it used to exist is occupied entirely by phase 5.

In phase 6, the 180° exchange springs in the regions of the same type no longer exist; instead, 90° exchange springs are formed in both types of regions. These 90° exchange springs are parallel to the interface and have opposite directions of rotation in different types of regions (see Section 6.4). Far from the interface, the antiferromagnetic order parameter is perpendicular to the magnetizations of the ferromagnetic layers.

If the surface energy of two 90° domain walls in the antiferromagnetic layer is lower than that of one 180° domain wall, then phase 4 must be replaced by another phase 4' in which the magnetizations of the ferromagnetic layers are mutually orthogonal and the antiferromagnetic interlayer is broken up into separate domains by 90° domain walls. Their number is two times larger than in the case shown in Figure 6.20 (see Figure 6.24). The relations given by Eqs. (6.67) and (6.68) that define the boundaries between the phases remain practically unchanged.

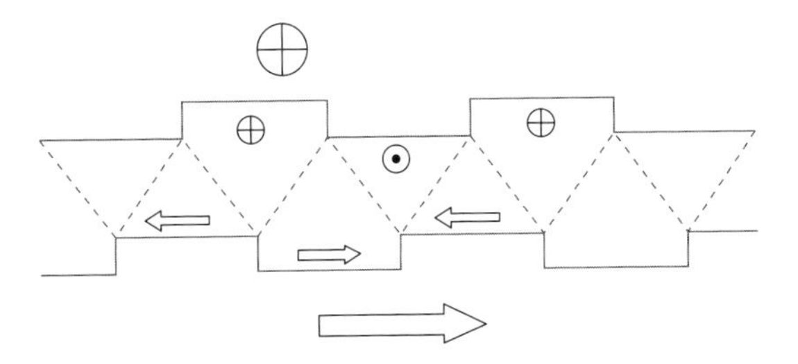

Figure 6.24 The 90° domain walls in the antiferromagnetic interlayer. The arrows indicate the direction of the order parameters.

6.6.3
Matching Experimental Data?

A nanodomain state has been observed in three-layer systems Fe/NiO/Fe [31] and Fe/CoO/Fe [32]; see also the review paper [33]. In contrast to the model observed above where all three layers had equal thicknesses, in the experiments cited, the lower (nearest to the substrate) Fe layer was fairly thick, while the thicknesses of antiferromagnetic interlayer and upper Fe layer varied over wide limits.

The experimental data appear explicable within the framework of the theory formulated above only with the supposition of (001) planes of nickel and cobalt oxides being uncompensated. In experiments on neutron scattering in CoO, two coexisting types of magnetic ordering were observed: AF-I and AF-II described by the wave vectors (0,0,1) and (1/2,1/2,1/2), respectively, with AF-I type ordering indeed resulting in uncompensated CoO(001) planes [34, 35]. However, as discussed in Chapter 7, Section 7.3.2.3, neutron diffraction investigations of exchange-biased Fe_3O_4/CoO multilayers showed no evidence of AF-I ordering [36, 37]. Similarly, this type of AF phase was never observed in NiO-based systems [38]. Therefore, the origin of the local uncompensation in CoO- or NiO-based multilayer magnetic structures seems to call for an explanation (see Chapters 5 and 7 ,Section 7.3.2.3).

If the system contains a thick ferromagnetic layer, one can assume with reasonable confidence that the latter always remains in a single-domain state. Instead of a single parameter, that is, the layer thickness a, as in Section 6.6.1.2, now we have two parameters: the thickness of antiferromagnetic interlayer d_{af} and the thickness of upper ferromagnetic layer d_f.

Referring to the simple evaluation in the framework of the exchange approximation similar to that performed in Section 6.6.1.2, for the nanodomain phase at $R > d_{af}$, the thickness of the domain wall in a thin ferromagnetic layer originated from an atomic step at any of the boundary planes, for $d_f > \gamma_{af} d_{af}$ equals

$$\delta_f \approx (d_f d_{af}/\gamma_{af})^{1/2} \tag{6.69}$$

For $d_f < \gamma_{af} d_{af}$, the thickness of the domain walls originating from the atomic steps at the interface between the thin ferromagnetic layer and the antiferromagnetic interlayer becomes equal to d_f/γ_{af}, as in the two-layer system (Eq. 6.46), while the domain walls originated from the atomic steps at the interface between the thick ferromagnetic layers and the antiferromagnetic interlayer have the thickness of the order of d_{af}, which continues to increase with increasing interlayer thickness.

The (R, d_{af}) and (R, d_f) phase diagrams of the asymmetric three-layer system are displayed in Figures 6.25 and 6.26. Phase 1 is that with the thin ferromagnetic layer subdivided into 180° nanodomains by the domain walls of equal thickness.

The new phase 5 is the phase with 180° nanodomains in the thin ferromagnetic layer, the domain walls having two different characteristic thicknesses.

Phase 2 is the Slonczewski phase with homogeneous and mutually perpendicular magnetizations of ferromagnetic layers and with exchange springs in the antiferromagnetic interlayer (corresponding to phase 3 in Figure 6.19).

Phase 3 is the one with collinear magnetizations of ferromagnetic layers, with spin vortices in the antiferromagnetic interlayer near the interfaces, and with the antiferromagnetic order parameters perpendicular to the magnetizations (corresponding to phase 5 in Figure 6.19).

A new phase (phase 4 in Figure 6.25) appears under the condition $d_{af} > R > d_f/\gamma_{af}$, in which the antiferromagnetic layer contains static spin vortices near the thick ferromagnetic layer surface, and the thin ferromagnetic layer is broken into nanodomains, the antiferromagnetic order parameter in the central part of the antiferromagnetic layer remaining homogeneous and perpendicular to the thick ferromagnetic layer magnetization. The magnetization vectors in 180° nanodomains are collinear with the antiferromagnetic order parameter and hence perpendicular to the magnetization of the thick ferromagnetic layer. Contrary to phase 1 in which the domain walls originate from the atomic steps on both interfaces, the domain walls in phase 4 are induced only by the steps on the

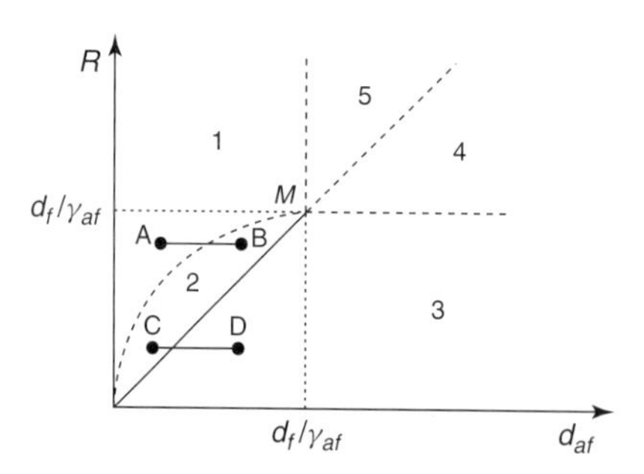

Figure 6.25 Phase diagram of asymmetric three-layer system: the phase with 180° domains in the thin ferromagnetic layer and domain walls of equal thickness (1); Slonczewski phase (2); the phase with static 90° spin vortices and with the antiferromagnetic order parameter directed perpendicular to the magnetizations of the ferromagnetic layers (3); the phase with 180° domains in the thin ferromagnetic layer and homogeneous antiferromagnetic order parameter far from interfaces (4); the phase with 180° domains in the thin ferromagnetic layer and two types of domain walls (5). Solid line relates to a first-order phase transition and M is the critical point. AB segment corresponds to the data reported in [32]; CD segment corresponds to the data from [39].

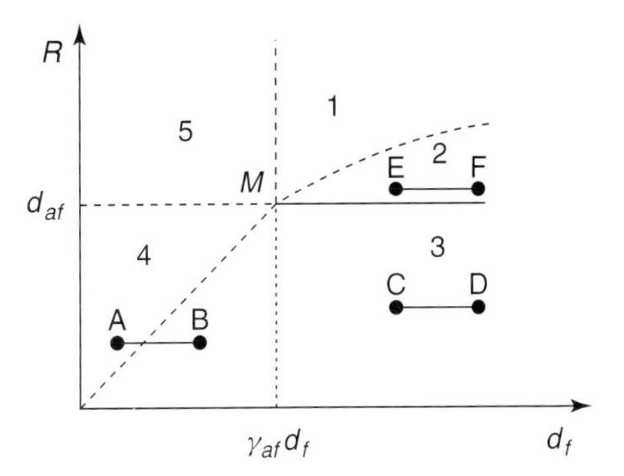

Figure 6.26 Phase diagram of asymmetric three-layer system: the phase with 180° domains in the thin ferromagnetic layer and domain walls of the equal thickness (1); Slonczewski phase (2); the phase with static 90° spin vortices and with the antiferromagnetic order parameter directed perpendicular to the magnetizations of the ferromagnetic layers (3); the phase with 180° domains in the thin ferromagnetic layer and homogeneous antiferromagnetic order parameter far from interfaces (4); the phase with 180° domains in the thin ferromagnetic layer and two types of domain walls (5). Solid line relates to a first-order phase transition and *M* is the critical point. AB segment corresponds to the data reported in [31]; CD and EF segments correspond to the data from [39].

interface between the oxide and the thin ferromagnetic layer. In this phase, as well as in the collinear phase, the interaction between the ferromagnetic layers through the antiferromagnetic interlayer is suppressed, and the problem reduces to two independent two-layer systems. In particular, the thickness of the domain wall in the thin ferromagnetic layer is given by Eq. (6.46).

The multilayer magnetic structure behavior observed in the experiments is in qualitative agreement with the phase diagrams derived from the theoretical model. The ranges of the parameters d_f and d_{af} corresponding, in our opinion, to the experimental data [31, 32, 39] are marked in Figures 6.25 and 6.26. For low interlayer thicknesses ($d_f < 1.3$ nm), no domains were observed in Fe/CoO/Fe system [32]. This experimental point is the only one not interpreted within our theory. However, if one allows for the presence of pinholes ignored in our model, it becomes possible to explain this point as well. For a higher interlayer thickness, the nanodomain state (phase 1) does occur, the thin ferromagnetic layer being broken into 180° domains, and then, with increasing interlayer thickness, this state transforms into phase 2 with mutually perpendicular orientation of the ferromagnetic layer magnetizations. In the case of Fe/CoO/Fe multilayer structure, the investigated range of the values of d_{af} does not reach the line of the phase transition to the collinear phase 3.

For the Fe/NiO/Fe multilayer structure, having a larger value of the parameter γ_{af} with respect to the Fe/CoO/Fe structure, the range of d_{af} values [39] begins in phase 2 and crosses the line of the phase transition to the collinear phase 3. In [32] the parameter d_f was varied at constant d_{af} value, from phase 4 to the collinear phase 3.

6.7 Conclusion

To summarize, we have shown that spin frustrations arising at the interfaces of a multilayer ferromagnet–antiferromagnet system determine many aspects of the behavior of magnetic order parameters in nanometer-thick layers.

We therefore ask both experimental physicists and technologists to exercise utmost care and diligence when measuring interface parameters, paying special attention to the characteristic distance R between the edges of neighboring atomic steps – a parameter that is of critical importance for the case when uncompensated atomic planes of the antiferromagnet are parallel to the interface. Unfortunately, this parameter has been given little attention by the researchers who chose to focus on the standard deviation of the interface from its mean position.

Only when the dependence of R on technological parameters is well established, will it become possible to select a technological route leading to a multilayered magnetic structure that corresponds to a certain region of one of the phase diagrams mentioned above, which should permit optimization of parameters of magnetoelectronic components and devices.

This will certainly require an *in situ* control of the R parameter in the process of growth of a multilayered structure. We wish success to those colleagues of ours who commit themselves to this difficult but extremely important endeavor.

References

1. Thompson, S.M. (2008) The discovery, development and future of GMR: the Nobel Prize 2007. *J. Phys. D: Appl. Phys.*, **41**, 093001 (20 p).
2. Hubert, A. (1974) *Theory of Domain Walls in Ordered Mediums*, Springer-Verlag, New York.
3. Levchenko, V.D., Morosov, A.I., Sigov, A.S., and Sigov, Yu.S. (1998) "Unusual" domain walls in multilayer systems: ferromagnet + layered antiferromagnet. *J. Exp. Theor. Phys.*, **87** (5), 985–990.
4. Levchenko, V.D., Morosov, A.I., and Sigov, A.S. (2000) Phase diagram of a thin ferromagnetic film on the surface of an antiferromagnet. *J. Exp. Theor. Phys. Lett.*, **71** (9), 373–376.
5. Morosov, A.I. and Sigov, A.S. (2004) New type of domain walls: domain walls caused by frustrations in multilayer magnetic nanostructures. *Phys. Solid State*, **46** (3), 395–410.
6. Slonczewski, J.C. (1995) Overview of interlayer exchange theory. *J. Magn. Magn. Mater.*, **150** (1), 13–24.
7. Morosov, A.I. (2008) Magnetic phase diagram of ferromagnet-antiferromagnet bilayer with a rough interface. *Phys. Solid State*, **50** (4), 703–708.
8. Stoeffler, D. and Gautier, F. (1995) Theoretical investigations of the magnetic behaviour of Cr monolayers deposited on a Fe(001) substrate: role of a mono-atomic step. *J. Magn. Magn. Mater.*, **147** (3), 260–278.
9. Fullerton, E.E., Sowers, C.H., and Bader, S.D. (1997) Interplay between biquadratic coupling and the Néel transition in $Fe/Cr_{94}Fe_6(001)$ superlattices. *Phys. Rev. B*, **56** (9), 5468–5473.
10. Berger, A. and Fullerton, E.E. (1997) Phase diagram of imperfect ferromagnetic/antiferromagnetic bilayers. *J. Magn. Magn. Mater.*, **165** (1-3), 471–474.
11. Koon, N.C. (1997) Calculations of exchange bias in thin films with ferromagnetic/antiferromagnetic interfaces. *Phys. Rev. Lett.*, **78** (25), 4865–4868.

12. Morosov, A.I. (2003) Magnetic structure of a compensated ferromagnet-antiferromagnet interface. *Phys. Solid State*, **45** (10), 1940–1943.
13. Morosov, A.I. and Rynkov, D.O. (2007) Magnetic structure of the interface between ferromagnet and antiferromagnet with parallel anisotropy axis. *Phys. Solid State*, **49** (10), 1940–1943.
14. Fullerton, E.E., Jiang, J.S., Grimsditch, M., Sowers, C.H., and Bader, S.D. (1998) Exchange-spring behavior in epitaxial hard/soft magnetic bilayers. *Phys. Rev. B*, **58** (18), 12193–12200.
15. Jiang, J.S., Fullerton, E.E., Sowers, C.H., Inomata, A., Bader, S., Shapiro, A.J., Shull, R.D., Gornakov, V.S., and Nikitenko, V.I. (1999) Spring magnet films. *IEEE Trans. Magn.*, **35** (5), 3229–3234.
16. Vlasko-Vlasov, V.K., Welp, U., Jiang, J.S., Miller, D.J., Grabtree, G.W., and Bader, S.D. (2001) Field induced biquadratic exchange in hard/soft ferromagnetic bilayers. *Phys. Rev. Lett.*, **86** (19), 4386–4389.
17. Landau, L.D. and Lifshitz, E.M. (1984) *Electrodynamics of Continuous Media*, Course of Theoretical Physics, Vol. 8, Pergamon, Oxford, § 43.
18. Berger, A. and Hopster, H. (1994) Magnetic properties of Fe films on Cr(100). *Phys. Rev. Lett.*, **73** (1), 193–196.
19. Escorcia-Aparicio, E.J., Choi, H.J., Ling, W.L., Kawakami, R.K., and Qiu, Z.Q. (1998) 90° magnetization switching in thin Fe films grown on stepped Cr(001). *Phys. Rev. Lett.*, **81** (10), 2144–2147.
20. Schlickum, U., Janke-Gilman, N., Wulfhekel, W., and Kirschner, J. (2004) Step-induced frustration of antiferromagnetic order in Mn on Fe(001). *Phys. Rev. Lett.*, **92**, 107203(4p).
21. Wulfhekel, W., Schlickum, U., and Kirschner, J. (2005) Topological frustrations in Mn films on Fe(001). *Microsc. Res. Tech.*, **66** (1), 105–116.
22. Morosov, A.I. and Sigov, A.S. (2009) Influence of single-ion anisotropy on the exchange bias in "ferromagnet-antiferromagnet" system, in *Giant Magnetoresistance: New Research* (eds A.D. Torres and D.A. Perez), NOWA Publishers.
23. Escorcia-Aparicio, E.J., Wolfe, J.H., Choi, H.J., Ling, W.L., Kawakami, R.K., and Qiu, Z.Q. (1999) Magnetic phases of thin Fe films grown on stepped Cr(001). *Phys. Rev. B*, **59** (18), 11892–11896.
24. Finazzi, M., Brambilla, A., Biagioni, P., Graf, J., Gweon, G.-H., Scholl, A., Lanzara, A., and Duò, L. (2006) Interface coupling transition in a thin epitaxial antiferromagnetic film interacting with a ferromagnetic substrate. *Phys. Rev. Lett.*, **97**, 097202(4p).
25. Morosov, A.I. and Sigov, A.S. (1999) Phase diagram of multilayers ferromagnet-layered antiferromagnet. *Phys. Solid State*, **41** (7), 1130–1137.
26. Levchenko, V.D., Morosov, A.I., and Sigov, A.S. (2002) Phase diagram of multilayer magnetic structures. *J. Exp. Theor. Phys.*, **94** (5), 985–992.
27. Morosov, A.I. and Morosov, I.A. (2008) Magnetic phase diagram of a ferromagnet-antiferromagnet-ferromagnet spin-valve structure with rough interfaces. *Phys. Solid State*, **50** (10), 1924–1930.
28. Paul, A. (2002) Effect of interface roughness on magnetic multilayers of Fe/Tb and Fe/Cr. *J. Magn. Magn. Mater.*, **240** (1-3), 497–500.
29. Shreyer, A., Ankner, J.F., Zeidler, Th., Zabel, H., Schafer, M., Wolf, J.A., Grunberg, P., and Majkrzak, C.F. (1995) Noncollinear and collinear magnetic structures in exchange coupled Fe/Cr(001) superlattices. *Phys. Rev. B*, **52** (22), 16066–16085.
30. Ustinov, V.V., Kirillova, M.M., Lobov, I.D., Maevskii, V.M., Mahnev, A.A., Minin, V.I., Romashev, L.N., Del', A.R., Semerikov, A.V., and Shreder, E.I. (1996) Optical, magnetooptical properties and giant magnetoresistance of Fe/Cr superlattices with noncollinear iron layers ordering. *J. Exp. Theor. Phys.*, **82** (2), 253–261.
31. Rougemaille, N., Portalupi, M., Brambilla, A., Biagioni, P., Lanzara, A., Finazzi, M., Schmid, A.K., and Duò, L. (2007) Exchange-induced frustration in Fe/NiO multilayers. *Phys. Rev. B*, **76**, 214425 (6 p).
32. Brambilla, A., Sessi, P., Cantoni, M., Finazzi, M., Rougemaille, N., Belkhou,

R., Vavassori, P., Duò, L., and Ciccacci, F. (2009) Frustration-driven micromagnetic structure in Fe/CoO/Fe thin film layered system. *Phys. Rev. B*, **79**, 172401 (4 p).

33. Finazzi, M., Duò, L., and Ciccacci, F. (2009) Magnetic properties of interface and multilayers based on thin antiferromagnetic oxide films. *Surf. Sci. Rep.*, **64** (4), 139–167.
34. Tomiyasu, K., Inami, T., and Ikeda, N. (2004) Magnetic structure of CoO studied by neutron and synchrotron x-ray diffraction. *Phys. Rev. B*, **70**, 184411 (6 p).
35. Inderhees, S.E., Borchers, J.A., Green, K.S., Kim, M.S., Sun, K., Strycker, G.L., and Aronson, M.C. (2008) Manipulating the magnetic structure of Co core/CoO shell nanoparticles: implications for controlling the exchange bias. *Phys. Rev. Lett.*, **101**, 117202 (4 p).
36. Ijiri, Y., Borchers, J.A., Erwin, R.W., Lee, S.-H., van der Zaag, P.J., and Wolf, R.M. (1998) Perpendicular coupling in exchange-biased Fe_3O_4/CoO superlattices. *Phys. Rev. Lett.*, **80** (3), 608–611.
37. Ijiri, Y., Schulthess, T.C., Borchers, J.A., van der Zaag, P.J., and Erwin, R.W. (2007) Link between perpendicular coupling and exchange biasing in Fe_3O_4/CoO multilayers. *Phys. Rev. Lett.*, **99**, 147201(4p).
38. Lind, D.M., Berry, S.D., Chern, G., Mathias, H., and Testardi, L.R. (1992) Growth and structural characterization of Fe_3O_4 and NiO thin films and superlattices grown by oxygen-plasma-assisted molecular-beam epitaxy. *Phys. Rev. B*, **45** (4p), 1838–1850.
39. Brambilla, A., Biagioni, P., Portalupi, M., Zani, M., Finazzi, M., and Duò, L. (2005) Magnetization reversal properties of Fe/NiO/Fe (001) trilayers. *Phys. Rev. B*, **72**, 174402 (9 p).

7
Antiferromagnetic–Ferromagnetic Oxide Multilayers: Fe_3O_4-Based Systems as a Model

P. J. van der Zaag and Julie A. Borchers

7.1
Introduction

Interfacial coupling between ferromagnetic (F) and antiferromagnetic (AF) layers in multilayer films can substantially alter the magnetic properties of the individual components, as discussed in Chapters 5 and 6. Throughout the past 20 years, intense research has focused on metallic systems with coupled layers and has led to unprecedented advancements in magnetic recording technology and other spintronic applications. More recently, focus has shifted to oxide-based materials as a result of discoveries of colossal magnetoresistance (CMR) in perovskite manganite films [1] (for a review see [2]), large tunneling magnetoresistance (TMR) through oxide barrier layers [3], and coupling among the electric, magnetic, and structural order parameters in multiferroic nanostructures [4]. In particular, the proposed device applications for the versatile multiferroic materials range from filters, oscillators, and phase shifters that can be tuned by magnetic fields to magnetoelectric generators and spin wave amplifiers that can be driven by electric field or current. Researchers in the field agree that realization of these applications requires a complete understanding of the magnetic interactions in these oxide multilayers as well as the ability to control and manipulate their characteristics on the nanoscale level. Though most oxide systems are quite complex, many of their properties can be understood within the framework of experimental and theoretical studies of more simplistic systems with well-characterized components such as ferrimagnetic Fe_3O_4 [5] and AF NiO or CoO. In addition, researchers in the field often take transition metal monoxides such as CoO, NiO, and FeO as model antiferromagnets to understand interface phenomena [6].

Specifically, the oxygen sublattice in the Fe_3O_4 spinel structure (see Figure 7.1) allows for growth of crystalline AF/F multilayers with sharp interfaces. Full analysis of the chemical magnetic structure of thin Fe_3O_4 films and especially of Fe_3O_4 interfaces is important for understanding related coupling studies in which oxides play a role. Nanostructures of current interest include multilayers with AF layers that yield exchange biasing as well as (all-oxide) tunnel junctions that exhibit enhanced magnetoresistance (MR) originating from the half metallic nature of

Magnetic Properties of Antiferromagnetic Oxide Materials.
Edited by Lamberto Duò, Marco Finazzi, and Franco Ciccacci

ISBN: 978-3-527-40881-8

Fe_3O_4 (see Figure 7.1b) [7]. As shown schematically in Figure 7.1(b), Fe_3O_4 has a gap for the majority Fe^{3+} spin band but the minority spin Fe^{2+} band crosses the gap leading to the well-known hopping conduction along the octahedral B-sites [5]. Moreover, even Fe_3O_4-layered nanoparticles are of current interest because of their potential applications in targeted drug delivery for cancer treatment.

This chapter will thus focus on experimental investigations of Fe_3O_4-based multilayers, as they are a model system for discussing key aspects of interface and chemical structure, their influence on anisotropy and interlayer exchange coupling, and their role in determining the resulting magnetic properties of AF/F composites. As in related AF/F oxide multilayers, exchange interactions in Fe_3O_4-based systems are mediated by superexchange via the oxygen sublattice and are thus quite sensitive to the local structural environment. The organization of this chapter is intended to highlight the close relationship between structure and magnetism in these exchange-coupled multilayers. Section 7.2 describes early ground-breaking research on the growth and structural characterization of Fe_3O_4-based multilayers. The section next reviews the detailed investigations of interfacial properties, including magnetic dead layers, interface anisotropy, and antiphase boundaries, that soon followed. In Section 7.3, the implications of these effects are explored within the context of magnetic exchange coupling in AF/F and related multilayer systems. The magnetoresistive and magnetooptical properties of these composite systems are discussed in Section 7.4. The last section provides a summary as well as an outlook for future investigations of exchange-coupled magnetic oxide systems.

7.2 Interface and Structural Effects

Seminal work on the growth of oxide magnetic multilayers was done by Terashima and Bando in the 1980s [8]. They were the first to grow high-quality superlattices using molecular-beam epitaxy (MBE) of ultrathin CoO and NiO on NaCl substrates. They realized, however, that magnetism in pure AF multilayers is difficult to characterize by conventional magnetic techniques due to the absence of a net magnetic moment. Their subsequent study dealt with CoO–Fe_3O_4 multilayers [9] as the magnetic interactions with ferro(i)magnetic layers had greater potential for applications. Many of the phenomena that fascinated researchers during the following decade were discovered in this system: A shifted hysteresis loop due to unidirectional anisotropy, a reduction of the saturation magnetization of the magnetic oxide with decreasing layer thickness, and an increase in the AF ordering temperature for very thin CoO layers in the multilayers.

Subsequently other groups started to grow oxide magnetic multilayers by MBE, notably Lind and coworkers [10–12], Wolf *et al.* [13], Hibma and coworkers [14, 15] and Chambers and colleagues [16–18]. Although in some of the initial studies evidence was found for intermixing at the interface [19], these groups all determined that by oxygen-enhanced MBE, AF–F oxide multilayers could be grown under

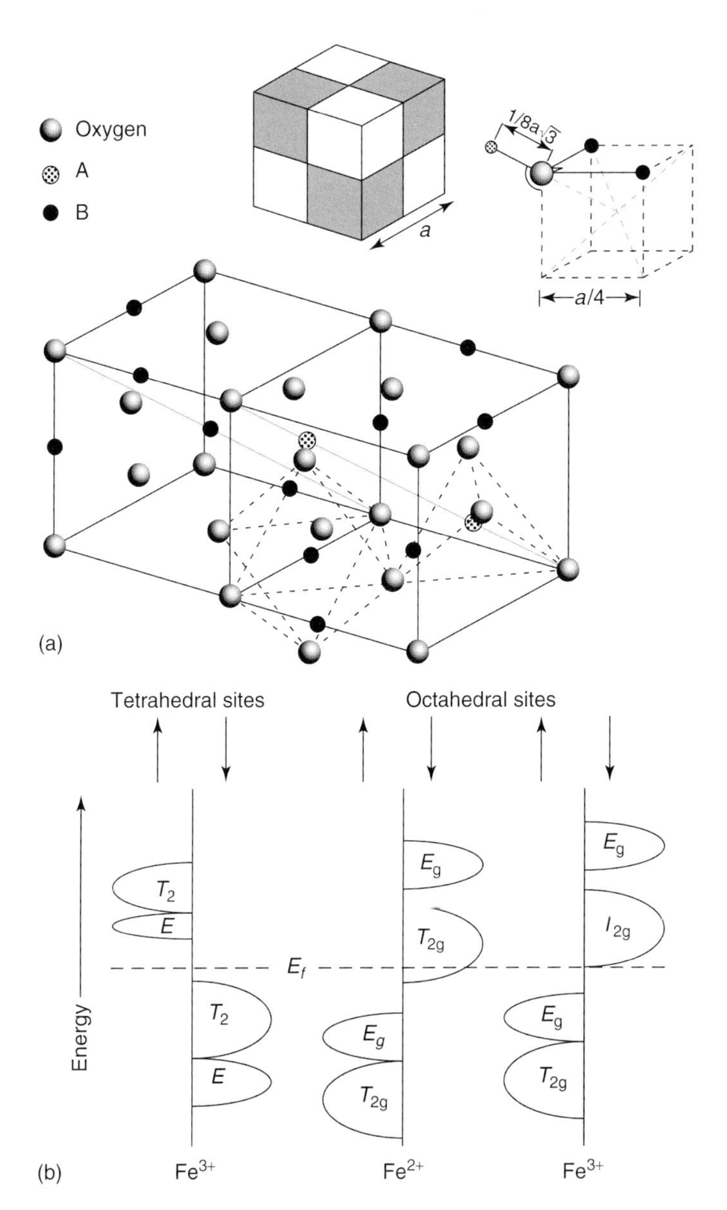

Figure 7.1 (a) The spinel lattice structure. The oxygen ions form a closed-packed face-centered cubic lattice, which has 64 tetrahedral (A)-sites and 32 octahedral (B)-sites. Only 16 of the A- and 8 of the B-sites are filled. For Fe_3O_4, Fe^{3+} ions occupy the A-sites and on the B-site both Fe^{2+} and Fe^{3+} ions are found. (b) Schematic representation of the electronic structure of the Fe-ions in Fe_3O_4. Note that the Fermi level, E_F, crosses the conduction band of the Fe^{2+} ions on the octahedral (B)-site.

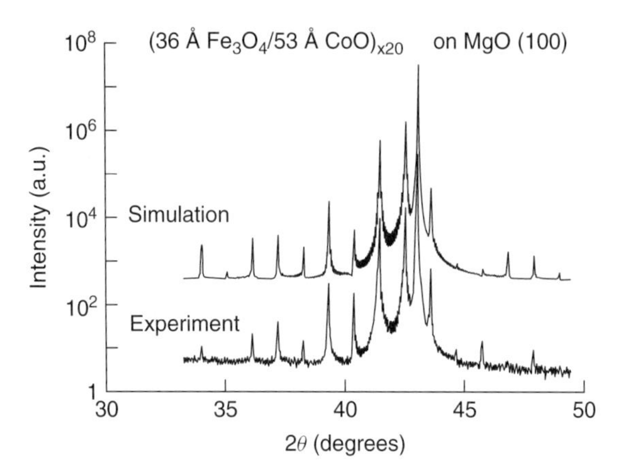

Figure 7.2 $\theta - 2\theta$ X-ray diffraction pattern for a (36 Å Fe_3O_4/53 Å CoO)$_{\times 20}$ multilayer grown on MgO(100). The top curve shows the simulated curve, which indicates that discrete layer thickness variation is less than 1 Å for the CoO and Fe_3O_4 layers in this case. Figure reproduced from [20].

similar growth conditions in a small pressure window (of 2×10^{-5} mbar [13]). The near absence of interfacial intermixing has been shown convincingly by both reflective high-energy electron diffraction (RHEED) and X-ray reflectometry [14], as well as by high-resolution X-ray diffraction [20] and neutron diffraction studies. X-ray diffraction revealed that single-crystalline superlattices could be obtained with rocking curve widths of $\approx 0.20^\circ$ and interface widths of 2 ± 1 Å [21]. Figure 7.2 shows the results of an X-ray diffraction (XRD) study for a (36 Å Fe_3O_4/53 Å CoO)$_{\times 20}$ multilayer grown on MgO(100). From the simulated spectrum, it was found that the roughness of the layers, included in the model through discrete layer thickness fluctuations, amounted to only 1 Å for CoO and Fe_3O_4 [20], attesting to the high quality of these AF–F multilayers.

7.2.1 Chemical and Structural Quality Effects

Although deposition techniques of oxide materials had advanced significantly due to work done on high T_C superconductors, growing mixed-valence oxide superlattices raised many challenges. First the properties of magnetic oxides are intimately dependent on cation distribution and stoichiometry, as this influences the magnetization and anisotropy [5]. Moreover, one could envision that these issues are particularly pronounced at the interfaces giving rise both to complications and new effects due to coupling between the layers. Consequently, it is important to assess the quality of the multilayers and to regulate growth at the interface in order to control the properties of the resulting multilayers. Terashima and Bando assessed the quality of their oxide multilayers by XRD and ^{57}Fe Mössbauer spectroscopy [9].

In their Mössbauer studies, they were able to study the free standing multilayers by ingeniously making use of the fact that their multilayers were grown on NaCl, which could be dissolved. They found that the Fe_3O_4 layers between 2.6 and 9 nm exhibited a hyperfine spectrum that was broadened compared to that of bulk Fe_3O_4. On the basis of these results, the authors correctly proposed a tendency toward superparamagnetism in the ultrathin Fe_3O_4 layers, as verified by later studies [22, 23]. Analysis of the possible types of interface structure, that is, type I–II or I–III (see Figure 7.3), showed excessive negative charge for type I–II, while type I–III has excessive positive charge [9]. Consequently, these authors predicted that the CoO–Fe_3O_4 interface structure would be unstable leading to cation redistribution at the Fe_3O_4 interface that is different from the bulk. This redistribution distorts the magnetic structure in the interior of the ultrathin Fe_3O_4 layer and thereby accounts for the Mössbauer data [8]. Alternatively the formation of vacancies could provide charge neutrality [13].

Assessing the stoichiometry of ultrathin mixed-valence oxide layers on an oxide substrate such as MgO, $SrTiO_3$, or Al_2O_3 is challenging. Determination of the stoichiometry through the Verwey metal–insulator transition temperature, T_v,

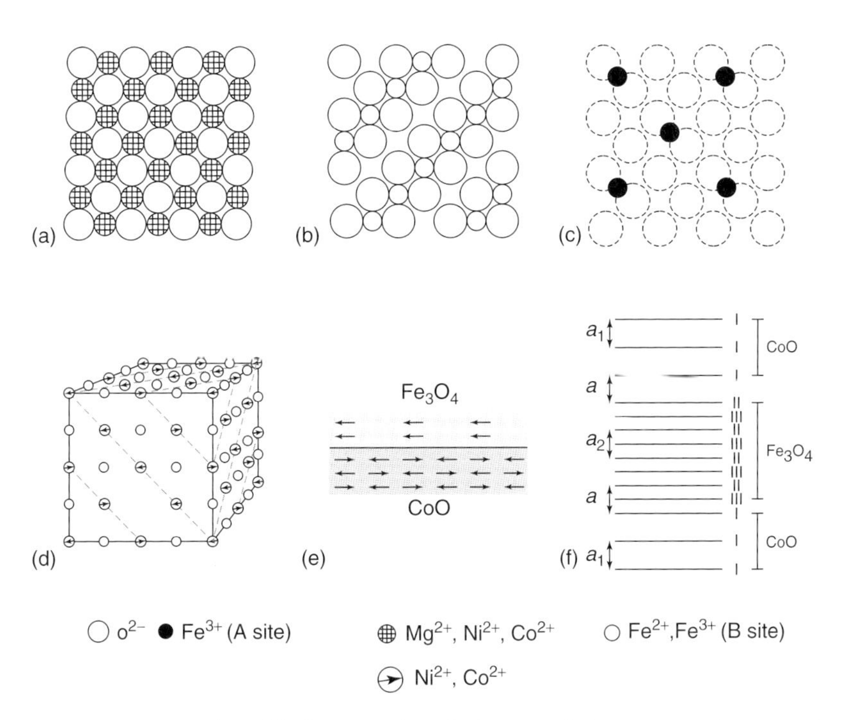

Figure 7.3 Structural model of the growth of Fe_3O_4 on a rock-salt crystal (e.g., MgO, NiO, CoO). Note that Fe_3O_4 has two different lattice planes, one in which the Fe^{2+} and Fe^{3+} ions fill the octahedral B-site in rows, along which the conduction occurs by hopping of electrons (see b) and another in which Fe^{3+} ions occupy A-sites (see c). Notice that Fe_3O_4 has a double magnetic lattice parameter with respect to CoO and NiO (see e).

occurring in Fe_3O_4 [5] is fraught with difficulty, since for epitaxially grown layers the strain, as well as differences in thermal expansion coefficients, will also influence T_v [24]. Thus only limited experimental work has been done to determine the stoichiometry of thin Fe_3O_4 layers. One approach taken has been to investigate the magnetooptical Kerr spectra, which has the advantage of being sensitive to only the top layer of the film [25]. (The top 2–4 nm of an unprotected film was found to be oxidized). A more traditional approach involves the use of ^{57}Fe Mössbauer spectroscopy for which the interior of the ultrathin layers was grown with enriched ^{57}Fe isotope [12]. In the latter study, it was shown that the stoichiometry of the Fe_3O_4 layer could be tuned over a broad range from 0.0 to 0.35 depending on the pressure of the NO_2, which is a very strong oxidizing agent. The more commonly used O_2 is a weaker oxidizing agent, and work in the Bando group using Mössbauer showed that only magnetite films $Fe_{3-x}O_4$ could be grown with $x \leq 0.06$ [26], where x is defined as the relative deviation of the stoichiometry from that of the ideal compound. Several groups have thus demonstrated that high-quality superlattices can be grown of oxide magnetic materials. However, when mixed-valence compounds are included, consideration should be given to some minor off-stoichiometry in the layers.

7.2.2
Interface Effects on Magnetic Properties

Before discussing the properties of ferro(i)magnetic–AF multilayers, it is useful and necessary to consider the interface properties of ferromagnetic layers deposited on nonmagnetic (NM) oxide layers. On this topic a significant amount of work has been done over the last decade, notably on the Fe_3O_4/MgO system. This system will therefore be taken as the model for this section from which the key aspects of the interface effects on the magnetic properties will be discussed. A noticeable amount of the work on this system has been done on Fe_3O_4/MgO multilayers with the aim of enhancing the magnetic contribution from the ultrathin ($\leq$5 nm) interface layers.

An understanding of the magnetic structure of thin Fe_3O_4 films and especially of Fe_3O_4 interfaces is crucial to the interpretation of coupling studies in which oxides play a role, such as when coupled to AF layers to yield exchange biasing or when embedded in multilayer structures to yield (all-oxide) tunnel junctions. The latter is of interest for applications owing to the half-metallic nature of Fe_3O_4. The scientific interest in the interface magnetism of the Fe_3O_4 layers themselves stems firstly from the discovery of the magnetic interface anisotropy favoring perpendicular anisotropy in oxide systems [27] and secondly from the anomalous behavior of the saturation magnetization of thin Fe_3O_4 layers [27, 28]. We start with the discussion of the latter phenomenon.

7.2.2.1 Reduced Magnetization and "Dead" Layers at the Interface

A number of studies have found that the saturation magnetization, M_s, of Fe_3O_4 layers decreases with Fe_3O_4 layer thickness to values below the bulk saturation magnetization of 497 kAm^{-1} for Fe_3O_4. This effect was initially reported by Parkin *et al.* for Fe_3O_4 layers deposited on polished silicon (111) crystals [29]. Further, studies in 1994 by Margulies *et al.* [28] showed that despite the application of a field of 7 T, the magnetization of Fe_3O_4 layers deposited by pulsed laser deposition (PLD) on MgO, as well as on other substrates such as $MgAl_2O_4$ and Al_2O_3, could not be saturated. Subsequent experiments on MBE-grown Fe_3O_4 layers deposited at 500 K also found anomalous magnetic properties [20, 27]. The key results of these and related studies on very well-defined Fe_3O_4/MgO layers are discussed here.

Van der Heijden *et al.* grew a series of Fe_3O_4/MgO multilayers by O_2-assisted MBE on MgO substrates in a UHV Balzers UMS 630 multichamber MBE system. Fe_3O_4 was deposited through e-gun evaporation from Fe targets in an O_2 ambient atmosphere at 2.8×10^{-5} mbar at 500 K [27]. Under the same conditions, MgO was deposited using Knudsen cells. Superconducting quantum interference device (SQUID) magnetometry was used to investigate at room temperature a series of Fe_3O_4/MgO multilayers on MgO(100) with Fe_3O_4 layer thickness varying between 1 and 50 nm and the MgO interlayer thickness chosen such that interlayer coupling effects were minimal [30]. The resulting magnetic moment normalized to Fe_3O_4 layer thickness (t) is shown in Figure 7.4(a) as a function of the Fe_3O_4 layer thickness, $t_{Fe_3O_4}$. These data show that for $t_{Fe_3O_4} \geq 30$ nm, the magnetic moment is, within the experimental accuracy, equivalent to the bulk moment of Fe_3O_4 at 300 K of 497 kAm^{-1} (dashed line in Figure 7.4a) [5]. However, below 30 nm the saturation moment, *even* in these high-quality samples, falls below the bulk value. Just as in earlier studies [29, 31], the behavior of the Fe_3O_4 layer was modeled using a so-called "dead" layer, which is a layer at the interface of the Fe_3O_4 layer with no net contribution to the overall magnetization. Thus, the overall magnetization of the Fe_3O_4 layer should adhere to the expression

$$M = M_s(\text{bulk}) - (2M_i/t) \tag{7.1}$$

where M_s(bulk) is the bulk magnetization, (M_i/M_s) is the thickness of the magnetically "dead" layer at the interface, t_d and the factor 2 is incorporated since each Fe_3O_4 layer has two interfaces with MgO. The analysis of the data becomes more straightforward by dividing by M_S and multiplying both sides of Eq. (1) by t yielding

$$t_M = t_{M_S(\text{bulk})} - 2t_d \tag{7.2}$$

In Figure 7.4(b), the data have been plotted using this format, and one can see that the data are well described by this equation between 1 and 50 nm. The inset in Figure 7.4(b) shows the same data in the range up to 10 nm. A positive x-axis intercept is found at 1.4 ± 0.6 nm, yielding a "dead" layer thickness t_d of 0.7 ± 0.3 nm. The larger dead layer thicknesses of 2.5 nm [29] and 4.0 nm [31] reported previously by others may be due to the nonoptimal growth conditions employed in these earlier studies, which relied on sputtering to prepare the Fe_3O_4 layer. Since the lattice constant of Fe_3O_4 is 8.38 Å [2], the result in Figure 7.4 might suggest that the magnetization of the Fe_3O_4 layer is disturbed over just the first lattice period and thus attests to the high-quality growth of these Fe_3O_4 layers by MBE. In the

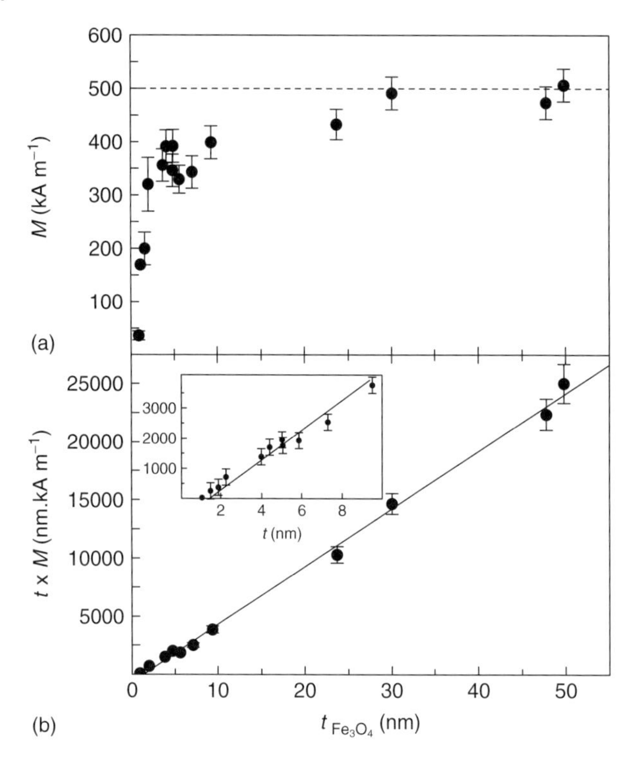

Figure 7.4 (a) Thickness dependence of the magnetic moment, M_S, of Fe_3O_4 grown on MgO(100) normalized to the total Fe_3O_4 volume of the sample at 300 K. (b) M multiplied by the Fe_3O_4 layer thickness, $t_{Fe_3O_4}$. In this figure, the solid line represents a linear fit to the data and the inset is an enlargement of small Fe_3O_4 layer thicknesses showing that M becomes zero at finite $t_{Fe_3O_4}$. Figure reproduced from [27].

study of ferrites, "dead" surface/interface layers have been invoked successfully to account for the anomalous saturation magnetization of ultrafine particles due to spin canting at the surface [32] and to explain the dependence of the magnetic permeability of bulk ferrites on the grain boundaries [33]. However, one has to bear in mind that a magnetically "dead" layer with $M_s = 0$ at the Fe_3O_4/MgO interface is an oversimplification. More likely the magnetization profile should be more rounded. Moreover, the above description provides limited insight into the exact nature of the magnetic state giving rise to this observed absence of magnetization at the Fe_3O_4 interface.

7.2.2.2 Anisotropy and Interface Anisotropy of Thin Fe_3O_4 Layers

The anisotropy in ultrathin MBE-grown Fe_3O_4 layers was first systematically investigated by van der Heijden *et al.* [34] and more recently by McGuigan *et al.* [35]. The key findings were again that for large layer thicknesses ($t \geq 30$ nm), the magnitude and temperature dependence of the magnetic anisotropy resembles the behavior of bulk Fe_3O_4 [34]. For lower thicknesses, the magneto–crystalline and

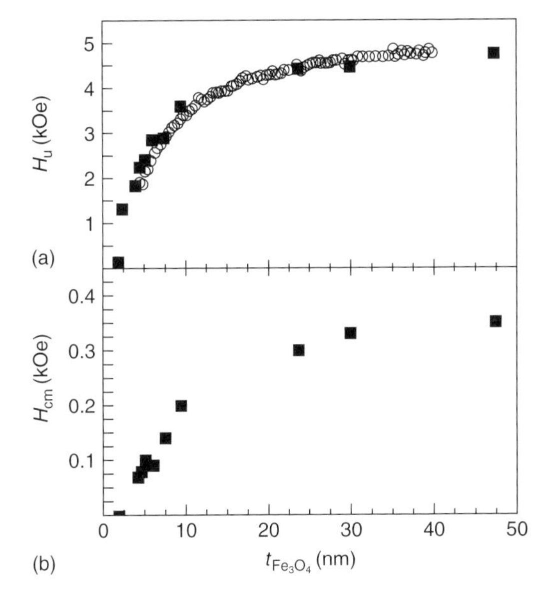

Figure 7.5 (a) The uniaxial out-of-plane field, H_u, and (b) the cubic magnetocrystalline anisotropy field, H_{cm}, as a function of Fe_3O_4 layer thickness, $t_{Fe_3O_4}$, obtained from ferromagnetic resonance experiments at room temperature (filled squares)for Fe_3O_4 grown on MgO(100). For comparison H_u obtained from magnetooptical Kerr effect MOKE experiments on a 0–40-nm Fe_3O_4 wedge at room temperature is also shown (○). Figure reproduced from [34] (© 1998 with permission from Elsevier).

the uniaxial out-of-plane anisotropies decreased monotonically with decreasing Fe_3O_4 layer thickness to finally vanish below 5 nm (see Figure 7.5 taken from [34]). Consequently, we focus the discussion on what happens for ultrathin layers possibly due to a contribution from interface anisotropy.

The study of the interface anisotropy of magnetic thin films was driven in large part by the relevance of this topic for magnetooptic recording media, which require a preferred axis of the magnetization perpendicular to the surface of the film plane. Thus far the study of perpendicular anisotropy had been focused exclusively on metallic (multi)layers. In the late 1980s, a number of studies were done on the properties of ferrite films deposited by sputtering, without focusing on obtaining perpendicular anisotropy in these films [36]. However, recent progress on controlled MBE growth of magnetic oxides film has allowed for a more detailed study of the surface contribution to the magnetic anisotropy that gives rise to the desired perpendicular magnetization.

From the study of metallic multilayers, it is well known that the out-of-plane anisotropy K of thin magnetic layers is composed of a bulk or volume contribution K_v and an interface or surface contribution K_s. The work on metallic magnetic multilayers has shown that the anisotropy of metallic systems can be described phenomenologically by the following equation [37]:

$$K = K_v + 2K_s/t \tag{7.3}$$

with t the layer thickness. The K_v contribution to K is composed of various terms including the shape anisotropy, the magnetocrystalline anisotropy, and the magnetoelastic anisotropy due to strained, coherent growth. For thin layers, the shape anisotropy, favoring in-plane preferential magnetization, can be overcome by the surface anisotropy contribution, which gives rise to a preferred magnetization contribution out of the film plane. This effect has been amply demonstrated for metallic systems [37], yet has been investigated for only one oxide system to date, Fe_3O_4 grown on a MgO substrate [27]. As discussed in the preceding section, there is no measurable magnetic moment contribution from the interface layer of Fe_3O_4 on MgO. Hence, Eq. (3) needs to be modified to account for this "dead"-layer effect by correcting the apparent thickness by t_d. Thus, in practice the relevant equation for K becomes

$$K = K_v + 2K_s/(t - 2t_d) \tag{7.4}$$

To study the interface anisotropy of Fe_3O_4, van der Heijden *et al.* [27] grew a wedged-shaped sample on a MgO(100) substrate. The first sample was composed of 0–40-nm Fe_3O_4/2-nm MgO/3-nm NiO. The MgO component of the capping layer ensures that a symmetrical situation is attained. The additional 3-nm NiO capping layer is needed to prevent reaction of MgO with H_2O and CO_2 during the measurements. To be able to detect any influence of this capping layer, only half of the wedged sample had the capping layer. Figure 7.6 schematically shows the layout of the samples studied. With a wedged-shaped sample, all the anisotropy measurements on various layer thicknesses can be performed on a single sample grown in a single growth run [38]. The slope of these wedges is 3-nm Fe_3O_4/mm, and consequently one can easily measure the hysteresis loop at a well-defined Fe_3O_4 layer thickness using the MOKE.

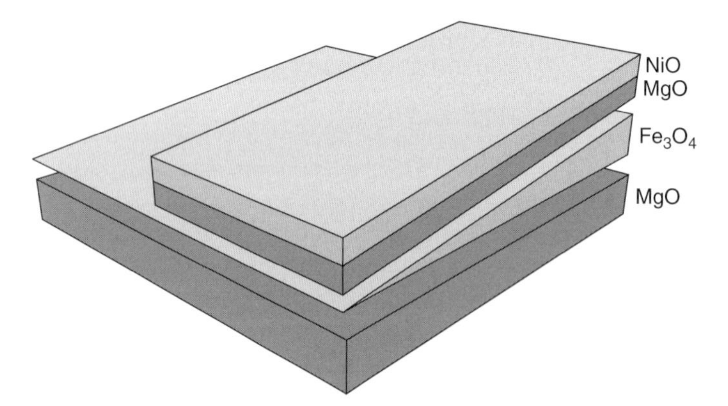

Figure 7.6 Schematic representation of the wedge-shaped sample used to study the interface anisotropy of Fe_3O_4 grown on MgO(100). The Fe_3O_4 wedge has a maximum thickness of 40 nm and a slope of 3.6 nm/mm. The wedge is partly covered by 2-nm MgO/3-nm^{-1} NiO to enable investigation of the influence of the top capping layer.

To determine the out-of-plane anisotropy, K, van der Heijden *et al.* [27] measured hysteresis loops along their wedged-shaped sample in both longitudinal and polar MOKE geometry with a magnetic field applied parallel and perpendicular to the film plane, respectively. Figure 7.5(a) shows the thickness dependence of the uniaxial out-of-plane anisotropy field, H_u, determined from the measured perpendicular saturation field, H_S. The results of longitudinal MOKE experiments show that the in-plane saturation magnetization is sufficiently small for Fe_3O_4 layer thicknesses larger than 5 nm (less than 10 kAm^{-1} [39]) compared to the perpendicular saturation field, H_s, and it can be neglected. Thus, K can be directly obtained from H_s for Fe_3O_4 thicknesses in excess of 5 nm using the equation [37]

$$K = -\tfrac{1}{2}\mu_0 M_s H_s \tag{7.5}$$

with μ_0 the vacuum permeability. Below 5 nm, the in-plane saturation is no longer negligible and should be incorporated when calculating K. Figure 7.7 shows the resulting thickness dependence of K multiplied by $(t - 2t_d)$ versus the effective magnetic Fe_3O_4 layer thickness $(t - 2t_d)$ [27]. For two points on this curve, at an Fe_3O_4 layer thickness of 9.3 and 23.7 nm, the value of K was deduced from SQUID magnetometry data for corresponding Fe_3O_4/MgO multilayers. In Figure 7.7, open data points were determined from the symmetric wedge, while the filled data points were determined from the part of the wedge that did not have any cap layer. The absence of a cap layer enables one to obtain data at lower Fe_3O_4 thickness as less light is absorbed by the cap layer. From this figure, one can immediately see that the phenomenological Eq. (7.3) (and Eq. (7.4)) also describes the data in this oxide system. The slope of the fit in Figure 7.7 gave values for the volume anisotropy K_v of $-0.123 \pm 0.0.10$ MJm^{-3} and -0.129 ± 0.010 MJm^{-3} for the wedged sample with and without a capping layer, respectively. These values were found to be in accord with the various terms contributing to K_v. Specifically, the shape anisotropy for the system studied is $-\frac{1}{2}\mu_0 M_s^2 = 0.155$ MJm^{-3}, the crystalline anisotropy at $K_1/4 = 0.003$ MJm^{-3} is small, and the pseudomorphic growth of Fe_3O_4 on MgO(100) results in a magnetoelastic anisotropy contribution of about 0.027 MJm^{-3}. The summation of these values yields a volume anisotropy contribution K_v that is in excellent agreement with the experimental data irrespective of the capping layer -0.125 MJm^{-3} [27].

However, the situation concerning K_s was less clear. The fit to the data (Figure 7.7) would suggest that at a Fe_3O_4 layer thickness $(t - 2t_d)$ of 3.5 nm or a total layer thickness of 5 nm, the surface anisotropy term that favors perpendicular anisotropy would balance the volume anisotropy term that favors in-plane anisotropy. Thus, at thicknesses below 5 nm, the system should exhibit positive anisotropy favoring perpendicular magnetization. However, experimentally this was not observed. The signal intensities of the MOKE experiment were insufficient to obtain reliable data below 5 and 10 nm for the noncapped and capped wedged sample. However, a SQUID magnetometry experiment on a (1.6-nm Fe_3O_4/2.5-nm MgO)$_{\times 20}$ multilayer capped with 3-nm NiO showed that hysteresis loops along the <100> in-plane axis and the film normal coincide. Ferromagnetic resonance (FMR) experiments confirmed that the out-of-plane, as well as in-plane, anisotropy vanishes below

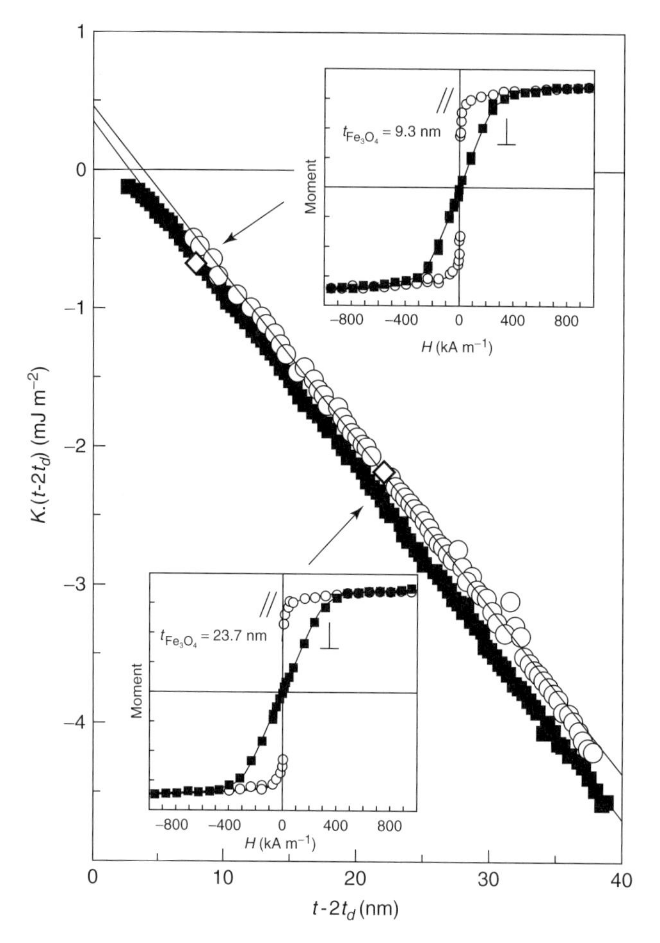

Figure 7.7 The thickness dependence of the out-of-plane anisotropy multiplied by the thickness obtained from the Fe_3O_4 wedge with the capping layer (◦) and without a cap layer (filled squares) grown on MgO(100) using the corrected layer thickness $t{-}2t_d$ (see text). Inset to figure: For two Fe_3O_4/MgO multilayers with Fe_3O_4 = 9.3 and 23.7 nm, respectively, the out-of-plane anisotropy has been calculated using the hysteresis loop determined by SQUID magnetometry with H applied along the [100] in-plane axis (◦) and applied perpendicular to the plane (filled square). Figure reproduced from [27].

a Fe_3O_4 layer thickness of 5 nm. This result led the authors to conclude that the "dead"-layer effect that they had observed somehow precludes a meaningful interpretation of the observed interface contribution to the anisotropy K_s in terms of a Néel-type mechanism [27].

7.2.2.3 **The Interface Structure: Antiphase Boundaries**

A key experiment to elucidate the nature of the magnetism of the Fe_3O_4 interface layers was the conversion electron Mössbauer spectroscopy (CEMS) experiment performed by Voogt *et al.* [23]. Using MBE techniques, these authors successfully grew Fe_3O_4/MgO multilayers in which 0.42-nm (i.e., 2 monolayers) Mössbauer probe layers with the ^{57}Fe Mössbauer active isotope were incorporated at both the Fe_3O_4 interfaces with the MgO layer or within the bulk of the Fe_3O_4 layer. In their series of samples, the Fe_3O_4 layer thicknesses were 1.8, 3.5, and 5.0 nm, respectively, and the MgO interlayer thickness was chosen to be 2.0 nm to exclude pinhole coupling between the Fe_3O_4 layers of the multilayer [30]. At first glance, the most surprising result of this room-temperature Mössbauer experiment was that the two hyperfine split sextets indicative of long-range ferromagnetic order in Fe_3O_4 decrease with decreasing Fe_3O_4 layer thickness regardless of the position of the probe layer (see Figure 7.8 [23]). The breakdown of the long-range order for Fe_3O_4 layer thicknesses less than 5 nm is clearly *not* an interface effect. The Mössbauer data clearly show a loss of long-range order as the two sextet peaks gradually deteriorate with decreasing layer thickness until they appear to coalesce into one peak for the 1.8 nm sample. Thus, this effect occurs throughout the entire layer and not just at the interface.

The two sextet components for the multilayer in which the Fe_3O_4 thickness is 5.3 nm correspond to the isolated Fe^{3+} and $Fe^{2.5+}$ ions in the octahedral sites. This result demonstrated that the structure of the Fe_3O_4 film is spinel-like in its entirety, and the presence of other iron-oxide phases at the interface, such as $Fe_{1-x}O$ or γ-Fe_2O_3, can be excluded. Moreover, the formation of a mixed $MgFe_2O_4$ ferrite at the interface in these high-quality MBE samples can also be excluded since $MgFe_2O_4$, with its high resistivity, does not have hopping conduction along the octahedral sites that would give rise to a $Fe^{2.5+}$ signal in the Mössbauer spectrum [23].

Drawing a comparison to CEMS studies of thicker Fe_3O_4 layers [40], the Mössbauer data indicate that the hyperfine fields of both the Fe^{3+} and $Fe^{2.5+}$

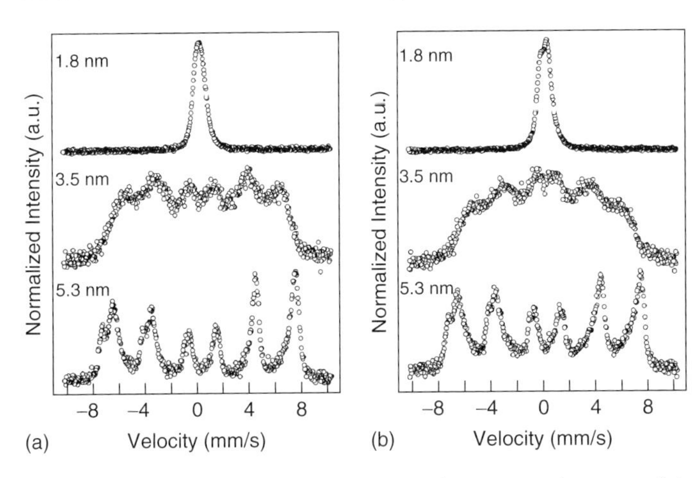

Figure 7.8 Room-temperature conversion electron Mossbauer spectra of (100) Fe_3O_4/MgO multilayers as a function of the Fe_3O_4 thickness (indicated in each spectrum) and the position of the ^{57}Fe probe layer: (a) In the center of the Fe_3O_4 layer and (b) at the Fe_3O_4/MgO interface. Figure reproduced with permission from [23] (© 1998 by the American Physical Society).

components become smaller as the Fe_3O_4 thickness is reduced. Even in the ferromagnetic phase, the magnetic moments of the Fe-ions are progressively fluctuating faster. Ultimately, this leads to the collapse of the sextet spectrum around 3.5 nm, while a transition to a fully paramagnetic state occurs at a layer thickness of 1.8-nm layer. Consequently, from these Mössbauer experiments, there is no evidence of any nonmagnetic or "dead" layer at the Fe_3O_4 interface.

Independent of each other, two groups proposed that the explanation of the anomalous magnetization behavior of thin Fe_3O_4 layers is caused by antiphase boundaries (APBs) [22, 23]. APBs develop because Fe_3O_4 has a lower symmetry and a larger unit cell than the underlying MgO substrate (and the MgO interlayer in the case of multilayers). Consequently, Fe_3O_4 islands can nucleate on a MgO substrate in eight unique ways. When nucleated on a substrate, these islands are not able to match up with each other in order to form a continuous Fe_3O_4 layer without lattice faults (Figure 7.9). Consequently, the Fe_3O_4 layer will have a continuous O_2-sublattice coincident with a fragmented, discontinuous cation sublattice [22, 23]. Such APBs have been observed directly in transmission electron microscopy (TEM) studies (see Figure 7.10) [22, 41, 42] as well as in scanning tunneling microscope (STM) studies (Figure 7.11) [43]. In Figure 7.11, the STM image of a $33 \times 33\,nm^2$ area of the surface of an MBE-grown Fe_3O_4 layer shows a regular network of bright features elongated along the <110> direction, which indicates a $p(1 \times 1)$ surface reconstruction. This reconstruction was interpreted as a clustering of tetrahedral ions, which have dangling bonds that rotate by 90° from one atomic plane to the next [43]. This 90° rotation is observed between two

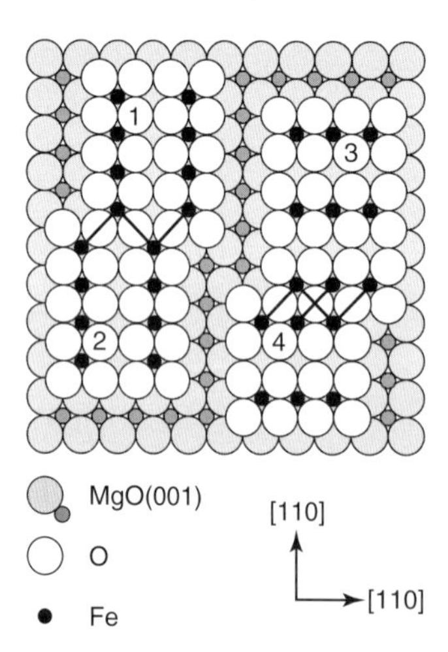

Figure 7.9 Schematic representation of the formation of antiphase boundaries (APBs) by the growth of Fe_3O_4 on a rock-salt structure, here, MgO(100). For clarity, one monolayer containing only octahedral Fe cations is shown. The coalescence of islands 1–2 and 3–4 leads to the formation of an APB as the islands are shifted with respect to each other. This shift is thought to give rise to 180^0 Fe–O–Fe superexchange paths, as indicated by the solid lines (© 1998 by the American Physical Society).

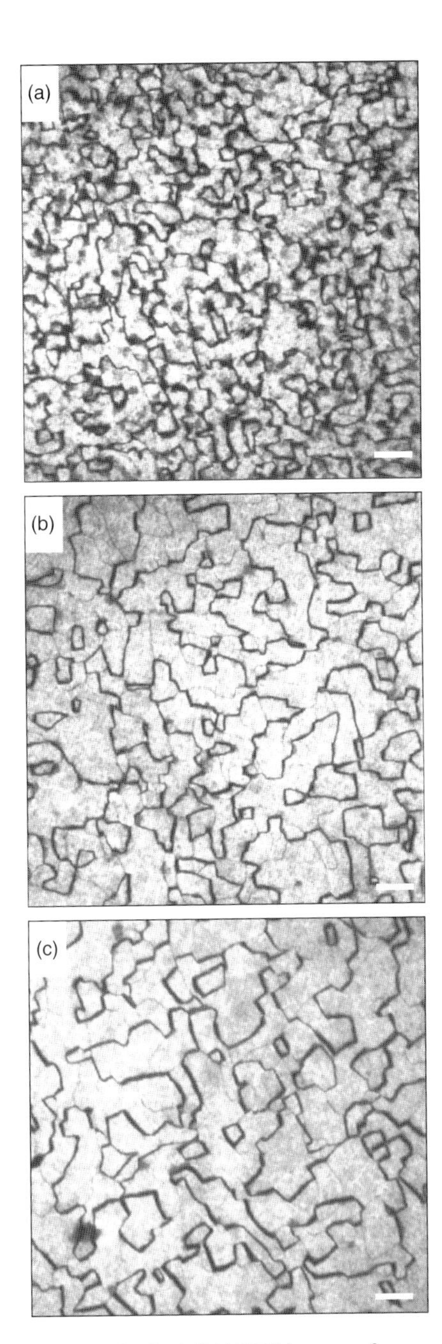

Figure 7.10 Dark-field TEM images of a 12-nm-thick Fe_3O_4 film (a) as-grown. The images in (b) and (c) were taken after postannealing at 300°C in an oxygen background pressure of 10^{-6} mbar for 1 and 2 h, respectively. The scale bar is 20 nm for all three images. Figure reproduced with permission from [42] (© 2003 by the American Physical Society).

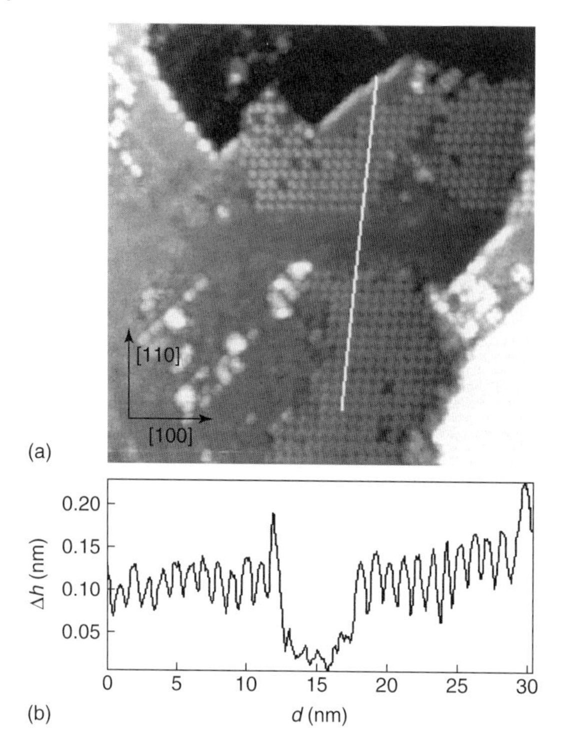

Figure 7.11 (a) Scanning tunneling microscope (STM) image of a $33 \times 33\,nm^2$ area of Fe_3O_4 grown on MgO(100) with a p(1×1) surface reconstruction showing two Fe_3O_4 islands rotated by 90° with respect to each other. (b) Line profile taken along the white line shown in (a). This line scan shows that the two Fe_3O_4 islands are at the same height on the surface. The two islands are separated by a disordered region taken to be an APB. Figure reproduced from [43] (© 1997 with permission from Elsevier).

adjacent terraces which are separated by a single step on one atomic layer (0.2 nm). However, Figure 7.11 also shows a disordered region between two Fe_3O_4 islands nucleated with a different (cationic) orientation relative to each other. These two islands appear to be separated by a disordered region of around 5 nm, which was taken to be an APB. Figure 7.10 shows a TEM image of antiphase boundaries in MBE-grown Fe_3O_4 layer at lower resolution, thereby offering a macroscopic view that allows one to determine the density of the APBs [42]. For this MBE-grown layer, the APB density or boundary length per unit area was determined to be $6.8 \times 10^{-3}\,nm^{-1}$, which is equivalent to an average antiphase domain (APD) size of $300\,nm^2$ [41]. This size is more than 2 orders of magnitude larger than the APD of $27.5\,nm^2$ found for sputtered layers [22, 28], indicating a strong dependence of APD size on growth conditions.

Although both the San Diego [22, 28] and Groningen [23] groups proposed that APBs play an essential role in explaining the anomalous magnetization behavior of ultrathin Fe_3O_4 layers, the models invoked to justify their conclusions differ. In addition to the CEMS studies on Fe_3O_4/MgO multilayers, Voogt *et al.*

performed SQUID studies at low temperature. They observed that the remanent magnetization at 5 K after field cooling in a 10 kOe (= 796 kAm^{-1}) field decreased rapidly with increasing temperature until it leveled off at what they defined as the blocking temperature [23]. They also determined that this blocking temperature decreased with film thickness with values of 40, 140, and 250 K for their 1.8, 3.5, and 5.3-nm thick layers, respectively. This magnetic behavior was attributed to superparamagnetism caused by magnetic domains within the Fe_3O_4 layer having dimensions of approximately $10 \times 10\,nm^2$. It was proposed that these domains originate from a new 180° Fe–O–Fe interaction across the APBs that is not present within Fe_3O_4. They surmised that these 180° Fe–O–Fe interactions are quite strong because a similar Fe–O–Fe superexchange interaction found in orthoferrites has a superexchange constant J of −25 K [44]. Since this new interaction at the APB is comparable to the dominant superexchange interaction between Fe on tetrahedral (A) and octahedral (B) sites within Fe_3O_4 (J_{AB} interaction of −23.4 K [5]), these authors assumed that the A–B coupling across each APB cancels the 180° Fe–O–Fe interaction, yielding an average net coupling strength of $3 \times 10^{19}\,m^{-2}$ for the AF superexchange interaction between neighboring domains within the Fe_3O_4 layer. This gives Rise to a barrier height $W \sim 3 \times 10$ eV per 10 nm of antiphase boundary for the fluctuation of the domain due to AF coupling across the APB. Comparing this calculation with the experimental data, the authors concluded that the influence of the superexchange barriers across the APBs must be largely suppressed due to frustration among the interdomain interactions. When Fe_3O_4 islands nucleate on the MgO substrate at random points during film growth, junctions between domains common to three domains will be in the majority. Each of these junctions have competing interdomain interactions in which superexchange barriers will effectively cancel each other out and enable the domain to fluctuate freely. In this situation, single-crystalline, epitaxial Fe_3O_4 films will become superparamagnetic.

In contrast, the San Diego group did not invoke superparamagnetism to explain how APBs account for the magnetism of thin Fe_3O_4 films [23]. These authors also concluded that new 180° Fe–O–Fe interactions occur across the APBs. In addition, these authors assumed that the number of A–A interactions increases along with the A–O–A angle relative to 180°. As a consequence, the interaction term also increases [22]. This assumption of an increase in A–A interactions may be incorrect as it relies on models of the APB in which the octahedral and tetrahedral ions are in close proximity. If one assumes that there must be charge neutrality along the APBs, these APBs are not likely to occur. Nevertheless, Margulies *et al.* [22] arrived at the same conclusion as Voogt *et al.* The altered exchange interaction across the APB causes the coupling between two adjacent domains to be AF since the 180° Fe–O–Fe coupling will overpower the A–B coupling. These authors used this model to explain their observation that their thick sputtered Fe_3O_4 films did not saturate even in extremely high fields up to 70 kOe (= 5572 kAm^{-1}). We note, however, that subsequent studies demonstrated that high-quality Fe_3O_4 films grown by MBE could be saturated (see Figure 7.4a) presumably due to a lower density of APBs.

The accepted explanation of how the APBs and the APDs influence the observed magnetic properties of Fe_3O_4 films originates largely from the work done by Hibma, Eerenstein, and coworkers. From resistivity measurements [45], these authors first found that the antiphase domain size, and thus the APB density, strongly increases with decreasing film thickness. Specifically, the antiphase domain size D was found to depend on film thickness, t, as [42].

$$D \propto \sqrt{t} \tag{7.6}$$

This effect also explained the thickness dependence of the magnetization and anisotropy (see Sections 7.2.2.1 and 7.2.2.2 as well). Subsequently, these authors [42] demonstrate that the antiphase domain size is not fixed but could be altered by annealing the Fe_3O_4 films after dissolving the MgO substrate to remove it. Figure 7.10(a) shows the dark-field TEM image of the as-grown 12-nm Fe_3O_4 film. Figures 7.10(b) and 7.10(c) show the effect of postannealing the film at 300°C for 1 and 2 h, respectively, in an oxygen atmosphere of 10^{-6} mbar. Clearly, the APB density and the antiphase domain size change. This result is quite important because the antiphase domain size previously was thought to be fixed by the initial growth conditions [41]. Eerenstein *et al.* showed that the activation energy for the diffusive motion of an APB is quite low at 26 ± 5 kJmol^{-1} (250 meV) [42], primarily because the oxygen sublattice is continuous at the APB and the cation sublattice is discontinuous. Since these cations can move into the mostly unoccupied interstices of the Fe_3O_4 lattice (Figure 7.1a) and since the rapid exchange of electrons between Fe^{2+} and Fe^{3+} above the Verwey temperature [5] prevents the buildup of long-range electrostatic interactions, this process does not require high energy and can occur at growth temperatures typical for Fe_3O_4 films. Consequently, the longer deposition times needed for thicker films give rise to a lower APB density. (These authors also showed that for MBE-based growth, the antiphase domain size does not depend on oxygen pressure.) Given that the critical nuclei for the films consist of both Fe and O atoms and that Fe diffuses faster, the island formation depends on the density of the slower O species. Differences in magnetic results that depend on growth conditions at various labs can thus be explained. (This model applies to both Fe_3O_4 films grown by MBE and sputtering, but not to pulsed laser deposited films as the atomic species are deposited at the same time in the latter.) In a subsequent paper [46], the Mössbauer investigations indicated that the magnetic fluctuations of the larger antiphase domains are most likely blocked while the smaller domains have magnetic moments that are superparamagnetic. A more refined analysis showed that only 50% of the coupling across APBs gives rise to AF coupling [47]. The authors thus could quantitatively explain the earlier results by Voogt *et al.* [23] on the observed blocking temperature dependence of the remanent magnetization.

APBs were thus directly observed both in sputtered and in MBE-grown films [22, 23, 41], and analysis of the magnetization revealed that new 180° Fe–O–Fe coupling occurs giving rise to AF coupling between domains originating from differently nucleated Fe_3O_4 islands. To gain more insight into the relevant model [22, 23], an experiment was performed to see if the superparamagnetic state predicted [23] could be observed. To investigate this, the hysteresis loop of a $(Fe_3O_4/MgO)_{\times 50}$ multilayer

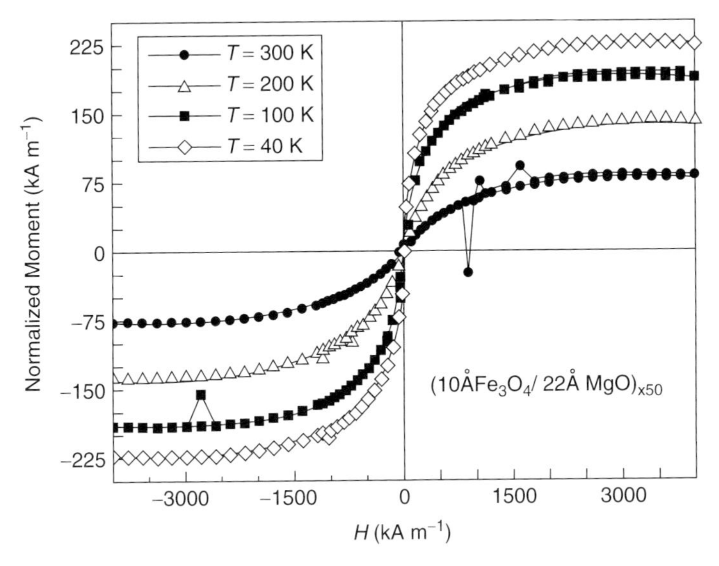

Figure 7.12 Hysteresis loops at various temperatures determined by SQUID magnetometry for a (10 Å Fe_3O_4/22 Å MgO)$_{\times 50}$ multilayer with a 30 Å NiO cap layer grown on MgO(100). *H* was applied along a [100] in-plane direction. The magnetic moment *M* is normalized to the total Fe_3O_4 volume of the multilayer after linear correction for the diamagnetic substrate contribution. Figure taken from [48].

with a 1-nm Fe_3O_4 and 2.2-nm MgO was measured by SQUID magnetometery [48], and the data are reproduced in Figure 7.12. These data clearly show saturation of the magnetization at all four temperatures investigated (40, 100, 200, and 300 K), where the observed moment decreases with increasing temperature. A key observation is that the temperature dependence of the hysteresis loop cannot be described by the Langevin function, which is expected for superparamagnetic behavior. One has to note that in measuring these hysteresis loops, a correction had to be made to the signal for the diamagnetic contribution of the MgO substrate. However, it was shown that it is not possible to adjust this linear correction such that a Langevin temperature dependence could be obtained. Thus, the proposed model based on superparamagnetic behavior of thin Fe_3O_4 layer induced by the 180° Fe–O–Fe coupling across the APBs does not seem to hold if magnetization measurements are considered.

There appears to be a contradiction between the magnetization measurements that find no evidence of a "dead" layer and the Mössbauer data that see evidence for motional narrowing of the Mössbauer spectra for films of less than 2.4-nm thickness [23, 46]. These differences can be reconciled by considering that both measurement techniques detect the magnetization at different timescales. Specifically, the Mössbauer timescale of 10^{-8} s is much shorter than the magnetization measurement scale of $\simeq$1 s. Consequently, fluctuations of the spins within the measurement time of Mössbauer give rise to motional narrowing or "superparamagnetism", while these spin fluctuations are not detectable in conventional magnetization measurements that consequently are sensitive to the missing magnetization. The layer thickness dependence can thus be understood in that the antiphase domain size D depends on the film thickness as $D \propto \sqrt{t}$, and the magnetic fluctuations of the spins in larger APDS are blocked or exceed the Mössbauer measurement time. Consequently, the *relative* contribution of the antiphase boundaries to the total Fe_3O_4 layer thickness decreases as the Fe_3O_4 layer becomes thicker, and the layers become easier to magnetize, as shown in Figure 7.4(a). A possible interpretation of the strictly linear dependence for the anisotropy in Figures 7.4(b) and 7.7 then is that the *absolute* contribution (i.e., amount) of the spins *in* the APBs is more or less constant giving the effect of a (small) "dead" fraction of spins which do not contribute to the magnetization signal. The absence of a so-called "dead" layer *at the Fe_3O_4 interface* bodes well for applications of Fe_4O_4 in oxide tunnel junctions (to be discussed in Section 7.4), as a true "dead" layer at the Fe_3O_4 interface with any NM tunnel barrier could severely limit the spin polarization available for tunneling and thereby reduce the attainable tunnel magnetoresistance ratio.

7.3 Magnetic Coupling Studies

Though the properties of magnetic oxide multilayers are determined largely by the nature of the magnetic coupling between the layers, the structural and magnetic characteristics of each individual component (as detailed in Section 7.2) strongly influence the behavior of the composite system. Because of the potential for spintronics applications (discussed in Section 7.4), recent research has focused primarily on AF–F oxide multilayers that exhibit exchange biasing, on F–F multilayers that act as exchange springs, and on F–F systems with NM interlayers that show tunneling MR. Simplistic multilayers composed of monoxide antiferromagnets, however, were the topic of the earliest investigations as their growth was straightforward. The magnetic phenomena discovered in these AF–AF and AF–NM systems, such as scaling of the ordering temperature, are described first in Section 7.3 as they provide a basis for understanding the complex behaviors observed later in F–AF multilayers, as discussed later in this section.

7.3.1 Antiferromagnetic Multilayers

7.3.1.1 AF–NM Multilayers: Finite-Size Scaling

Finite-size scaling, which is an important field of study in physics, refers to changes in the physical properties of solids resulting from a reduction in a dimension below a certain characteristic length scale. Early studies of this effect in magnetic $MnFe_2O_4$ nanoparticles prepared by wet-chemical methods [49, 50] were fraught with difficulty due to nonequilibrium cation distributions [51] and surface chemistry/oxidation effects [52]. Recently, a systematic investigation of related nanoparticles confirmed unequivocally that these structural issues were actually responsible for the observed deviations in T_N that were initially attributed to finite-size scaling [53]. Advances in the growth of magnetic oxide multilayers and thin films by sputtering and MBE have opened a more controlled approach for characterizing finite-size scaling effects on magnetic properties such as the ordering temperature T_N. Since these oxides are strongly correlated systems, the correlation length ζ is short and finite-size scaling should occur only in ultrathin films. Indeed, investigations of metallic magnetic films preceding the oxide studies established that finite scaling of the ordering temperature occurs only for ultrathin films that are 5–6 monolayers thick [54].

Figure 7.13 summarizes the key results of measurements of the CoO ordering temperature T_N and its dependence on layer thickness in CoO–NM oxide multilayers. This figure includes data on both CoO/MgO multilayers prepared by MBE [21, 55, 56] and CoO/SiO_2 multilayers grown by reactive sputtering [57, 58]. Specifically, Figure 7.13(b) shows the ordering temperature determined from susceptibility measurements of the CoO/SiO_2 multilayers. These data suggest that the

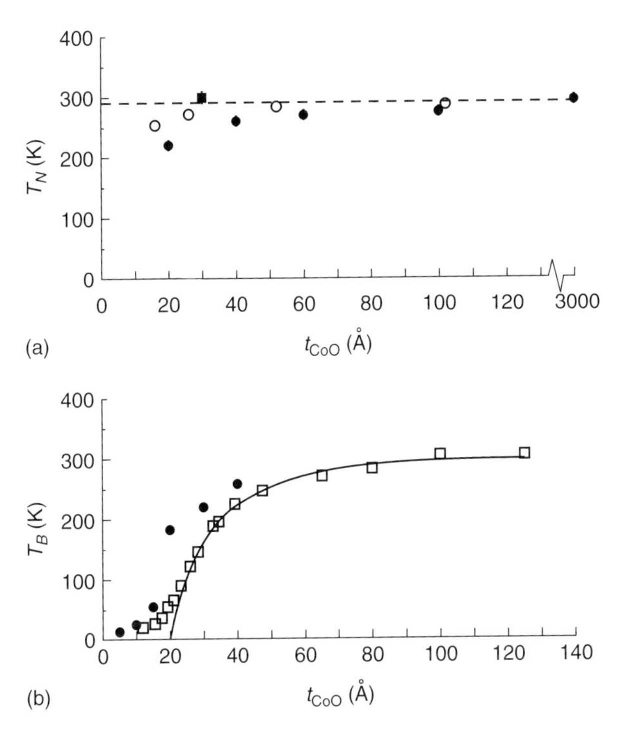

Figure 7.13 (a) Néel temperature, T_N, data taken thermodynamic measurement in (21, 55, 56). The ◦ data points are based on CoO/MgO multilayers, while the • data points are based on CoO/SiO_2 multilayers [55]. The filled square is derive from Figure 7.14. (b) 'Ordering' temperature reflecting the blocking temperature, T_B, determined from susceptibility measurements in [55] (•) and [57] (□) for CoO as a function of CoO layer thickness, t_{CoO}, (see text for details).

CoO bulk T_N of 291 K is reached only at CoO thicknesses greater than 100 Å $\simeq$ 25 monolayers. In contrast, neutron diffraction studies of high-quality *single-crystalline* $(30\,\text{Å CoO}/30\,\text{Å MgO})_{\times 333}$ and $(30\,\text{Å CoO}/18\,\text{Å MgO})_{\times 50}$ multilayers demonstrated that T_N of the CoO layers remains bulk-like [21, 56]. The plot of the intensity of the $(\frac{1}{2}\frac{1}{2}\frac{1}{2})$ AF reflection as a function of temperature for the $(30\,\text{Å CoO}/18\,\text{Å MgO})_{\times 50}$ multilayer (Figure 7.14) clearly shows that $T_N = 300 \pm 10$ K. The correlation length deduced from the width of the $(\frac{1}{2}\frac{1}{2}\frac{1}{2})$ reflection for this multilayer is 38 Å, thus ruling out the possibility that interlayer coupling among the CoO layers across the MgO layers leads to an enhancement of the ordering temperature. Moreover, Abarra *et al.* studied a series of sputtered CoO/MgO multilayers by specific heat and observed a very limited reduction of T_N (◦ data points in Figure 7.13a) [55].

The apparent discrepancy between the behavior of the ordering temperature of CoO/MgO multilayers (Figure 7.13a) and CoO/SiO_2 multilayers (Figure 7.13b) was addressed by Tang *et al.* [58] who demonstrated that the structural integrity of the CoO/MgO multilayers was better than that of the CoO/SiO_2 multilayers even though both were grown by reactive sputtering. Given that SiO_2 is amorphous, it was not surprising that in their CoO/SiO_2 multilayers the structure of the CoO was also amorphous for layer thicknesses less than 20 Å. In contrast, their CoO/MgO multilayers grown on MgO substrates exhibited higher ordering temperatures

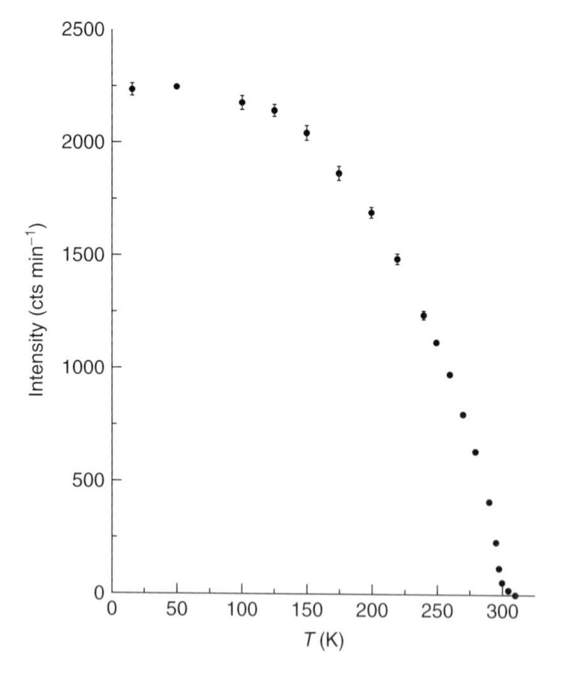

Figure 7.14 The intensity, I, of the $(\frac{1}{2}\frac{1}{2}\frac{1}{2})$ reflection as a function of temperature, T, for a $(30\,\text{Å CoO}/30\,\text{Å MgO})_{\times 333}$ multilayer grown on MgO(100). Note that I drops to zero at around 305 $\pm$10 K, corresponding to the bulk Néel temperature of CoO.

because lattice matching of the CoO and MgO rock-salt structures results in the growth of polycrystalline multilayers with large grain sizes [58]. Moreover, for their CoO/MgO samples, the authors also identified apparent contradictions among values of the "ordering" temperature determined by measurements of the specific heat in comparison to values obtained from analysis of the decay time of the thermoremanent moment or from a comparison of the zero- and field-cooled magnetization. The ordering temperatures that they determined for the CoO/SiO_2 multilayers by specific heat measurements are plotted in Figure 7.13(a) (• data points). The "ordering temperature" found from measurements of the thermoremanent magnetization and the zero- and field-cooled magnetization is actually a blocking temperature [58], and the trend with CoO layer thickness matches that shown in Figure 7.13(b) for CoO/SiO_2 multilayers. The explanation for the discrepancy between the data in Figure 7.13(a) and 7.13(b) may be that magnetic susceptibility measurements for thin films are unduly influenced by uncompensated surface spins [55], while neutron diffraction and specific heat directly probe the bulk of the AF material. Consequently, taking into account the quality of the samples studied, limited evidence exists, as of yet, for finite scaling of T_N for CoO layers as thin as $\approx$25 Å (e.g., 6 monolayers). It would be highly desirable to extend these studies to include *single-crystalline*, MBE-grown samples with CoO layer thicknesses $\leq$30 Å.

The ordering temperature of MBE-grown, ultrathin NiO single layers on MgO was also studied using magnetic X-ray dichroism (MCD) [59], and it was shown to be reduced below T_N of bulk NiO (520 K) [60]. Specifically, T_N was found to be 470, 430, and 295 K for NiO thicknesses of 20, 10, and 5 monolayers, respectively [59]. Note that for CoO/MgO and NiO/MgO sputtered multilayers, T_N for NiO is considerably more suppressed than for CoO layers of comparable thickness [55]. To date, there exists no explanation for the origin of the differences in behavior observed among ultrathin metallic films [54], CoO multilayers in which finite scaling effects are observed only below 5–6 monolayers (if at all), and NiO multilayers in which finite-size scaling occurs in much thicker layers.

7.3.1.2 AF–AF Multilayers: Exchange Coupling

As described in Section 7.2, the first CoO–NiO AF–AF multilayers were grown by MBE by Terashima and Bando [61]. These authors were also the first to study oxide multilayers using neutron diffraction techniques, which directly probe the AF order. In each of their multilayers, the CoO and NiO layers had individual thicknesses less than 20 Å, and they behaved as a *single* magnetic unit with a *single* T_N that varied linearly with the relative CoO/NiO content [62]. The XRD results for multilayers with a bilayer thickness $\leq$50 Å were consistent and showed a single diffraction peak that yielded a lattice parameter intermediate between that of the individual CoO and NiO components [61]. The interest in these AF–AF multilayers was further stimulated by the work of Carey and Berkowitz, who showed that AF NiO or $Co_xNi_{1-x}O$ could be used as a pinning layer to exchange bias (see Section 7.3.2) a permalloy overlayer in read heads [63]. Exchange biasing [64, 65] was also observed in related systems involving CoO–NiO multilayers, and these multilayers were subsequently studied by neutron diffraction [66, 67] and specific heat techniques [55]. Characterization of the ordering temperature as a function of NiO layer thickness, t_{NiO}, and CoO layer thickness, t_{CoO}, yielded the following picture:

1) Multilayers with t_{CoO} and $t_{NiO} \leq 20$ Å behave as a single entity with T_N determined by the average NiO : CoO ratio [55, 62, 66].
2) Multilayers with $t_{CoO} \leq 30$ Å, $t_{NiO} \leq 50$ Å no longer behave as a single antiferromagnet and show two transition temperatures [55, 66].
3) For t_{NiO}, $t_{CoO} \geq 60$ Å, the layers behave as individual layers with transition temperatures approaching their bulk values [62, 65, 66].

This trend is clearly illustrated in Figure 7.15 [65] that shows the blocking temperature, T_B (i.e., the temperature at which the exchange biasing [68] ceases to exist) plotted as a function of CoO layer thickness for a series of $Ni_{80}Fe_{20}$/CoO–NiO trilayers with the NiO thickness > 450 Å and the intermediary CoO thickness varied from 10 to 55 Å (see inset in Figure 7.15). The three different regimes identified above are apparent with a transition region extending over a CoO thickness range from 30 to 50 Å.

The scaling of T_N can be understood within the context of the magnetic proximity effect that gives rise to long-range spin alignment. Using a mean-field approach [69],

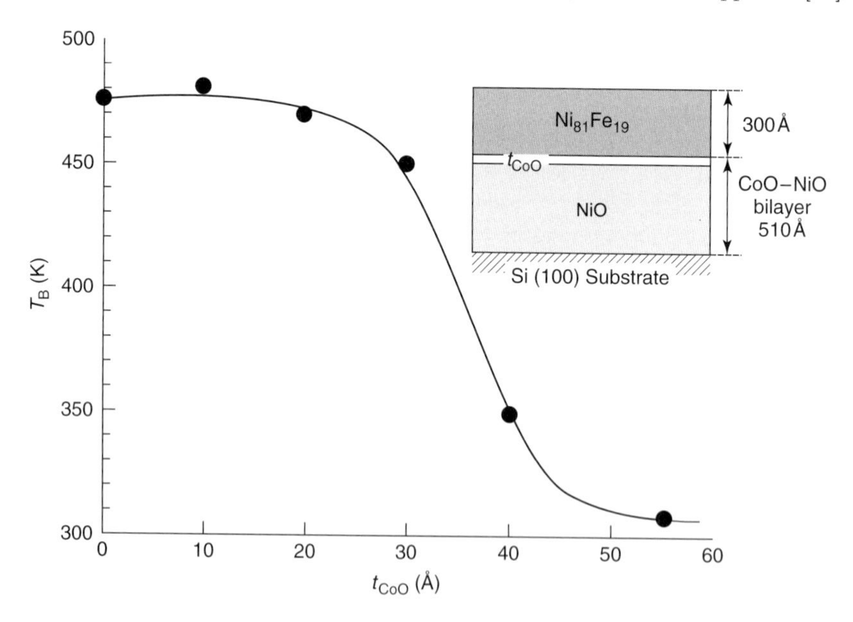

Figure 7.15 Blocking temperature, T_B, versus CoO layer thickness, t_{CoO}, for thick NiO/thin CoO/$Ni_{81}Fe_{19}$ exchange couples. The drawn line indicates variation in TB. T_N for NiO/$Ni_{81}Fe_{19}$ is 475 K and for CoO/$Ni_{81}Fe_{19}$ is 305 K. The inset shows the sample structure used. Figure reproduced from [65].(© 1993 by the American Institute of Physics).

a single bilayer is modeled as N_{NiO} planes of NiO and N_{CoO} planes of CoO. The magnetic structure of the bilayer is determined from a mean-field equation that includes only nearest-neighbor interactions. For the ith spin in the bilayer

$$< S_i > = B_S \left(\frac{J_i < S_{i-1} > + J_i < S_{i+1} >}{k_B T} \right) \tag{7.7}$$

where B_S is the Brillouin function. The exchange parameter J_{NiO} and J_{CoO} were determined from the T_N of bulk NiO and CoO, and J_i at the interface was taken as the average of J_{NiO} and J_{CoO}. This model reproduced qualitatively the thickness dependence of T_N as determined from neutron measurements [66], and it also predicted that the moment decays smoothly through the interface to near zero in the center of the CoO layer at temperatures just below T_N for NiO. According to the model, the Co spins near the interface are polarized by exchange coupling to Ni moments, and this polarization gives rise to, for example, the increase of the CoO ordering temperature of approximately 80 K observed for a (43 Å NiO/29 Å CoO)$_{\times 100}$ multilayer [66]. The results of these neutron diffraction experiments are consistent with ordering temperature measurements for other multilayers involving ionic antiferromagnets such as $FeF_2/CoFe_2$ [70] and Fe_3O_4/NiO [71, 72].

Remarkably, in the CoO/NiO multilayers described above, the measured magnetic coherence length along the growth axis (i.e., perpendicular to the growth plane) is 250 Å at 300 K and 170 Å at 450 K [66], which exceeds the individual NiO layer thickness even at temperatures (450 K) at which the CoO is effectively disordered [66]. The coherence of the NiO spins at temperatures greater than the apparent T_N of the CoO suggests that the magnetic exchange interactions may be of longer range than previously believed in these AF oxide systems.

7.3.2 Antiferromagnetic–Ferromagnetic Coupling

7.3.2.1 Exchange Anisotropy

Exchange anisotropy or exchange biasing was discovered by Meiklejohn and Bean in 1956 upon field cooling Co particles with surface oxidation [68]. They observed a hallmark signature of exchange biasing: a hysteresis loop shifted along the field axis for temperatures below the AF Néel temperature, T_N. Figure 7.16 shows this effect for a [111] oriented Fe_3O_4/CoO bilayer grown on α-Al_2O_3(0001) upon field cooling through the CoO T_N. The shift of the hysteresis loop along the field axis is defined as H_{eb} and is indicated in Figure 7.16. The extent of the shift, H_{eb}, depends on temperature as shown in the figure. The temperature at which the biasing vanishes is defined as the blocking temperature T_B with $T_B \leq T_N$. Note that in this case H_{eb} depends linearly on T. For recent reviews of this effect that include basic models, refer to [73, 74] in addition to Chapter 5 [75]. A discussion of various theories for biasing can be found in the reviews of [76, 77] as well as in Chapter 6 [78].

From simplistic energetic models, one can derive an equation for the expected magnitude of the field shift, H_{eb}, by balancing the Zeeman field with the exchange

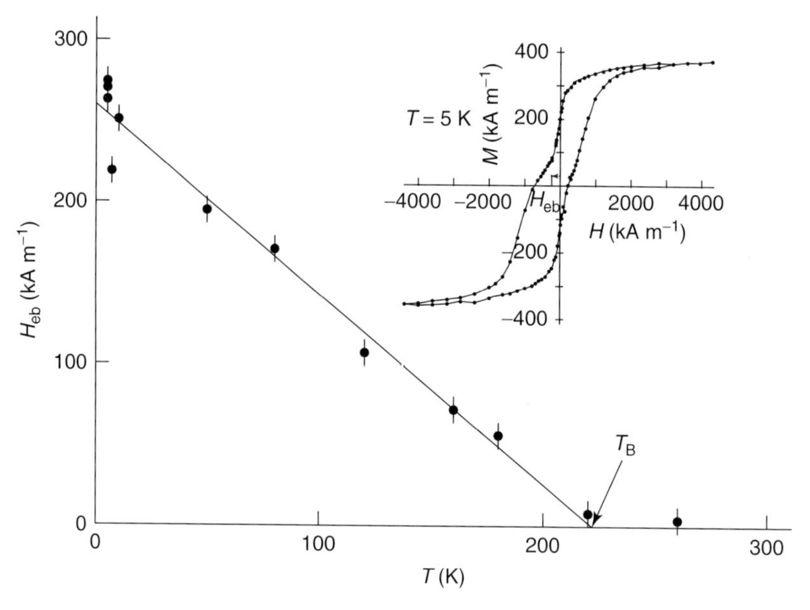

Figure 7.16 The right-hand side shows the hysteresis loop of a [111] oriented 125 Å Fe_3O_4/33 Å CoO bilayer grown on α-Al_2O_3(0001) at 5 K after cooling from 350 K in a field of 4400 kAm^{-1}. As a result, the hysteresis loop shifts along the field axis by H_{eb} as indicated in the figure. The exchange-biasing field H_{eb} versus temperature T depends linearly on T and vanishes at the blocking temperature T_B. In this case, $T_B = 220 \pm 10$ K. Figure reproduced with permission from [56] (© 2000 by the American Physical Society).

coupling across the interface:

$$H_{eb} = \frac{n2J_{ex}S_F S_A F}{a^2 \mu_0 M_F t_F} \tag{7.8}$$

where a is the lattice parameter, n/a^2 is the number of exchange-coupled bonds across the AF–F interface per unit area, J_{ex} is the exchange constant, S_i the spin of either the F or AF, μ_0 is the vacuum permeability, M_F is the magnetization of the F layer, and t_F is the thickness of the F layer. An important aspect of Eq. (7.8) is that it predicts that $H_{eb} \propto t_F$, indicating that exchange biasing is an interface effect. This dependence has been verified for many systems exhibiting exchange bias. For the oxides discussed here, this has been demonstrated to be the case in [79]. Unfortunately, the model described by Eq. (7.8) also predicts that the absolute value of H_{eb} is 2 orders of magnitude larger than the H_{eb} values measured in metallic systems such as $Ni_{80}Fe_{20}$/FeMn. This discrepancy led to the development of models for exchange biasing in which domain-wall formation in the AF layer reduces the expected H_{eb} [80, 81].

Since a general discussion of the experimental results on exchange biasing using AF oxide bilayers is provided in Section 5.3 of Chapter 5 [75], the section here focuses exclusively on the exchange-biasing phenomenon in oxide multilayers. The advantage of these materials, in general, is that the analysis is straightforward because the (super)exchange interaction is local and well understood [5], and values for the exchange interaction across the interface can be found in the literature [82]. Moreover, as discussed in Section 7.2.1, the growth of high-quality, single-crystalline multilayers avoids complications from extrinsic structural effects such as small grain sizes and rough interfaces. In addition, neutron diffraction measurements, designed to directly elucidate the magnetic structure within the AF layers, can easily be applied to crystalline oxide superlattices with multiple bilayer repeats, as opposed to individual polycrystalline AF layers with thicknesses at which exchange biasing occurs (typically < 100 Å) which do not generate sufficient scattered intensity. As in the previous sections of this chapter, we focus our discussion on MBE-grown systems using magnetite, with particular emphasis on Fe_3O_4/NiO and Fe_3O_4/CoO, which have been extensively studied due to their matching oxide sublattices. The implications of these investigations for other oxide systems, such as the perovskites and manganites grown by PLD, are clear.

Table 7.1 lists examples of the various Fe_3O_4/CoO and $Fe_3O_4/Co_xFe_{2-x}O_4$ multilayers which have been shown through the years to exhibit exchange biasing. (We note that Fe_3O_4/Co_3O_4, which is a F/AF system comprised of two spinels, is not included in the table as no exchange biasing was observed [83].) Since theoretical models generally provide only zero-temperature results, one should consider H_{eb} at $T = 0$ K when comparing theoretical and experimental results. To eliminate differences in layer thickness among the various systems studied, the exchange anisotropy constant $K_{\mathrm{eb}}^{(0)}$ has been calculated in Table 7.1 for those all-oxide systems for which low-temperature biasing data are available

$$K_{\mathrm{eb}}^{(0)} = \mu_0 M_S H_{\mathrm{eb}}^{(0)} t_F \tag{7.9}$$

As we mentioned before in reference to $Ni_{80}Fe_{20}$/FeMn multilayers grown by MBE, the magnitude of the difference between the measured and calculated (via Eq. 7.8) exchange anisotropy *even* at low temperature is quite large (i.e., nearly 2 orders

Table 7.1 List of the magnitude of the exchange anisotropy in various all-oxide exchange-biased systems for which low-temperature data are available. Δ is the ratio of the biasing calculated from Eq. (7.8) and the experimentally found value for biasing at low temperature, $H_{\mathrm{eb}}^{(0)}$. Data for the classic metallic $Ni_{80}Fe_{20}$/FeMn system grown by MBE are included for comparison

System	Orientation	Substrate	Deposition	$K_{\mathrm{eb}}^{(0)}$ (mJm^{-2})	Δ	References
Fe_3O_4/CoO	(100)	$SrTiO_3$(100)	MBE	2.11	8	[84]
Fe_3O_4/CoO	(111)	α-Al_2O_3(0001)	MBE	1.43	35	[84]
Fe_3O_4/CoO	(100)	MgO(100)	PLD	1.2 ± 0.3	14	[79]
Fe_3O_4/CoO	(111)	MgO(111)	PLD	1.1 ± 0.3	47	[79]
$Fe_3O_4/Co_xFe_{2-x}O_4$	(100)	MgO(100)	PLD	1.8 ± 0.4		[79]
$Fe_3O_4/Co_xFe_{2-x}O_4$	(111)	MgO(111)	PLD	1.0 ± 0.2		[79]
$Ni_{80}Fe_{20}$/FeMn	(111)	Cu(111)	MBE	0.22	73	[84]

of magnitude), as observed and reported many times for related metallic systems. In comparison, the biasing measured for all-oxide multilayers is significantly larger, and the experimental and theoretical anisotropies differ by less than an order of magnitude. The classical interpretation of exchange biasing thus gives values of the bias field that are, at least, the right order of magnitude for AF–F oxide systems. The contrast between the behavior of (111) and (100) oriented bilayers may be understood within the context of differences in the construction [84] (and reconstruction) of their interfacial planes (as discussed, in part, in Section 7.3.2). It is also interesting to note that the results for Fe_3O_4/CoO multilayers grown by PLD are consistent with those for multilayers grown by MBE, albeit with slightly lower biasing.

7.3.2.2 Dependence on Antiferromagnetic Thickness

An explanation for the characteristic dependence of H_{eb} on AF layer thickness is an essential feature of any model for exchange biasing. In particular, Malozemoff made a very specific prediction that biasing only occurs below a critical thickness $t_{A,crit1}$ of the AF layer (see Figure 7.5 of [81]). For Fe_3O_4/CoO bilayers, the experimentally determined dependence of H_{eb} on t_{CoO} has been plotted in Figure 7.17 both for [111] Fe_3O_4/CoO bilayers grown on α-Al_2O_3(0001) substrates as well as for [100] bilayers grown on $SrTiO_3$(100) and MgO(100) substrates [84, 85]. In these sample series, the CoO layer thickness, t_{CoO}, was varied while the Fe_3O_4 layer thickness, $t_{Fe_3O_4}$, was held constant at 120 Å. At first glance, the observed dependence for two of the sample series strongly resembles that predicted by the Malozemoff theory (*viz.*, a regime where $H_{eb} \propto 1/t_{AF}$, followed by a regime in which H_{eb} is constant). Upon closer inspection of Figure 7.17(a), blatant inconsistencies are apparent as noted previously in [84]. Specifically, Malozemoff's theory includes an expression for the critical thickness $t_{A,crit1}$ above which biasing should disappear:

$$t_{A,crit1} = \frac{f_i\sqrt{A/K}}{4\sqrt{\pi}} \tag{7.10}$$

where f_i is a parameter of order 1 [81], A is the exchange stiffness, and K is the uniaxial anisotropy constant of the antiferromagnet. Using parameters for bulk CoO ($K_{CoO} = 1.1 \times 10^7$ Jm^{-3} [86]), one finds that $t_{A,crit1} \leq 2$ Å, which is clearly at odds with the data shown in Figure 7.17(a). In fact, this calculated value for $t_{A,crit1}$ more closely matches the CoO layer thickness of $\approx$4 Å corresponding to the *onset* of exchange bias at low temperatures. As the roughness associated with deposition of the CoO underlayer is of this order, it is likely that the onset of biasing is determined entirely by factors related to the growth in contrast to Malozemoff's predictions. In comparison to related ionic systems such as $MnFe_2$/Fe and FeF_2/Fe [87] in which T_N is less than 100 K, interpretation of the onset of biasing in CoO-based systems is more straightforward as H_{eb} was measured at temperatures well below T_N of CoO. Finally, we note that the constant regime of H_{eb} in the Malozemoff theory is an artifact of the nonanalytical expression used for the random-field energy, which disappears when an analytic expression is used [84].

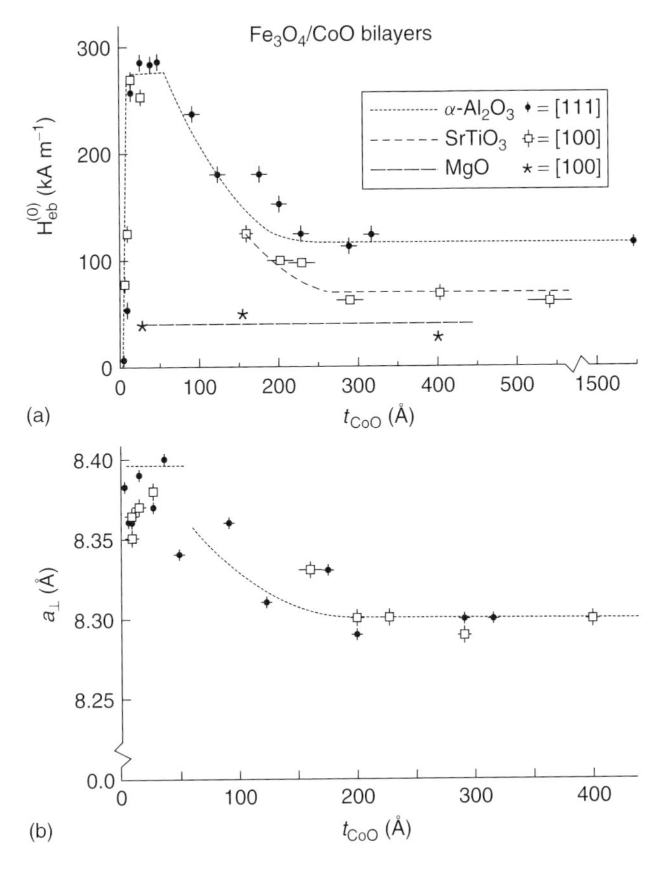

Figure 7.17 (a) The low-temperature exchange-biasing field, $H_{eb}^{(0)}$, versus the CoO AF layer thickness, t_{CoO}, for Fe_3O_4/CoO bilayers with a Fe_3O_4 layer thickness of ≈120 Å grown on three different substrates α-Al_2O_3(0001), $SrTiO_3$(100), and MgO(100). For details on the t_{CoO}-dependence, see text. (b) The Fe_3O_4 perpendicular lattice parameter, $a_\perp$, versus t_{CoO} for the Fe_3O_4/CoO bilayers grown on α-Al_2O_3 and $SrTiO_3$.

In Figure 7.17(b), the perpendicular lattice parameter $a_\perp$ of the Fe_3O_4 layer as measured by X-ray diffraction is plotted as a function of t_{CoO} for the [111] Fe_3O_4/CoO bilayers grown on α-Al_2O_3(0001) and for the [100] bilayers grown on $SrTiO_3$(100) [85]. A striking parallel to the variation of H_{eb} with t_{CoO} (Figure 7.17a) is evident indicating that the dependence of $a_\perp$ on t_{CoO} is also a direct consequence of growth-induced variations among these samples. Specifically, Figure 7.17(b) shows that three regimes exist for the growth of Fe_3O_4/CoO bilayers with critical CoO thicknesses of approximately 100 Å and 200 Å. For $t_{CoO} \leq 100$ Å, the perpendicular lattice parameter, $a_\perp$, is less than or equal to the Fe_3O_4 lattice parameter of 8.396 Å [5]. For $t_{CoO} \geq 200$ Å, $a_\perp = 8.30$ Å as a direct consequence of lattice-matched growth of Fe_3O_4 on the thick CoO underlayer. Since the lattice constant of CoO for the double oxygen sublattice is 8.52 Å, epitaxial growth of the Fe_3O_4 oxygen sublattice on the CoO(100) sublattice gives rise to an in-plane lattice stress of 1.48%. Using the elastic constants of Fe_3O_4 [88], the corresponding strain σ is 4.29 GPa and the resulting perpendicular lattice parameter is 8.30 Å for Fe_3O_4, as observed. Clearly, the CoO underlayer strains the Fe_3O_4 layer in this thickness regime for both the [111] and [100] bilayers.

For $t_{CoO} \leq 100$ Å, the situation is reversed, and the relatively thick Fe_3O_4 layer with its bulk-like lattice parameter (Figure 7.17b) strains the thin CoO layer. Correspondingly, H_{eb} exhibits no dependence on t_{AF} in this regime (Figure 7.17a). The observed t_{AF} dependence of H_{eb} thus appears to originate from induced strain that increases the anisotropy of the CoO layer and subsequently increases H_{eb}. The thickness-independent biasing data for [100] Fe_3O_4/CoO bilayers on MgO in Figure 7.17(a) further support this conclusion. In this case, the CoO is closely lattice matched to the MgO(100) substrate. The Fe_3O_4 layer, rather than the CoO layer, is thus strained for all t_{AF} considered. Also, biasing fields obtained for [111] 66 Å Fe_3O_4/NiO bilayers grown on α-Al_2O_3(0001) were a factor of four stronger than those observed for [100] bilayers grown on MgO(100), much to the surprise of the authors [89]. While these results are seemingly at variance with the general trends shown in Table 7.1, these differences can now be understood in terms of strain effects.

The recent results obtained by Hibma and coworkers on CoO thin films have revealed that variations in the CoO strain can alter the CoO single-ion anisotropy by as much as 4.8 meV, from 3.4 meV to −1.7 meV [90]. Qualitatively, the data for the Fe_3O_4/CoO bilayers shown in Figure 7.17 can be explained by these same effects. The enhancement of H_{eb} in Fe_3O_4/CoO bilayers by a factor of three may result from strain induced by the Fe_3O_4 overlayer on the CoO. This strain would influence the Co^{2+} single-ion anisotropy, which can be enhanced by a factor of three from its bulk value to a value of $\sim 7 \times 10^6$ Jm^{-3} in CoO thin films due to the 4% tensile stress induced by epitaxial growth on MnO(100), according to the data of [90]. Although we have argued here that the Malozemoff model for the AF thickness dependence of the bias field may not be appropriate for this Fe_3O_4/CoO system, the observed dependence of H_{eb} on K_{CoO} seems to be consistent with the general trend predicted by models for exchange biasing that are based on domain-wall formation (including Malozemoff's) [80, 81]

$$H_{eb} \propto \sqrt{A_{AF} K_{AF}} \tag{7.11}$$

Additional support for these domain-wall models is provided by investigations of the temperature dependence of the anisotropy. For antiferromagnets with cubic anisotropy such as CoO and NiO, K_{AF} varies with T as [63, 91]

$$K_{AF} \propto (1 - T/T_N)^2 \tag{7.12}$$

Combining Eqs. (7.11) and (7.12) yields

$$H_{eb} \propto (1 - T/T_N) \tag{7.13}$$

giving the linear dependence of H_{eb} on T typically seen for Fe_3O_4/CoO bilayers irrespective of orientation [56, 79, 84, 92] (see Figure 7.16).

The formation of AF domain walls as predicted by these models has been investigated using high angle neutron diffraction by Borchers *et al.* in Fe_3O_4/NiO multilayers [93], and the formation of F domain walls was probed using polarized neutron reflectometry (PNR) by Ball *et al.* in Fe_3O_4/NiO and Fe_3O_4/CoO multilayers [94]. As described in detail in the next section, diffraction measurements of the (111) magnetic reflection provided direct information about the NiO AF order as a function of temperature and field in these samples. Specifically, these experiments [93] revealed that the observed exchange biasing is associated with domain walls that "freeze" into the NiO layers during field cooling. These domains were relatively small (i.e., 40 nm in-plane and 80 nm along the growth axis) and their size did not vary with subsequent field cycling. In contrast, the sizes of the AF domains both parallel and perpendicular to the growth plane were larger after cooling in zero field to 78 K, but decreased systematically when the field magnitude was increased. PNR measurements of similar Fe_3O_4/NiO bilayers provided additional evidence of domain-wall formation in the exchange-biased state but within the ferromagnetic, rather than the AF, layer [94]. We note that PNR is sensitive to the depth dependence of the magnetization of ferromagnetic layers along the growth axis. Using this technique, the authors determined that the Fe_3O_4 magnetization was depleted in a saturating field applied opposite to the cooling direction, and they suggested that this reduction could be a consequence of domain walls parallel to the growth plane that form within the ferromagnetic layer. As the anisotropy of NiO is reasonably large, it is possible that domain walls form both in the F and AF layers, as indicated by these combined neutron measurements, consistent with the general predictions of the Malozemoff model [81]. However, the dependence of the bias field on the AF anisotropy magnitude remains untested in these systems.

7.3.2.3 **Perpendicular Coupling**

Owing to the similar oxygen sublattices of the spinel Fe_3O_4 and the rock-salt NiO and CoO, the AF order in these epitaxial oxide multilayers can be probed directly with neutron scattering techniques. Neutron diffraction has thus proven to be a powerful tool for obtaining a deeper understanding of exchange biasing, especially in light of its sensitivity to nanoscale magnetic structures in buried layers. The complex interplay between the F and AF structures in (100) Fe_3O_4/CoO superlattices that exhibit exchange biasing was revealed by a series of neutron diffraction studies that tracked the temperature and field dependence of a series of magnetic reflections [21, 95, 96]. Figure 7.18 shows a scan perpendicular to the growth axis through the (111) reflection (indexed relative to the Fe_3O_4 lattice) for a $(100\,Å\ Fe_3O_4/30\,Å\ CoO)_{\times 50}$ multilayer [96]. The resultant lineshape can easily be separated into contributions originating from the individual CoO and Fe_3O_4 layers. Specifically, the large, narrow component corresponds to the CoO AF reflection, and the AF order parameter can be extracted directly from measurements of its intensity as a function of temperature. The narrow peak sits on top of a broad reflection corresponding to the (111) Fe_3O_4 reflection that has both magnetic and structural contributions. The full-width of these reflections

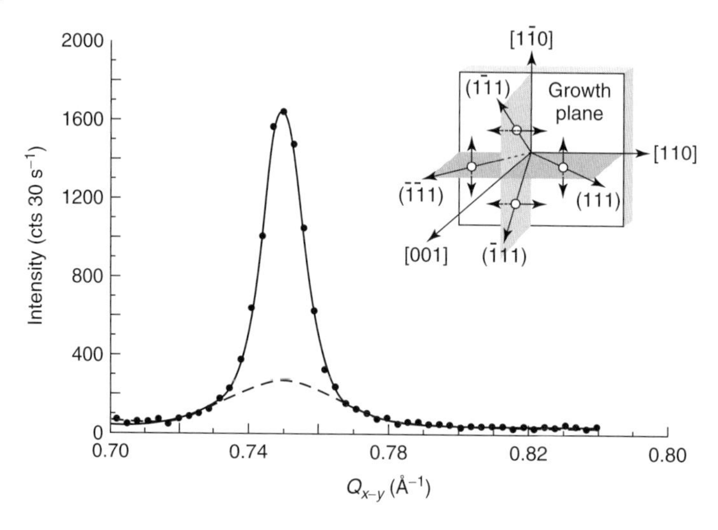

Figure 7.18 Neutron diffraction scan of the (111) reflection (indexed relative to the Fe_3O_4 lattice) within the growth plane for a (100 Å Fe_3O_4/30 Å CoO)$_{\times 50}$ multilayer film grown on MgO at 78 K. The reflection has a two component line shape. The Fe_3O_4 component, indicated by the dashed line, is broadened due to the presence of APBs and is invariant with temperature and field (see Section 7.2.2.3). The inset illustrates the AF spin structure consisting of the four labeled {111} domains with spins alternating in the directions indicated by the double-tipped arrows. Figure reproduced with permission from [96] (© 2007 by the American Physical Society).

is inversely proportional to the coherence length of the structural grains and/or magnetic domains, and the source of the broadening of the Fe_3O_4 component was initially identified in related studies of Fe_3O_4/NiO multilayers [71, 72] to be structural stacking faults. It is now clear that these Fe_3O_4 stacking faults are actually identical to the antiphase boundaries that were discussed in Section 7.2.2.3 as an explanation for the anomalous behavior of ultrathin Fe_3O_4 layers grown on MgO.

Figure 7.19 shows the dependence of the (111) reflection on field preparation for two multilayers with different CoO layer thicknesses, (100 Å Fe_3O_4/30 Å CoO)$_{\times 50}$ and (100 Å Fe_3O_4/100 Å CoO)$_{\times 50}$. The filled data points (•) correspond to the (111) reflection after zero field cooling to 78 K. The relative intensity of the narrow CoO component was *increased* (△ data points) by cooling in a field ($H = 14$ kOe) parallel to the in-plane [110] direction (Figure 7.18 inset) from 320 K. Field cooling along the perpendicular in-plane [1$\bar{1}$0] axis resulted in a *decrease* of the (111) CoO intensity (∘ data points in Figure 7.19) [21, 95]. Further diffraction studies showed that in the latter case, the intensities of the complementary CoO ($\bar{1}$11) and (1$\bar{1}$1) reflections (Figure 7.18 inset) simultaneously *increased*. The intensity of each of these four {111} reflections is proportional to the relative population of the four corresponding magnetic domains, each of which has a different AF spin propagation direction (Figure 7.18 inset). The data in Figure 7.19 thus demonstrate that the population

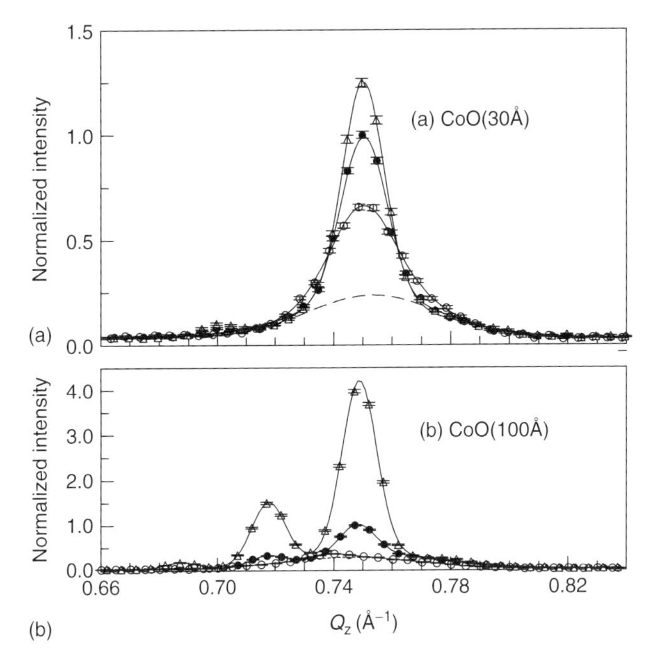

Figure 7.19 Scans of the (111) reflection along the [001] growth axis direction (scattering vector Q_z) for (a) (100 Å Fe_3O_4/30 Å $CoO)_{\times 50}$ and (b) (100 Å Fe_3O_4/100 Å $CoO)_{\times 50}$ taken at 78 K at $H = 0$. The symbols •, ◦ and △ indicate data taken after zero-field cooling (initial state), field cooling ($H = 1120\,\mathrm{KAm^{-1}}$) from 320 K in the $[1\bar{1}0]$ direction, and field cooling ($H = 1120\,\mathrm{KAm^{-1}}$) from 320 K in the [110] directly, respectively. The data have been normalized to the initial scans. The peaks have been fitted with a Gaussian line profile. Figure reproduced with permission from [21] (© 1998 by the American Physical Society).

of the four different CoO AF domains strongly depends on the direction of the cooling field, even though CoO has no net moment. (As the broad background component due to APBs in the Fe_3O_4 does not change with field, the field-induced changes in Figure 7.19 are solely due to the CoO spin reorientations.) Information about the CoO easy-axis directions within each of the four domains was obtained from neutron diffraction with polarization analysis, and it proved to be the key to determining the relative orientation of the Fe_3O_4 and CoO spins. As schematically represented in the inset of Figure 7.18, the polarized beam data revealed that CoO spins in the (111) and $(\bar{1}\bar{1}1)$ domains always lie along the $[1\bar{1}0]$ in-plane direction, while CoO spins in the $(1\bar{1}1)$ and $(\bar{1}11)$ domains are always oriented parallel to the [110] in-plane direction. Combined with the field dependence measurements in Figure 7.19, the authors concluded that field cooling preferentially drives the CoO spins into domains in which they are aligned *perpendicular* to the applied cooling field and to the Fe_3O_4 spins [21, 95]. In short, the coupling between the CoO and Fe_3O_4 spins is perpendicular in these exchange-biased superlattices.

Since early models claimed that only collinear coupling between the AF and F can give rise to exchange biasing, the key question for Fe_3O_4/CoO superlattices

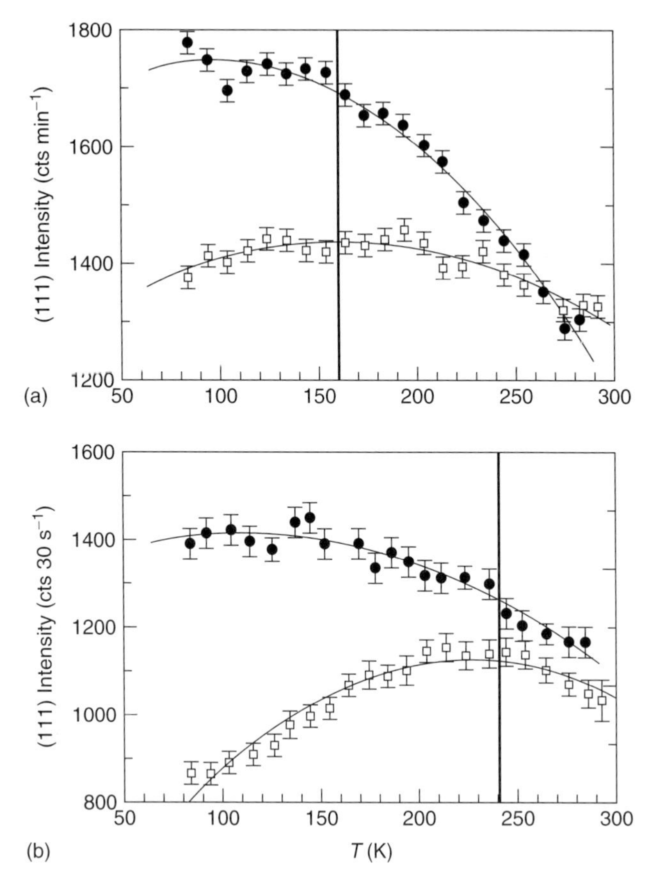

Figure 7.20 Changes in the intensity of the (111) CoO reflection intensity as a function of increasing temperature, T, for two multilayers (a) (100 Å Fe_3O_4/17 Å CoO)$_{\times 50}$ and (b) (100 Å Fe_3O_4/30 Å CoO)$_{\times 50}$. The ● data points were taken after cooling from room temperature in zero field, while the □ data points were obtained in zero field after cooling the sample in a 4000 kAm^{-1} field from room temperature to 78 K. The solid lines are guides to the eye. Vertical lines indicate the blocking temperature T_B independently determined by magnetometry. Figure reproduced with permission from [96] (© 2007 by the American Physical Society).

is whether the observed perpendicular coupling is responsible for the exchange biasing or is merely a "spectator." The former supposition is strongly supported by the following experimental evidence. First, the CoO spins did not reorient upon field cooling when the Fe_3O_4 layers were replaced by nonmagnetic MgO layers in a (30 Å MgO/30 Å CoO)$_{\times 333}$ multilayer [21, 95]. Second, the onset of perpendicular coupling in several Fe_3O_4/CoO superlattices occurred at the *same* temperature as the onset of exchange biasing [96] (i.e., at the so-called blocking temperature T_B). To demonstrate this effect, Figure 7.20 [96] shows the intensity of the (111) CoO reflection as a function of temperature upon heating for two multilayers, (100 Å Fe_3O_4/17 Å CoO)$_{\times 50}$ and (100 Å Fe_3O_4/30 Å CoO)$_{\times 50}$, that were cooled in zero field (filled data points) and in a 5 T field applied along the $[1\bar{1}0]$ axis (open data points). After cooling in zero field, the (111) intensity gradually decreases with increasing temperature for both samples in a manner consistent with a typical order parameter for an antiferromagnet. In contrast, after field cooling the initial (111) intensity is significantly reduced because the CoO spins have been driven perpendicular to the field into either the $(\bar{1}11)$ or $(1\bar{1}1)$ domain. Upon heating, the intensity of the (111) reflection *increases* with temperature until it reaches a broad maximum at a temperature that matches the independently determined T_B (i.e., vertical lines in Figure 7.20) for each sample. Above this maximum, the intensities from all four of the {111} AF domains were shown to be equal, indicating a randomization of the AF spins [96]. This definitive result directly links the onset of biasing below T_B with the freezing of the perpendicular F–AF spin alignment. Both effects thus have a common origin in the all-oxide Fe_3O_4/CoO system.

Coincident with and also following the discovery of perpendicular coupling in Fe_3O_4/CoO multilayers, this effect was also reported for several other exchange-biased systems including (100) Fe_3O_4/NiO [97, 98] and (110) Fe/FeF_2 [99]. In contrast to the behavior of Fe_3O_4/CoO and Fe_3O_4/NiO superlattices, the F layer in (110) $Ni_{80}Fe_{20}$/FeMn multilayers [100] oriented itself perpendicular to the AF layer, giving rise to uniaxial, rather than to unidirectional, anisotropy. The behavior of this particular system is more consistent with early phenomenological models of the AF–F coupling at the interfaces. Specifically, if the interfacial plane of the AF has a net uncompensated moment, then the alignment of the F relative to the AF is expected to be collinear and to exhibit exchange bias in the absence of pronounced roughness. If the interfacial plane is compensated, then the relative AF–F alignment is expected to be perpendicular and to show no bias. Within this context, thought-provoking results were obtained by Krug *et al.* as part of their X-ray magnetic circular dichroism (XMCD) and X-ray magnetic linear dichroism (XMLD) investigation of compensated (110) and (001) interfaces and the uncompensated (111) interfaces of epitaxial NiO films grown on single-crystal Fe_3O_4 [98]. Specifically, these authors observed that the (001) interface exhibited perpendicular coupling in agreement with results for CoO/Fe_3O_4 superlattices, yet collinear coupling was found for *both* the (111) and (110) NiO interfaces. The collinear coupling in the nominally compensated (110) structures was explained to be a consequence of strained growth. Specifically, magnetoelastic effects favor {111} AF domains out of the sample plane with interface structures that are uncompensated. In contrast, it was determined in exchange-biasing studies of Fe_3O_4/CoO [56, 84] that the dependence of H_{eb} on temperature and on t_{CoO} is similar for (111) and (100) oriented multilayers, apart from a difference of $0.7\,MJm^{-2}$ in the absolute value of the exchange anisotropy constant $K_{eb}^{(0)}$ (see Table 7.1). This agreement does not make sense because the magnetic surface structures of these two orientations differ drastically, as shown in Figure 7.1 [84] and as observed by Krug [98]. However, it is possible that the antiphase domains (see Section 7.2) or even just the strain resulting from the growth of Fe_3O_4 on the rock-salt structure may be responsible for a net compensated (111) interface, giving rise to perpendicular coupling at least for this Fe_3O_4/CoO system.

In the related ionic system Fe/FeF_2, perpendicular coupling has also been observed, and the trends are again not intuitive. For a single AF crystal of FeF_2 onto which a 20-nm Fe layer was deposited, perpendicular coupling was observed for the compensated (110) surface, as well as for the uncompensated (100) surface once the system was cooled [101]. Surprisingly, the biasing is strongest for a twinned (110) FeF_2 film in which the F spins are aligned relative to the AF spins at an angle of 45°, presumably due to frustration [101]. The explanation for the orientation dependence of the coupling and biasing in this, as well as in the Fe_3O_4/CoO, system should be key to fully understanding the exchange-biasing effect.

To this end, Koon proposed a theoretical model for the exchange-biased state in which the AF spins align perpendicular to the F spins [102] as a result of frustrated exchange at the compensated interfaces [102, 103]. However, for Heisenberg exchange between F–AF spins across the interface, such a spin–flop coupling is uniaxial and thus cannot lead to unidirectional exchange anisotropy [104]. A possible mechanism for breaking this symmetry is the Dzyaloshinsky-Moriya interaction [96] which was considered as a possible origin for exchange biasing in Fe_3O_4/CoO multilayers with APBs.

Finazzi has also studied interface coupling for F/AF bilayers [105] that are compensated and uncompensated. Since direct exchange is the focus of his model, however, this theory may not directly be applicable to Fe_3O_4/CoO and Fe_3O_4/NiO multilayers in which the exchange interaction is due to superexchange. Finazzi's idea was to examine the crossover between compensated to uncompensated interfaces, and he arrived at a phase diagram for collinear and perpendicular coupling with axes of u and H_{eb}/mJ_{AF}. The parameter u is a measure for the density of uncompensated moments ($u = 0$ for a fully compensated interface, $u = 1$ for a fully uncompensated interface), and H_{eb}/mJ_{AF} is a measure of the exchange coupling across the interface normalized to the spin–spin interaction in the antiferromagnet. (If J_{AF} is $\gg H_{eb}$, the exchange coupling across the interface will have limited influence on the AF spin structure.) Following Finazzi's model qualitatively, one would expect that uncompensated interfaces with $u = 1$, such as the (111) surface in the Fe_3O_4/NiO system, to exhibit collinear coupling as observed by Krug *et al.* [98]. However, the compensated ($u = 0$) (100) interface should exhibit perpendicular coupling consistent with the Koon model for exchange biasing [102], as observed for (100) Fe_3O_4/NiO [98] and Fe_3O_4/CoO [21]. It is possible that variations in interface roughness and the density of APBs from one sample set to the next leads to deviations of u from the ideal values of 0 and 1 for the (100) and (111) surfaces respectively, and thus accounts for some of the differences apparent in measured values of H_{eb} (see Table 7.1).

7.3.2.4 **Reduction of the Blocking Temperature**

In many AF–F systems, the exchange-biasing field has been shown to decrease with increasing temperature, but the temperature T_B at which H_{eb} vanishes also exhibits a dependence upon the AF layer thickness. As an example, Figure 7.16 shows that the bias field H_{eb} systematically decreases with temperature for a [111] oriented Fe_3O_4/CoO multilayer with $t_{CoO} = 33$ Å [56]. For this particular multilayer, the temperature T_B at which H_{eb} vanishes is equal to 230 ± 10 K, which is lower than T_N of 291 K for bulk CoO. Further studies revealed that T_B is similarly reduced below T_N for all Fe_3O_4/CoO bilayers in which $t_{CoO} < 50$ Å, irrespective of crystalline orientation and substrate [56, 84]. In fact, this relationship between T_B and T_N is ubiquitous as it has been observed for a wide assortment of oxide and metallic exchange-biased systems with thin AF layers [84, 106–113]. This behavior was perhaps first reported for thin FeMn layers in $Ni_{80}Fe_{20}$/FeMn bilayers by Parkin and Speriosu who proposed that the reduction of T_B relative to the ordering temperature T_N originates from finite-size scaling in these thin AF layers [106]. These authors proposed the following equation to describe this variation of T_B with AF layer thickness, t_{AF}:

$$\frac{T_N - T_B(t_{AF})}{T_N} \propto t_{AF}^{-\delta} \tag{7.14}$$

where the values found for δ typically are in the range of 1.2–1.6. Theoretically, one would expect δ values of 1.56 or 1.42 depending on whether the Ising or Heisenberg model is applied [87].

Many of the exchange-biased systems studied to date are, however, polycrystalline in which local variations of the crystal size could introduce an additional time dependence of T_B [114] that might contribute to its reduction [115]. These complications are avoided by using the same single-crystalline, epitaxial AF/F samples for both magnetization measurements of T_B [84] and neutron diffraction measurements of T_N [56], the former of which is exemplified by the data in Figure 7.21. To date the all-oxide Fe_3O_4/CoO is the only system for which both these parameters have been *directly measured*, as shown in Figure 7.21 [56]. For this system, the dependence of T_B on CoO thickness closely follows expectations. Specifically, T_B decreases systematically with CoO layer thickness for $t_{AF} < 50$ Å irrespective of the crystalline orientation [111] or [100] of the substrate. (Both $SrTiO_3$ and MgO substrates were used for the (100)-oriented superlattices.) The key result, however, is that the ordering temperature T_N of the CoO clearly increases with increasing CoO layer thickness (◇ data points in Figure 7.21). This observation is consistent with the induced magnetic order reported for other all-oxide multilayers [66, 72], and it can again be understood within the framework of a mean-field model [69] in which the ordering temperature of the two constituents, here Fe_3O_4 and CoO, approach each other as a function of relative layer thickness (See discussion in Section 7.3.1). Further, the scaling of T_N is not intrinsic to the individual thin CoO layers as evidenced by the observation that T_N is neither reduced nor enhanced in an uncoupled $(30\,Å\ CoO/30\,Å\ MgO)_{\times 333}$ multilayer measured by neutron diffraction techniques (Figure 7.14). Further support for the observed enhancement of T_N is provided by a recent nuclear resonance scattering (NRS) study of a 1.6-nm-thick AF FeO layer sandwiched between two Fe layers [116]. The T_N measured for the FeO layer was as high as 800 ± 30 K and greatly exceeded T_N of 198 K for bulk FeO. The blocking temperature, however, was independently determined to be < 50 K from bulk magnetization measurements. The disparity between T_N and T_B is consistent with the results reported for Fe_3O_4/CoO.

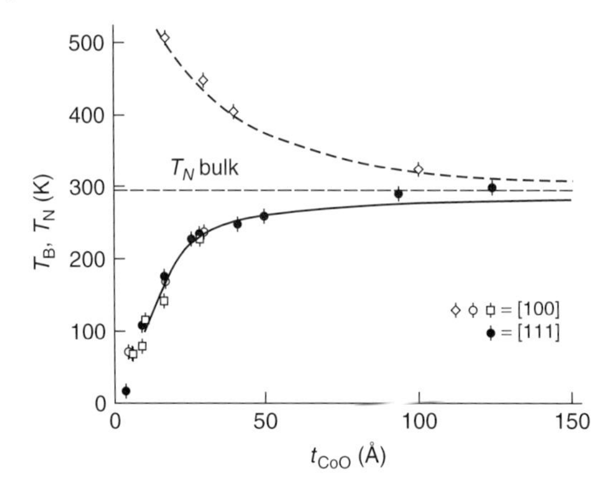

Figure 7.21 The measured Néel temperature, T_N (◇) and blocking temperature, T_B, (●, ◦, and □) for CoO versus the CoO layer thickness, t_{CoO}, for the Fe_3O_4/CoO system. Note the divergence of the two curves, indicating that the measured reduction of T_B is not due to a reduction of the ordering temperature at low t_{CoO}. The curve drawn through the T_B data points is the theoretical curve for t_{AF} dependence of T_B for CoO according to the theory of Lang *et al.* [117]. Note the excellent correspondence (see text for details). Figure adapted with permission from [56] (© 2000 by the American Physical Society).

The results shown in Figure 7.21 can be understood by separately considering the physical mechanisms that are responsible for driving the transitions at T_N and T_B. Specifically, T_N is the ordering temperature above which the CoO layer becomes paramagnetic, and it thus poses an upper limit for the temperature at which biasing can occur. In contrast, exchange biasing is caused by changes in the anisotropy of the AF layers, as clearly demonstrated by the fact that higher bias fields are obtained for higher anisotropy materials such as CoO or NiO [63]. The rate of reduction of H_{eb} with increasing temperature, and hence T_B itself, should thus be coupled to the anisotropy of the AF material. Consequently, a comprehensive theory for exchange biasing in oxide systems should be able to explain not only the magnitude of H_{eb}, which has been the focus of many of the more recent theories [80, 81], but also the temperature dependence of H_{eb} and the difference between T_B and T_N.

A promising theory [117] does follow such an approach and concentrates on the weakening of the spin–spin interactions in the antiferromagnet that originates from lattice vibrations within the ultrathin layers. This weakening accounts for the reduction observed in T_B when the variation of anisotropy with t_{AF} is considered and leads to the following equation for the dependence of T_B on the AF layer thickness t_{AF}:

$$T_B\left(t_{AF}\right) = T_N(\exp(-2S_{vib}(\infty)/[3R(t_{AF}/2a - 1)])) \tag{7.15}$$

where $S_{vib}(\infty)$ is the vibrational entropy of melting of the AF material, R is the ideal gas constant, and a is the lattice parameter [117] and where we have taken $T_B(\infty)$ as T_N. In Figure 7.21, the theoretically predictive curve based on Eq. (7.15) using the relevant parameters for CoO ($T_N = 291$ K, $a = 0.426$ nm, and $S_{vib}(\infty) = 6.789\,\mathrm{J mol^{-1}}$ [118]) has been drawn. Note the excellent agreement between the T_B

data for Fe_3O_4/CoO data [56] and this theory which is based on a straightforward model without adjustable parameters. Also for several other AF materials, quantitative agreement of this theory with experimental results has thus far been obtained [117].

A final detail is that Eq. (7.15) based on the theory by Lang *et al.* only provides a valid description of the data if $t_{AF} > 2a = 8.52\,\text{Å}$ for CoO (as otherwise T_B would increase again due to the change in sign of the denominator). Although it is reasonable to assume that two unit cells of CoO would be needed for biasing to occur, the data in Figures 7.17(a) and 7.21 clearly show that biasing exists also for $t_{CoO} < 10\,\text{Å}$ [84] and a detectable T_B can be determined. Hence, the theory should be extended to include the regime for t_{AF} between 0 and $2a$ for a complete resolution of this problem.

In conclusion, direct experimental observation has shown that the reduction of T_B with t_{AF} does not necessarily originate from a corresponding reduction of T_N of the antiferromagnet, consistent with the results discussed in Section 7.3.1. Thus T_B and T_N are not directly related to each other and must be considered separately in any experimental investigation and theoretical interpretation. Moreover, a promising theory is now available that relates the reduction in T_B to a weakening of the spin–spin interaction in the sublattice of the antiferromagnet. This theory recognizes that the role of finite-size scaling may be less significant than previously believed.

7.3.3 Coupling across Intermediary Layers

Thus far we have discussed coupling between AF–AF layers and between F–AF layers comprised of oxide materials in which the magnetic interactions originate from superexchange. Another important type of magnetic interaction involves the coupling between magnetic layers across an intermediary layer. This interaction has been probed extensively in metallic systems as it is relevant for optimizing layer thickness to maximize the giant magnetoresistance (GMR) effect that is the basis of operation for read heads in magnetic storage devices. While all-oxide systems are insulating and do not exhibit GMR, investigations of coupling across intermediary layers enhance the general understanding of superexchange interactions, and these novel geometries hold great promise for spintronic and microwave device applications. We start the discussion by first considering the coupling between ferromagnetic oxide layers separated by a nonmagnetic (i.e., paramagnetic) intermediary layer.

7.3.3.1 Coupling across a Nonmagnetic Layer

The most detailed study to date on interlayer coupling in an all-oxide system focused on two Fe_3O_4 layers separated by MgO [30]. Using a wedge-shaped sample geometry (shown in Figure 7.22 top [30]), the authors probed the magnetization of these layers with MOKE over a wide range of MgO interlayer thicknesses, all in a single sample. The $Co_xFe_{3-x}O_4$ layer at the bottom was coupled to the lower Fe_3O_4 layer in order to enhance its coercivity and thus to distinguish it from the

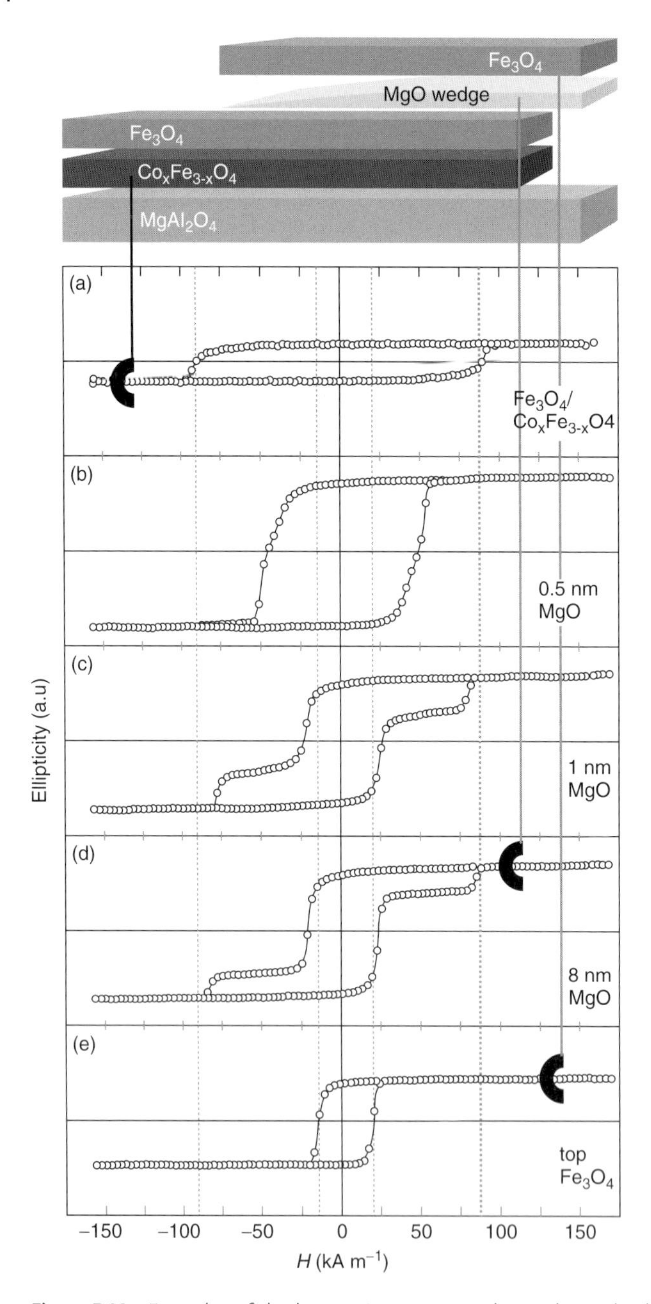

Figure 7.22 Examples of the hysteresis loops obtained along the MgO wedge used in the study of the interlayer coupling between two Fe_3O_4 layers. From top to bottom the measurements were performed from left to right on the wedged-sample depicted at the top of the figure. The measurements were done in the longitudinal MOKE geometry with H along a [110] axis. The figure is adapted from [30].

top Fe_3O_4 layer. (The penetration length of light is such that the entire structure was measured.) The MgO layer was offset at the beginning and end of the wedge, allowing for characterization of the uncoupled top and bottom Fe_3O_4 layers. The interlayer exchange coupling strength J_{iec} was determined by measuring minor hysteresis loops and extracting the shift of the loop with respect to $H = 0$. Assuming that the Fe_3O_4 layers are in a single domain state, one can deduce J_{iec} from [30]

$$J_{iec} = H_{shift}\,\mu_o M_S^A t^A \tag{7.16}$$

in which μ_0 is the vacuum permeability, M_S^A is the saturation magnetization of the top (free) Fe_3O_4 layer, and t^A is the thickness of this layer.

The interlayer exchange coupling was investigated at room temperature using two samples: 33-nm $Co_{0.17}Fe_{2.83}O_4$/32.5-nm Fe_3O_4/ 0–8.3-nm wedge MgO/21.5-nm Fe_3O_4 (Figure 7.23a) and 30-nm $Co_{0.2}Fe_{2.8}O_4$/30-nm Fe_3O_4/ 2–45-nm wedge MgO/20-nm Fe_3O_4 (Figure 7.23a inset). Typical MOKE data [30] are shown in the lower part of Figure 7.22. The behavior expected for the individual Fe_3O_4 layers is observed at the extremes, while hysteresis loops measured in the regions with the intermediary layer have steps. At an MgO layer thickness of 0.5 nm, a single, square hysteresis loop is observed indicative of strong ferromagnetic coupling. Figure 7.23(a) shows the MgO interlayer thickness (t_{MgO}) dependence of the interlayer exchange coupling constant, J_{iec}, derived from these data. From this plot, one can immediately identify two distinct regimes for the interlayer coupling: $t_{MgO} < 1.2$ nm and $t_{MgO} \geq 1.2$ nm. STM studies of a 40-nm Fe_3O_4 layer grown on MgO(100) by the same research group revealed that the Fe_3O_4 consisted of terraces separated by steps with vertical height variations of one to four oxygen planes, that is, 0.2–0.8 nm [43]. Consequently, for MgO layers ≤ 1.2 nm deposited on a Fe_3O_4 layer, incomplete coverage resulted in pinhole coupling [30] between the Fe_3O_4 layers on either side of the MgO layer. For $t_{MgO} \geq 1.2$ nm, the authors determined that the interlayer coupling in Figure 7.23(a) can be explained by so-called orange-peel coupling (i.e., magnetostatic coupling due to correlated interface roughness) between the two Fe_3O_4 layers [30]. The inset to Figure 7.23(a) shows the t_{MgO} dependence of J_{iec} in the regime of thick MgO interlayers. At these length scales, the interlayer coupling strength is of the order of $\simeq 10\ \mu Jm^{-2}$. The curves shown in both the inset and Figure 7.23(a) represent fits to the expression derived by Néel for this coupling [119]:

$$J_{iec} = \frac{1}{2\sqrt{2}}\pi p h^2 \mu_0 M_S^2 e^{-p\sqrt{2}t} \tag{7.17}$$

where $2\pi/p$ is the periodicity of the interface roughness, σ, in both the x and y directions which has height variations h (with $\sigma = h\sin(px)\sin(py)$), and where t is the spacer layer thickness. As shown in Figure 7.23(a) inset, a good fit is obtained with the parameters $2\pi/p = (39 \pm 2) \times 10$ nm and $2h = 3.1 \pm 0.1$ nm [30], which are consistent with the values of the average terrace length and step height variation, 25 nm and 0.3 nm respectively, deduced from STM studies (see Figure 7.11) [43].

These authors also studied the temperature dependence of J_{iec} in a uniform sample of 27.5-nm $Co_{0.2}Fe_{2.8}O_4$/30-nm Fe_3O_4/5-nm MgO/20-nm Fe_3O_4 grown on

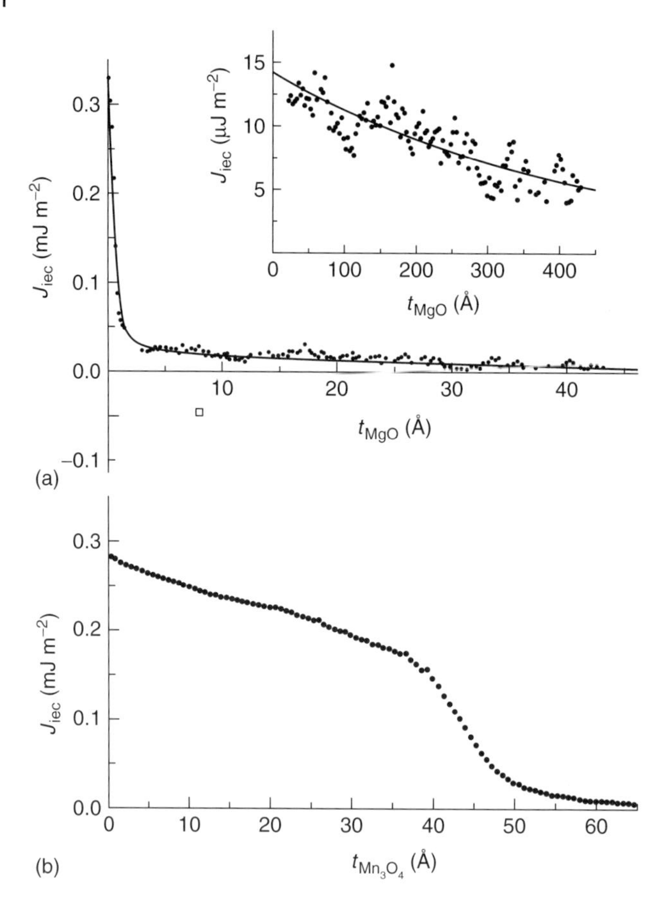

Figure 7.23 Interlayer exchange coupling constant, J_{iec}, at room temperature as a function of spacer thickness for (a) MgO and (b) Mn_3O_4. Data taken from (30, 120, 121).(The □ data points were derived from [120]). Note the very high-interlayer exchange coupling for Mn_3O_4 up to 50 Å, which precludes use of this paramagnetic barrier in tunnel junctions. Details about the fits of the J_{iec} across MgO are discussed in the text.

(001) $MgAl_2O_4$, and they determined that it was $\propto M_S^2$ as predicted by Eq. (7.17). The theoretically relevant interlayer coupling at $T = 0\,K$, $J_{iec}^{(0)}$, was found to be $22\,\mu\,Jm^{-2}$ in the "orange-peel" coupling regime [30]. This result is consistent with the value of J_{iec} for Fe_3O_4/MgO/Fe heteroexpitaxial structures grown by PLD which was determined to be $30\,\mu Jm^{-2}$ for $t_{MgO} = 2\,nm$ [122]. The agreement between the two values is reasonable given that the roughness of the top Fe_3O_4 layer for PLD growth may be slightly larger than the roughness for MBE growth.

In more recent studies of MBE-grown Fe_3O_4/MgO/Fe_3O_4 trilayers with $t_{MgO} <$ 1.2 nm, Shvets and coworkers [120] found that the Fe_3O_4 interlayer coupling for a trilayer with an MgO thickness of 0.8 nm is AF with $J_{iec} = -46\,\mu J/m^2$. In reference to the discrepancy between their results and those of [30], the authors asserted that their use of a better quality, unpolished substrate with lower roughness enabled their discovery of AF interlayer exchange coupling [120] in lieu of pinhole coupling [30]. Antiferromagnetic interlayer coupling of similar magnitude has also been reported for fully epitaxial Fe/MgO/Fe trilayers with t_{MgO} between 0.5 and 0.7 nm [123]. Wu *et al.* performed calculations which indicated that the presence of AF coupling in Fe_3O_4/MgO/Fe_3O_4 seems related to the relative oxygen state of the Fe_3O_4 electrodes [120] rather than to vacancies in the MgO tunneling barrier [123].

In a manner analogous to the initial Fe_3O_4/MgO/Fe_3O_4 trilayer study, interlayer exchange coupling has been investigated at room temperature in Fe_3O_4/Mn_3O_4/Fe_3O_4 using wedged-shaped samples [121]. Since the T_C of Mn_3O_4 is low (42 K) [124], the Mn_3O_4 interlayer was initially expected to exhibit paramagnetic behavior. The data shown in Figure 7.23(b) [121] indicate instead that strong ferromagnetic interlayer coupling, in contrast to that observed in MgO, persists up to Mn_3O_4 thicknesses of 50 Å. One might speculate that the Mn_3O_4 ordering temperature is enhanced due to local polarization by the Fe_3O_4 at the interfaces in light of prior results on proximity magnetism in oxide multilayers (see Section 7.3.1 [66, 71]). However, there was no experimental evidence of an increase in the Mn_3O_4 T_C toward 300 K. In view of the 3% lattice mismatch between Mn_3O_4 and Fe_3O_4, it is more likely that the coupling originates from pinholes that span a distance of up to 5 nm. Indeed, RHEED studies indicate that the roughness stemming from the growth of Mn_3O_4 on Fe_3O_4 is much larger than that resulting from the deposition of MgO on Fe_3O_4 [121].

When considered together, these experimental results suggest that weak interlayer exchange coupling can clearly be obtained across MgO. In particular the use of smooth, flat substrates can minimize interlayer interactions originating from extrinsic effects (such as pinholes) at thicknesses below 1 nm. As this length scale is quite relevant for the operation of tunnel junctions with magnetic oxide components (discussed in Section 7.4), further study of this AF coupling in the low $t_{MgO} < 1.5$ nm regime is essential for understanding its origin and for optimizing the interface quality in order to control it.

7.3.3.2 Coupling across an Antiferromagnetic Layer

The intrinsic coupling between two Fe_3O_4 layers across an AF layer (i.e., NiO) was investigated at 300 K using two wedge-shaped samples [125]: 20-nm Fe_3O_4/0–2-nm NiO/20 nm Fe_3O_4 and 25-nm Fe_3O_4/0–9-nm NiO/25-nm Fe_3O_4. (Note that these samples were not field cooled, and the focus of this investigation thus differs from that of the exchange-biasing Fe_3O_4/NiO studies described in Section 7.3.2) For NiO layer thickness (t_{NiO}) less than 0.7 nm, the measured saturation magnetization was large indicating the presence of "pinhole" coupling, if any at all, arising from ferromagnetic bridges similar to those observed in MgO interlayers (see Section 7.3.3.1). The magnetic behavior of the wedges in the regime $t_{NiO} \geq 0.7$ nm, however, was quite different. As an example, in the MOKE data (Figure 7.24) measured for the 20-nm Fe_3O_4/1.4-nm NiO/20-nm Fe_3O_4 trilayer in an applied field H parallel to the in-plane [110] axis, the parallel remanent magnetization was reduced to 0.5 M_S and equaled the perpendicular component of the magnetization. For H parallel to

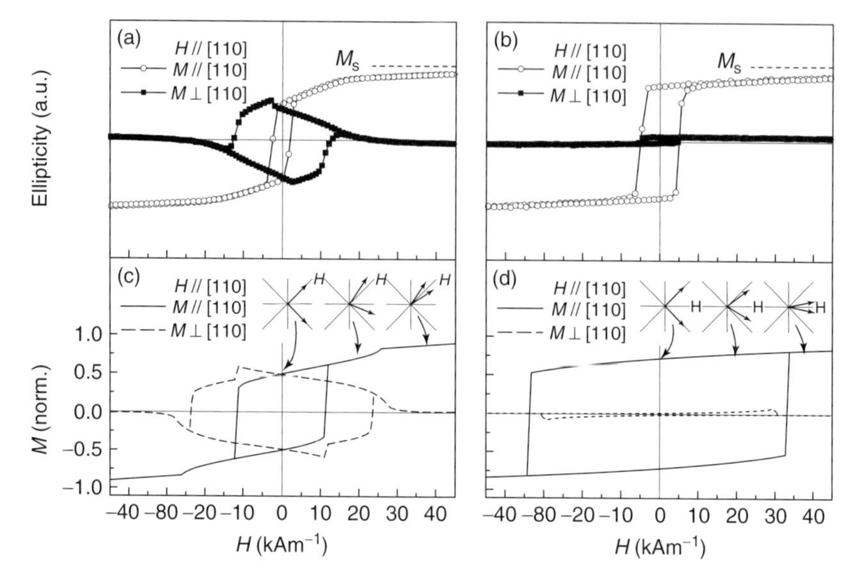

Figure 7.24 Hysteresis loop of a 20-nm Fe_3O_4/1.4-nm NiO/20 nm Fe_3O_4 trilayer measured by MOKE for coupling studies across an AF NiO interlayer. (a) Corresponds to H parallel to [110] and (b) corresponds to H parallel to [100]. In panels (c) and (d) the corresponding calculated hysteresis loops for two magnetic layers with an in-plane cubic anisotropy and mutual 90° coupling for (c) H applied along [110] and (d) H applied along [100]. The inset in these panels shows the configuration of the magnetic moments, represented by arrows, at different positions along the hysteresis loop. Figure reproduced with permission from [125] (© 1999 by the American Physical Society).

[100], the zero-field magnetization was 0.7 M_S, and virtually no perpendicular magnetization component was observed. Qualitatively this behavior can be understood if the magnetizations of both Fe_3O_4 layers are oriented perpendicular to each other along the mutually orthogonal [110] and $[\bar{1}10]$ axes. A more quantitative agreement with the field-dependent magnetization was achieved using the following model for the angular dependence per unit area [125]:

$$\begin{aligned} E &= -\mu_o M_S t_1 H\cos(\phi_1 - \theta) - \mu_o M_S t_2 H\cos(\phi_2 - \theta) \\ &\quad + K_1 t_1 \cos^2\phi_1 \sin^2\phi_1 + K_1 t_2 \cos^2\phi_2 \sin^2\phi_2 \\ &\quad + J_2 \cos^2(\phi_1 - \phi_2) \end{aligned} \tag{7.18}$$

in which the bulk saturation magnetization of Fe_3O_4 M_S equals 496 kAm^{-1}, the cubic magnetocrystalline anisotropy K_1 equals $-9\,\text{kJm}^{-3}$ [30], t_i is the thickness of the ith layer, and ϕ_1, ϕ_2, and θ are the angles of the magnetic moments M_1, M_2, and the applied field respectively. The last term in Eq. (7.18) is the phenomenological biquadratic exchange coupling with coupling constant J_2. Good agreement with the hysteresis loops was found for $J_2 = 0.35 \pm 0.08\,\text{MJm}^{-2}$ (Figure 7.24c). For H applied along [110], which is parallel to M_1 and perpendicular to M_2, the response of the system is intriguing as the rigid 90° moment unit first rotates to a spin–flop configuration symmetric about the field H and then collapses.

Note that the magnitude of the interlayer coupling constant is a factor of 1000 larger than the magnitude of interlayer coupling due to uncorrelated roughness (see Section 7.3.3.1 [30]), and the latter can thus be ignored. The likely microscopic origin of this biquadratic coupling is the lateral variation of the AF interlayer thickness as it locally gives rise to competition between F and AF interlayer interactions [126]. Strictly speaking this model requires an uncompensated AF–F interface, but the NiO(100) surface is compensated. As described throughout this chapter, the Fe_3O_4 and NiO (or CoO) interface is complex due to differences in the magnetic sublattices [84] as well as structural APBs in the Fe_3O_4. After taking into account features that are unique to this system, it is possible that the Slonczewski model may still be applicable.

In this system, the biquadratic coupling vanishes for $t_{NiO} > 5.4$ nm. This result can be understood from Eq. (7.18) since perpendicular alignment of M_1 and M_2 is only maintained when J_2, which favors perpendicular alignment, is greater than $t_i K_1/2 = 0.09\ \mathrm{MJm^{-2}}$. In the Slonczewski model, the coupling parameter C, which is related to J_2 in Eq. (7.16), is inversely proportional to t_{AF} [126]. Since $J_2 = 0.35\ \mathrm{MJm^{-2}}$ at $t_{NiO} = 5.5$ nm [125], one can thus expect the perpendicular coupling to cease at or above a NiO thickness of 5.5 nm, as is indeed the case [125].

7.3.4
Perpendicular Anisotropy

While more a consequence of growth conditions than of interlayer exchange interactions, perpendicular anisotropy was observed in related Fe_3O_4-CoO systems, such as (100) Fe_3O_4/CoO bilayers on MgO and (100) $Co_{0.07}Fe_{2.93}O_4$/CoO on MgO, and should thus be mentioned here. Using wedge-shaped samples, Bloemen *et al.* have shown that for Fe_3O_4 grown on a CoO underlayer, a preferred perpendicular orientation occurs when the CoO layer thickness exceeds 200 Å and the Fe_3O_4 thickness is comparable (i.e., less than 280 Å) (see Figure 7.25) [20]. Evidently, the preferred perpendicular orientation of Fe_3O_4 with respect to the film plane does not originate from an interface anisotropy contribution, K_s, as we have seen in Section 7.2.2.2 that this is absent in Fe_3O_4. The observed thickness dependence of the perpendicular anisotropy is instead due to the changing growth mode of CoO. For $t_{CoO} \leq 200$ Å, the CoO film growth on the MgO substrate is strained. For larger thickness, it becomes energetically favorable for the CoO layer to relax, form misfit dislocations at the MgO–CoO interface, and adopt its bulk lattice parameter. Subsequently, Fe_3O_4 layers of modest thickness grow coherently on the strained CoO base layer, which is about 1.5% larger than the Fe_3O_4 lattice (see Section 7.3.2.2). Because of the enhanced magnetostrictive contribution, the strained growth leads to a perpendicular preferred direction. With increasing Fe_3O_4 thickness, the strain relaxes and the magnetostrictive contribution diminishes, leading to a preferred in-plane magnetization direction [20]. It is notable that this effect can be enhanced by adding Co to the magnetite. Because of the large

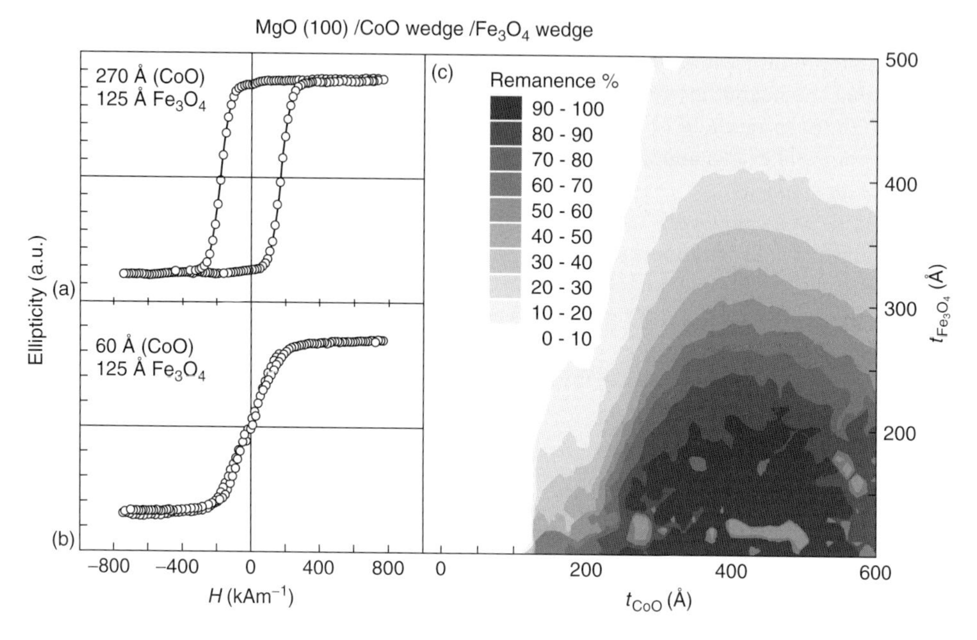

Figure 7.25 Magnetooptical Kerr (MOKE) data on the remanence in a double wedge $CoO–Fe_3O_4$ grown on MgO(100) to probe the perpendicular anisotropy. In (a) and (b) individual hysteresis loops are shown measured with the *H*-field applied perpendicular to the film plane for conditions indicated in the figure. (c) Contour plots of perpendicular remanence obtained from 4800 polar hysteresis loops, similar to those shown in (a) and (b) for CoO layer thickness, t_{CoO}, between 1 and 600 Å and Fe_3O_4 layer thickness, $t_{Fe_3O_4}$, between 0 and 500 Å. The relationship between the gray scale and remanence is given by the central legend. Figure taken from [20].

single-ion anisotropy of Co^{2+}, the magnetostrictive constants for Co-substituted magnetite are larger by 2 orders of magnitude, leading to a preferred magnetization direction perpendicular to the plane that is stable over a much larger thickness regime [20]. Not much work has been done on this topic, apart from the work by Horng *et al.* who examined the magnetic anisotropies, including perpendicular anisotropy, of Fe_3O_4 and $CoFe_2O_4$ films grown on MgO(001) [127].

Current metallic spintronic devices, such as hard disk drives, already utilize magnetic multilayer geometries that promote perpendicular anisotropy coincident with exchange coupling to AF base layers or through NM interlayers. Magnetic oxides nanostructures are poised to replace these devices in many applications because their magnetic properties are even richer and more robust. The AF order in ultrathin CoO layers, for example, persists to thicknesses down to at least 30 Å, and the length scale of the magnetic interaction in AF–AF multilayers is definitely larger than naively expected for superexchange. AF–F oxide multilayers exhibit exchange biasing that is stronger than that observed in most metallic systems, and the ordering and blocking temperatures of these structures can be tuned simply by varying the relative composition of the bilayer. Because of the complex nature of the lattice structures in these oxides, frustration drives 90° coupling to produce exchange biasing or to promote strong coupling across AF interlayers. This interaction can be stronger than that achieved in the more common collinear coupling geometry. The intriguing array of magnetic interactions in these multilayers leads to an equally unique set of properties, as discussed in the next section. The challenge remains as to how to best exploit them in future devices, such as magnetic tunnel junctions.

7.4 Properties of Coupled Systems

7.4.1 Magnetoresistance Effects

Much of the work discussed in the previous section is geared toward the ultimate goal of making tunnel junctions in which Fe_3O_4 is used as an electrode material. In this section, we discuss the properties of tunnel junctions (i.e., multilayers) in which Fe_3O_4 has been incorporated. The MR of individual Fe_3O_4 thin films has been discussed in articles by Ziese [128], and a general overview of the use of oxide layers in tunnel junctions to enable oxide spintronics has appeared recently [129]. Hence, this section focuses on Fe_3O_4-based systems specifically, building on what has been discussed in Sections 7.2 and 7.3 of this chapter.

Tunnel MR is defined as the variation in resistance between the parallel (R_{pp}) and antiparallel (R_{ap}) states of the magnetic electrodes divided by the resistance of the antiparallel state. Because of magnetostatic energy, the state is antiparallel in zero-field and parallel in high fields, which means

$$\mathrm{TMR} = \frac{R_{pp} - R_{ap}}{R_{ap}}\,\% = \frac{R(H) - R(0)}{R(0)}\,\% \tag{7.19}$$

Hence, the TMR is positive if $R_{pp} > R_{ap}$ and negative when $R_{ap} > R_{pp}$. The TMR can be related to the spin polarization P_1 and P_2 of both electrodes by the Julière formula [130]. Thus,

$$\mathrm{TMR} = \frac{2P_1P_2}{1 - P_1P_2}\,\% \tag{7.20}$$

In this equation, the spin polarization P_i at the ith electrode is given by the normalized difference between the densities of the majority spin, $N_{i\uparrow}$ and the minority spin, $N_{i\downarrow}$ at the Fermi level E_F

$$P_i = \frac{N_{i\uparrow}(E_F) - N_{i\downarrow}(E_F)}{N_{i\uparrow}(E_F) + N_{i\downarrow}(E_F)} \tag{7.21}$$

The spin polarization of a material is positive when the majority spin at E_F is parallel to the bulk magnetization direction and negative when the minority spin at E_F is parallel. Given the theoretical expectation that Fe_3O_4 is a half-metallic ferromagnet [131] with only minority spin polarization at the Fermi level [7],

Eqs. (7.20) and (7.21) predict an infinite TMR if two Fe_3O_4 electrodes are used in a tunnel junction. We note, however, that experiments have shown that the spin polarization determined from tunneling experiments is not an intrinsic property of the electrode material but instead depends on the barrier [132]. The high T_C of Fe_3O_4 of 858 K [5] provides a key advantage over other half-metallic oxides including the manganites (with a parent compound of $LaMnO_3$) in which $T_C < 300$ K precluding room-temperature applications. In addition, the corrosion resistance of ferrites make a Fe_3O_4-based tunnel junction ideal for chemically demanding applications such as in-gap read heads for in-contact magnetic recording [133].

7.4.1.1 Tunnel Junctions using Fe_3O_4–MgO

The initial reports on TMR in tunnel junctions comprised of Fe_3O_4–MgO were quite disappointing. A low-field MR effect corresponding to the switching field of the electrodes was found for both a PLD-grown Fe_3O_4/50 Å MgO/Fe_3O_4 junction [134] and an MBE-grown CoO/Fe_3O_4/60 Å MgO/Fe_3O_4 junction [121], but the observed magnitude of the MR was $\leq 0.5\%$, which was much less than expected based on the considerations given above. For the PLD growth of these tunnel junctions, the importance of producing the MgO tunnel barrier at high-growth temperatures (473 K) has been acknowledged [135].

Recently, much higher MR effects of $\simeq -10\%$ have been reported by several research groups [136–138]. Specifically, TMR values of −8.5% at 300 K and −22% at 80 K were measured for a 120 Å Fe/80 Å Fe_3O_4/20 Å MgO/100 Å Co junction grown epitaxially on MgO(001) by sputtering [138]. The structural features of the multilayer stack that accounted for this improvement relative to previous results include the following:

1) Low mean roughness of $\approx$0.3 nm.
2) A thin 20 Å MgO tunnel barrier since the tunnel current decreases exponentially with barrier thickness. As discussed in Section 7.3.3.1 and shown in Figure 7.23(a), this thickness is adequate to decouple the magnetic response of both electrodes.
3) A thin Fe_3O_4 layer ($\leq$10 nm) to suppress the Verwey transition and to stabilize the insulating state at low T, permitting measurements over a large T-range.
4) A reduction of the APB density by employing an Fe underlayer to improve Fe_3O_4 growth quality [139].

The importance of the last structural improvement is highlighted by the results obtained on Fe_3O_4/60 Å MgO/CoFe/IrMn/Ta junctions grown on Al_2O_3(0001) by Kado *et al.* [136, 137]. These authors found a clear relationship between the observed TMR ratio and the resistance-area product RA (see Figure 7.26 [137]). Tunnel junctions with fewer imperfections should show a higher RA, and hence the negative MR ratio for these junctions should be considered to be intrinsic since it is coincident with a high RA value. These studies revealed that electron transport through defects in the tunneling barrier or at the interface with the tunnel barrier may result in a reversal of the sign of the TMR [136]. This discovery helped to resolve the debate in the literature about why certain studies found positive rather

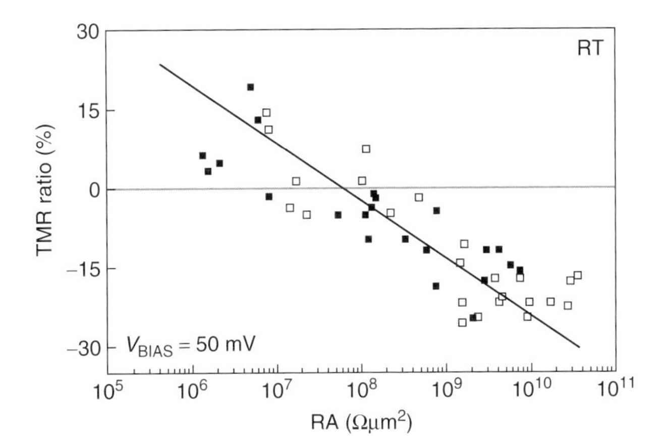

Figure 7.26 Relationship between the resistance-area product, *RA*, and the observed MR ratio for Fe_3O_4/5-nm MgO/1-nm Al_2O_3/CoFe magnetic tunnel junction at room temperature. Note that for high values of *RA* only negative MR is observed reaching a value of −26%. The different symbols refer to samples from different lots. Figure reproduced with permission from [137] (© 2008 by the American Institute of Physics).

than negative TMR as was expected based on the negative spin polarization of Fe_3O_4. The performance of the tunnel junction (Figure 7.26) was further improved to yield a TMR of −26% at room temperature by adding an amorphous 10 Å Al_2O_3 layer that reduced the interfacial roughness from 5.5 nm for the Fe_3O_4/60 Å MgO tunnel barrier to 3.7 nm for the Fe_3O_4/50 Å MgO/10 Å Al_2O_3 stack. Finally, Kado determined that the TMR can be enhanced even more by depositing the MgO barrier at a higher temperature.

7.4.1.2 Tunnel Junctions using Fe_3O_4–AlO_x

A number of groups [140–143] have built upon the work done on traditional tunnel junctions by using an AlO_x barrier. An MR of 13% at room temperature, increasing to 43% at 4.2 K, was first reported for a sputtered glass/Co/Al_2O_3/$Fe_{3-x}O_4$/Al structure with a 20 Å Al_2O_3 tunnel barrier and a very thin 1.5–2-nm $Fe_{3-x}O_4$ layer [140]. Other groups have not significantly improved upon this initial result, reporting TMR values varying between −11% [142] and +3% [143] at room temperature. Overall the MR effect found in this system does not exceed the −26% effect achieved in the Fe_3O_4–MgO system (see Section 7.4.1.1).

7.4.1.3 Tunnel Junctions using Fe_3O_4–oxide–LSMO

Fe_3O_4–oxide–$La_{0.7}Sr_{0.3}MO_3$ (LSMO) tunnel junctions have been fabricated by PLD with paramagnetic barriers ($CoCr_2O_4$, $FeGa_2O_4$) [144–146] and nonmagnetic tunnel barriers (Mg_2TiO_4, $SrTiO_3$) [146, 147], and the reports vary significantly. For Fe_3O_4/$SrTiO_3$/LSMO, Chen and Ziese observed no correlation between the switching of the electrodes and the junction MR [147]. On the basis of their analysis of the voltage dependence of the junction conductance measured at temperatures ranging from 110 to 160 K [147], they concluded that indirect tunneling occurs via defect states. This problem originates from differences in deposition parameters used for Fe_3O_4 on the one hand and $SrTiO_3$ and LSMO on the other. The low oxygen partial pressure, pO_2, needed for the deposition of the Fe_3O_4 electrode, may lead to the introduction of defects via deoxygenation into the $SrTiO_3$ barrier. Subsequent deoxygenation of the LSMO electrode may also explain the low T_C of the LSMO and the vanishing of the MR above 135 K [147].

However, Suzuki and coworkers did observe a strong correlation between the MR and the switching of the magnetic electrodes in tunnel junctions comprised of $Fe_3O_4/CoCr_2O_4$(CCO)/LSMO grown on $SrTiO_3(110)$ substrates. Their initial results yielded the negative MR (−25% at 60 K [144, 145]) that was expected based on the negative spin polarization of Fe_3O_4 and the positive spin polarization of LSMO. The voltage, V, dependence of the conductance of $\sim V^{4/3}$ [144] showed that the dominant contribution was from inelastic hopping through the barrier since it was too thick (60 Å) to support direct tunneling. Surprisingly, the hopping mechanism through the CCO still preserved the spin information [144]. Subsequent studies by the same group on $Fe_3O_4/FeGa_2O_4$/LSMO and Fe_3O_4/Mg_2TiO_4/LSMO found essentially the same result with MR values of −11% at 70 K and −26% at 60 K, respectively. Again, for most voltage biases the tunneling across the barrier was indirect. Only at low bias (<50 mV) was the conductance proportional to V^2 [146] which corresponded to direct or resonant tunneling [148].

7.4.1.4 Tunnel Junctions using a $CoFe_2O_4$ Spin Filter

An alternative approach, which has been pursued recently, is to use $CoFe_2O_4$ layer as a spin filter layer [149, 150]. The first system considered was $CoFe_2O_4/MgAl_2O_4/Fe_3O_4$ grown by PLD on $MgAl_2O_4$ (001). The $MgAl_2O_4$ layer serves to magnetically decouple the Fe_3O_4 and the $CoFe_2O_4$ layers. The idea was to exploit the difference in energy levels in the conduction band of a magnetic insulator ($CoFe_2O_4$), leading to a difference in the tunnel barrier heights for the spin-up and spin-down electrons. Consequently, the tunneling probability for one of the spin directions should be higher than for the other, giving a MR effect in the tunneling current. For this device structure, the current–voltage curves were found to depend on the relative orientation of M of the $CoFe_2O_4$ and the Fe_3O_4 layers. The electrical properties of these devices were characterized by conductive atomic force microscopy (CAFM), from which it was deduced that the net spin polarization exceeded 70% for most of the junctions studied [149]. Moussy and coworkers optimized a related device structure comprised of $Pt(111)/CoFe_2O_4/\gamma\text{-}Al_2O_3(111)/Co(0001)$ epitaxially grown by oxygen-plasma-assisted MBE [150–152]. In this device, they observed a spin polarization of −3% at 290 K and −18% at 2 K. This latter value could be optimized further to −26% by using a superconducting Al spin analyzer to create the structure $Pt(111)/CoFe_2O_4/\gamma\text{-}Al_2O_3(111)/Al$ [153]. This result demonstrates that stronger oxidation conditions during growth have a positive effect on the $CoFe_2O_4$ barrier possibly because the oxygen vacancies in the $CoFe_2O_4$ layer create point defects in the band gap that lower the effective barrier height for the tunneling electrons [153].

In summary, the main problem with Fe_3O_4–oxide–LSMO tunnel junctions is that the room-temperature MR effect is negligible due to the low T_C of LSMO. Another disadvantage of this system is that it has to be grown with LSMO as the bottom electrode because of the high-deposition temperature of 700–800°C that is required [146, 147]. Consequently, APBs (see Section 7.2.2.3) may form at the Fe_3O_4–oxide barrier interface when Fe_3O_4 is grown as the *top* layer, and this inherent disorder may further reduce the MR. The advantage of both Fe_3O_4–MgO and Fe_3O_4–AlO_x is that they can both be grown by sputtering [138, 140]. Specifically, higher room-temperature TMR values of up to −26% have been reported for Fe_3O_4–MgO-based tunnel junctions [137]. It is clear that the best MR results for Fe_3O_4/MgO/3d transition metal junctions are obtained when the Fe_3O_4 electrode is at the *bottom* [136–138]. Investigations by Eerenstein and Hibma [45] demonstrated that this growth geometry avoids problems such as disorder and spin mixing associated with the APBs because they are annealed out for $t_{Fe_3O_4} \geq$ 5 nm (see Section 7.2.2.3). Shvets and coworkers improved the interlayer coupling further by using substrates with very low roughness [120]. Future studies of the influence of roughness in Fe_3O_4/MgO/3d transition metal junctions and related systems thus might lead to additional improvements in parameters such as the switching field, making them more suitable for various device applications.

7.4.2
Magnetooptical Effects

Though $CoFe_2O_4$ has a large Kerr effect [36], early attempts to use it as an optical recording medium failed because of high noise. Given the similarities between Fe_3O_4 and $CoFe_2O_4$ as well as the local nature of the intervalence charge transfer transitions responsible for the optical and magnetooptical (MO) properties [154] in Fe_3O_4, the choice to investigate the magnetooptical Kerr effects of Fe_3O_4–CoO bi- and multilayers [155] was obvious. Consistent with expectations, studies revealed that the AF CoO layer in (100) oriented bilayers does not contribute to the observed Kerr effect. In (111)-oriented bilayers, however, the Kerr rotation in a broad band around 1.7 eV increased but only for $t_{CoO} \leq 12.3$ nm. For even thinner CoO layers between 3.1 and 4.4 nm, this band increased, as well as another band near 2.7 eV [155]. The contribution of the Fe_3O_4 layers to the spectra did not differ significantly from the conventional bulk spectra, apart from some slight oxidation at the top of the Fe_3O_4 layers that was discovered when the bilayers were examined [155]. These results seem to be consistent with ultraviolet photoelectron spectroscopy (UPS) measurements of CoO(100) grown on Fe_3O_4(100) in which the electronic structure of only the first monolayer on both sides of the CoO–Fe_3O_4 interface was found to be disturbed [156].

In view of the relevance of Fe_3O_4 in magnetoelectric heterostructures, Cheng *et al.* investigated the magnetooptical properties of epitaxial Fe_3O_4 grown on MgO, $SrTiO_3$, and $BaTiO_3$ [157]. The authors found a reduction of the intensity of the magnetooptical transition compared to that of bulk Fe_3O_4, and they determined that the saturation value of the MOKE rotation angle was directly proportional to the lattice misfit. The latter observation was attributed to an increase in APBs with misfit strain [157]. Perhaps the most intriguing nanostructure with the best potential for magnetooptical applications involved embedding Fe_3O_4 in 3D opal photonic crystals. Both the polar Kerr effect and the Faraday effect showed a strong enhancement near the photonic band gap at $\simeq 1.8\,\text{eV}$ [158].

7.5
Conclusions and Outlook

The last 20 years have produced significant advances in our understanding of AF–F oxide multilayers. Using either MBE or PLD, these systems can now be routinely made on a variety of substrates including MgO, $SrTiO_3$, $MgAl_2O_4$, and α-Al_2O_3. The controlled growth consistently results in high-quality crystalline superstructures with few complicating structural defects, and the study of AF–F interactions for various interface configurations, both compensated and uncompensated, is thus straightforward. From these investigations we have learned the following:

1) The anomalous properties of ultrathin Fe_3O_4 can be explained by the occurrence of APBS, which are a fundamental consequence of growing a lower symmetry oxide film on a higher symmetry oxide substrate. Fortunately, for potential applications of these layers, the density of antiphase boundaries is not a fixed property of the material and can be manipulated by tightly controlling growth and annealing conditions.
2) Although a comprehensive theory for exchange biasing in oxide multilayers is still not available, a promising model based on the weakening of the spin–spin interaction explains the anomalous dependence of the blocking temperature T_B on the AF layer thickness, which contrasts with the dependence of the Néel temperature T_N on AF thickness. This model and other models, however, should be expanded further to include the influence of growth-induced effects, as experiments demonstrate that they can have a strong influence on the properties of the AF oxides and hence on the observed biasing. For example, experiments clearly show that strain can alter the AF anisotropy and subsequently, the AF thickness dependence of the biasing field.
3) From the oxide multilayer studies, it is apparent that spin frustration can give rise to perpendicular coupling in trilayers at zero field across an AF oxide interlayer and in AF–F multilayers where the onset of the coupling is linked to the onset of exchange biasing.

The results on magnetic oxide multilayers have also led to the design and development of magnetic tunnel junctions comprised of magnetic oxide layer stacks. It has even been proven possible to incorporate these layers in tunnel junctions with oxide barrier layers as thin as a couple of tens of Ångstroms. These structures rely upon some of the experimental innovations developed to highlight interlayer interactions in early studies of oxide multilayers (i.e., exchange coupling of Fe_3O_4 to magnetically harder layers such as $Co_xFe_{2-x}O_4$ as an alternative to exchange biasing). The progress in this field has been impressive: within a decade the reported *room-temperature* TMR effect for Fe_3O_4–MgO-based tunnel junctions has risen from less than 0.5% to 26%, an increase of a factor of 50. It would seem that further improvement would be possible with better understanding of the effect of interface roughness and variations in the electronic states at (magnetic) oxide interfaces on spin-polarized transport. Notably, the Fe_3O_4/MgO/3d-transition metal is an excellent candidate for further study given that it exhibits TMR at room temperature in contrast to other systems.

Presently, new research directions on materials such as multiferroics are emerging, building on the work discussed in this chapter. In particular, the combination of magnetic oxides with semiconductors is attracting much attention. For instance, magnetic oxide materials can be combined with semiconductors to form Schottky diodes [159, 160], and thin nanowires with Fe_3O_4-based layers exhibit new MR effects due to domain-wall scattering of polarized electrons [161]. A particularly interesting trend in realizing and studying *semiconductor* spintronic device structures is the combination of semiconductors such as ZnO with Fe_3O_4, which has a high spin polarization and small conductivity mismatch with (oxide) semiconductors, in an all-oxide ferromagnetic/semiconductor heteroepitaxial structure (see for example [162]). We hope that this chapter thus provides a firm basis and a complete background on magnetic oxide multilayers, as it is essential for understanding and manipulating the fundamental interactions that drive the physics in these new oxide multilayer structures.

Acknowledgments

We would like to thank our friends and colleagues with whom we have worked on this topic over the years. In alphabetical order: Ross Erwin, Lou-Fé Feiner, Yumi Ijiri, Wim de Jonge, David Lind, Thomas Schulthess, and Ronald Wolf. It was a great pleasure working with them. Special thanks are due to Paul van der Heijden and Willem Fontijn, who worked on this topic for their PhD and were instrumental to many of the results reported. (Eindhoven/Gaithersburg, June 2009).

References

1. von Helmolt, R., Wecker, J., Holzapfel, B., Schultz, L., and Samwer, K. (1993) Giant negative magnetoresistance in perovskite $La_{2/3}Ba_{1/3}MnO_x$. *Phys. Rev. Lett.*, **71** (14), 2331–2333.
2. Coey, J.M.D., Viret, M., and von Molnar, S. (1999) Mixed valence manganites. *Adv. Phys.*, **48** (2), 167–293.
3. Moodera, J.S., Kinder, L.S., Wong, T.M., and Merservey, R. (1995) Large tunnel magneto-resistance at room-temperature in ferromagnetic thin film tunnel junctions. *Phys. Rev. Lett.*, **74** (16), 3273–3276.
4. Ramesh, R. and Spaldin, N.A. (2007) Multiferroics: progress and prospects in thin films. *Nat. Mat.*, **6** (1), 21–29.
5. Smit, J. and Wijn, H.P.J. (1959) *Ferrites*, Philips Technical Library, Eindhoven.

6. Finazzi, M., Duò, L., and Ciccacci, F. (2009) Magnetic properties of interfaces and multilayers based on thin antiferromagnetic oxide films. *Surf. Sci. Reports*, **64** (4), 139–168.
7. Yanase, A. and Siratori, K. (1984) Band structure in the high temperature phase of Fe_3O_4. *J. Phys. Soc. Jpn.*, **53** (1), 312–317.
8. Terashima, T. and Bando, Y. (1984) Formation of artificial superlattices comprised of ultrathin layers of CoO and NiO by reactive evaporation. *J. Appl. Phys.*, **56** (12), 3445–3450.
9. Terashima, T. and Bando, Y. (1987) Formation and magnetic properties of artificial superlattice of CoO-Fe_3O_4. *Thin Solid Films*, **152** (3), 455–463.
10. Lind, D.M., Berry, S.D., Chern, G., Mathias, H., and Testardi, L.R. (1992) Growth and structural characterization of Fe_3O_4 and NiO thin films and superlattices grown by oxygen-plasma-assisted molecular-beam epitaxy. *Phys. Rev. B*, **45** (4), 1838–1850.
11. Chern, G., Berry, S.D., Lind, D.M., Mathias, H., and Testardi, L.R. (1992) Electrical transport properties of Fe_3O_4/NiO superlattices. *Phys. Rev. B*, **45** (7), 3644–3651.
12. Lind, D.M., Tay, S.P., Berry, S.D., Borchers, J.A., and Erwin, R.W. (1993) Structural and magnetic ordering in iron-oxide nickel-oxide multilayers by X-ray and neutron diffraction. *J. Appl. Phys.*, **73** (10), 6886–6891.
13. Wolf, R.M., De Veirman, A.E.M., van der Sluis, P., van der Zaag, P.J., and aan de Stegge, J.B.F. (1994) MBE growth of CoO and Fe_3O_4 films and CoO/Fe_3O_4 multilayers. *Mat. Res. Soc. Symp. Proc.*, **341**, 23–28.
14. James, M.A., Voogt, F.C., Niesen, L., Rogojanu, O.C., and Hibma, T. (1998) The role of interfacial structure in determining magnetic behaviour in MBE-grown Fe_3O_4-MgO multilayers on MgO (001). *Surf. Science*, **402–404** (1–3), 332–336.
15. Voogt, F.C., Fuji, T., Smulders, P.J.M., Niessen, L., James, M.A., and Hibma, T. (1999) NO_2-assisted molecular-beam epitaxy of Fe_3O_4, $Fe_{3-\delta}O_4$, and γ-Fe_2O_3 thin films on MgO(100). *Phys. Rev. B*, **60** (15), 11193–11206.
16. Gao, Y. and Chambers, S.A. (1997) Heteroepitaxial growth of α-Fe_2O_3, γ-Fe_2O_3 and Fe_3O_4 thin films by oxygen-plasma-assisted molecular beam epitaxy. *J. Cryst. Growth.*, **174** (1–4), 446–454.
17. Kim, Y.J., Gao, Y., and Chambers, S.A. (1997) Selective growth and characterization of pure, epitaxial α-Fe_2O_3 (0001) and Fe_3O_4 (001) films by plasma-assisted molecular beam epitaxy. *Surf. Sci.*, **371** (2–3), 358–370.
18. Chambers, S.A. (2000) Epitaxial growth of thin film oxides. *Surf. Sci. Repts.*, **39** (5–6), 105–180.
19. de Boer, D.K.G., Leenaers, A.J.G., and Wolf, R.M. (1995) X-ray reflectometry from samples with rough interfaces - an oxidic multilayer study. *J. Phys. D: Appl. Phys.*, **28** (4A), A227–A230.
20. Bloemen, P.J.H., van der Heijden, P.A.A., Wolf, R.M., aan de Stegge, J., Kohlepp, J.T., Reinders, A., Jungblut, R.M., van der Zaag, P.J., and de Jonge, W.J.M. (1996) Magnetic and structural properties of MBE-grown oxidic multilayers. *Mater. Res. Soc. Symp. Proc.*, **401**, 485–499.
21. Ijiri, Y., Borchers, J.A., Erwin, R.W., Lee, S.-H., van der Zaag, P.J., and Wolf, R.M. (1998) Perpendicular coupling in exchange biased Fe_3O_4/CoO superlattices. *Phys. Rev. Lett.*, **80** (3), 608–611.
22. Margulies, D.T., Parker, F.T., Rudee, M.L., Spada, F.E., Chapman, J.N., Aitchison, P., and Berkowitz, A.E. (1997) Origin of the anomalous magnetic behaviour in single crystal Fe_3O_4 films. *Phys. Rev. Lett.*, **79** (25), 5162–5165.
23. Voogt, F.C., Palstra, T.T.M., Niessen, L., Rogojanu, O.C., James, M.A., and Hibma, T. (1998) Superparamagnetic behavior of structural domains in epitaxial ultrathin magnetite films. *Phys. Rev. B*, **57** (14), R8107–R8110.
24. van der Zaag, P.J., Fontijn, W.F.J., Gaspard, P., Wolf, R.M., Brabers, V.A.M., van der Veerdonk, R.J.M., and van der Heijden, P.A.A. (1996) A study of the magneto-optical Kerr spectra

of bulk and ultrathin Fe_3O_4. *J. Appl. Phys.*, **79** (8), 5936–5938.

25. Fontijn, W.F.J., van der Heijden, P.A.A., Voogt, F.C., Hibma, T., and van der Zaag, P.J. (1997) Comparison of a stoichiometric anlaysi of $Fe_{3-\delta}O_4$ layers by magneto-optical Kerr spectroscopy with Mössbauer data. *J. Magn. Magn. Mater.*, **165** (1–3), 401–404.
26. Fujii, T., Takano, M., Katano, R., Bando, Y., and Isozumi, Y. (1990) CEMS study of the growth and properties of Fe_3O_4 films. *J. Cryst Growth.*, **99** (1–4), 606–610.
27. van der Heijden, P.A.A., Bloemen, P.J.H., Gaines, J.M., van Eemeren, J.T.W.M., Wolf, R.M., van der Zaag, P.J., and de Jonge, W.J.M. (1996) Magnetic interface anisotropy of MBE-grown ultra-thin (001) Fe_3O_4 layers. *J. Magn. Magn. Mater.*, **159** (3), L293–L298.
28. (a) Margulies, D.T., Parker, F.T., and Berkowitz, A.E. (1994) Magnetic anomalies in single-crystal Fe_3O_4 thin films. *J. Appl. Phys.*, **75** (10) 6097–6099; (b) Margulies, D.T., Parker, F.T., Spada, F.E., and Berkowitz, A.E. (1994) Anisotropy in epitaxial Fe_3O_4 and $NiFe_2O_4$ films. *Mater. Res. Soc. Symp. Proc.*, **341** 553–558; (c) Margulies, D.T., Parker, F.T., Spada, F.E., Li, J., Sinclair, R., and Berkowitz, A.E. (1996) Anomalous moment and anisotropy behavior in Fe_3O_4 films. *Phys. Rev. B*, **53** (14) 9175–9187.
29. Parkin, S.S.P., Sigbee, R., Felici, R., and Felcher, G.P. (1986) Observation of magnetic dead layers at the surface of iron oxide films. *Appl. Phys. Lett.*, **48** (9), 604–606.
30. van der Heijden, P.A.A., Bloemen, P.J.H., Metselaar, J.M., Wolf, R.M., Gaines, J.M., van Eemeren, J.T.W.M., van der Zaag, P.J., and de Jonge, W.J.M. (1997) Interlayer coupling between Fe_3O_4 layers separated by an insulating nonmagnetic MgO layer. *Phys. Rev. B*, **55** (17), 11569–11575.
31. Sakamoto, M., Asano, M., Fujii, T., Nanba, T., Osaka, A., Miura, Y., and Takada, J. (1992) Preparation of Fe_3O_4/MgO artificial superlattices by activated reactive evaporation method, in *Ferrites: Proceedings of the Sixth International Conference on Ferrites(ICF6)* (eds T. Yamaguchi and M. Abe), The Japan Society and Powder and Powder Metallurgy, Tokyo, pp. 872–875.
32. Coey, J.M.D. (1971) Noncollinear spin arrangement in ultrafine ferrimagnetic crystallites. *Phys. Rev. Lett.*, **27** (17), 1140–1142.
33. (a) Visser, E.G. and Johnson, M.T. (1991) A novel interpretation of the complex permeability in polycrystalline ferrites. *J. Magn. Magn. Matter.*, **101** (1–3), 143–147; (b) van der Zaag, P.J., Ruigrok, J.J.M., Noorderemeer, A., van Delden, M.H.W.M., Por, P.T., Rekveldt, M.Th., Donnet, D.M., and Chapman, J.N. (1993) The initial permeability of polycrystalline MnZn-ferrites: the influence of domain ad microstructure. *J. Appl. Phys.*, **74** (6) 4085–4095.
34. van der Heijden, P.A.A., van Opstal, M.G., Swüste, C.H.W., Bloemen, P.J.H., Gaines, J.M., and de Jonge, W.J.M. (1998) A ferromagnetic resonance study on ultra-thin Fe_3O_4 layers grown on (001) MgO. *J. Magn. Magn. Mater.*, **182** (1–2), 71–80.
35. McGuigan, L., Barklie, R.C., Sofin, R.G.S., Arora, S.K., and Shvets, I.V. (2008) In-plane anisotropies in Fe_3O_4 films on vicinal MgO(100). *Phys. Rev. B* **77** (17), 174424-1–17442-9.
36. Martens, J.W.D. and Voermans, A.B. (1984) Cobalt ferrite thin-films for magneto-optical recording. *IEEE Trans. Magn.*, **MAG-20** (5), 1007–1012.
37. de Jonge, W.J.M., Bloemen, P.J.H., and den Broeder, F.J.A. (1994) Experimental investigations of magnetic anisotropy, in *Ultrathin Magnetic Structures*, vol. 1 (eds J.A.C. Bland and B. Heinrich), Springer Verlag, Berlin/Heidelberg, pp. 65–90.
38. This technique was introduced by Purcell, S.T., Folkerts, W., Johnson, M.T., McGee, N.W.E., Jager, K., aan Stegge, J., Zeper, W.B., and Hoving, W. (1991) Oscillations with a period of 2 Cr monolayers in the antiferromagnetic exchange coupling in a (001) Fe/Cr/Fe sandwich structure. *Phys. Rev. Lett.*, **67** (7), 903–906.

39. van der Heijden, P.A.A., Hammink, J.J., Bloemen, P.J.H., Wolf, R.M., van Opstal, M.G., van der Zaag, P.J., and de Jonge, W.J.M. (1995) Magnetic properties of epitaxial MBE grown thin Fe_3O_4 films on MgO(100). *Mater. Res. Soc. Symp. Soc.*, **384**, 27–32.
40. Fujii, T., Takano, M., Katano, R., Isozumi, Y., and Bando, Y. (1994) Surface and interface properties of epitaxial Fe_3O_4 films studied by Mössbauer spectroscopy. *J. Magn. Magn. Mater.*, **130** (1–3), 267–274.
41. Hibma, T., Voogt, F.C., Niessen, L., van der Heijden, P.A.A., de Jonge, W.J.M., Donkers, J.J.T.M., and van der Zaag, P.J. (1999) Anti-phase domains and magnetism in epitaxial magnetite layers. *J. Appl. Phys.*, **85** (8), 5291–5293.
42. Eerenstein, W., Palstra, T.T.M., Hibma, T., and Celotto, S. (2003) Diffusive motion of antiphase boundaries in Fe_3O_4. *Phys. Rev. B*, **68** (1), 014428-1–014428-7.
43. Gaines, J.M., Bloemen, P.J.H., Kohlepp, J.T., Bulle-Lieuwma, C.W.T., Wolf, R.M., Reinders, A., Jungblut, R.M., van der Heijden, P.A.A., van Eemeren, J.T.W.M., aan de Stegge, J., and de Jonge, W.J.M. (1997) A STM study of Fe_3O_4 (100) grown by molecular beam epitaxy. *Surf. Sci.*, **373** (1), 85–94.
44. (a) Motida, K. and Miyahara, S. (1970) On the 90° exchange interactions between cation (Cr^{3+}, Mn^{2+}, Fe^{3+} and Ni^{2+}) in oxides. *J. Phys. Soc. Jpn.*, **28** (5) 1188–1196; (b) Boekema, C. and Sawatzky, G.A. (1972) Covalency effects and hyperfine interactions in the rare earth orthoferrites. *Int. J. Magn.* **3** (4) 341–348.
45. Eerenstein, W., Palstra, T.T.M., Hibma, T., and Celotto, S. (2002) Origin of the increased resistivity in epitaxial Fe_3O_4 layers. *Phys. Rev. B*, **66** (20), 201101(R).
46. Eerenstein, W., Hibma, T., and Celotto, S. (2004) Mechanism for supermagnetic behavior in epitaxial Fe_3O_4 films. *Phys. Rev. B*, **70** (18), 184404-1–184404-6.
47. Celotto, S., Eerenstein, W., and Hibma, T. (2003) Characterization of antiphase boundaries in epitaxial magnetite films. *Eur. Phys. J. B*, **36** (2), 271279.
48. van der Heijden, P.A.A. (1998) Magnetic properties of oxide-based thin films and multilayers. Chapter 5. Ph.D. dissertation, Eindhoven University of Technology.
49. Tang, Z.X., Sorensen, C.M., Klabunde, K.J., and Hadjipanayis, C.G. (1991) Size dependent Curie temperature in nanoscale $MnFe_2O_4$ particles. *Phys. Rev. Lett.*, **67** (25), 3602–3605.
50. Kulkarni, G.U., Kannan, K.R., Arunarkavalli, T., and Rao, C.N.R. (1994) Particle-size effects on the value of T_C of $MnFe_2O_4$: Evidence for finite-size scaling. *Phys. Rev. B*, **49** (1), 724–727.
51. van der Zaag, P.J., Johnson, M.T., Noordermeer, A., and Bongers, P.F. (1992) Comment on "Size dependent Curie temperature in nanoscale $MnFe_2O_4$ particles". *Phys. Rev. Lett.*, **68** (20), 3112.
52. van der Zaag, P.J., Brabers, V.A.M., Johnson, M.T., Noordermeer, A., and Bongers, P.F. (1995) Comment on "Particle-size effects on the value of T_C of $MnFe_2O_4$: Evidence for finite-size scaling". *Phys. Rev. B*, **51** (17), 12009–12010.
53. Yang, A., Chinnasamy, C.N., Greneche, J.M., Chen, Y., Yoon, S.D., Hsu, K., Vittoria, C., and Harris, V.G. (2009) Large tunability of Néel temperature by growth-rate-induced cation inversion in Mn-ferrite nanoparticles. *Appl. Phys. Lett.*, **94** (11), 113109-1–113109-3.
54. (a) Schneider, C.M., Bressler, P., Schuster, P., Kirschner, J., de Miguel, J.J., and Miranda, R. (1990) Curie temperature of ultrathin films of fcc cobalt epitaxially grown on atomically flat Cu(100) surfaces. *Phys. Rev. Lett.*, **64** (9) 1059–1062; (b) de Miguel, J.J., Cebollada, A., Gallego, J.M., Miranda, R., Schneider, C.M., Schuster, P., and Kirschner, J. (1991) Influence of the growth conditions on the magnetic properties of fcc cobalt films: from monolayers to superlattices. *J. Magn. Magn. Mater.*, **93**, 1–9.

55. Abarra, E.N., Takano, K., Hellman, F., and Berkowitz, A.E. (1996) Thermodynamic measurements of magnetic ordering in antiferromagnetic superlattices. *Phys. Rev. Lett.*, **77** (16), 3451–3454.

56. van der Zaag, P.J., Ijiri, Y., Borchers, J.A., Feiner, L.F., Wolf, R.M., Gaines, J.M., Erwin, R.W., and Verheijen, M.A. (2001) Difference between the blocking and Néel temperature in the exchange biased Fe_3O_4/CoO system. *Phys. Rev. Lett.*, **84** (26), 6102–6105.

57. Ambrose, T. and Chien, C.L. (1996) Finite-size effects and uncompensated magnetization in thin antiferromagnetic CoO layers. *Phys. Rev. Lett.*, **76** (10), 1743.

58. Tang, Y.J., Smith, D.J., Zink, B.L., Hellman, F., and Berkowitz, A.E. (2003) Finite size effects on the moment and ordering temperature in antiferromagnetic CoO layers. *Phys. Rev. B*, **67**, 054408-1–054408-7.

59. Alders, D., Tjeng, L.H., Voogt, F.C., Hibma, T., Sawatzky, G.A., Chen, C.T., Vogel, J., Sacchi, M., and Iacobucci, S. (1998) Temperature and thickness dependence of magnetic moments in NiO epitaxial films. *Phys. Rev. B*, **57** (18), 11623–11631.

60. Roth, W.L. (1958) Multispin axis structures for antiferromagnets. *Phys. Rev.*, **111** (3), 772–781.

61. Terashima, T. and Bando, Y. (1984) Formation of artificial superlattice composed of ultrathin layers of CoO and NiO by reactive evaporation. *J. Appl. Phys.*, **58** (12), 3445–3450.

62. Takano, M., Terashima, T., Bando, Y., and Ikeda, H. (1987) Neutron diffraction study of artificial CoO-NiO superlattices. *Appl. Phys. Lett.*, **51** (3), 205–206.

63. Carey, M.J. and Berkowitz, A.E. (1992) Exchange anisotropy in coupled film of $Ni_{81}Fe_{19}$ with NiO and $Co_xNi_{1-x}O$. *Appl. Phys. Lett.*, **60** (24), 3060–3062.

64. Carey, M.J., Berkowitz, A.E., Borchers, J.A., and Erwin, R.W. (1993) Strong interlayer coupling in CoO/NiO antiferromagnetic superlattices. *Phys. Rev. B*, **47** (15), 9952–9955.

65. Carey, M.J. and Berkowitz, A.E. (1993) CoO-NiO superlattices: interlayer interactions and exchange anistropy with $Ni_{81}Fe_{19}$. *J. Appl. Phys.*, **73** (10), 6892–6896.

66. Borchers, J.A., Carey, M.J., Erwin, R.W., Majkrzak, C.F., and Berkowitz, A.E. (1993) Spatially modulated antiferromagnetic order in CoO/NiO superlattices. *Phys. Rev. Lett.*, **70** (12), 1878–1881.

67. Borchers, J.A., Carey, M.J., Berkowitz, A.E., Erwin, R.W., and Majkrzak, C.F. (1993) Propagation of antiferromagnetic order across paramagnetic layer of CoO/NiO superlattices. *J. Appl. Phys.*, **73** (10), 6898–6900.

68. (a) Meiklejohn, W.H. and Bean, C.P. (1956) New magnetic anisotropy. *Phys. Rev.*, **102** (14) 1413; (b) ibidem (1957) **105** (3), 904–913.

69. Wang, R.W. and Mills, D.L. (1992) Onset of long-range order in superlattices - mean-field theory. *Phys. Rev. B*, **46** (18), 11681–11687.

70. Ramos, C.A., Lederman, D., King, A.R., and Jaccarino, V. (1990) New antiferromagnetic insulator superlattices: Structural and magnetic characterization of $(FeF_2)_m(CoF_2)_n$. *Phys. Rev. Lett.*, **65** (23), 2913–2915.

71. Borchers, J.A., Erwin, R.W., Berry, S.D., Lind, D.M., Lochner, E., and Shaw, K.A. (1994) Magnetic structure determination for Fe_3O_4/NiO superlattces. *Appl. Phys. Lett.*, **64** (3), 381–383.

72. Borchers, J.A., Erwin, R.W., Berry, S.D., Lind, D.M., Ankner, J.F., Lochner, E., Shaw, K.A., and Hilton, D. (1995) Long-range magnetic order in Fe_3O_4 superlattices. *Phys. Rev. B*, **51** (13), 8276–8286.

73. Nogués, J. and Schuller, I.K. (1999) Exchange anistropy. *J. Magn. Magn. Mater.*, **192** (2), 203–232.

74. Berkowitz, A.E. and Takano, K. (1999) Exchange anistropy - a review. *J. Magn. Magn. Mater.*, **200** (1–3), 552–570.

75. Fecioru-Morariu, M., Nowak, U., and Güntherodt, G. (2010) Exchange bias by antiferromagnetic oxides, in *Magnetic Properties of Antiferromagnetic Oxide Materials: Surfaces, Interfaces and*

Thin Films, (eds L. Duò, M. Finazzi, and F. Ciccacci), Wiley-VCH, Weinheim. pp. 143–190.

76. Kiwi, M. (2001) Exchange bias theory. *J. Magn. Magn. Mater.*, **234** (3), 584–595.
77. Stamps, R.L. (2000) Mechanisms for exchange bias. *J. Phys. D: Appl. Phys.*, **33** (23), R247–R268.
78. Morosov, A.I. (2010) Theory of ferromagnetic-antiferromagnetic interface coupling, in *Magnetic Properties of Antiferromagnetic Oxide Materials: Surfaces, Interfaces and Thin Films* (eds L. Duò, M. Finazzi, and F. Ciccacci), Wiley-VCH, Weinheim. pp. 191–238.
79. Kleint, C.A., Krause, M.K., Höhne, R., Walter, T., Semmelhack, H.C., Lorenz, M., and Esquinazi, P. (1998) Exchange anisotropy in epitaxial Fe_3O_4/CoO and $Fe_3O_4/Co_xFe_{3-x}O_4$ bilayers grown by pulsed laser deposition. *J. Appl. Phys.*, **84** (9), 5097–5104.
80. Mauri, D., Siegmann, H.C., Bagus, P.S., and Kay, E. (1987) Simple model for thin ferromagnetic-films exchange coupled to an antiferromagnetc substrate. *J. Appl. Phys.*, **62** (7), 3047–3049.
81. Malozemoff, A.P. (1988) Heisenber-to-Ising crossover in a random-field model with uniaxial anisotropy. *Phys. Rev. B*, **37** (13), 7673–7679.
82. Broese van Groenou, A., Bongers, P.F., and Stuyts, A.L. (1968/1969) Magnetism, microstructure and crystal chemistry of spinel ferrites. *Mater. Sci. Eng.*, **3** (6), 317–392.
83. Sohma, M., Kawaguchi, K., and Manago, T. (2001) Strain-induced magnetic anisotropy in epitaxial Fe_3O_4/Co_3O_4 multilayers. *J. Appl. Phys.*, **89** (5), 2843–2846.
84. van der Zaag, P.J., Ball, A.R., Feiner, L.F., Wolf, R.M., and van der Heijden, P.A.A. (1996) Exchange biasing in MBE grown Fe_3O_4/CoO bilayers: the antiferromagnetic layer thickness dependence. *J. Appl. Phys.*, **79** (8), 5103–5105.
85. van der Zaag, P.J., Feiner, L.F., Wolf, R.M., Borchers, J.A., to be published.
86. Kanamori, J. (1963) in *Magnetism*, vol. 1 (eds G.T. Rado and H. Suhl), Academic, New York, pp. 198–199.
87. Lund, M.S., Macedo, W.A.A., Liu, K., Nogués, J., Schuller, I.K., and Leighton, C. (2002) Effect of anisotropy on the critical antiferromagnet thickness in exchange-biased bilayers. *Phys. Rev. B*, **66** (5), 054422-1–054422-7.
88. Li, Z. and Fisher, E.S. (1990) Single-crystal elastic constants of zinc ferrite ($ZnFe_2O_4$). *J. Mater. Sci. Lett.*, **9** (7), 759–760.
89. (a) Gatal, C. and Snoeck, E. (2005) Epitaxial growth and exchange coupling in NiO-Fe_3O_4 bilayers deposited on MgO(001) and Al_2O_3(0001). *J. Magn. Magn. Mater.*, **272–276** Suppl 1, e823–e824; (b) Gatal, C., Snoek, E., Serin, V., and Fert, A.R. (2005) Epitaxial growth and magnetic exchange anisotropy in Fe_3O_4/NiO bilayers grown on MgO (001) and Al_2O_3(0001). *Eur. Phys. J. B*, **45** (2), 157–168.
90. Csiszar, S.I., Haverkort, M.W., Hu, Z., Tanaka, A., Hsieh, H.H., Lin, H.-J., Chen, C.T., Hibma, T., and Tjeng, L.H. (2005) Controlling orbital moment and spin orientation in CoO layers by strain. *Phys. Rev. Lett.*, **95** (18) 187205-1–187205-4.
91. Malozemoff, A.P. (1988) Mechanisms of exchange anisotropy (invited). *J. Appl. Phys.*, **63** (8), 3874–3879.
92. van der Zaag, P.J., Wolf, R.M., Ball, A.R., Bordel, C., Feiner, L.F., and Jungblut, R. (1995) A study of the magnitude of exchange biasing in [111] Fe_3O_4/CoO bilayers. *J. Magn. Magn. Mater.*, **148** (1–2), 346–348.
93. Borchers, J.A., Ijiri, Y., Lind, D.M., Ivanov, P.G., Erwin, R.W., Qasba, A., Lee, S.H., O'Donovan, K.V., and Dender, D.C. (2000) Detection of field-dependent antiferromagnetic domains in exchange-biased Fe_3O_4/NiO superlattices. *Appl. Phys. Lett.*, **77** (25), 4187–4189.
94. (a) Ball, A.R., Fredrikze, H., Lind, D.M., Wolf, R.M., Bloemen, P.J.H., Rekveldt, M.Th., and van der Zaag, P.J. (1996) Polarized neutron diffraction studies of magnetic oxidic Fe_3O_4/NiO

and Fe_3O_4/CoO multilayers. *Physica B*, **221** (1–4) 388–392; (b) Ball, A.R., Leenaers, A.J.G., van der Zaag, P.J., Shaw, K.A., Singer, B., Lind, D.M., Fredrikze, H., and Rekveldt, M.Th. (1996) Polarized neutron reflectometry study of an exchange biased Fe_3O_4/NiO multilayer. *Appl. Phys. Lett.* **69** (10) 1489–1491.

95. Ijiri, Y., Borchers, J.A., Erwin, R.W., Lee, S.-H., van der Zaag, P.J., and Wolf, R.M. (1998) Role of the antiferromagnet in exchange-biased Fe_3O_4/CoO superlattices (invited). *J. Appl. Phys.*, **83** (11), 6882–6887.

96. Ijiri, Y., Schulthess, T.C., Borchers, J.A., van der Zaag, P.J., and Erwin, R.W. (2007) Link between perpendicular coupling and exchange biasing in Fe_3O_4/CoO multilayers. *Phys. Rev. Lett.*, **99** (14), 147201-1–147201-4.

97. Borchers, J.A., Ijiri, Y., Lind, D.M., Ivanov, P.G., Erwin, R.W., Lee, S.-H., and Majkrzak, C.F. (1999) Polarized neutron diffraction studies of exchange coupled Fe_3O_4/NiO superlattices. *J. Appl. Phys.*, **85** (8), 5883–5885.

98. Krug, I.P., Hillebrecht, F.U., Haverkort, M.W., Tanaka, A., Tjeng, L.H., Gomonay, H., Fraile-Rodriguez, A., Nolting, F., Cramm, S., and Schneider, C.M. (2008) Impact of interface orientation on magnetic coupling in highly ordered systems: a case study of the low-indexed Fe_3O_4/NiO interfaces. *Phys. Rev. B*, **78** (6), 064427-1–064427-4.

99. Moran, T.J., Nogués, J., Lederman, D., and Schuller, I.K. (1998) Perpendicular coupling at Fe-FeF_2 interfaces. *Appl. Phys. Lett.*, **72** (5), 617–619.

100. Jungblut, R., Coehoorn, R., Johnson, M.T., Sauer, Ch., van der Zaag, P.J., Ball, A.R., Rijks, Th.G.S.M., aan de Stegge, J., and Reinders, A. (1995) (Invited Paper) Exchange biasing in MBE grown $Ni_{80}Fe_{20}/Fe_{50}Mn_{50}$ bilayers. *J. Magn. Magn. Mater.*, **148** (1–2), 300–306.

101. Nogués, J., Moran, T.J., Lederman, D., Schuller, I.K., and Rao, K.V. (1999) Role of interfacial structure on exchange-biased FeF_2-Fe. *Phys. Rev. B*, **59** (10), 6984–6993.

102. Koon, N.C. (1998) Calculations of exchange bias in thin films with ferromagnetic/antiferromagnetic interfaces. *Phys. Rev. Lett.*, **78** (25), 4865–4868.

103. Hinchey, L.L. and Mills, D.L. (1986) Magnetic properties of ferromagnet-antiferromagnet superlattice structures with mixed-spin antiferromagnetic sheets. *Phys. Rev. B*, **34**, 1689.

104. Schulthess, T.C. and Butler, W.H. (1998) Consequences of spin-flop coupling in exchange biased films. *Phys. Rev. Lett.*, **81** (20), 4516–4519.

105. Finazzi, M. (2004) Interface coupling in a ferromagnetic/ antiferromagnetic bilayer. *Phys. Rev. B*, **69** (4), 064405.

106. Parkin, S.S.P. and Speriosu, V.S. (1990) Exchange anisotropy in $Ni_{81}Fe_{19}/Fe_{50}Mn_{50}$ multilayered structures: evidence for finite-size scaling in ultrathin antiferromagnetic layers, in *Magnetic Properties of Low Dimensional Systems II*, Springer Proceedings in Physics, vol. 50 (eds L.M. Falicov, F. Mejia-Lira, and J.L. Moran-Lopez), Springer, Berlin, pp. 110–120.

107. Devasahayam, A.J. and Kryder, H.H. (1999) The dependence of the antiferromagnet/ferromagnet blocking temperature on antiferromagnet thickness and deposition conditions. *J. Appl. Phys.*, **85** (8), 5519–5521.

108. Fuke, H.N., Saito, K., Yoshikawa, M., Iwasaki, H., and Sahashi, M. (1999) Influence of crystal structure and oxygen content on exchange coupling properties of IrMn/CoFe spin-valve films. *Appl. Phys. Lett.*, **75** (23), 3680–3682.

109. Sang, H., Du, Y.W., and Chien, C.L. (1999) Exchange coupling in $Fe_{50}Mn_{50}$ bilayer: dependence on antiferromagnetic layer thickness. *J. Appl. Phys.*, **85** (8), 4931–4933.

110. van Driel, J., de Boer, F.R., Lenssen, K.-M.H., and Coehoorn, R. (2000) Exchange biasing in $Ir_{19}Mn_{81}$: dependence on temperature, microstructure and antiferromagnetic layer thickness. *J. Appl. Phys.*, **88** (22), 975–982.

111. Xi, H.W., White, R.M., Gao, Z., and Mao, S.N. (2002) Antiferromagnetic thickness dependence of blocking temperature in exchange

coupled polycrystalline ferromagnet/antiferromagnet bilayers. *J. Appl. Phys.*, **92** (8), 4828–4830.

112. Baltz, V., Sort, J., Landis, S., Rodmacq, B., and Dieny, B. (2005) Tailoring size effects on the exchange bias in ferromagnet-antiferromagnet < 100nm nanostructures. *Phys. Rev. Lett.*, **94** (11), 117201-1–117201-4.

113. Rickart, M., Guedes, A., Ventura, J., Sousa, J.B., and Freitas, P.P. (2005) Blocking temperature in exchange coupled MnPt/CoFe bilayers and synthetic antiferromagnets. *J. Appl. Phys.*, **97** (10), 10K110-1–10K110-3.

114. van der Heijden, P.A.A., Maas, T.F.M.M., de Jonge, W.J.M., Kools, J.C.S., Roozeboom, F., and van der Zaag, P.J. (1998) Thermally assisted reversal of the exchange biasing in NiO and FeMn based systems. *Appl. Phys. Lett.*, **72** (4), 492–494.

115. Fulcomer, E. and Charap, S.H. (1972) Thermal fluctuations aftereffect model for some systems with ferromagnetic-antiferromagnetic coupling. *J. Appl. Phys.*, **43** (10), 4190–4199.

116. Couet, S., Schlage, K., Rüffer, R., Stankov, S., Diederich, Th., Laenens, B., and Röhlsberger, R. (2009) Stabilization of antiferromagnetic order in FeO nanolayers. *Phys. Rev. Lett.*, **103** (9), 097201-1–097201-4.

117. Lang, X.Y., Zheng, W.T., and Jiang, Q. (2007) Dependence of the blocking temperature in exchange biased ferromagnetic/antiferromagnetic bilayers on the thickness of the antiferromagnetic layer. *Nanotechnology*, **18** (15), 155701-1–155701-6.

118. Lang, X.Y., Zheng, W.T., and Jiang, Q. (2006) Size and interface effects on ferromagnetic and antiferromagnetic transition temperatures. *Phys. Rev. B*, **73** (22), 224444-1–224444-8.

119. Néel, L. (1962) Magnetisme-Sur un probleme de magnetostatique relatif de couches minces ferromagnetique. *Compt. Rendus.*, **255** (14), 1545; (b) Néel, L. (1962) Magnetisme-Sur un nouveau mode de couplgae entre les animantations de duex couches minces ferromagnetique. *Compt. Rendus*, **255** (15), 1676.

120. Wu, H.-C., Arora, S.K., Mryasov, O.N., and Shvets, I.V. (2008) Antiferromagnetic interlayer exchange coupling between Fe_3O_4 layers across a nonmagnetic MgO dielectric layer. *Appl. Phys. Lett.*, **92** (18), 182502-1–182502-3.

121. van der Zaag, P.J., Bloemen, P.J.H., Gaines, J.M., Wolf, R.M., van der Heijden, P.A.A., van de Veerdonk, R.J.M., and de Jonge, W.J.M. (2000) On the construction of an Fe_3O_4 based all-oxide spin valve. *J. Magn. Magn. Mater.*, **211** (1–3), 301–308.

122. Orna, J., Morellon, L., Algarabel, P.A., Pardo, J.A., Sangio, S., Magen, C., Snoeck, E., De Theresa, J.M., and Ibarra, M.R. (2008) Fe_3O_4/MgO/Fe heteroepitaxial structures for magnetic tunnel junctions. *IEEE Trans. Magn.*, **44** (11), 2862–2864.

123. Faure-Vincent, J., Tiusan, C., Bellouard, C., Popova, E., Hehn, M., Montaigne, F., and Suhl, A. (2002) Interlayer magnetic coupling interactions of two ferromagnetic layers by spin polarized tunneling. *Phys. Rev. Lett.*, **89** (10), 107206-1–107206-4.

124. Brabers, V.A.M., and Whall, T.E., (1992) Hausmannite, Mn_3O_4 in. *Landolt-Börnstein, New Series*, vol. III 27 h, (ed. H.P.J wijn) Springer Verlag, Berlin/Heidelberg, pp. 86–99.

125. van der Heijden, P.A.A., Swüste, C.H.W., de Jonge, W.J.M., Gaines, J.M., van Eemeren, J.T.W.M., and Schep, K.M. (1999) Evidence for roughness driven 90° coupling in Fe_3O_4/NiO/Fe_3O_4 trilayers. *Phys. Rev. Lett.*, **82** (5), 1020–1023.

126. Slonczewski, J. (1995) Overview of interlayer exchange theory. *J. Magn. Magn. Mater.*, **150** (1), 13–24.

127. Horng, L., Chern, G., Chen, M.C., Kang, P.C., and Lee, D.S. (2004) Magnetic anisotropic properties of Fe_3O_4 and $CoFe_2O_4$ ferrite epitaxy thin films. *J. Magn. Magn. Mater.*, **270** (3), 389–396.

128. (a) Ziese, M. and Blythe, H.J. (2000) Magnetoresistance of magnetite. *J. Phys.: Condens. Matter.*, **12** (1)

13–28; (b) Ziese, M. (2002) Extrinsic magnetotransport phenomena in ferromagnetic oxides. *Rep. Prog. Phys.*, **65** (2) 143–249.

129. Bibes, M. and Barthélémy, A. (2007) Oxide spintronics. *IEEE Trans. Electron. Dev.*, **54** (5), 1003–1023.

130. Juliere, M. (1975) Tunneling between ferromagnetic films. *Phys. Lett.*, **54** (3), 225–226.

131. de Groot, R.A., Mueller, F.M., van Engelen, P.G., and Buschow, K.H.J. (1983) New class of materials: half-metallic ferromagnets. *Phys. Rev. Lett.*, **50** (25), 2024–2027.

132. de Teresa, J.M., Barthélémy, A., Fert, A., Contour, J.-P., Montaigne, F., and Seneor, P. (1999) Role of the metal-oxide interface in determining the spin polarization of magnetic tunnel junctions. *Science*, **286** (5439), 507–509.

133. van der Zaag, P.J., Ruigrok, J.J.M., and Gillies, M.F. (1998) New options in thin film recording heads through ferrite layers. *Philips J. Res.*, **51** (1), 173–195.

134. Li, X.W., Gupta, A., Xiao, G., Gian, W., and Dravid, V.P. (1998) Fabrication and properties of heteroepitaxial magnetite (Fe_3O_4) tunnel junctions. *Appl. Phys. Lett.*, **73** (22), 3282–3284.

135. Kiyomura, T., Maruo, Y., and Gomi, M. (2000) Electrical properties of MgO insulating layers in spin-dependent tunneling junctions using Fe_3O_4. *J. Appl. Phys.*, **88** (8), 4768–4771.

136. Kado, T., Saito, H., and Ando, K. (2007) Room-temperature magnetoresistance in magnetic tunnel junctions with Fe_3O_4 electrode. *J. Appl. Phys.*, **101** (9), 09J511-1–09J511-3.

137. Kado, T. (2008) Large room-temperature inverse magnetoresistance in tunnel junctions with a Fe_3O_4 electrode. *Appl. Phys. Lett.*, **92** (9), 092502-1–092502-3.

138. Greullet, F., Snoeck, E., Tiusan, C., Hehn, M., Lacour, D., Lenoble, O., Magen, C., and Calmels, L. (2008) Large inverse magnetoresistance in fully epitaxial Fe/Fe_3O_4/MgO/Co magnetic tunnel junctions. *Appl. Phys. Lett.*, **92** (5), 053508-1–053508-3.

139. Magen, C., Snoeck, E., Lüders, U., and Bobo, J.F. (2008) Effect of metallic buffer layers on the antiphase boundary density of epitaxial Fe_3O_4. *J. Appl. Phys.*, **104** (13), 013913-1–013913-7.

140. Seneor, P., Fert, A., Maurice, J.-L., Montaigne, F., Petroff, F., and Vaurès, A. (1999) Large magnetoreistance in tunnel junction with an iron oxide electrode. *Appl. Phys. Lett.*, **74** (26), 4017–4019.

141. Matsuda, H., Tekeuchi, M., Adachi, H., Hiramoto, M., Matsukawa, N., Odagawa, A., Sersune, K., and Sakakima, H. (2002) Fabrication and magnetoresistance properties of spin-dependent tunnel junction using epitaxial Fe_3O_4 films. *Jpn. J. Appl. Phys.*, **41** (4A), L387–L390.

142. Yoon, K.S., Koo, J.H., Do, Y.H., Kim, K.W., Kim, C.O. and Hong, J.P. (2005) Performance of $Fe_3O_4/AlO_x/CoFe$ magnetic tunnel junctions based on half-metallic Fe_3O_4 electrodes. *J. Magn. Magn. Mater.*, **285** (1–2), 125–129.

143. Bataille, A.M., Mattana, R., Seneor, P., Tagliaferri, A., Gota, S., Bouzehouane, K., Deranlot, C., Guittet, M.-J., Moussy, J.-B., de Nadaï, C., Brookes, N.B., Petreoff, F., and Gautier-Soyer, M. (2007) On the spin polarization at the Fe_3O_4/γ-Al_2O_3 interface probed by spin-resolved photoemission and spin-dependent tunneling. *J. Magn. Magn. Mater.*, **316** (2), e963–e965.

144. Hu, G. and Suzuki, Y. (2002) Negative spin polarization of Fe_3O_4 in magnetite/manganite-based junctions. *Phys. Rev. Lett.*, **89** (27), 276601-1–276601-4.

145. Hu, G., Chopdekar, R., and Suzuki, Y. (2003) Observation of inverse magnetoresistance in epitaxial magnetite/manganite junctions. *J. Appl. Phys.*, **93** (10), 7516–7519.

146. Alldredge, L.M.B., Chopdekar, R.V., Nelson-Cheeseman, B.B., and Suzuki, Y. (2006) Spin-polarized conduction in oxide magnetic tunnel junctions with magnetic and nonmagnetic barrier layers. *Appl. Phys. Lett.*, **89** (18), 182504-1–182504-3.

147. Chen, Y.F. and Ziese, M. (2007) Magnetotransport properties of

Fe_3O_4-$La_{0.7}Sr_{0.3}MnO_3$ junctions. *J. Phys. D: Appl. Phys.*, **40** (11), 3271–3276.

148. (a) Glazman, L.I. and Matveev, K.A. (1988) Inelastic tunneling across thin amorphous films. *Zh. Eksp. Theor.*, **94** (6), 332–343; (b) *Sov Phys. JETP*, **67** (6), 1276–1282.

149. Chapline, M.G. and Wang, S.X. (2006) Room-temperature spin filtering in a $CoFe_2O_4/MgAl_2O_4/Fe_3O_4$ magnetic tunnel barrier. *Phys. Rev. B*, **74** (1), 014418-1–014418-8.

150. Ramos, A.V., Guittet, M.-J., Moussy, J.-B., Manttana, R., Deranlot, C., Petroff, F., and Gatel, C. (2007) Room temperature spin filtering in epitaxial cobalt-ferrite tunnel junctions. *Appl. Phys. Lett.*, **91** (12), 122107-1–122107-3.

151. Ramos, A.V., Moussy, J.-B., Guittet, M.-J., Gutier-Soyer, M., Gatel, C., Bayle-Guillemaud, P., Warot-Fonrose, B., and Snoeck, E. (2007) Influence of a metallic or oxide top layer in epitaxial magnetic bilayers containing $CoFe_{24}$(111) tunnel barriers. *Phys. Rev. B*, **75** (22), 224421-1–224421-8.

152. Mocuta, C., Barbier, A., Ramos, A.V., Guittet, M.-J., Moussy, J.-B., Stanescu, S., Gatel, C., Mattana, R., Deranlot, C., and Petroff, F. (2009) Crystalline structure of oxide-based epitaxial tunnel junctions: the effect of optical lithography studied by X-ray microdiffraction. *Eur. Phys. J. Special Topics*, **167**, 53–58.

153. Ramos, A.V., Santos, T.S., Miao, G.X. Guitet, M.-J., Moussy, J.-B., and Moodera, J.S. (2008) Influence of oxidation on the spin-filtering properties of $CoFe_2O_4$ and the resultant spin polarization. *Phys. Rev. B.*, **78** (18), 180402(R)-1–180402(R)-4.

154. Fontijn, W.F.J., van der Zaag, P.J., Devillers, M.A.C., Brabers, V.A.M., and Metselaar, R. (1997) Optical and magneto-optical polar Kerr spectra of Fe_3O_4 and Mg^{2+} or Al^{3+} substituted Fe_3O_4. *Phys. Rev. B*, **56** (9), 5432–5442.

155. Fontijn, W.F.J. (1998) Magneto-optical spectroscopy of spinel type ferrites in bulk materials and layered structures. Chapter 8. Ph.D. dissertation, Eindhoven University of Technology.

156. Wang, H.Q., Altman, E.I., and Henrich, V.E. (2008) Interfacial properties between CoO(100) and Fe_3O_4(100). *Phys. Rev. B.*, **77** (8), 085313–085311, 085313–085310.

157. Cheng, J., Sterbinsky, G.E., and Wessels, B.W. (2008) Magnetic and magneto-optical properties of heteroepitaxial magnetite thin films. *J. Cryst. Gwth.*, **310** (16), 3730–3734.

158. Pavlov, V.V., Usachev, P.A., Pisarev, R.V., Kurdyukov, D.A., Kaplan, S.F., Kimel, A.V., Kirilyuk, A., and Rasing, Th. (2008) Enhancement of optical and magneto-optical effects in three-dimensional opal/Fe_3O_4 magnetic photonic crystals. *Appl. Phys. Lett.*, **93** (7), 072502–072501, 072502–072503.

159. Watts, S.M., Boothman, C., van Dijken, S., and Coey, J.M.D. (2005) *Appl. Phys. Lett.*, **86** (21), 212108–212101, 212108–212103.

160. Yan, H., Zhang, M., and Yan, H. (2009) Electrical transport, magnetic properties of the half-metallic Fe_3O_4-based Schottky diode. *J. Magn. Magn. Mater.*, **321** (15), 2340–2344.

161. Goto, K., Tanaka, H., and Kawai, T. (2009) Controlled fabrication of epitaxial $(Fe,Mn)_3O_4$ artificial nanowire structures and their electric and magnetic properties. *Nano Lett.*, **9** (5), 1962–1966.

162. Nielsen, A., Brandlmaier, A., Althammer, M., Kaiser, W., Opel, M., Simon, J., Mader, W., Goenenwein, S.T.B., and Gross, R. (2008) All oxide ferromagnet/semiconductor epitaxial heterostructure. *Appl. Phys. Lett.*, **93** (16), 162510-1–162510-3.

8
Micromagnetic Structure – Imaging Antiferromagnetic Domains using Soft X-Ray Microscopy

Hendrik Ohldag

8.1
Introduction

The existence of ferromagnetism has been known for thousands of years, which has led to the development of devices that significantly contributed to the advancement of our society. We would not be where we are today (literally) but for the compass or electromagnets and generators. More recently, ferromagnetic materials have been used extensively in high-tech applications such as magnetic memories and sensors. Antiferromagnetism, however, came into existence some 60 years ago when it was discovered by Néel [1], who pioneered the early understanding of these exotic and elusive magnets [2]. Antiferromagnets are elusive because they do not possess a macroscopic magnetic moment detectable with conventional macroscopic magnetic probes such as superconducting quantum interference devices (SQUIDs), vibrating sample magnetometry (VSM), magneto-optical Kerr effect (MOKE), or magnetic force microscopy (MFM). The absence of a macroscopic moment makes it considerably more challenging to obtain information about the antiferromagnetic order or the details regarding the micromagnetic antiferromagnetic domain structure, as it is necessary to directly probe the magnetic order or the electronic properties of the material on an atomic scale. Early studies on antiferromagnetic spin order, therefore, used spin-resolved neutron scattering techniques capable of mapping out the symmetry of the spin system in bulk , for example, NiO and CoO [3–5]. More recently, synchrotron-based X-ray dichroism spectroscopy, scattering, and microscopy approaches have been established as valuable tools for a better understanding of these fascinating materials. It is seen that soft X-ray dichroism spectromicroscopy is a perfect tool for achieving this goal. It allows for element-by-element characterization of ferro- and antiferromagnetic properties in a complex sample. It is also sensitive to the sample's chemical state, because it addresses the local electronic properties of each element. It also allows for excellent surface and interface sensitivity, which is crucial for modern-device applications. A more detailed description of X-ray dichroism is given in Chapter 3. Here, only a brief description of the features of this technique, which are relevant to this chapter, is given.

Magnetic Properties of Antiferromagnetic Oxide Materials.
Edited by Lamberto Duò, Marco Finazzi, and Franco Ciccacci

ISBN: 978-3-527-40881-8

From a scientific and engineering point of view, it is the absence of a macroscopic magnetic moment that makes an antiferromagnet a perfect candidate for providing a magnetic reference in magnetic sensors, as its compensated magnetic structure is not sensitive to external fields. The foundation for modern magnetic sensors was laid in 1956 by Meiklejohn and Bean [6], who found that a ferromagnet in contact with an antiferromagnet exhibits a preferred magnetic direction or unidirectional anisotropy. The magnetic moment of such a ferromagnet will always point in a particular predefined direction when the external magnetic field is removed; hence, it is suitable as a magnetic reference in a magnetic sensor. It is found that the unidirectional anisotropy originates from the exchange interaction between an antiferromagnet and a ferromagnet across their interface, as this determines the magnetic reversal behavior of the entire structure. Consequently, the characterization of the antiferromagnetic order at surfaces and interfaces, the visualization of its magnetic domain structure, and in particular the determination of the direction of the antiferromagnetic spin axis within each domain are crucial to obtain a profound understanding of the origins of magnetic order in an antiferromagnet and to improve our understanding of complex phenomena such as antiferromagnetic/ferromagnetic-exchange coupling and anisotropy [7]. Hence the section first comprises a brief description of the antiferromagnetic domain structure of NiO, which serves as a model antiferromagnet. However, the methods and results described in the following are not restricted to NiO alone but can be applied to other antiferromagnetic and even ferrimagnetic oxides. For example, recent X-ray microscopy studies of multiferroic $BiFeO_3$ have greatly improved our understanding of such multifunctional materials. In addition to multiferroics, many base compounds for high-temperature superconductors and colossal magnetoresistance materials are antiferromagnetic oxides, whose crystallographic, electronic, transport, and magnetic properties are closely linked.

8.1.1 Origin of Antiferromagnetic Domains

The formation of magnetic domains is a typical phenomenon in magnetically ordered systems in general. However, the origin of the formation of a domain in an antiferromagnet is different from that in a ferromagnet due to the magnetically compensated nature of the spin structure. In a ferromagnet, the magnetostatic energy stored in the stray field created by the net magnetic moment increases the total energy significantly, so the system decomposes consequently into magnetic domains to decrease the magnetostatic energy. Ideally, the magnetic flux then remains closed within the sample and the exchange energy needed to rotate spins against each other from one domain to another can be afforded. For a more detailed overview, see the book by Hubert and Schäfer [8]. Energy terms related to magnetic flux closure, however, do not contribute significantly to the magnetic energy of an antiferromagnet. The first contribution to the magnetic energy, which both systems have in common, is the *magnetocrystalline anisotropy* energy. The effect of the magnetocrystalline anisotropy is that the spins follow the orbital moment and align parallel to certain crystal axes, which are given by the electronic structure and the symmetry of the crystal. Although the interaction between shape and magnetocrystalline anisotropy determines the easy axes in a ferromagnet, higher order contributions have to be considered in an antiferromagnet, for example, the *magnetoelastic anisotropy*. The combination of magnetoelastic, short-range dipolar, and magnetocrystalline energies determines the easy axes in an antiferromagnet. If more than one easy axis is possible or if lattice defects change the elastic energy of the system, the antiferromagnet will decompose in several magnetic domains. Gomonay *et al.* [9] recently presented a theoretical investigation, in which the correlation between magnetostriction and antiferromagnetic domain formation in the bulk and at surfaces was studied. The authors found that, similar to magnetic charges at ferromagnetic surfaces, "elastic charges" at antiferromagnetic surfaces cause an "elastic stray field", which determines the domain structure.

To ultimately describe the antiferromagnetic domain pattern at the surface of NiO, the well-known bulk structure is first reviewed here. In NiO, nickel and oxygen atoms form a rock-salt-type crystal. Figure 8.1a shows the principal symmetry. Nickel atoms are represented by small spheres forming an fcc sublattice. They are connected by oxygen atoms, represented by larger spheres, along the [100] direction. The oxygen 2p orbital overlaps with a 3d orbital of each of the neighboring nickel atoms, leading to the so-called superexchange interaction. As per the Pauli principle, this interaction between the two otherwise independent electrons causes the antiparallel alignment of their spins. The three different [100] axes define the *intrinsic antiferromagnetic propagation axis*. However, the symmetry of the spin order is not completely defined yet, because the coupling along a particular [110] direction is still ambiguous.

From Figure 8.1a, it is seen that to obtain a magnetically compensated cell, the spin arrangement along three of the [110] axes needs to exhibit an antiferromagnetic order, whereas the other three need to show ferromagnetic order. Together with the condition of antiferromagnetic order along [100], one finds that only four different permutations are possible, neglecting spin inversion. The characteristic feature for each of these permutations is that atoms within a particular (111) plane show ferromagnetic order, while adjacent (111) planes are aligned antiparallel. A macroscopic crystal may now exhibit different spin symmetries in different regions of the crystal. In this case, the spin lattice is *twinned* and the different spin symmetry domains are referred to as *T(win)-domains*. Table 8.1 lists the four possible T-domains as defined by their ferromagnetic plane or their antiferromagnetic propagation axes. Two T-domains adjoining each other will form an *antiferromagnetic domain wall* parallel to a certain crystal plane. The wall plane is defined by the magnetic axes that the two domains have in common and the possibilities are listed in Table 8.1. For example, T_1 and T_2 share the ferromagnetic axis $[01\bar{1}]$ and the antiferromagnetic axis [011]. These two axes together define a (100) plane, which now separates a (111)-ordered region from a $(\bar{1}11)$-ordered region. The (100) plane is defined by two common magnetic axes. There are two more domain walls possible between T_1 and T_2 defined by only one of the common [110] axis together with one of the [100] intrinsic antiferromagnetic axis, which all antiferromagnetic domains have in

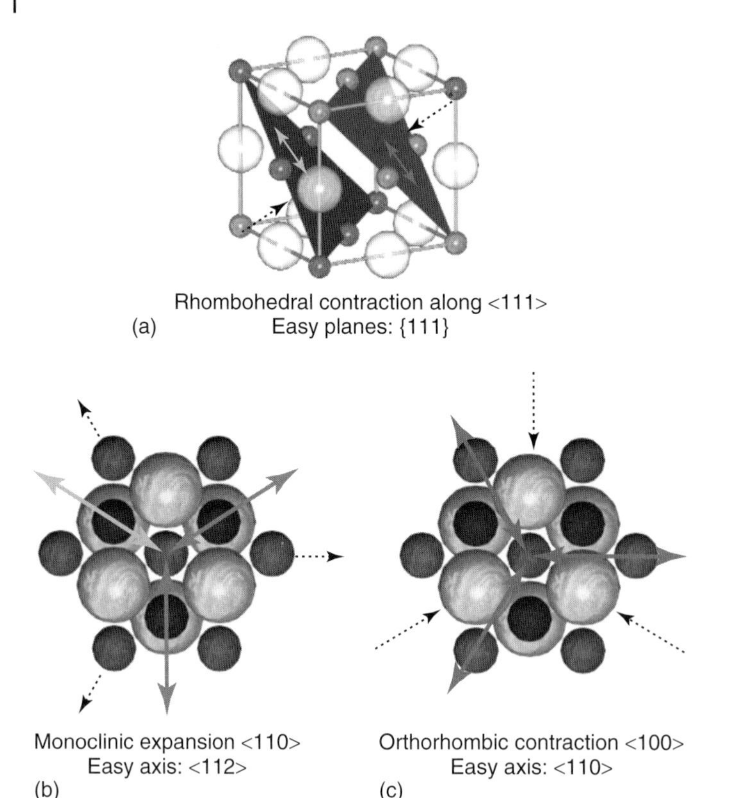

Figure 8.1 (a) A lattice cell in NiO. Ni atoms are represented by small spheres and O atoms by larger spheres. The spins of each atom within a particular (111) plane exhibit ferromagnetic order and are aligned parallel to the plane. At the same time, a rhombohedral lattice contraction along [111] is observed due to magnetostrictive effects. (b, c) View of a (111) plane. A monoclinic expansion favors spins to align along [112] into the direction of the next nearest neighbor (b). An orthorhombic contraction along [100], on the other hand, favors a [110] easy axis toward the next neighbor (c).

common. In this example, these are the (110) and $(1\bar{1}0)$ planes. A complete list of possible T-domain walls in NiO is given in Table 8.2.

So far, only the spin symmetry of the complete antiferromagnetic crystal has been addressed. Whether the *crystallographic T(win)-domains* of the spin lattice also represent *magnetic S(pin)-domains* will depend on whether different T-domains favor a different magnetic spin axis. A detailed discussion of the origin of antiferromagnetic domains in NiO and their relation to magnetoelastic deformation as the origin of antiferromagnetic domain formation has been presented by Yamada *et al.* [10–12]. The authors calculated the magnetostatic energy of NiO, allowing a complete deviation from the perfect cubic lattice in terms of rhombohedral ([111]), monoclinic ($[1\bar{1}0]$), and orthorhombic ([001]) deformation. The first conclusion in [10] is that the short-range dipolar interaction between ordered (111)-planes contributes the most to the magnetostatic energy. This forces the spins parallel to the

Table 8.1 Possible antiferromagnetic symmetries named T-domains, identified by the (111)-plane that exhibits ferromagnetic order.

Name	FM plane	AFM axis	FM_1 axis (T-wall)	FM_2 axis (S-Domain)
		[110]	$[1\bar{1}0]$	$[11\bar{2}]$
T_1	(111)	[101]	$[10\bar{1}]$	$[1\bar{2}1]$
		[011]	$[01\bar{1}]$	$[\bar{2}11]$
		$[1\bar{1}0]$	[110]	$[1\bar{1}2]$
T_2	$(\bar{1}11)$	$[10\bar{1}]$	[101]	$[12\bar{1}]$
		[011]	$[01\bar{1}]$	[211]
		$[1\bar{1}0]$	[110]	$[\bar{1}12]$
T_3	$(1\bar{1}1)$	[101]	$[10\bar{1}]$	[121]
		$[01\bar{1}]$	[011]	$[21\bar{1}]$
		[110]	$[1\bar{1}0]$	[112]
T_4	$(11\bar{1}]$	$[10\bar{1}]$	[101]	$[\bar{1}21]$
		$[01\bar{1}]$	[011]	$[2\bar{1}1]$

The ferromagnetic and antiferromagnetic propagation axes are given in the columns. Note that one of the ferromagnetic propagation axis coincides with the spin axis realized in antiferromagnetic spin domains, while the other one is realized in walls between the twin domains.

Table 8.2 Possible T-walls between T-domains T_i and T_j.

T_1/T_2	T_1/T_3	T_1/T_4
$[01\bar{1}][011]$	$[10\bar{1}][101]$	$[1\bar{1}0][110]$
(100)(011)	(010)(101)	(001)(110)
T_2/T_3	T_2/T_4	T_3/T_4
$[110][1\bar{1}0]$	$[101][10\bar{1}]$	$[011][01\bar{1}]$
$(001)(01\bar{1})$	$(010)(10\bar{1})$	(100)[011]

The first row denotes the common magnetic axes. Planes listed in the second row describe high-symmetry favorable domain walls.

(111) plane and leads to a rhombohedral contraction along [111] of $\frac{\Delta l}{l} \sim 10^{-4}$. The anisotropy energy along [100] is comparably small and the spins can easily rotate into the (111) plane. The situation is illustrated in Figure 8.1(a). Whether an additional uniaxial anisotropy exists within the (111)-plane depends on the existence of additional lattice distortions, which may lift the degeneracy between the equivalent [112] (the second next neighbor) and [110] (the nearest neighbor) directions. This is illustrated in Figure 8.1(b) and (c), which shows a view along [111] of the atoms of two (111) planes, one above the other. The calculation presented by Yamada *et al.*

suggested that an orthorhombic contraction along [100] favors [110] as an easy axis, whereas the monoclinic expansion along [110] favors [112]. Both the situations are shown in the figure. By comparing their calculations to (bulk) domain patterns generated under stress and external field, the authors conclude that a monoclinic expansion of the lattice is actually realized in NiO and that the easy spin axis in a T-domain is [112]. They conclude that the spins within a particular T-domain are aligned parallel to the plane and point toward one of the three possible next nearest neighbors, forming three different S(pin)-domains within each of the four possible T(win)-domains. These 12 different domains in NiO are listed in the last column of Table 8.1 and they fully describe the possible antiferromagnetic (S)pin domains in NiO.

8.1.2
Soft X-Ray Spectroscopy

The experiments described in this chapter were performed using synchrotron radiation. Synchrotron radiation is an excellent source of tunable and polarized X-rays in the energy range between 200 and 1000 eV. An introduction to synchrotron radiation is given in the book by Attwood [13]. Here it is sufficient to note that the horizontal deflection of the electron beam stored in the synchrotron by strong beam bending magnets leads to the emittance of polarized X-rays over a broad energy range. These X-rays are guided toward an experimental endstation using the so-called beamline. The beamline consists of mirrors to guide and focus the X-rays as well as a grating optic to select a small energy band. The energy resolution of a soft X-ray beamline can be in the range of $1000 < \frac{E}{\Delta E} < 10\,000$. Although the center of the X-ray beam is linearly polarized in the horizontal direction, the upper and lower parts of the beam consist of left and right circularly polarized X-rays. One can choose the polarization for the experiment by placing apertures into the beam at the appropriate positions. This allows for polarization-dependent (dichroism) *soft X-ray absorption spectroscopy* (*XAS*) and *soft X-ray microscopy* studies to characterize antiferromagnetic as well as ferromagnetic order.

The energy of soft X-rays is such that the absorption of a soft X-ray photon can lead to the excitation of 2p core-level electrons into empty 3d final states in transition metals such as Mn, Fe, Co, Ni, and Cu. The intensity of this transition, and hence the line shape of the absorption resonance, is strongly influenced by the chemical environment and magnetic symmetry of each species, since it is given by the overlap of the 2p and the 3d orbitals. The X-ray absorption cross section is measured by either detecting the intensity of the transmitted X-rays using the *Lambert–Beer* law or by collecting the electron yield emitted from the sample. The second approach is referred to as *total electron yield* (*TEY*) technique. The absorption of an incident X-ray by a core-level electron and the recombination processes that cause the electron to be emitted from the sample are depicted in Figure 8.2. The incoming photon produces a photoelectron and the resulting core hole recombines by emitting an Auger electron from the valence band. The photoelectron and the Auger electron may now either be emitted into the vacuum or, due to their limited

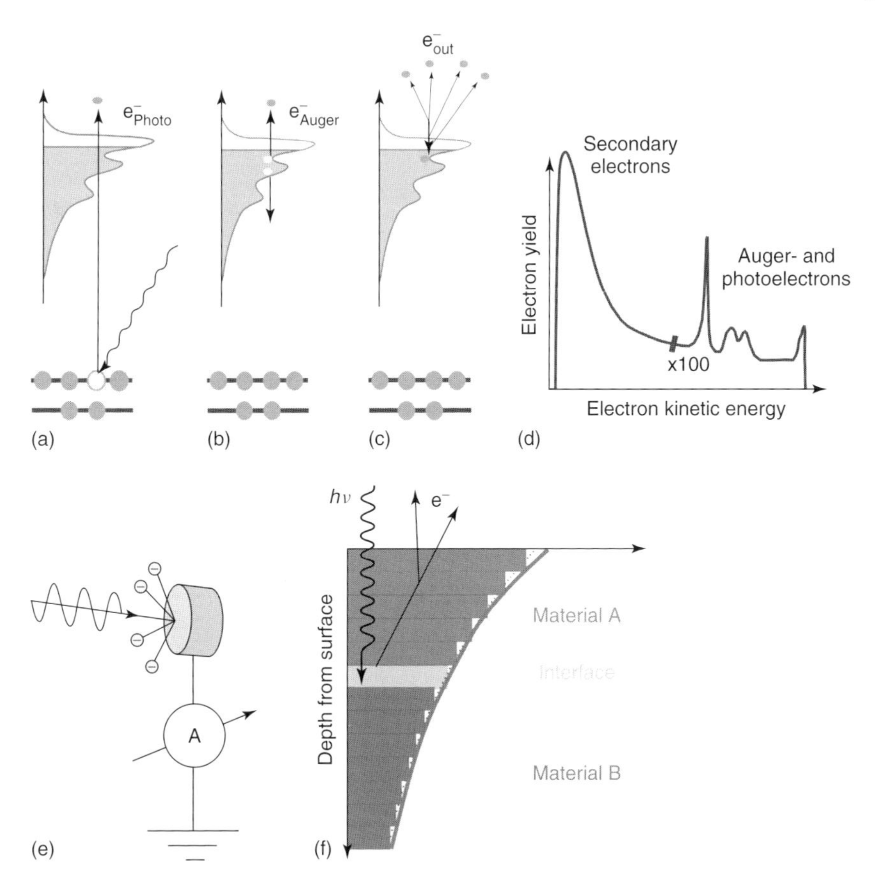

Figure 8.2 Example of the X-ray absorption process at the L-resonance of a 3d metal. (a) If the photon energy of the incoming X-ray is equal to or larger than the binding energy of the L-electron, the X-ray can be absorbed and a core-level electron is excited into an empty final state. (b) The resulting core hole may be filled by a valence electron (down arrow), leading to the emission of an Auger electron from the valence band. (c) Both the Auger- and the photoelectrons lose kinetic energy due to inelastic scattering and produce numerous low-energy secondary electrons. The energy spectrum of all electrons escaping the sample is shown in (d), consisting mostly (90%) of electrons with a kinetic energy less than 5eV produced by inelastic scattering of the initial Auger- and photoelectrons. (e) The current between the sample and the ground potential is proportional to the electron yield generated by X-ray absorption process. (f) The contribution of deeper layers to the total electron yield (TEY) signal decays exponentially with the distance z to the surface. This makes the characterization of surfaces and buried interfaces possible.

mean-free path, lose their energy by inelastic scattering. In the latter case, a larger number of low-energy secondary electrons per initial photoelectron are generated. Because of their larger mean-free path, these secondary electrons will eventually be emitted into the vacuum. If the sample is grounded through a current amplifier, the electron current that is necessary to compensate for the emitted charge can be measured (Figure 8.2e). Typical sample currents obtained using synchrotron radiation are between 0.1 and 10 nA.

The TEY is proportional to the absorption coefficient $\mu(E)$. At normal X-ray incidence, the TEY dN_e from a layer of thickness dz at depth z is related to the X-ray absorption $\mu(E)$ by

$$dN_e \sim \mu(E) \exp[-z\mu(E)] \exp\left[\frac{-z}{\lambda}\right] dz \tag{8.1}$$

The second exponential term denotes the probability for an electron generated at a distance z from the surface to leave the sample. The electron mean-free path λ is a combination of the mean-free path of the initial photo- and Auger electrons (very short <1 nm) as well as the secondary low-energy electrons (long >10 nm). Spectroscopic information is obtained only from layers that contribute to the electron yield. Experimental values for the effective λ vary between 2 and 5 nm. The first exponential function describes the decay of the X-ray intensity as it penetrates the material. Because the decay length of the X-rays is, in general, much larger than the electron escape length, this term is usually neglected and the X-ray intensity is assumed to be constant within the probing depth of the TEY method[1]. Integrating Eq. 8.1 yields the direct proportionality between TEY and $\mu(E)$.

The depth profile of the TEY method is shown in Figure 8.2. The sketch shows the exponential decay of the contribution of each atomic layer in a stack of two materials to the TEY. It is assumed that λ is constant over the entire volume. Deeper layers contribute with a smaller electron yield to the total yield than surface layers. A rough estimate yields that 82% of the electron yield originates from the first 2 nm. For this reason, the TEY technique is considered a surface-sensitive approach. If material B is buried deeper than about 10 nm from the surface, it is usually not detected in a TEY experiment. The electron yield arising from the buried layer B is suppressed by layer A, but the relative contribution of the top layer of material B to the total signal arising from material B is independent of the thickness of top layer, material A. Again, assuming a typical λ of 2 nm and an atomic layer thickness of 0.2 nm, the contribution of the top layer of material B to the electron yield of material B is always between 8 and 10%. This means that if the chemical properties of the top layer of material B are different from the other layers in the bulk of material B, this contribution can be identified. For a more detailed description, see [14].

8.1.3
Photoemission Electron Microscope

X-ray absorption cross section can be obtained using the electron yield method. Specifically, this detection method can be used in two different ways. First, one can spatially average over the illuminated sample area and collect spectroscopic

1) This assumption is even appropriate right at the absorption resonance where the decay length might be as short as 10nm.

information (XAS). Second, one can use a photoemission electron microscope (PEEM) to magnify and image the lateral distribution of the electron yield. The lateral variation of elements and their chemistry as well as their ferromagnetic and antiferromagnetic order can be resolved using PEEM. Review papers focusing on magnetic imaging methods are presented, for example, by Stöhr *et al.* [15–17], by Hillebrecht [18], and in articles by Kuch *et al.* and Scholl *et al.* in "Magnetic nanostructures" edited by Hopster [19, 20]. A general overview about PEEM has been given by Bauer [21].

The idea of cathode lens microscopy was developed in the first half of the last century. Driven by the increased use of synchrotron radiation in the late 1980s and the success of electron-yield-based absorption spectroscopy, an ultrahigh-vacuum compatible and synchrotron-based PEEM was proposed by Tonner and Harp [22] and a few years later by Engel *et al.* [23]. Figure 8.3 shows a sketch of a two-lens PEEM. The irradiation of the sample surface with monochromatic X-rays leads to the emission of electrons, as described in the previous section. If the absorption coefficient μ varies laterally, so will the electron yield. The excited electrons are

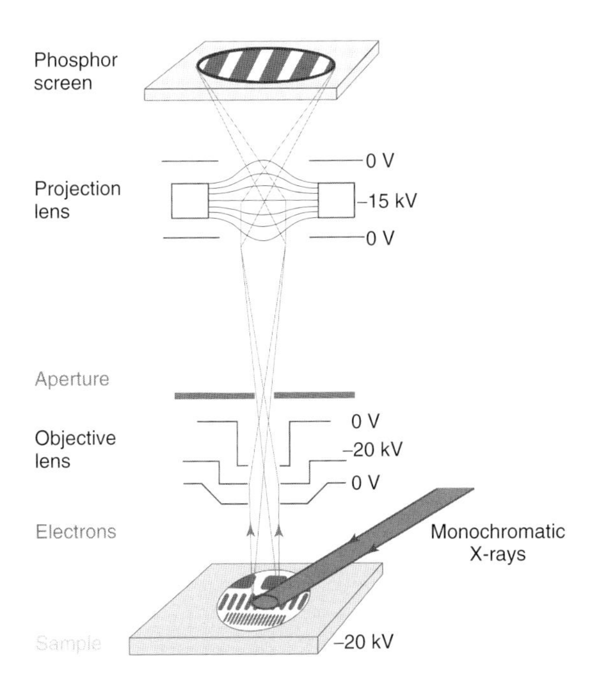

Figure 8.3 Irradiation of a surface with (monochromatic) X-rays leads to the emission of secondary electrons. The spatial distribution of the electron yield is magnified onto a screen by electrostatic lenses. A typical voltage of $10\,\mathrm{kV\,mm^{-1}}$ is applied to the sample (cathode) to accelerate the electrons toward the microscope column. Energy and angular selection of the electrons passing through the microscope column is achieved by an aperture in the back focal plane of the objective lens.

accelerated toward the microscope. An electrostatic objective lens[2] and a projective lens will magnify and image the electron yield on a two-dimensional electron detector, for example, a phosphor screen or a microchannel plate. The visible light image generated by the channel plate and screen can then be recorded using a CCD camera. A common feature of all cathode lens microscopes is that the sample itself, as the cathode, is part of the optical system. The sketched microscope consists of two triode electrostatic lenses. The outer electrodes are at the same potential as the microscope housing, whereas the inner electrode is at high negative voltage. The equipotential lines produced by such an arrangement resemble the shape of thick concave optical lens. They are shown within the projection lens.

Two different microscopes were used for the experiments described in this chapter. One was the commercial STAIB-PEEM [24] microscope. The other was the PEEM2 instrument [25] built at the Advanced Light Source (ALS) in Berkeley, CA, USA. The typical spatial resolution obtained with the STAIB-PEEM using synchrotron radiation was 120–150 nm and about 50–70 nm was obtained using PEEM2. Using the PEEM2 setup, XAS spectra from small areas (less than 1 μm^2) on the sample (*XAS Spectromicroscopy*) can easily be obtained by synchronizing beamline and microscope control software. While the X-ray energy is increased step by step through an absorption resonance, a PEEM image is acquired at each energy. The result is the so-called image stack. After the actual images have been acquired at the synchrotron, the XAS spectra can be extracted using suitable image processing software.

8.1.4 Soft X-Ray Dichroism

Magnetic information is extracted from soft X-ray spectra by varying the angle between the polarization vector and the magnetic axis (or the direction) of the studied sample. The ability to acquire high-quality spectra from single ferromagnetic or antiferromagnetic domains makes PEEM a powerful tool for imaging the magnetic domain structure of these systems [26–28]. An overview of recent experiments can be found in [17].

In the case of a ferromagnetic sample, the line shape of the spectrum will depend on the relative orientation between the helicity vector of the incoming circularly polarized X-rays and the direction of the magnetic moment (X-ray magnetic circular dichroism, XMCD). This is shown in the sketch in the upper left of Figure 8.4. The magnetization vector of each domain includes a different angle with the helicity of the X-rays and, therefore, each of the X-ray absorption spectra obtained from each domain (upper right panel) exhibits a different Fe L_3/L_2 ratio and images acquired at the two resonances show four different gray levels with opposite contrast at

2) Sometimes a magnetic objective lens, with smaller aberrations, is used in a PEEM. However, the magnetic stray field of such an objective lens may perturb the magnetic configuration of the sample.

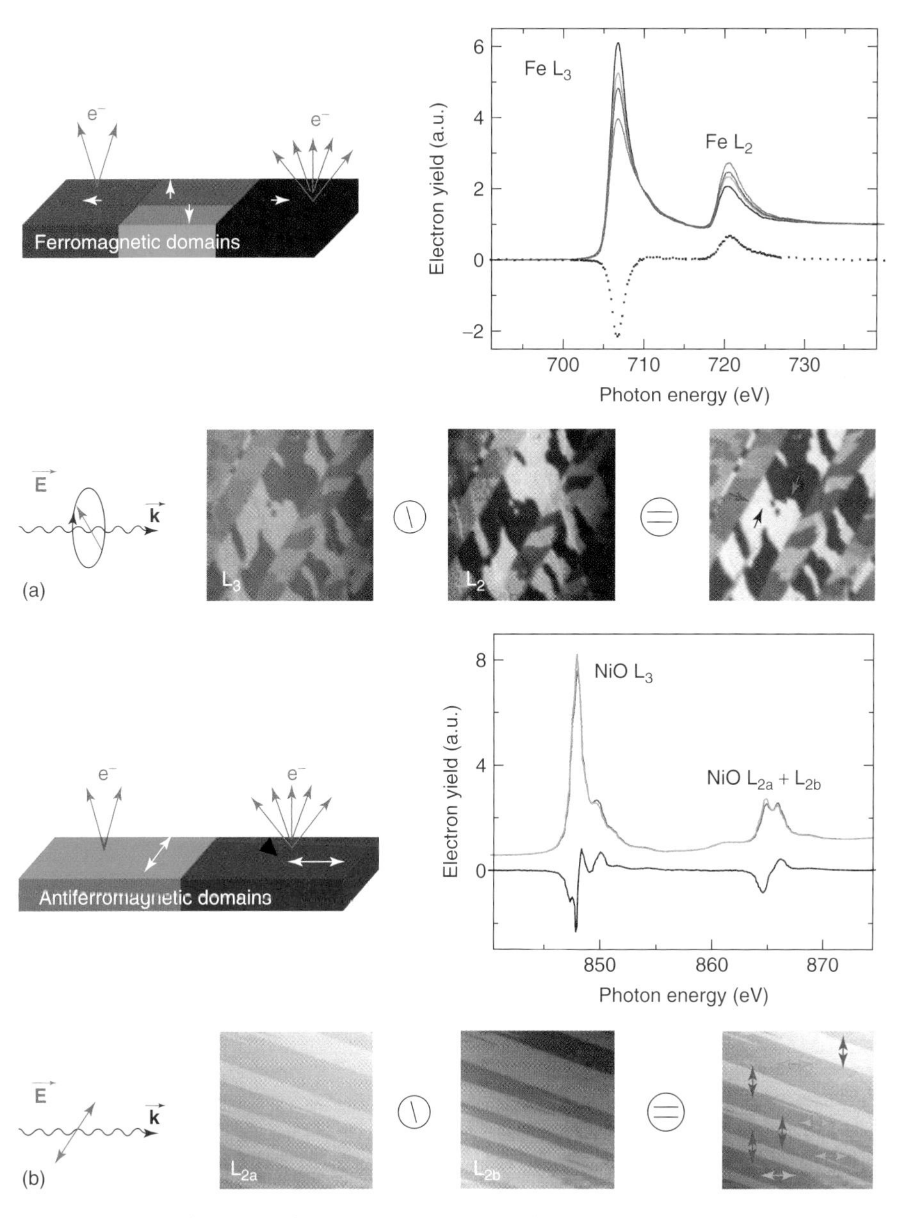

Figure 8.4 (a) L-edge XAS and XMCD spectra of Fe as well as PEEM images obtained at the absorption resonances. (b) L-edge XAS and XMLD spectra of NiO as well as PEEM images at the relevant energies. Note that the NiO L_2 absorption peak is split into two peaks labeled L_{2a} and L_{2b}. Magnetic domain images using XMCD and XMLD are obtained by dividing two images with the same background (such as topography) but opposite magnetic contribution.

the two resonances[3]. The final XMCD image on the right is derived by dividing the single images and it represents a map of the XMCD intensity, that is, the Fe L_3/L_2 ratio and, thus, the direction of the magnetic moment. The further advantage of the calculation of the image ratio is that electron yield variations caused by inhomogeneous illumination or topographic features will not contribute to the final image. The basic symmetry of a domain pattern can already be deduced from the XMCD image itself, especially if the experimental setup allows one to rotate the sample so that several images in different geometries can be acquired. However, a precise determination of the direction of the magnetization in each domain from the XMCD intensity involves an appropriate background subtraction and normalization that cannot be easily achieved in a PEEM image. For this purpose, the spectra calculated from image stacks acquired in each of the ferromagnetic domains are used. The absolute XMCD intensity within each domain can be extracted this way and the angle of the magnetic moment can be determined by comparing the experimentally derived values with appropriate references [29].

A similar correlation between the shape of the absorption spectra and the angle between the polarization vector and antiferromagnetic axis is observed when linear polarization is used to obtain antiferromagnetic contrast (X-ray magnetic linear dichroism, XMLD); see, for example, [27, 30–35]. The same procedures for image acquisition and data analysis as described for XMCD are applied in the case of XMLD. The procedures to acquire an XMLD image and the XMLD spectra resulting from image stacks are shown in Figure 8.4b. For all antiferromagnetic domain images of NiO shown in this chapter, two images are acquired at within the Ni L_2 resonance, which is split into two peaks labeled L_{2a} and L_{2b}. The XMLD spectrum acquired at these two energies exhibits an opposite XMLD effect and can, therefore, be used for XMLD imaging. In addition, because the isotropic absorption intensity of the two peaks is very similar, the same exposure time can be chosen for the two images. This leads to a very efficient suppression of the nonmagnetic background and allows for the acquisition of generally beautiful images. The final XMLD image shown at the bottom right represents the ratio of the two images with opposite contrast. Again as for the case of XMCD, useful information about the principal symmetry of the domain pattern and the distribution of antiferromagnetic axes can be extracted from XMLD images obtained in different geometries or with different orientations of the linear polarization vector. Nevertheless, an exact determination of the antiferromagnetic axis can only be achieved by acquiring absorption spectra in each antiferromagnetc domain, analyzing the microspectroscopic data and extracting the concrete XMLD spectrum. At this point, it needs to be noted that it is crucial to know the shape of the XMLD spectrum at the Ni L_3 resonance. It had always been assumed that the absorption intensity at the NiO L_{2a} peak is maximum when the electric field vector is perpendicular to the spin axis of the antiferromagnet. However, Arenholz and coworkers have recently shown that this

3) The incoming circularly polarized X-ray preferentially excites core-level electrons with one particular spin, which serves as a spin-polarized probe of the spin-split-unoccupied electronic structure of the ferromagnet.

assumption is not correct, in general, and that the correct interpretation of the NiO XMLD spectra depends on the particular experimental geometry. The authors found that the shape of the XMLD at the L_3 resonance holds the key to the correct interpretation of the L_2 XMLD [36, 37]. In light of this, some of the initial findings based on the original interpretation of the XMLD effect [27, 28] had to be revisited and corrected [38].

8.2
Antiferromagnetic Domain Imaging using PEEM

In this section, an example of how XMLD can be used to image the antiferromagnetic domain structure at NiO(001) surfaces is given. It will be evident that the domain structure at the surface can be fully described by the bulk domain structure introduced in Section 8.1.1. However, we also see that the observed surface domain pattern does depend on the preparation conditions of the surface. The antiferromagnetic surfaces, discussed in this section, are NiO(001) that were cleaved *ex situ* before introducing them to the ultrahigh-vacuum apparatus. Before the surfaces were imaged using the PEEM microscope, they were annealed for several hours at 600 K above T_N (525 K) in an oxygen atmosphere of 10^{-6} torr to remove surface contaminations. All these samples showed large antiferromagnetic domains with domain walls congruent with the T-domains and walls listed in Table 8.2. One T-domain usually consists of only one particular S-domain. However, if the surfaces are annealed at lower temperatures (Section 8.2.2) below T_N, a rich S-domain substructure will be observed. If the surfaces are annealed above T_N, the lattice can relax and release the stress that has been introduced during the cleavage process. This, in turn, leads to annihilation of the smaller magnetic S-domains and growth of the larger T-domains. The temperature dependence of the domain pattern is addressed again in Section 8.4.

8.2.1
Imaging Antiferromagnetic Domains and Domain Walls

For the experiments presented in this section, variable X-ray polarization was essential. In particular, the ability to change the angle between the polarization vector of the incident X-rays and the direction of the magnetic axis was the key. The geometry that applies to all experiments is shown in Figure 8.5a. The angle of incidence (θ) is 30°. The incoming X-rays are either linearly or circularly polarized. In the case of linear polarization, the only component of the polarization that is present is the horizontal one, $E_{\|}$, parallel to the sample surface. For circular polarization, an additional component is present, which is referred to as $E_{\perp}$ here. $E_{\perp}$ and the surfaces normal include a fixed angle that is identical to the angle of incidence. Note that the vertical polarization component also shows a projection onto the surface plane and that needs to be taken into account for a detailed analysis of the XMLD contrast. The sample can be rotated around its surface normal such

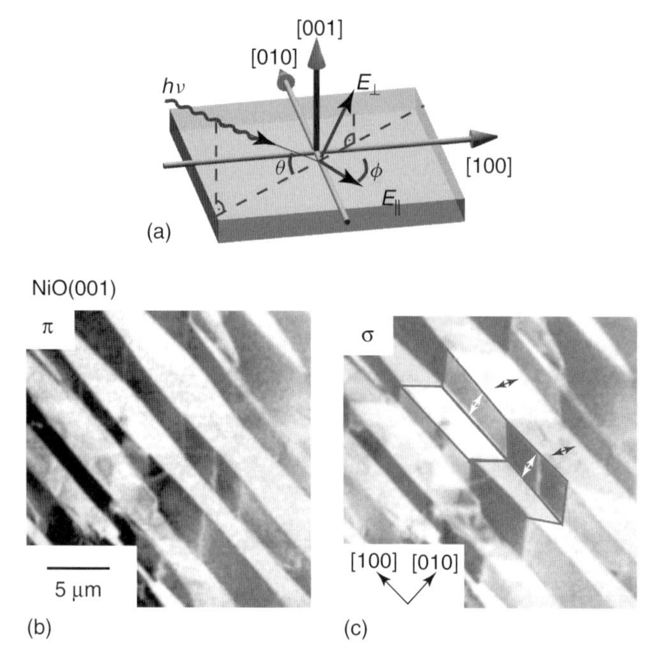

Figure 8.5 (a) X-rays are incident onto a NiO(001) surface at an angle θ. The incident polarization can be split in two main polarization components: $E_{\parallel}$ and $E_{\perp}$. In addition, the sample can be rotated around its normal by an angle ϕ, which is determined by the angle between $E_{\parallel}$ and a particular in-plane sample direction. Note that the angle of incidence θ also appears between $E_{\parallel}$ and the surface normal.Two antiferromagnetic domain images acquired using linear polarization π (b) and mixed σ polarization (c) at the Ni L_2 resonance. The crystal axes and the dimensions of the image are indicated in the insets. Only (100) domain walls are observed in (b), whereas (c) also shows (110) walls. The direction of the antiferromagnetic axis parallel to [121] is indicated by arrows in (c).

that the $E_{\parallel}$ component includes any angle ϕ with a particular crystallographic in-plane axis, for example, [100].

Figure 8.5b and 8.5c shows an example of two antiferromagnetic domain images obtained with the ALS PEEM2 using linear π and circular (or mixed) σ polarization. Both the images were acquired at the same spot on the sample. The sample was a NiO(001) single crystal that had been annealed in an oxygen atmosphere of 10^{-6} torr for several hours after cleaving and before imaging. Figure 8.5b shows the distribution of the antiferromagnetic axes imaged with linear polarization (π). Because the electric field vector $E_{\parallel}$ lies completely in the surface, only the in-plane projection of the antiferromagnetic axis contributes to the observed contrast. The pattern itself is governed by stripe-shaped antiferromagnetic domains parallel to [100] directions and separated by (100) walls. We can, therefore, assume that we observe a pure T-domain pattern as described in the previous section. In addition, to address the out-of-plane contribution of the antiferromagnetic domain pattern, we then "switched on" the out-of-plane component of the polarization by using σ-polarized X-rays. The resulting PEEM images now show four different antiferromagnetic domains. Each in-plane domain is split into two domains with different projections of their antiferromagnetic axis on the vertical polarization vector $E_{\perp}$. In other words, each of the antiferromagnetic in-plane domains separated by domain walls along (100) is split into two different antiferromagnetic out-of-plane domains. These two out-of-plane domains are separated by (110) walls. The direction of the magnetic axis in each antiferromagnetic domain has been determined from the line shape of the X-ray absorption spectra acquired in each AF domain and is indicated in the image by arrows. We find that only S-domains are observed with a large projection of the antiferromagnetic spin axis onto the sample surface plane. The spin direction in the four T-domains is given by the four possible [121] directions.

As mentioned in the introduction, the samples had been annealed above T_N at 600 K in vacuum after cleaving. Annealing causes the atomic arrangement at the surface to relax. As a result, no sub-S-domain structure is observed and only a T-domain structure with highly symmetric domain walls that altogether form the herring bone pattern as shown in Figure 8.5 is observed. Nevertheless, it is also evident that the surface is not completely relaxed. In particular, in Figure 8.5a, we see that the (100)-walls are not completely straight and parallel to the [100] direction; instead they are slightly curved. This may be simply attributed to the way the sample is mounted in the microscope, or even more likely, due to the (still) relatively low annealing temperature of 600 K, which is not high enough to sufficiently heal defects in the bulk of the samples, causing a deformation of the domain wall planes.

To further probe whether the assignment of the antiferromagnetic axis is correct, we focus on the domain walls between the observed T-domains. Domain walls in antiferromagnets are often described on the same basis as the domain walls in ferromagnetic materials. The analogy arises from the fact that on a macroscopic scale, both structures can be characterized by a single vector, that is, the magnetization for ferromagnets or the difference between the two sublattice magnetization for the antiferromagnet. For antiferromagnets, however, a wider configuration of domain walls is possible than for ferromagnets. This is because for antiferromagnetic crystals magnetic ordering is not only related to the appearance of a new magnetic vector but also to a change in the crystal translational symmetry as pointed out previously. As the properties of domain walls provide a key for understanding the magnetic microstructure [8], experimental determinations of their properties are extremely important.

Figure 8.6 shows three different images of the domain structure acquired with different orientations of the incoming X-rays. These images were obtained from the same sample as before but from a different area on the surface. Figure 8.6a was obtained in the same geometry (using σ-polarized X-rays) as the one in Figure 8.5, but the image has been rotated such that the [100] direction is vertical. The other two images were obtained using both polarizations (σ and π) after rotation of the sample by 25°. Figure 8.6a also shows four different contrast

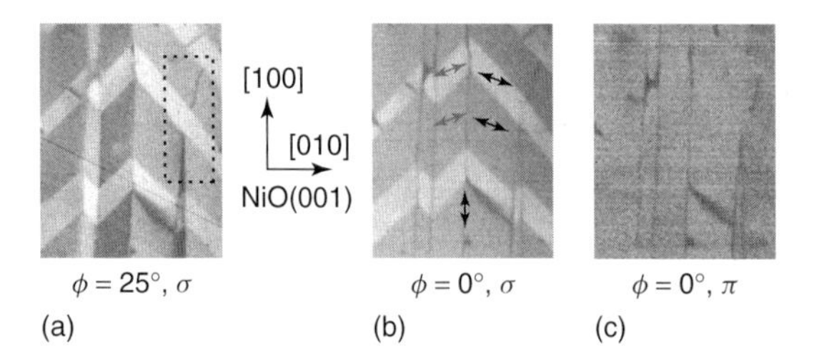

Figure 8.6 (a) The image has been acquired with σ-polarized X-ray, incident at 25° with respect to the in-plane [100] direction on a NiO(001) surface. As shown in Figure 8.5, all four different T-domains are visible. (b, c) The two images were taken with the incident direction parallel to the [100] axis but with different polarizations to reveal out-of-plane as well as in-plane antiferromagnetic domains. The horizontal dimension of the image is 8 μm. All images were obtained at the Ni L_2 resonance.

levels, forming the typical herring bone pattern. However, we also observe an interesting deviation from this pattern. One subset of alternating T-domains forming the herring bone vanishes in a singularity (see highlighted area in Figure 8.6). Surprisingly, the domain walls that separate the original T-domains from the adjacent domains on the left and on the right continue beyond the singularity upward in the image representing an antiferromagnetic analog to a ferromagnetic 360° wall, which even extends beyond a few alternations of the remaining herring bone.

After rotation of the sample, the contrast between domains vanishes completely when the image is obtained with pure linear polarization parallel to [010]. This behavior indicates that the projection of the antiferromagnetic axis onto the [010] direction is the same for all four domains and corroborates our previous findings that all T-domains are formed by S-domains belonging to one subclass. All T-domains are made up of [121]-type S-domains as indicated by the arrows. It also becomes evident that the observed (110) domain walls separate domains with identical in-plane axes but different out-of-plane orientations of the spin axis, for example, [121] and [$12\bar{1}$]. The only difference that remains in the domain image (Figure 8.6b and 8.6c) is the domain wall contrast. T-domains in Figure 8.6 exhibit a spin axis with the same projection on the [010] axis or the [001] axis. However, the spin axis in each domain is either rotated clockwise or counterclockwise to the same extent with respect to the wall plane so that the domains themselves do not create any XMLD contrast. Only the domain wall itself, where the antiferromagnetic spin axis rotates from one domain to the other, can be observed using XMLD contrast.

Finally, the structure of the domain wall is discussed. Antiferromagnetic domain walls are often much wider than typical ferromagnetic domain walls, which are only a few nanometers wide. For example, in a 180° ferromagnetic domain wall, the direction of the spins within the wall will be reversed. The length scale on which this can happen is essentially the length of a spin wave in a ferromagnet. Wider domain walls can also appear; however, they are most often observed in nonequilibrium situations. The walls in antiferromagnetic NiO are more of the order of 120 nm, in the case of the 360° wall even close to 200 nm, so they can be easily resolved in experiments [39]. Antiferromagnetic domain walls are wider because the formation of a domain wall requires that the different lattice contractions on each side of the domain wall be compensated for in the wall plane. This involves physically moving atoms against their equilibrium positions, which can only be accomplished over several hundreds of lattice sites. A more detailed description of the theory behind antiferromagnetic domain walls on NiO and their correlation with magnetoelastic fields can be found in [9, 39][4].

8.2.2
Magnetism and Crystallography

The final step in analyzing the antiferromagnetic domain structure of NiO is to correlate the observed magnetic domains with the underlying crystallographic domains. So far, we have only assumed that the observed domain pattern in Figure 8.5 is a one-to-one match with the T-domain structure, but we do not have a concrete experimental proof for such an assumption. For this reason, it is important to revisit Figure 8.1. As can be seen, each Ni atom possesses a magnetic moment that is antiparallel exchange coupled with its second nearest neighbors along [100]. The oxygen atom, however, does not possess a magnetic moment. It is surrounded by the same number of "spin-up" Ni atoms as "spin-down" Ni atoms. For this reason, the oxygen X-ray absorption spectrum should not show a magnetic linear dichroism. However, we also saw that the crystal undergoes a rhombohedral distortion, which leads to a deviation away from the simple cubic coordination of the oxygen atom. It is well known that such a noncubic atomic arrangement will, in turn, lead to the observation of an X-ray linear dichroism (XLD) effect, since, simply put, the optical properties (e.g., absorption coefficient) are now dependent on the crystal direction. This type of linear dichroism is widely used for the orientational characterization of polymers for example [16, 17], and has previously been found to be applicable also to T-domains in NiO [40].

This additional type of dichroism can now be used to separately address the antiferromagnetic and crystallographic domains in NiO(001). Figure 8.7a shows an antiferromagnetic domain image, acquired from a sample that was annealed at 450 K after cleaving. A very rich domain structure is observed, and it is still possible to identify some domain walls that seem to run straight along defined crystal axis. However, most of the domain walls exhibit a strong curvature and are not correlated with crystallographic axis as it would be expected for pure S-domain walls. To ascertain the T- and S-domain walls, we show XLD images obtained at the oxygen K resonance [38]. Figure 8.7b and 8.7c was acquired using π polarized X-rays and σ-polarized light, respectively. The difference between the two images is the same as in the case of the Ni XMLD images in Figure 8.5; one image reveals

4) The paper by Weber and coworkers is based on the original XMLD interpretation [39]. Although the absolute direction of the spins as assumed in this work might not be correct, the relative orientation is still true so that the findings and conclusions are still valid.

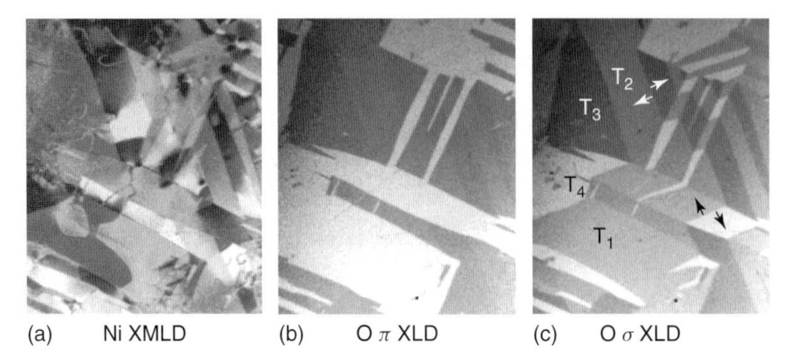

Figure 8.7 A complex domain structure consisting of twin and spin domains is observed using Ni L-edge XMLD (a) on a cleaved NiO(001) surface. Oxygen X-ray linear dichroism (XLD), which is only sensitive to the T-domain structure, is then used to image the underlying T-domain pattern. (b, c) O K-edge XLD images obtained with π-polarized X-rays show T-domains with different projection of the contraction axis onto the surface plane (indicated in (c) by arrows), while images obtained with σ polarization reveal the arrangement of all four different T-domains. The horizontal extent of the images is 12 μm.

only T-domains with a different in-plane component of the contraction axis, whereas the other one shows the different out-of-plane components. Consequently, Figure 8.7c shows four different gray scales corresponding to four different T-domains separated by (100) and (110) walls. We may assign a particular T-domain to each observed crystallographic domain by employing our knowledge about the T-domain structure in the bulk presented in Table 8.2. The T-domains and their corresponding in-plane projection of the contraction axis are shown in Figure 8.7c.

Later in the chapter, this approach is used to address the origin of the spin orientation that is observed at the surface of NiO upon deposition of a ferromagnetic layer. It is a very powerful technique, in particular, in a microscope, since it allows to correlate magnetic and crystallographic order and, even more important, it enables us to observe how changes in the crystallographic order will lead to changes in the magnetic domain structure. This is, in particular, relevant for studies of systems where the correlation between different degrees of freedom is of interest, like mulitferroic materials for instance.

8.3
Antiferromagnetic Domains in Exchange-Coupled Systems

We now focus our attention on the evolution of the antiferromagnetic domain pattern after deposition of a ferromagnetic layer. Antiferromagnetic–ferromagnetic-exchange anisotropy plays an important role in modern-device applications, since the exchange coupling between the ferromagnet and any excess moment in the antiferromagnet – caused by symmetry breaking at the interface or through defects

– can lead to an effect called *exchange bias*. As long as these excess moments are strongly coupled to the antiferromagnetic matrix such that they are not affected by (conventional) external fields, they can, in turn, introduce a well-defined magnetization direction into the ferromagnet through exchange coupling. Such a bilayer will always exhibit its preferred magnetic direction (unidirectional anisotropy) when the external magnetic field is removed and, hence, serve as a magnetic reference in a magnetic storage or sensor device. The orientation of the unidirectional anisotropy can be defined during growth or by annealing the system above the Néel temperature (T_N) in a magnetic field. Independent of the existence of a unidirectional anisotropy, the exchange phenomena between the two layers will affect the observed uniaxial anisotropy of the ferromagnetic layer. In both cases, it is necessary to obtain a fundamental understanding of the micromagnetic structure. The domain structure is determined by the anisotropy energies. In addition to the interface itself and crystallographic defects, the domain walls and crystallographic boundaries themselves are origins of the uncompensated moments that contribute to the exchange coupling between antiferromagnets and ferromagnets.

8.3.1 Antiferromagnetic–Ferrmagnetic-Exchange Coupling

The general characteristic of the domain pattern in exchange-coupled Fe/NiO and Co/NiO is described in this section. Figure 8.8 shows images of the ferromagnetic as well as the antiferromagnetic domain patterns after deposition of 10–12 ML of Fe or Co on NiO(001). The ability to tune the X-ray energy either to the NiO L_{2a}/L_{2b} or Co/Fe L_3/L_2 absorption resonance and to vary the polarization of the X-rays allows one to study the magnetic structure of the top ferromagnetic layer as well as the buried NiO surface. Although the NiO signal is suppressed by a factor of 4–5 due to the limited escape length of the secondary electrons, the domain pattern of the buried antiferromagnet can still be easily identified. The lateral scale is rather large in order to illustrate that the observations are not restricted to a small local area on the sample.

Both XMLD images in Figure 8.8 exhibit two gray levels, representing two different antiferromagnetic domains. These domains are separated by (100) or (010) walls. The XMCD images, on the other hand, show four different gray scales corresponding to four different ferromagnetic domains. These four ferromagnetic domains split up into two different subgroups with each subgroup following one particular antiferromagnetic domain, such that antiferromagnetic (100) walls in the XMLD images are coincident with ferromagnetic (100) walls in the XMCD images. Considering only these walls, the ferromagnetic domains form an exact replica of the antiferromagnetic domains.

The contrast arising from the different magnetic domains is now analyzed qualitatively. For this purpose, legends are shown in Figure 8.8, which correlate the observed dichroism intensity with the magnetic orientation. Image areas showing strong bright and dark XMCD intensity represent domains where the projection of the spins onto the helicity of the light is large, meaning that the magnetic moments

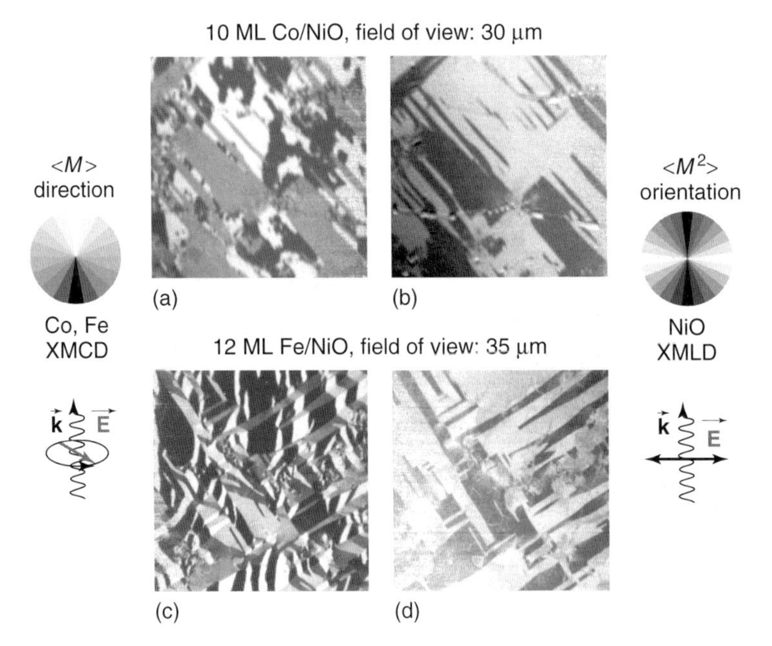

Figure 8.8 XMCD images (a) and XMLD (b) images of 10 ML of Co/NiO and 12 ML of Fe/NiO(c, d) obtained using the respective L absorption resonance. X-rays are incident along the vertical image direction. The legends on the side correlate the observed gray levels to the direction of the magnetic moment in each ferromagnetic domain and the orientation of the magnetic axis in each antiferromagnetic moment. Perpendicular alignment between ferromagnetic moment and antiferromagnetic axis is observed.

are pointing up or down in the image plane. The more grayish domains exhibit smaller projections onto the polarization vector, indicating that their moments are aligned close to the left and right in the image plane. In the XMLD images, antiferromagnetic domains with a more vertical orientation of the magnetic axis are shown in dark gray scales compared to horizontal domains shown in brighter shades. A qualitative conclusion can already be made about the orientation of the exchange coupling between ferromagnetic Fe or Co and the antiferromagnetic NiO(001) surface. The perpendicular alignment between ferromagnetic directions and antiferromagnetic axes is observed in this case. Results obtained from (thick) Fe/NiO and Co/NiO are qualitatively identical. For this reason, results obtained only from the Co/NiO(001) sample are discussed in the following.

Figure 8.9 shows how the direction of the magnetic moments in the ferromagnetic Co domains and in the antiferromagnetic NiO domains is determined. Here, the sample was rotated around the surface normal so that the in-plane component of the electric field vector was parallel to [012], [010], or [011]. The observed XMCD contrast is maximal for one subset of ferromagnetic domains and vanishes for the other subset, when the X-rays were incident along the [110] direction, showing directly that the magnetization in the ferromagnetic domains is aligned along [110]. The same is true for the contrast arising from the antiferromagnetic domains. The

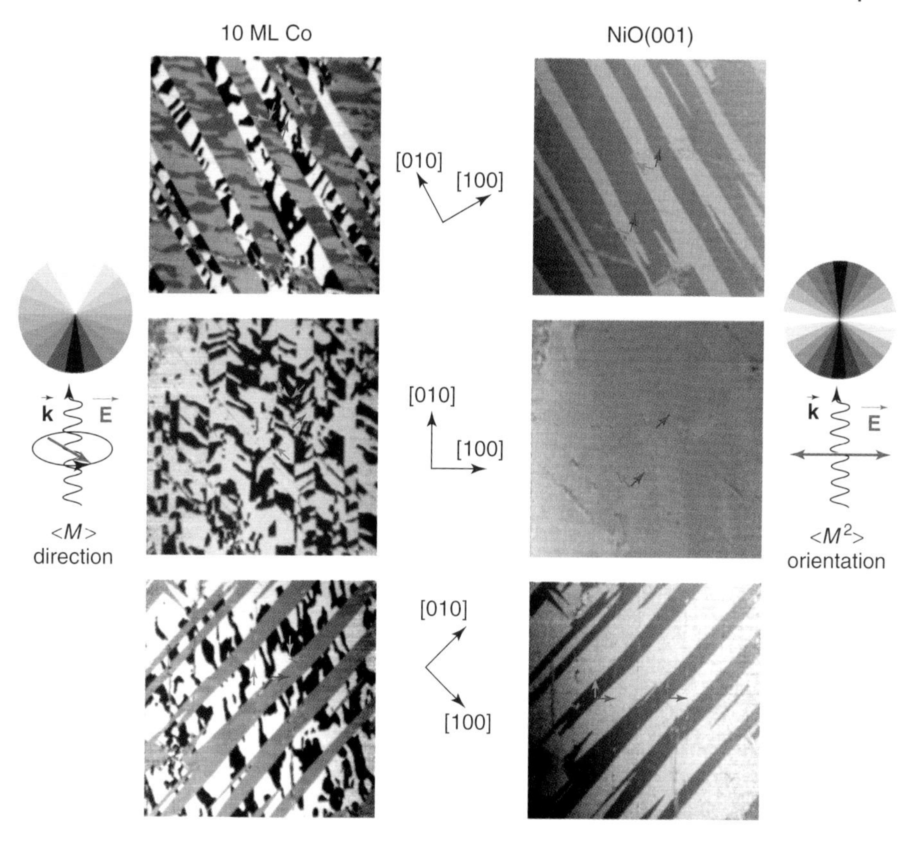

Figure 8.9 XMLD and XMCD images of 2.5 nm on Co/NiO(001) obtained in several different geometries at the Co and Ni L absorption resonance. From the azimuthal dependence of the observed contrast, it can be concluded that magnetic moment in the Co domains is parallel to [110] as indicated by the arrows in the figure. The magnetic axes in NiO domains are oriented perpendicular to the ferromagnetic moments. The field of view is 30 μm.

contrast in the [110] orientation is larger than for the [012] geometry, indicating that the magnetic axis is now closer to [110] than to [121] as before. This is then further corroborated by local X-ray absorption spectra obtained in each domain, which exhibit a value of the XMLD contrast that is very close to what can be expected to be the maximum for NiO. Hence, it appears that the antiferromagnetic spin axis has been rotated from the [121] to the [110] direction upon deposition of the ferromagnetic layer as demonstrated in detail in Figure 8.10. Figure 8.10a shows an antiferromagnetic domain pattern in NiO before Co deposition and Figure 8.10b and 8.10c shows the antiferromagnetic and ferromagnetic domains after deposition of Co. It can be clearly seen that the complex T-domain pattern that is observed before Co deposition with four different T-domains forming a zigzag-like structure is greatly simplified upon deposition of the ferromagnet. Only two domains are still visible, the (110) walls vanish, and only the (100) walls remain.

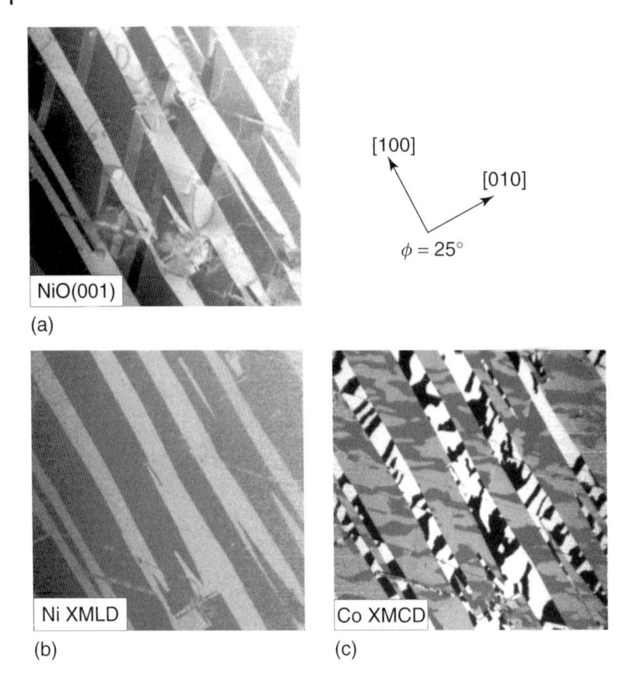

Figure 8.10 (a) Ni L-edge XMLD image of AF NiO(001), which again shows a zigzag-like domain pattern with some mixture of S-domains. We show XMLD images of NiO (b) and XMCD images of Co (c) after deposition of 2.5 nm of Co, obtained at the respective L absorption resonances. The field of view is 30 μm.

Table 8.3 Thermodynamic properties of selected 3d-metals and their oxides.

	Electronegativity	E^0 [V]
Fe	1.6	−0.47
Co	1.7	−0.27
Ni	1.8	−0.23

The table lists the electronegativity for Fe, Co, and Ni as well as their reduction potential $M^+ + e^-$.

8.3.2
Magnetic Domains at Interfaces of Antiferromagnets with Ferromagnets

To fully understand what happens upon deposition of a ferromagnetic metal onto an antiferromagnetic oxide surface, we briefly look at the chemistry of such an interface. One may already intuitively expect that an interface between a metal and an oxide is not sharp, because chemical forces that are not negligible at $T > 0$ might drive oxygen atoms to diffuse from the oxide into the metal layer. Table 8.3

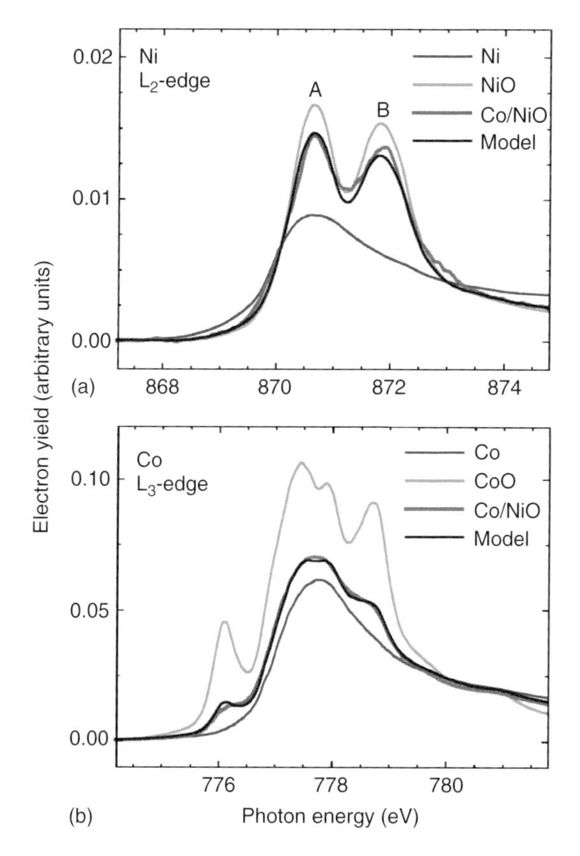

Figure 8.11 Absorption spectra obtained from (1 nm)Co/(1 nm)NiO in Co/NiO (red). These are compared to the reference metal (blue) and oxide (green) spectra. The black spectra show a fit to the (red) Co/NiO spectra assuming that 1 ML of Co is oxidized and 1 ML of NiO is reduced at the interface (reproduced with permission from Ref. 28).

lists the electronegativity and reduction potential of the three investigated species to illustrate this assumption. The electronegativity is a measure for the ability of a particular species to attract electrons. Fe and Co exhibit a lower electronegativity than Ni, which means that in the presence of O, Fe and Co will be predominantly oxidized compared to Ni. This is also reflected in the reduction potential, which is the lowest for Ni. These elemental data show the general tendency of Co or Fe to reduce NiO if they are in close contact.

The chemical reaction at the interface can be visualized by employing the elemental and chemical sensitivity of XAS. For this reason, Co L_3 and Ni L_2 XAS spectra were obtained from a (1 nm)Co/(1 nm)NiO sample as shown in Figure 8.11. The reference spectra for the metal represented in blue and for the bare oxide surface in green are also shown. The spectra obtained from the Co/NiO sample is shown in red. The XAS intensity for the Ni L_2 edge appears to be decreased compared to the oxide sample, indicating the increasing metallic character of Ni atoms in the film. On the other hand, the spectrum obtained at the Co L_3 resonance exhibits a multiplet character and increased peak intensity. Both observations are an indicator of the more oxidic character of the original metallic Co film.

The lineshapes observed for the unknown samples suggest that their XAS spectra can be represented as a superposition of the pure metal and the oxide spectra. To corroborate this assumption, the last spectrum in the graph (black) shows a linear combination of metal and oxide spectra to best fit the unknown spectra. An excellent agreement is achieved demonstrating that indeed at the interface a certain amount of Co is oxidized to CoO and NiO is reduced at the same time to Ni. The thickness of the interfacial layer consisting of $CoNiO_x$ can be derived from the weighting factors used for the fit. Considering an electron escape length of 2.5 nm, one estimates that about a monolayer of metallic Ni is formed on top of the NiO and a monolayer of CoO at the bottom of Co. Therefore, the thickness of the interfacial $CoNiO_x$ layer is estimated to be about 2 ML [14, 28].

For Fe/NiO, a stronger initial interfacial redox reaction has been reported [14] due to the larger difference in the reduction potential. However, the identification of the particular oxide which is realized at the interface is more difficult, because iron forms different stable oxides such as Fe_2O_3, Fe_3O_4, or FeO at room temperature. For this reason, it is not surprising that the coupling that is observed between NiO and Fe depends on the amount of Fe that is available for the reaction.

Figure 8.12 shows XMCD images of ferromagnetic Co and Ni as well as an XMLD image of antiferromagnetic NiO. Images were acquired in two different orientations of the sample. The legends on the left- and right-hand side of the figure show the direction of the incident X-rays, the polarization vectors, and the correlation of image gray levels with the orientation of the magnetic moment within each domain. The Co XMCD images reveal three different contrast levels representing parallel, antiparallel, and perpendicular alignment of the spin moment with respect to the helicity of the light. In addition, XMLD contrast arising from antiferromagnetic domains in NiO can be seen and the perpendicular alignment between antiferromagnetic NiO and ferromagnetic Co is again observed. By employing the Ni XMCD effect, the interfacial spin polarization finally becomes visible. Again three different contrast levels are observed as in the Co XMCD images. The spatial distribution of the ferromagnetic Ni domains is a complete replica of the Co ferromagnetic domain pattern, demonstrating that the Ni XMCD contrast is the result of metallic Ni moments ferromagnetically exchange coupled to the Co domains. However, the relative contrast levels are different in the Co and the Ni XMCD images. This is due to the fact that the very small XMCD signal (0.3%) is still superimposed on the XMLD effect, which does not exactly cancel out[5].

5) In principle, the XMLD asymmetry is identical at the L_3 and L_2 NiO resonances. Howvever, since the line shape is slightly different, it does not cancel out completely.

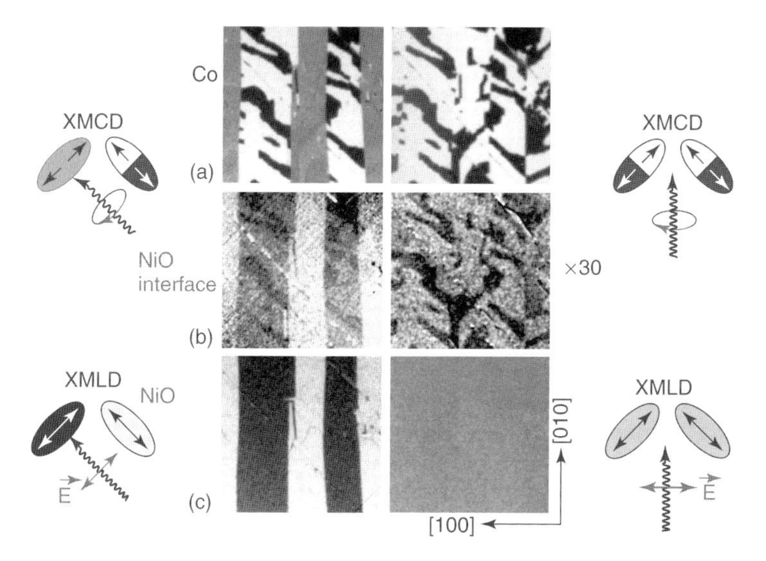

Figure 8.12 XMCD images of the ferromagnetic Co layer (a) and the uncompensated interfacial Ni moments (b) as well as XMLD images of the antiferromagnetic domains of the NiO(001) surface (c) Images were obtained at the L absorption resonances in two different geometries to show the parallel alignment of uncompensated interfacial spins and ferromagnetic Co spins.

For this reason, a second set of images were acquired in another geometry in which no XMLD contrast was expected. These are shown on the right-hand side of Figure 8.12. The angle between antiferromagnetic axes and electric field vector is the same for both antiferromagnetic domains and they cannot be distinguished. In this geometry, the Ni XMCD contrast is not superimposed on the residual XMLD contrast and the observed domain pattern for the interfacial Ni moments forms an *exact* replica of the domain pattern observed for the Co film, and parallel coupling between interfacial Ni and Co along [110] is also observed. The thickness of the $CoNiO_x$-like layer is estimated from the size of the XMCD contrast. The XMCD contrast between two domains calculated from the image intensity is 0.5% as shown in Figure 8.12. Taking into account the particular geometry of the experiment, the corrected XMCD asymmetry between opposite domains can be calculated to be 1% in the *as-prepared* state. The maximum XMCD value expected for metallic Ni is 16% [41]. Assuming that the electron escape length in NiO is 2.5 nm, the first monolayer contributes to the total signal to the extent of 8%. Hence, an XMCD signal of 1.3% can be expected from a full monolayer of Ni on top of NiO. This leads to the conclusion that about 0.75 ML of Ni or 1.5 ML of $CoNiO_x$ is produced as a result of the interfacial chemical reaction. Assuming layer-by-layer reduction of the NiO, the following general expression is derived to calculate the thickness of the metallic nickel layer from the corrected XMCD difference I_{XMCD}, which is the XMCD ratio observed in the image as defined in Figure 8.4.

$$x = -2.5\ \mathrm{nm} \ln\left(1 - \frac{I_{XMCD}}{16\%}\right) \tag{8.2}$$

A thickness of 0.75 ML for the *as-prepared* sample is in accordance with results obtained on thin film samples by Regan *et al.* [14]. The observed thickness of the $CoNiO_x$-like interfacial layer is similar for single crystalline or polycrystalline samples. This shows that the thickness of the interfacial layer can be extracted from the size of the XMCD contrast arising from the metallic Ni atoms located at the interface as well as from a linear combination of (nonmagnetic) metal and oxide reference spectra.

The question remains whether the interfacial CoO is antiferromagnetic or forms a magnetically homogenous ferro(i)magnetic layer together with the Ni atoms. In other words, the question is whether the final sample structure is $Co/CoNiO_x/NiO$ or Co/CoO/Ni/NiO. Although T_N of bulk CoO is below room temperature (290 K), a thin film of CoO will show magnetic order since it is exchange coupled to the other magnetic layers with a much higher ordering temperature [42]. To demonstrate this, we show XMLD images from 2 ML of CoO directly deposited on NiO as shown in Figure 8.13. The CoO layer was deposited on a clean NiO surface in a 10^{-6} mbar oxygen atmosphere using a Co electron beam evaporator at a deposition rate of about 0.5 ML per minute. Comparing the XMLD images of CoO and NiO, the consequences of the exchange coupling between the two layers and the antiferromagnetic order of the thin CoO film at room temperature can be clearly seen. The CoO domain pattern is identical to the NiO domain pattern. The XMLD asymmetry between adjacent CoO domains is about 2%. A similar effect was never observed in the $CoNiO_x$ interfaces, although, in principle, dichroism effects of less than 0.5% can be observed. This leads to the conclusion that the interfacial CoO is not antiferromagnetically ordered but rather forms a single magnetic layer together with the interfacial Ni metal spins that does not exhibit an XMLD effect as CoO.

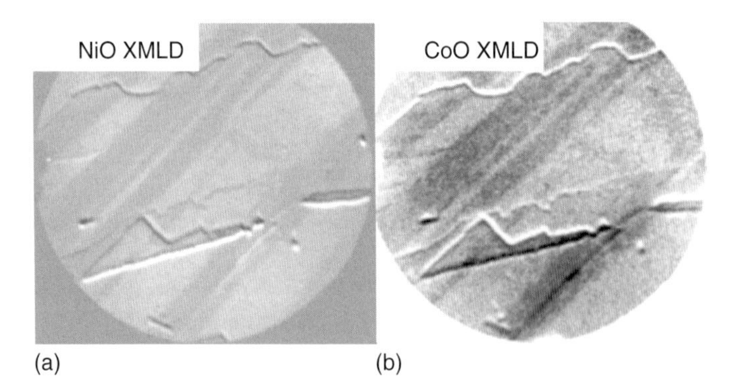

Figure 8.13 XMLD images of antiferromagnetic domains in a 2-ML CoO/NiO bilayer. The field of view in these images acquired with the STAIB-PEEM is 150 μm. An identical T-domain pattern with domain walls along (100) planes can be identified in both images. The images were obtained at the Ni (a) and Co L (b) absorption resonances.

8.3.3
Origin of Spin Reorientation

We have seen that a ferromagnetic layer of Co or Fe exhibits a perpendicular alignment with the underlying antiferromagnetic axis at the (100) plane. This is because the NiO(001) surface is magnetically compensated, which means that the direction of the spins on adjacent lattice sites at the surface is opposite, so that the ideal surface does not exhibit a magnetic moment. We refer to the (100) surface as a magnetically compensated surface, in contrast to (111) terminated surface which exhibits a macroscopic magnetic moment, since the (111) planes in NiO are ferromagnetically ordered. The observed perpendicular alignment between the two layers can be understood if we consider that a single Co atom on a NiO surface is surrounded by as many Ni atoms in a "spin-up" as in a "spin-down" state. Assuming that all the next nearest Ni atoms are at roughly the same distance to the Co atom, the size, and in particular the sign, of the exchange energy between the Co site and all the surrounding Ni sites will be identical. For the purpose of this argument, we may assume that it is ferromagnetic. However, since half of the Ni spins point in the opposite direction, the Co site cannot align itself with one half of the spins without paying the energy to align opposite to the other half. For this reason, it will find its minimum energy position by perpendicular alignment. This frustrated alignment is often referred to as *spin-flop* coupling and is a direct consequence of the abrupt change in the sign of the magnetic exchange energy across the compensated antiferromagnetic/ferromagnetic interface.

The origin of the reorientation of the antiferromagnet is more delicate to investigate. Initial reports on the coupling between Co and NiO based on the original interpretation of the XMLD effect concluded that the coupling is parallel and that it is mediated by the uncompensated Ni spins. The magnetization of the Co layer needs to be aligned parallel to the surface plane, because the ferromagnetic film has a strong shape anisotropy to minimize any stray field emerging from the sample. Since the antiferromagnetic axis of the compensated surface was believed to be coupled parallel to the ferromagnetic layer through uncompensated Ni moments, the antiferromagnetic axis should also be aligned parallel to the surface plane. However, now we know that the coupling between the two layers is not mediated by exchange coupling but rather by spin-flop coupling corresponding to the magnetically frustrated interface. This leaves us with the question as to why the axis of the antiferromagnet rotates into the plane of the interface. Again, the ability to address crystallographic and magnetic domains separately proves to be an invaluable tool to solve this question.

Figure 8.14 shows Ni XMLD, Co XMCD, and O XLD images of a bare NiO(001) surface and of a Co/NiO bilayer. The first row shows a similar spin and twin domain arrangement as before, except for the fact that only three twin but four spin domains are observed in this particular geometry, because the X-rays are incident along the [110] direction. Upon deposition of a thin Co film (2.5 nm) and formation of a thin-interface-mixed oxide layer, only 2 S-domains remain at the NiO surface due to the reorientation of the antiferromagnetic axis. For example,

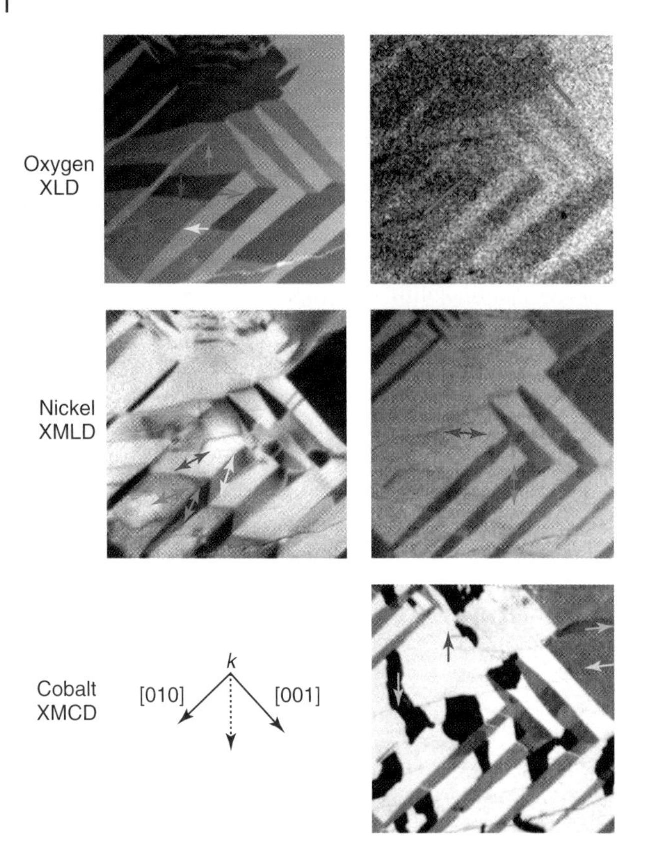

Figure 8.14 The left row shows dichroism images obtained for the bare surface while the right row shows images obtained from the same spot after deposition of a thin ferromagnetic Co film. The orientation of the antiferromagnetic spin axis and the direction of the ferromagnetic moments in each domain are indicated in the Ni L-edge XMLD and Co L-edge XMCD images. Upon Co deposition, (100) domain walls vanish in the antiferromagnetic and in the O K-edge XLD images (reproduced with permission from [38]).

we find that the domain wall between the "red" and the "blue" antiferromagnetic domains vanishes, whereas the domain wall between the red and the green domain remains. If the uncompensated Ni moments caused by the chemical reduction at the interface are not the reason for the reorientation, then maybe the oxidized Co sites play a role. For this purpose, we take a close look at the contrast between the corresponding T-domain walls observed with O XLD. The overall contrast in this image is now strongly reduced due to the Co layer on top that reduces the electron yield arising from the NiO layer.

A detailed analysis of the contrast across the domain walls reveals a qualitatively similar development of the O XLD contrast compared to the Ni XMLD contrast. Although the contrast across the red/blue domain wall does not completely vanish, it is still much strongly reduced (by about a factor of 3) than the contrast across the red/green domain wall after deposition of the ferromagnetic Co layer. This observation indicates that the crystallographic T-domains undergo a structural reorientation in a similar manner as the antiferromagnetic domains undergo a spin reorientation. Such a behavior can be readily explained with an interfacial lattice distortion of NiO toward a CoO-like structure. The (001) surface of CoO would exhibit only two T-domains as observed here due to the tetragonal distortion of the CoO lattice in contrast to the rhombohedral distortion of the NiO lattice [43]. This means that the observed rotation of the axis from [121] to [110] is caused by the change in magnetic anisotropy from a situation that is typical for NiO toward a CoO-like symmetry with an easy magnetic axis of [110] [38].

8.4 Temperature Dependence of the Antiferromagnetic Domain Structure

In this section, the question addressed is how the observed domain pattern on NiO(001) and Co/NiO(001) evolves if the temperature is increased above T_N. The behavior close to T_N is of special interest, because it is a common observation that the exchange bias field disappears at the so-called blocking temperature T_B significantly below T_N [44]. It will be demonstrated that the ability to distinguish between antiferromagnetic and ferromagnetic order gives interesting new insights into the origin of T_B and how it can be related to the antiferromagnetic domain structure. For this experiment, 30–40 XMLD or XMCD images were acquired while the sample was slowly annealed from room temperature to 600 K. XMCD images are the results of images acquired with circular polarization at the Co L_3 and L_2 absorbtion edges. XMLD images were obtained using linear polarized X-rays at the NiO L_{2a} and L_{2b} resonance.

The dependence of the dichroism signal on the magnetic properties is reviewed first. In the case of NiO, this can be done in a straightforward manner, because the temperature dependence of the XMLD in NiO depends only on the magnetic state of the sample (see also Chapter 2). We present a series of XMLD images obtained in a fixed experimental geometry, which allows us to neglect contributions from anisotropic XMLD [37] to the total XMLD signal. The XMLD intensity depends on the thermodynamic expectation values of the magnetic moment $<M>$ for the ferromagnet or its square $<M^2>$ in the case of an antiferromagnet. The second term denotes the dependence on the angle between the relevant polarization vector and the magnetic direction or axis.

$$I_{\mathrm{XMCD}} \sim < M > (\boldsymbol{\sigma} \cdot \mathbf{M}) \tag{8.3}$$

$$I_{\mathrm{XMLD}} \sim < M^2 > (\mathbf{u}_E \cdot \mathbf{M})^2 \tag{8.4}$$

I_{XMCD} and I_{XMLD} refer to the XMCD and XMLD intensities observed in the images as defined in Figure 8.4 and $\mathbf{u}_E$ describes a unit vector collinear to the electric field vector of the incoming X-ray beam. If a dichroism image is acquired, the intensity of every image point is given by these expressions, and changes in the dichroism intensity may be attributed to two different effects:

1) the thermodynamic average of the magnetization $<M>$ or its square $<M^2>$ changes and
2) the direction of $<M>$ or its square $<M^2>$ changes.

A decrease in the thermodynamic average will be observed if the temperature is increased. For the temperature range considered here (300–600 K), the average of the magnetic moment $<M_{Co}>$ of the ferromagnet will decrease linearly by a very small amount, because the Curie temperature is much higher (1398 K). The temperature dependence of $<M^2>$ is more complicated and is denoted in the following, derived from a molecular field approximation [45]:

$$< M^2 > = g^2 \mu_B^2 J(J+1)) - g\mu < M > \coth(g\mu_B H/2kT) \tag{8.5}$$

In this equation, g denotes the gyromagnetic ratio, J the quantum number of the total spin (1 for a $3d^8$ system)[6], μ_B the Bohr magneton, k the Boltzmann constant, H the molecular field and $<M>$ the expectation value of the moment of the antiferromagnet. It can be shown that above $T > 0.6T_N$, the thermodynamic expectation value $<M^2>$ decreases linear with temperature. This behavior is shown, for example, by the straight line as shown in Figure 8.15. If the experiment does *not* show a linear decrease in the XMLD signal, consequently the second origin for a change in the XMLD intensity has to be considered and a reorientation of the magnetic moment or the magnetic axis needs to be assumed. In the following, a distinct deviation from the linear dependence is observed in NiO and Co/NiO. The deviation reaches its maximum in a temperature region that is typically attributed to the vanishing exchange anisotropy (0.85–0.95 of T_N), which occurs at the blocking temperature (T_B) of the system [44, 46].

Temperature-dependent XMLD images were obtained from a cleaved NiO(001) surface. For each XMLD image, the average XMLD intensity, I_{XMLD}, of the image has been calculated. The result is divided by the value obtained above, T_N, for normalization purposes. In addition, the standard deviation has been calculated from the image data. These two values are now plotted versus the relative sample temperature, T/T_N (Figure 8.15). For comparison, a straight line representing the expected linear decrease for I_{XMLD} is also shown. The temperature dependence of the XMLD intensity deviates significantly from the expected linear decrease. The maximum deviation occurs between 85 and 95% of T_N. Following the considerations in the previous section, this indicates that the direction of the antiferromagnetic axis averaged over the image area changes during the annealing process. The assumption is corroborated by the behavior of the standard deviation, which exhibits a maximum at this temperature range, because the XMLD intensities

6) Note that, in general, g refers to the Landé factor and J is the total angular momentum. However, since the orbital momentum is quenched in 3d metals, we can use these simplified terms instead.

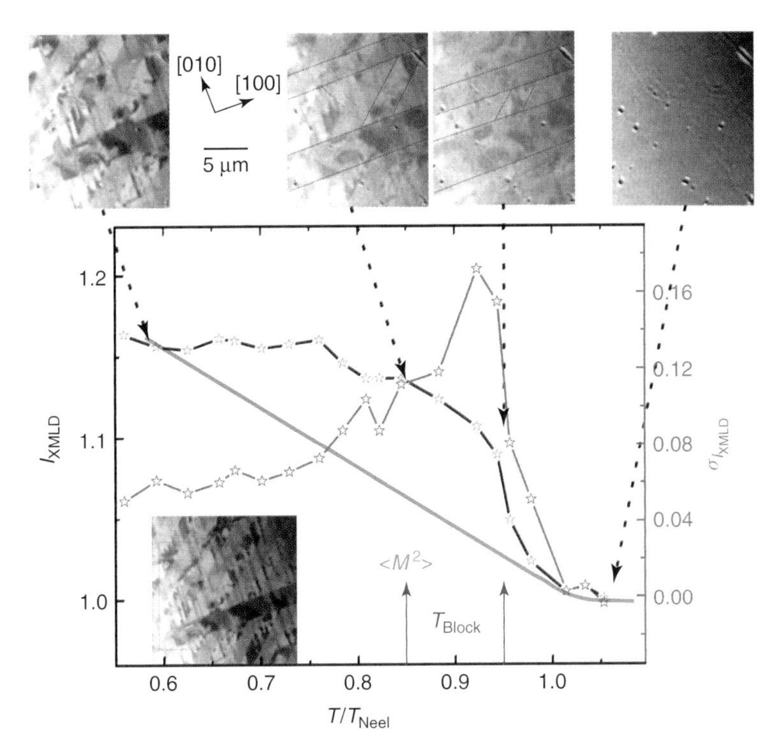

Figure 8.15 The plot shows the dependence of the average XMLD intensity obtained from all pixels in the acquired images and the standard deviation of their intensity distribution. The straight line indicates the expected linear decrease in $<M^2>$ (see the text). The maximum deviation from the linear decrease coincides with a maximum in the standard deviation around T_B. Images for some characteristic temperature values are shown. The black lines in these images serve to visually enhance domain wall features.

in the image are distributed in a larger range around the average value than at room temperature.

This is demonstrated by four selected images from this temperature series. Except for the fact that the overall contrast is reduced, there are only small differences in the domain patterns observed at 300 and 420 K ($0.8 \times T_N$). A small increase in the temperature, that is, to 460 K ($0.87 \times T_N$) leads to drastic changes in the domain arrangement. Most of the (110) walls, which could still be observed at 420 K, disappeared and a new (100) wall appeared. At 540 K, above T_N no XMLD contrast is observed. Annealing the NiO(001) surface does not only lead to a decrease in $<M^2>$ but also leads to a reorganization and reorientation of the antiferromagnetic domain pattern. As a possible reason for the reorganization, consider the fact that the magnetic anisotropy in NiO(001) is closely linked to the crystallographic structure and deformations thereof (Section 8.1.1). As the temperature increases, the deformation of the lattice is directly affected, which leads to changes in the magnetic domain pattern. The observed reorganization processes are especially

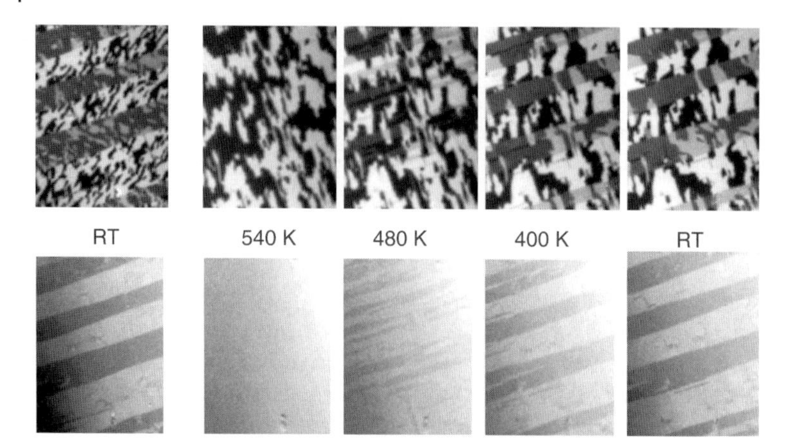

Figure 8.16 Images of the antiferromagnetic and the ferromagnetic domain structure while cooling the sample through $T_N = 525$ K. Four different ferromagnetic and two different antiferromagnetic domains can be identified below T_N. The antiferromagnetic domain pattern disappears above 525 K and is re-established at room temperature after cooling without significant changes. However, a different domain pattern is observed right below T_N. The field of view and the sample orientation are similar to Figure 8.15. The images were obtained at the L absorption resonances of Co and Ni.

pronounced around $0.9 \times T_N$, a temperature that is usually attributed to the blocking temperature of NiO(001) [44].

The same experiment can now be performed after deposition of 10 ML of Co onto a cleaved and annealed NiO(001) surface (see Figure 8.16). The XMCD as well as XMLD images were obtained during annealing and cooling. At room temperature, the same domain patterns as before are observed. Above T_N, the antiferromagnetic contrast vanishes and only two ferromagnetic domains remain. Judging from the XMCD contrast, Co domains with spin directions either left or right in the image plane have vanished in favor of domains with either spins up or down. Second, most of the ferromagnetic domain walls have changed their orientation from (110) at room temperature to a plane that is parallel to the vertical direction in the image. A possible reason for this preferential alignment of magnetic moments within the domains and the domain walls may be the small magnetic field originating from the filament used to anneal the sample. Note that the ferromagnetic domain pattern observed above T_N is partly preserved throughout the cooling process in areas with vertical orientation of the antiferromagnetic axis (Figure 8.16).

The experiments were performed using a NiO single crystal that had gone through several annealing cycles earlier so that the overall characteristics of the domain pattern had already been set. We can, therefore, expect that the initial domain pattern should be reinstated at room temperature, as observed in this case. During the cooling process, however, the antiferromagnetic domain pattern observed at room temperature is not directly reinstated below T_N and additional domains appear around 80% of T_N. At these temperatures, the antiferromagnetic domain pattern can be affected by the ferromagnetic layer on top, while at room temperature the domain pattern of the antiferromagnetic surface has been rigidly established and is anchored deeply in the bulk of the antiferromagnetic single crystal. Similar to the findings on the bare surface, this indicates that reorganization of the antiferromagnetic domain pattern takes place at around 85% of T_N. For this reason, we can correlate the blocking temperature of the NiO(001) surface directly with the temperature at which the competition between antiferromagnetic domain formation driven by exchange with the antiferromagnetic bulk domain structure, on one side, and ferromagnetic domains, on the other side, is maximal. In the case of an exchange-coupled antiferromagnetic/ferromagnetic bilayer, this will be the temperature at which any fixed uncompensated moments, which were originally imprinted into the antiferromagnet, will be annihilated. At this temperature, one will also observe a maximum in the macroscopic uniaxial anisotropy observed through the coercivity of the ferromagnetic layer in the coupled system. At the blocking temperature, not only the magnetization of the ferromagnetic layer has to be reversed, but also a significant amount of domains in the antiferromagnetic layer is affected by the reversal. The reversal of the antiferromagnetic layer will, in general, lead to the formation of an antiferromagnetic exchange spring as predicted by Mauri and Siegmann [47] and observed using synchrotron techniques by Scholl *et al.* [48] and using neutron-based techniques by Roy *et al.* [49]

8.5 Antiferromagnetic Domains and Exchange Bias

In this section, we discuss two different exchange bias systems in which our understanding of the antiferromagnetic microstructure will help us to understand the temperature dependence of the antiferromagnetic/ferromagnetic-exchange anisotropy and the correlation between antiferromagnetic domain size and the exchange bias effect on a microscopic scale.

8.5.1 A Quick Look at a Fluoride

The striking observation made during the temperature-dependent experiments is that the antiferromagnetic domain structure in a single crystal is not static upon annealing. The domain pattern does not simply vanish upon annealing due to the reduction of $<M^2>$, but undergoes topological changes. This behavior is common for ferromagnetic systems but had not yet been observed for antiferromagnets. The fact that exchange bias disappears significantly below T_N, especially for NiO(001), can now be understood, as mentioned. Although the samples grown on single-crystal surfaces do not exhibit exchange bias, these results present strong evidence and represent the very first experimental observation for the fact that the antiferromagnetic domain structure on NiO(001) rearranges significantly below the ordering temperature.

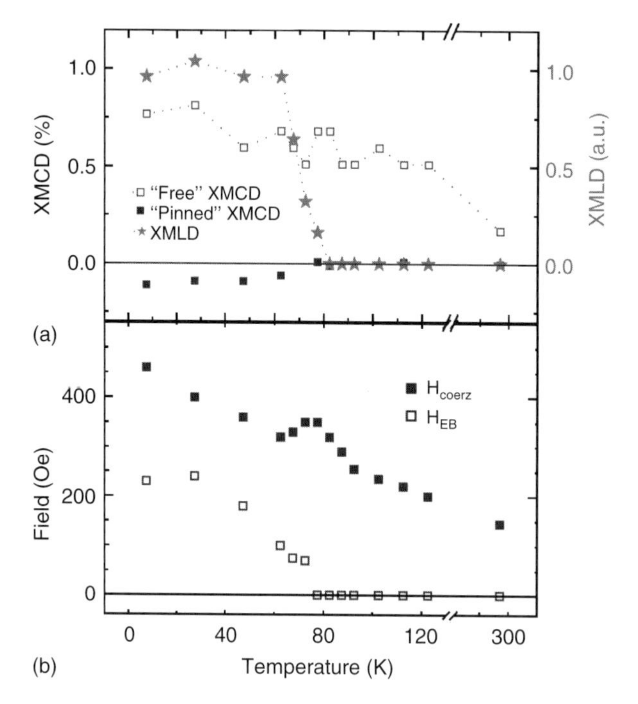

Figure 8.17 (b) This figure shows the coercive and the exchange bias field of a $Co/FeF_2(110)$ sample that exhibits a macroscopic unidirectional anisotropy. The exchange bias vanishes at $T_N = 78$ K. (a) This shows the temperature dependence of the XMLD effect as well as the size of the XMCD effect of frozen or pinned uncompensated moments as well as the so-called "free" uncompensated moments (reproduced with permission from [50]).

To explore this behavior in more detail and to show its relevance for the exchange bias effect, we briefly discuss a study of an exchange biased Co/FeF_2 sample [50]. Note that for FeF_2, T_B and T_N are very close to each other. This is shown in Figure 8.17b, where the temperature dependence of the exchange bias field is plotted, and we find that H_{EB} vanishes at $T_N = T_B = 78$ K. We also observe that the XMCD signal, which originates from the so-called frozen or "pinned" uncompensated moments, vanishes at 78 K, together with the XMLD signal of the antiferromagnetic FeF_2. These pinned moments do not follow an external field and preserve their orientation even if the ferromagnetic layer on top is completely reversed. The pinned moments at the interface between Co and FeF_2 are directly linked to the antiferromagnetic spin structure in FeF_2 and "use" the antiferromagnetic spins as anchor, since the antiferromagnet itself is not sensitive to external fields. Once the antiferromagnetic order vanishes above T_N, the pinned spins are not observed anymore. In addition, we also observe an XMCD signal arising from uncompensated "free" moments, which are located in the interface near region. These moments are relatively strongly coupled to the ferromagnetic Co layer and follow the magnetization of the Co layer. The temperature dependence shows that even at room temperature they give rise to an XMCD signal. The temperature dependence of this XMCD signal decreases only gradually following the temperature dependence of the magnetic moment of the Co layer. This observation provides a possible explanation regarding why, again, we do not observe a linear decrease in the XMLD signal but a rather steep sudden decrease around T_N, as shown in Figure 8.17a. The magnetic order in the topmost layers of the antiferromagnet is stabilized by exchange coupling to the ferromagnetic layer on top and hence can persist at higher temperatures compared to the bare surface. However, once the thermal excitation of the spin arrangement in the bulk of the thick $FeFe_2$ layer increases significantly, the antiferromagnetic order in the interface region breaks up rapidly [51].

8.5.2 Magnetic Reversal Mechanism on the Microscopic Scale

The final section discusses the importance of the micromagnetic structure of the antiferromagnet for our understanding of the exchange bias effect. In the previous section, we have seen that the temperature dependence of the domain pattern in coupled antiferromagnetic/ferromagnetic bilayers can be understood from a micromagnetic point of view. We now discuss the reversal behavior of different ferromagnetic regions on an exchange-coupled $Co/LaFeO_3$ sample. Nolting *et al.* [26] and Scholl *et al.* [52] have shown earlier that in $Co/LaFeO_3$, like in NiO, the ferromagnetic domain structure is an exact replica of the underlying antiferromagnetic domain structure. Figure 8.18a shows XMCD images of ferromagnetic domains on such a $Co/LaFeO_3$ sample. The images were acquired in remanence after applying a magnetic field along the vertical direction in the images. We note that the sample has not been treated any further by field cooling to introduce a horizontal shift of the hysteresis loop[7].

After application of 220 Oe, in either the positive or the negative direction, all ferromagnetic domains, which exhibit an easy axis along the vertical direction, are aligned with the external field. They either exhibit a black or a white XMCD contrast. The gray areas represent ferromagnetic domains, which are aligned along the horizontal direction and are therefore not affected by the vertical magnetic field. If, however, a smaller field of only 100 Oe is applied, we see that only roughly half of the domains reverse their orientation. Local hysteresis loops are obtained by monitoring the XMCD intensity in each domain, which depends on the magnitude of the applied field. The resulting so-called switching loops are shown Figure 8.18b and 8.18c. The loop obtained by averaging over the entire area of interest is symmetric with respect to zero applied field, meaning that the sample does not show exchange bias on a macroscopic scale. Single loops, however, obtained from a single ferromagnetic domain are not symmetric as indicated by

7) For more details on how to prepare samples that exhibit a macroscopic exchange bias effect, see Chapter 5.

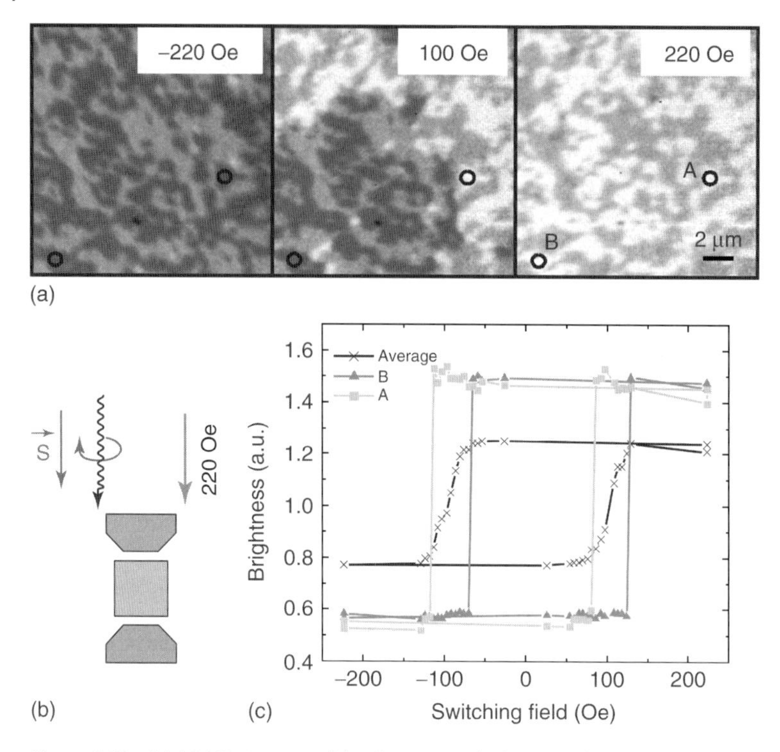

Figure 8.18 (a) XMCD images of the ferromagnetic Co layer in a Co/$LaFeO_3$ bilayer film after application of an (vertical) in-plane magnetic field of −220 Oe, +100 Oe, and +220 Oe. The images were obtained in remanence using PEEM at the Co L resonance. The arrows in (b) indicate the direction of the magnetic moment in each domain. The spatial variation of the exchange bias field is shown in (c). Two of the three hysteresis loops shown in this graph were obtained at locations (A) and (B) in the images marked with a circle and the third loop is the result of averaging over the entire image. (Courtesy of A. Scholl.)

the two examples in the graph. Both loops show an exchange bias effect, where one is shifted to the left and the other is shifted to the right. This means that an antiferromagnetic/ferromagnetic exchange-coupled sample will show exchange bias on a microscopic level in the as-prepared state. However, the magnitude and the sign of the exchange bias fields show a wide distribution, so that no macroscopic effect is observed. Macroscopically, only an increase in coercivity compared to a single ferromagnetic layer can be detected.

Scholl and coworkers, who reported this study in [52], went one step further and analyzed the correlation between the domain size and the absolute magnitude of the exchange bias field. It had been proposed earlier by Malozemoff [53] as well as by Takano *et al.* [54] that the exchange bias field should increase with a decrease in domain or grain size, since uncompensated moments than can be frozen are mostly located at domain and grain boundaries. In this case, the observed ferromagnetic domains are nicely correlated with the antiferromagnetic domains underneath. And indeed the authors found that the larger the underlying antiferromagnetic domain, the smaller the absolute magnitude of the loop shift of the ferromagnetic domain.

8.6 Summary and Outlook

The sections in this chapter addressed the antiferromagnetic domain structure of bare antiferromagnetic surfaces, antiferromagnetic surfaces that are coupled to thin ferromagnetic layers, their temperature-dependent behavior, and the role of the antiferromagnetic domain structure in exchange bias. The chapter addresses the fact that a detailed, quantitative description of the micromagnetic structure of antiferromagnetic surfaces and interfaces is crucial for the understanding of antiferromagnetic ordering and magnetic phenomena of complex coupled systems. We demonstrated that it is possible to characterize the antiferromagnetic domain patterns at surfaces in detail and that because of the correlation between the magnetoelastic energy and antiferromagnetic domains, the surface domain pattern depends on the surface preparation. For this purpose, it is crucial that the correlation between magnetic and crystallographic domains can be made visible by means of nonmagnetic linear dichroism at the oxygen resonance compared to magnetic linear dichroism at the Ni resonance. The use of soft X-ray dichroism spectromicroscopy was even more powerful for the coupled antiferromagnetic/ferromagnetic bilayer samples. Here, the element sensitivity and the fact that ferromagnetic and antiferromagnetic order can be distinguished using linear or circular polarized X-rays allowed for a complete chemical and crystallographic characterization of the interface. We could see that an interfacial chemical reaction leads to the formation of an additional layer at the Co/NiO interface that has a more CoO-like crystallographic structure and, hence, favors a reorientation of the antiferromagnetic spin axis toward a situation that resembles the anisotropy of CoO. Finally, we investigated the magnetic behavior of antiferromagnetic/ferromagnetic bilayer upon annealing above T_N. We saw that the vanishing exchange coupling between ferromagnetic and antiferromagnetic domains varies domain by domain, providing another instrumental clue toward the understanding of the exchange bias effect. Although we mainly presented results obtained for NiO(001) surfaces, the findings reported here are applicable to other antiferromagnetic oxides as well – and even to antiferromagnetic fluorides. Considering the increasing role that antiferromagnetic oxides play in modern devices and materials that are of significant interest to the scientific community today, like high T_C superconductors, multiferroic , and colossal magnetoresistance materials, it seems obvious that an understanding of the antiferromagnetic microstructure can also provide crucial insight into the electronic, mechanical, and transport properties of such materials. This chapter has provided a glimpse into the possibilities that are available toward a deeper understanding of antiferromagnetic oxides using soft X-ray dichroism spectromicroscopy. Today, synchrotron radiation is available worldwide, and so the tools described here are available to a much larger community of researchers than it was some 10 years ago when most of the projects described here were initiated. For this reason, it is well conceivable that synchrotron-based spectroscopy and microscopy will significantly affect the field of research on antiferromagnetic oxides in the future.

References

1. Néel, L. (1948) *Ann. Phys. (Paris)*, **3**, 137.
2. Néel, L. (1967) *Ann. Phys.*, **2**, 61.
3. Roth, W.L. (1958) *Phys. Rev.*, **111**, 772.
4. Roth, W.L. (1960) *J. Appl. Phys.*, **31**, 2000.
5. Slack, G.A. (1960) *J. Appl. Phys.*, **31**, 1571.
6. Meiklejohn, W.H. and Bean, C.P. (1956) *Phys. Rev.*, **102**, 1413.
7. Finazzi, M., Duó, L., and Ciccacci, F. (2009) *Surf. Sci. Reps.*, **64**, 139.
8. Hubert, A. and Schäfer, R. (1998) *Magnetic Domains*, Springer, Heidelberg.
9. Gomonay, H. and Loktev, V.M. (2002) *J. Phys. Cond. Mat.*, **14**, 3959.
10. Yamada, T., Saito, S., and Shimomura, Y. (1966) *J. Phys. Soc. Japan*, **21**, 664.
11. Yamada, T., Saito, S., and Shimomura, Y. (1966) *J. Phys. Soc. Japan*, **21**, 672.
12. Yamada, T. (1966) *J. Phys. Soc. Japan*, **21**, 650.
13. Attwood, D. (1999) *Soft X-Rays and Extreme Ultraviolet Radiation: Principles and Applications*, Cambridge University Press, Cambridge.
14. Regan, T.J., Ohldag, H., Stamm, C., Nolting, F., Lüning, J., and Stöhr, J. (2001) *Phys. Rev. B*, **64**, 214422.
15. Stöhr, J., Padmore, H.A., Anders, S., Stammler, T., and Scheinfein, M.R. (1998) *Surf. Rev. Lett.*, **5**, 1297.
16. Stöhr, J. (1992) *NEXAFS Spectroscopy*, vol. 25, Springer Series in Surface Sciences, Springer, Heidelberg.
17. Stöhr, J. and Siegmann, H.C. (2006) *Magnetism*, vol. 152, Springer Series in Solid State Sciences, Springer, Heidelberg.
18. Hillebrecht, F.U. (2001) *J. Phys. Cond. Mat.*, **13**, 11163.
19. Kuch, W. in *Magnetic Microscopy of Nanostructures* (eds H. Hopster and H.P. Oepen), Springer Verlag, Berlin, to be published. 1–28.
20. Scholl, A., Ohldag, H., Nolting, F., Anders, S., and Stöhr, J. Study of ferromagnet-antiferromagnet interfaces using x-ray peem, in *Magnetic Microscopy of Nanostructures* (eds H. Hopster and H.P. Oepen), Springer Verlag, Berlin, to be published. 29–50.
21. Bauer, E. (2001) *J. Phys. Cond. Mat.*, **13**, 11391.
22. Tonner, B.P. and Harp, G.R. (1988) *Rev. Sci. Instrum.*, **59**, 853.
23. Engel, W., Kordesch, M.E., Rothermund, H.H., Kubala, S., and Oertzen, Av. (1991) *Ultramicroscopy*, **36**, 148.
24. STAIB Instruments, PEEM 350 operating manual.
25. Anders, S., Padmore, H.A., Duarte, R.M., Renner, T., Stammler, T., Scholl, A., Scheinfein, M.R., Stöhr, J., Séve, L., and Sinkovic, B. (1999) *Rev. Sci. Instrum.*, **70**, 3973.
26. Nolting, F., Scholl, A., Stöhr, J., Seo, J.W., Fompeyrine, J., Siegwart, H., Locquet, J.-P., Anders, S., Lüning, J., Fullerton, E.E., Toney, M.F., Scheinfein, M.R., and Padmore, H.A. (2000) *Nature*, **405**, 767.
27. Ohldag, H., Scholl, A., Nolting, F., Anders, S., Hillebrecht, F.U., and Stöhr, J. (2001) *Phys. Rev. Lett.*, **86**, 2878.
28. Ohldag, H., Regan, T.J., Stöhr, J., Scholl, A., Nolting, F., Lüning, J., Stamm, C., Anders, S., and White, R.L. (2001) *Phys. Rev. Lett.*, **87**, 247201.
29. Chen, C.T., Idzerda, Y.U., Lin, H.-J., Smith, N.V., Meigs, G., Chaban, E., Ho, G.H., Pellegrin, E., and Sette, F. (1995) *Phys. Rev. Lett.*, **75**, 152.
30. Alders, D., Vogel, J., Levelut, C., Peacor, S.D., Hibma, T., Sacchi, M., Tjeng, L.H., Chen, C.T., van der Laan, G., Thole, B.T., and Sawatzky, G.A. (1995) *Europhys. Lett.*, **32**, 259.
31. Alders, D., Tjeng, L.H., Voogt, F.C., Hibma, T., Sawatzky, G.A., Chen, C.T.,

Vogel, J., Sacchi, M., and Iacobucci, S. (1998) *Phys. Rev. B*, **57**, 11623.

32. Spanke, D., Solinus, V., Knabben, D., Hillebrecht, F.U., Ciccacci, F., Gregoratti, L., and Marsi, M. (1998) *Phys. Rev. B*, **58**, 5201.
33. Stöhr, J., Scholl, A., Regan, T.J., Anders, S., Lüning, J., Scheinfein, M.R., Padmore, H.A., and White, R.L. (1999) *Phys. Rev. Lett.*, **83**, 1862.
34. Scholl, A., Stöhr, J., Lüning, J., Seo, J.W., Fompeyrine, J., Siegwart, H., Locquet, J.-P., Nolting, F., Anders, S., Fullerton, E.E., Scheinfein, M.R., and Padmore, H.A. (2000) Observation of antiferromagnetic domains in epitaxial thin films. *Science*, **287**, 1014.
35. Hillebrecht, F.U., Ohldag, H., Weber, N.B., Bethke, C., Weiss, M., Mick, U., and Bahrdt, J. (2001) *Phys. Rev. Lett.*, **86**, 3419.
36. Arenholz, E. and van der Laan, G. (2006) *Phys. Rev. B*, **74**, 094407.
37. Arenholz, E., Chopdekar, R.V., van der Laan, G., and Suzuki, Y. (2007) *Phys. Rev. Lett.*, **98**, 197201.
38. Ohldag, H., van der Laan, G., and Arenholz, E. (2009) *Phys. Rev. B*, **79**, 052403.
39. Weber, N.B., Ohldag, H., Gomonaj, H., and Hillebrecht, F.U. (2003) *Phys. Rev. Lett.*, **91**, 237205.
40. Kinoshita, T., Wakita, T., Sun, H., Tohyama, T., Harasawa, A., Kiwata, H., Hillebrecht, F.U., Ono, K., Matsushima, T., and Oshima, M. (2004) *Jour. Phys. Soc. Jap.*, **73**, 2932.
41. Chen, C.T., Smith, N.V., and Sette, F. (1991) *Phys. Rev. B*, **43**, 6785.
42. Carey, M.J. and Berkowitz, A.E. (1992) *Appl. Phys. Lett.*, **60**, 3060.
43. Silinsky, P. and Seehra, M.S. (1981) *Phys. Rev. B*, **24**, 419.
44. Nogués, J. and Schuller, I.K. (1999) *J. Magn. Magn. Mater.*, **192**, 203.
45. Regan, T.J. (2001) X-Ray absorption spectroscopy and microscopy study of ferro- and antiferromagneti thin films with application to exchange anisotropy, PhD thesis, Stanford University.
46. Berkowitz, A.E. and Takano, K. (1999) *J. Magn. Magn. Mater.*, **200**, 552.
47. Mauri, D., Siegmann, H.C., Bagus, P.S., and Kay, E. (1987) *J. Appl. Phys.*, **62**, 3047.
48. Scholl, A., Liberati, M., Arenholz, E., Ohldag, H., and Stöhr, J. (2004) *Phys. Rev. Lett.*, **92**, 247201.
49. Roy, S., Fitzsimmons, M.R., Park, S., Dorn, M., Petracic, O., Roshchin, I.V., Li, Z.-P., Batlle, X., Morales, R., Misra, A., Zhang, X., Chesnel, K., Kortright, J.B., Sinha, S.K., and Schuller, I.K. (2005) *Phys. Rev. Lett.*, **95**, 047201.
50. Ohldag, H., Shi, H., Arenholz, E., Stöhr, J., and Lederman, D. (2006) *Phys. Rev. Lett.*, **96**, 027203.
51. Grimsditch, M., Hoffmann, A., Vavassori, P., Shi, H., and Lederman, D. (2003) *Phys. Rev. Lett.*, **90**, 257201.
52. Scholl, A., Nolting, F., Seo, J.W., Ohldag, H., Stöhr, J., Raoux, S., Locquet, J.-P., and Fompeyrine, J. (2004) *Appl. Phys. Lett.*, **85**, 4085.
53. Malozemoff, A.P. (1987) *Phys. Rev. B*, **35**, 3679.
54. Takano, K., Kodama, R.H., Berkowitz, A.E., Cao, W., and Thomas, G. (1997) *Phys. Rev. Lett.*, **79**, 1130.

Index

Magnetic Properties of Antiferromagnetic Oxide Materials.
Edited by Lamberto Duò, Marco Finazzi, and Franco Ciccacci

ISBN: 978-3-527-40881-8